This book is due for return on or before the last date shown below.

FACILITIES DESIGN

DESIGN

Fourth Edition

FACILITIES DESIGN

DESIGN

Fourth Edition

Sunderesh S. Heragu

CRC Press
Taylor & Francis Group
Boca Raton London New York

CRC Press is an imprint of the
Taylor & Francis Group, an **informa** business

CRC Press
Taylor & Francis Group
6000 Broken Sound Parkway NW, Suite 300
Boca Raton, FL 33487-2742

Printed on acid-free paper
Version Date: 20151102

International Standard Book Number-13: 978-1-4987-3289-5 (Pack - Book and Ebook)

Library of Congress Cataloging-in-Publication Data

Names: Heragu, Sunderesh S., author.
Title: Facilities design / Sunderesh S. Heragu.
Description: Fourth edition. | Boca Raton : Taylor & Francis, CRC Press, 2016.
Identifiers: LCCN 2015042523 | ISBN 9781498732895 (alk. paper)
Subjects: LCSH: Factories--Design and construction. | Plant layout.
Classification: LCC TS177 .H47 2016 | DDC 658.2/3--dc23
LC record available at http://lccn.loc.gov/2015042523

Visit the Taylor & Francis Web site at
http://www.taylorandfrancis.com

and the CRC Press Web site at
http://www.crcpress.com

*This book is dedicated to the memory of my father,
Heragu Ramanuja Sesharanga Iyengar, who taught
me to work with reason and not excuse.*

Contents

Preface

Manufacturing and service companies spend a significant amount of time and money in designing or redesigning their facilities. It is an extremely important function that must be addressed before products are produced or services rendered. A poor design costs a company a significant amount of resources and results in poor quality, low employee morale, and customer dissatisfaction. This book deals with the proper design, layout, and location of facilities. The first edition of this book was published in 1996.

This fourth edition is different from the previous editions in the following ways. Dated materials have been removed and new material introduced. Case studies covering layout, logistics, supply chain, warehousing, and materials handling are included in each chapter. New software for layout—Layout-iQ™—is included in the book. Several new material-handling equipment and systems are introduced. State-of-the-art topics in materials handling, warehousing, and logistics not found elsewhere are presented in this book. The author believes this book has an ideal blend of theory and practice.

The topics and chapters are reorganized as follows: Topics covered in Chapters 1 through 9 are suitable for undergraduate students. Chapters 10 through 13 are suitable for a graduate class, which can be included with Chapters 1, 6, and 8 in a semester-long graduate class.

Chapter 1 introduces the reader to the design and planning problems encountered in manufacturing and service systems. Various types of layout problems are discussed in Chapter 2: product analysis and equipment selection, as well as a discussion on personnel and space requirement analyses. Chapter 3 addresses process and material flow analysis. This chapter details the typical types of data required for process and material flow analysis and a method for data development and generation. This chapter also describes the commonly used tools for presenting layout designs and includes a detailed, real-world case study for data development and generation.

Of the seven chapters that deal with facility layout, two—Chapters 4 and 5—are devoted to the layout problem. Chapter 4 discusses traditional models for facility layout including the popular systematic layout planning (SLP) model in detail. It includes a layout project involving the SLP model and lists others that are of historical interest. Chapter 5 discusses algorithms for the layout problem. It also includes modern software used for layout in practice.

Chapter 6 covers group technology and cellular manufacturing at the elementary level. It also includes a project on machine grouping and layout that builds on the case study in Chapter 3. A detailed case study involving the use of PFAST for a real-world problem is also presented in this chapter. Chapter 7 covers materials handling in detail. It includes a case study, taxonomy of material-handling devices, an educational module titled "10 Principles of Materials Handling," different types of material-handling devices with illustrations, and several design and planning models and algorithms. Chapter 8 shows an automated storage and retrieval system in action and discusses warehouse functions, different types of automated material-handling devices used in warehouses with illustrations, and design and operational models used for warehouse planning and management, as well as a discussion on RFID and a warehouse design, planning, and analysis software module.

Chapter 9 pertains to facility location and logistics management problems. Its emphasis is on simple-to-moderate mathematical models and includes a detailed case study. Chapter 10 covers advanced mathematical programming models for facility layout and is suitable only for a graduate class. Similarly, Chapter 11 covers advanced algorithms—optimal and heuristic—for layout problems. It also includes a discussion on next-generation factory layouts. Chapter 12 discusses more complex models for the facility location problem that are suitable primarily for the graduate audience. Analytical queuing and queuing network models are discussed in Chapter 13. The MPA software is available at http://sundere.okstate.edu/downloadable-software-programs-and-data-files.

MPA is a software package that can be used in the solution of queuing models. A discussion of simulation as it applies to material handling and facility layout is also provided in Chapter 13.

A typical one-semester undergraduate course may include Chapters 1 through 9. A graduate-level course could include Chapter 1 along with Chapters 7 through 13. Of course, other combinations for the graduate and undergraduate courses are possible depending upon the instructor's interests and the student's preparation.

This book comes with extensive supporting material, including software, data files for many of the numerical examples discussed in almost all the chapters, and PowerPoint files for 12 chapters. These are available at the following website: http://sundere.okstate.edu/book.

Acknowledgments

In a way, this page was the hardest part of the book to write. I have so many people to be thankful for—too numerous to list. However, I would be remiss if I did not show my deep appreciation to my wife Rita and son Keshav for their patience, support, and understanding as I worked through this fourth edition. Our Goldendoodle, Rohan, remained by my side much of the time during the preparation of this book.

Years ago, when I was pursuing my PhD in the field of facilities layout, I picked up the book by Francis and White (1974). I was so impressed by this book that it motivated me to write the first edition of *Facilities Design*. I have also benefited from other textbooks in this field, especially the first editions of the books by Askin and Standridge, Viswanadham and Narahari, and Tompkins and White. A number of people at private and government organizations in the United States and abroad—Karen Auguston, Jerry Crucetti, Charles Donaghey, Les Gould, Rick Korchak, Mark Lattimore, Ronald Mantel, Tim O'Donnell, Detlef Spee, Doug Stone, John Usher, M.A. van der Lande, and Dick Ward—have provided useful material that has been included in this book. Their support is deeply appreciated. The following reviewers provided very helpful comments and influenced the content, level, and organization of the first edition of this textbook: Sadashiv Adiga, Han Bao, Yavuz Bozer, Margaret Brandeau, Louis Brennan, Mike Diesenroth, Charles Donaghey, Marc Goetschalckx, Jeff Goldberg, Manjunath Kamath, Bill Kennedy, Panos Kouvelis, Ed McDowell, Udatta Palekar, Roderick Reasor, Nanua Singh, Joe Svestka, and John Usher. I am also thankful to my former colleagues Charles Malmborg and Robert Graves and my former students Ja-Shen Chen, Jing Jia, Siva Kakuturi, Byung-In Kim, Jasmit Kochhar, Chris Lucarelli, Gang Meng, Srinivasan Rajagopalan, Srinivasan Ramaswamy, Sandeep Venugopalan, and Ruben Mora for their help with parts of the book.

I also thank Bailey Whitman, a junior at Oklahoma State University, for helping me with this edition. I am also grateful to Professors Joe Mize and Ken Case for the influence they have had on my career, directly and indirectly.

Despite my best efforts to keep the book error free, it is likely that the readers of the book will find some. All errors and omissions remain my responsibility. However, I trust you will find the book well written, up-to-date, and useful in the study of modeling and analysis of facility design, layout, location, material handling, and warehousing/storage problems. I welcome comments from readers and would appreciate bringing any errors or omissions to my attention via e-mail (sunderesh.heragu@okstate.edu).

This book is dedicated to the memory of my father Heragu Ramanuja Sesharanga Iyengar, who taught me to *work with reason and not excuse*.

Sunderesh S. Heragu

Author

Sunderesh S. Heragu is regents professor, Donald and Cathey Humphreys Chair, and head of the School of Industrial Engineering and Management at Oklahoma State University, Stillwater, Oklahoma. Previously, he was professor and the Mary Lee and George F. Duthie Chair in engineering logistics in the Industrial Engineering Department at the University of Louisville, where he was also the director of the Logistics and Distribution Institute, Louisville, Kentucky. He was on the faculty of the Decision Sciences and Engineering Systems (now Industrial and Systems Engineering Department) at Rensselaer Polytechnic Institute from 1991 to 2004. He has also taught at the State University of New York, Plattsburgh, and held visiting appointments at State University of New York, Buffalo, the Technical University of Eindhoven, and the University of Twente, the Netherlands.

He completed his BE in mechanical engineering at Malnad College of Engineering, Hassan, India (affiliated to the University of Mysore); MBA at the University of Saskatchewan, Saskatoon, Canada, and PhD in industrial engineering at the University of Manitoba, Winnipeg, Canada.

His current research interests are in the development of real-time decision support systems for emergency preparedness for the healthcare, public health, and emergency service sectors; modeling and analysis of drive-through mass vaccination clinic; supply chain management; design of next-generation factory layouts; intelligent agent modeling of automated warehouse systems; application of RFID technology to improve intraplant and interplant logistics; and the integration of design and planning activities in advanced logistical systems. He has been principal investigator (PI) or co-PI of projects totaling more than $20 million funded by the National Science Foundation, the Department of Homeland Security, the Defense Logistics Agency, and private industry in many of the areas mentioned. His previous research has focused on the application of deterministic and stochastic mathematical optimization models and/or knowledge-based techniques to supply chain management, logistics, facility location, layout, material flow network, order picking in automated warehouses, scheduling, cellular manufacturing, and group technology problems. He has authored or coauthored more than 250 articles, many of which have appeared in journals, magazines, books, and conference proceedings. He has taught several courses in operations research, service logistics, inventory control, production and operations management, quality control and facilities design, planning, and control.

Dr. Heragu received the David F. Baker Distinguished Research Award from the Institute of Industrial Engineers (IIE). He is also a fellow of the IIE, has won the IIE Award for Technical Innovation, the IIE Transactions on Design and Manufacturing Award and IIE Transactions Award for Best Paper published in "Feature Applications" for a paper coauthored with his student, the Gold Award of Excellence for leadership in facilities planning and design, and the New York State and United University Professions New Faculty Development Award—a campus-wide award at SUNY–Plattsburgh. In addition, master's and doctoral theses done under his direction have received the Delmar W. Karger Award for outstanding master's and doctoral thesis. His biography appears in several *Who's Who* compilations.

Dr. Heragu has been invited to talk at several conferences and serves or has served on the editorial boards of the *International Journal of Automation and Logistics* (as advisor), the *International Journal of Lean Enterprise Research*, the *International Journal of Operations and Quantitative Management*, the *Asia Pacific Journal of Operational Research*, the *European Journal of Industrial Engineering*, and the *International Journal of Industrial Engineering*.

Dr. Heragu was senior vice-president for publications and a member of the Board of Trustees for IIE, an ABET examiner for Systems and Industrial Engineering programs, the Facilities Layout and Material Handling Department editor for *IIE Transactions*, the director of IIE's Facilities

Planning and Design Division; is a senior member of the Institute for Electrical and Electronics Engineers (IEEE), the Institute for Operations Research and Management Sciences (INFORMS), the Production and Operations Management Society (POMS), and the Society for Manufacturing Engineers (SME); and is a past member of the College-Industry Council on Material Handling Education (CICMHE).

1 Introduction to Facility Design

1.1 CASE STUDY

Ann Inc. was founded in 1954 as Ann Taylor. It offers two brands—Ann Taylor and Loft—and is a retailer specializing in women's apparel, shoes, and accessories. It operates over 1,000 retail stores in most states in the United States and Canada. To stock and restock its retail stores, Ann Inc. has one warehouse located in Louisville, KY.

Ann Inc. located its distribution center (DC) in Louisville to take advantage of the United Parcel Service (UPS) shipping network. The air-hub location for UPS is in Louisville, KY. The 256,000 square feet DC receives supplies from all over the world. More than 60% of the inventory is cross-docked. In other words, these items come into the DC from a factory in the Far East via container ships, rail, and road already packaged and labeled to be dispatched to one of the 1,000 retail stores, are unloaded from the receiving truck, go through a sortation system, and are loaded to a shipping truck without the item spending any time on a shelf in the DC. Some items like the cartons shown in Figure 1.1 do, but notice from Figures 1.2 and 1.3 that the garments come in from the factory already in a clothes hanger, go through a sortation system (Figures 1.4 through 1.6), and are transported through the supply chain all the way to a retail store on a hanger!

Some basic design and operational questions that arise in facilities design relative to a DC include the following:

- Where should the DC be located?
- What percentage of the goods handled should be cross-docked?
- What material handling systems should be installed for sorting and moving the goods from inbound trucks to outbound trucks?
- How many part-time workers to assign in each month of the year?

We will answer the aforementioned questions with respect to Ann Inc. As mentioned previously, Ann Inc. decided to locate its warehouse in Louisville, KY, so that they could utilize UPS' extensive distribution network (with its only air-hub located in Louisville) and be able to ship packages to stores overnight. Because the retail stores are located in fashionable shopping districts where the cost to rent or lease retail space is relatively high, these stores wish to use more of their available space for item display than to maintain store inventory. As a result, these stores require replenishment of their inventory two or three times a week. It is the primary function of the DC to ensure the required items are delivered to each of the 1,000 stores as requested on time even though the factories may be located in multiple countries all over the world. Because of the air-hub location, Ann Inc. can ship as late as 10 p.m. on a given day and be able to deliver to any of its U.S. stores the very next day!

Because storing items in a warehouse is a non-value-added activity, Ann Inc. also likes to minimize the time cartons containing high-value items spend on storage racks and thus opts to cross-dock most—more than 60%—of its items even though they may be coming from as far away countries as China.

Like most retailers, Ann Inc. faces significant demand fluctuation. Its DC activity tends to peak in the period of October–December. Thus, to keep labor costs down, it relies on seasonal employees to fulfill the orders. In order to ensure the availability of seasonal employees, Ann Inc. offers competitive wages as well as annual health-care and retirement benefits. Seasonal employment with

FIGURE 1.1   Cartons stored in Ann, Inc. distribution center.

FIGURE 1.2   High-value, high-margin dresses on clothes hangers unloaded from a receiving truck.

FIGURE 1.3   Clothes on hangers traveling through the DC.

FIGURE 1.4 "Bom-bay" sorter deposits specific ordered items into a box.

FIGURE 1.5 Conveyor-based sortation system.

FIGURE 1.6 Tilt-tray sortation system.

good year-round benefits is attractive to two-income households, where one person has a steady job and the family needs health insurance and retirement benefits, but without both earners having to commit to full-time employment. Of course, if an employee wants full-time employment, Ann Inc. will accommodate his or her request, but the employee must be willing to work in all areas of the DC, not just in order picking.

As can be seen from the aforementioned example, facilities design covers a myriad of design and operational problems, and each must be solved efficiently for the organization to succeed as a whole. The purpose of this book is to look at key design and operational problems that arise in manufacturing and service systems alike.

1.2   INTRODUCTION

Facilities can be broadly defined as buildings where people utilize material, machines, and other resources to make a tangible product or provide a service. Due to various forces—internal and external—it is important that a facility be properly managed so that the stated purpose is achieved in a manner that several objectives are satisfied. Examples of such factors include producing a product or providing a service at a relatively lower cost, higher quality and using the least amount of natural resources. In order to manage facilities* so that the objectives, which are often conflicting with one another, are attained, it is important to understand the underlying decision problems faced in such systems.

Although the design, layout, planning, and location of facilities have been formally studied and researched as a discipline since the mid-1950s, organizations as far back as 4000 BC have used these ideas in planning their tasks. For example, the Egyptians had to choose a location for erecting pyramids based on astrological calculations. When the Romans built the Coliseum, some form of a blueprint and detailed layout drawing must have been developed. The industrial revolution saw the transition from custom production to mass production. The assembly line concept implemented by Ford and the work division concept introduced by Gilbreth led to the acceptance of mass production techniques in the factories. The emergence of *Operations Research* as a field in the 1950s led to the development of analytical models for facilities design and planning. In fact, ever since the quadratic assignment problem was introduced by Koopmans and Beckman (1957) to model macro-level location problems (e.g., location of warehouses across a large geographic region) and micro-level location problems (e.g., layout of departments in a factory), there have been significant efforts at solving these models and developing software for the factory layout problem. The systematic layout planning (SLP) technique introduced in the late 1950s is a technique that is used by consultants even today. Many software packages, for example, the Factory suite of programs and Layout-iQ, discussed in Chapter 5 are modeled on the SLP. Computerized relative allocation of facilities technique (CRAFT) was one of the layout software packages developed for finding good-quality layout solutions. These and some other developments in the field of facilities design and planning are documented in Table 1.1.

In recent years, the manufacturing and service industries have witnessed significant developments. The increase in the number and types of automated systems in the two industries is a testament to the developments taking place. However, these developments have taken place at the expense of system design problems. Design problems have become even more complex, and designers and users of automated systems have developed new tools to cope with these problems. Manufacturing or service system design, which is a hierarchical combination of several problems, is a complex activity. It involves solving a number of design and planning problems arranged in a hierarchy—determining products to be manufactured or services to be provided, manufacturing

---

* Facilities can also mean service facilities where no tangible product is produced. Much of the discussion in this section is relevant to service systems as well.

TABLE 1.1

Chronological List of Facilities Planning and Design Activities

Date	Event
4000 BC	Egyptians developed expertise in finding suitable locations for pyramids according to their astrological calculations.
100 BC–AD 100	Romans developed full-fledged methods for the construction of temples, arenas, and other buildings. Detailed planning of public and residential buildings.
1700–1900	Industrial revolution period.
1910	First industrial engineering textbook *Factory Organization and Administration* published by Hugo Diemer.
1913	First moving automotive assembly line introduced by *Henry Ford*.
1954	Quadratic assignment problem for micro- and macro-level location problems introduced by Koopmans and Beckman.
1955–1995	Optimal and heuristic algorithms for the quadratic assignment problem.
1959	Systematic layout planning approach introduced by Muther.
1963	Computerized relative allocation of facilities technique (CRAFT) introduced by Armour and Buffa.
Early 1980s	The flexible manufacturing system concept is introduced, and attention shifts toward achieving plant-wide flexibility via medium-volume, medium-variety production using cellular manufacturing techniques.
1985–present	Modern software for facilities design problems.
1990s–present	Research on new layout concepts including dynamic layouts, robust layouts, and reconfigurable layouts introduced to support mass customization techniques.

or service processes to be used, number and types of manufacturing or service equipment that are capable of performing the required processes, etc. Additionally, in the case of manufacturing systems, preliminary process plan development, determining the tooling and fixture requirements, layout of manufacturing cells and machines, material handling methods to be used, and number and types of specific material handling devices capable of performing the required material handling moves are some of the more important design questions that need to be addressed (see Figure 1.7).

Solving the manufacturing cell determination and cell layout problems is generally required only for manufacturing systems that produce a large number of components and for which manufacturing activities can be decomposed into almost mutually independent cells. For mass production or continuous production systems, we may bypass these two problems. It should be emphasized that the design and planning problems shown in Figure 1.7 are by no means exhaustive and the problems need not necessarily be solved in the order shown. Often, it may be necessary to backtrack or iterate between two or more problems.

Clearly, there are many more decision problems, for example, marketing, advertising, packaging, warehousing, and financial. Ideally, all these problems would have to be looked at simultaneously in order to arrive at a good decision. However, each of these is difficult to solve when considered separately. Hence, the approach taken most commonly is to solve the problems sequentially—design problems first and planning next. Sometimes it is necessary to iterate between two or more problems because the decision taken on one problem has an impact on that of another and vice versa.

Design problems are those that are solved rather infrequently. They are typically addressed once in 3–5 years or more because it is costly to alter a design decision frequently. For example, once a plant is built in a certain geographic area, it is difficult to justify moving the plant to another location within 1 or 2 years of the first decision because the fixed costs would not have been fully recovered within this time period. Similarly, if heavy-duty machines have already been laid out in a

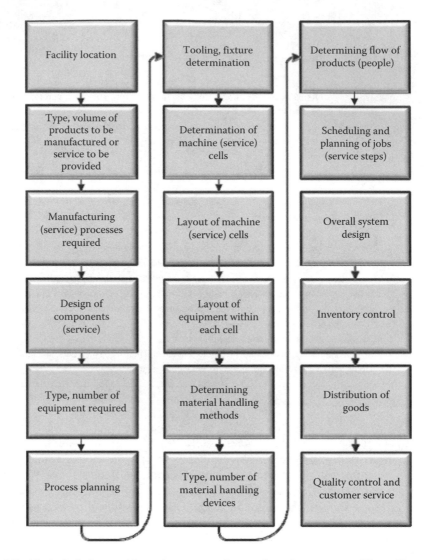

FIGURE 1.7 Typical design problems in automated manufacturing systems. (From Heragu, S.S. and Lucarelli, C.M., Automated manufacturing system design using analytical techniques, in: Irwin, J.D., ed., *The Industrial Electronics Handbook*, A Volume of the Electrical Engineering Handbook Series, CRC Press Inc., Boca Raton, FL, 1996.)

plant, we have to live with that layout for a period of at least 3–5 years because it is time consuming and expensive to change the facility layout frequently. Because we do not have the luxury to change design decisions often, it is extremely important that in making these decisions, we consider their long-term impact and take into consideration manufacturing or service activities that are to take place not only in the short run but also in the long run.

On the other hand, decisions on planning problems are made relatively frequently because changing such decisions is not as costly as changing design decisions. The scheduling of jobs coming into a shop floor and inventory planning can be done as frequently as 1 week, if necessary. Whereas decisions on some planning problems are made on a daily or weekly basis, some others are done on a monthly or quarterly basis. Thus, the time horizon for planning decisions can be short term or intermediate term.

1.3   FACILITY LAYOUT

In Section 1.2, we listed several facility design and planning problems in automated manufacturing systems. An important consideration in the design of both manufacturing and service systems is the physical arrangement or layout of departments. The remainder of this chapter focuses on the facility layout problem.

Studies on manufacturing systems indicate that 30%–75% of a product's cost can be attributed to material handling expenses (Sule, 1991). Tompkins and White (1984) mention that material handling activities account for 20%–50% of a manufacturing company's total operating budget. Thus, if their departments are arranged optimally, manufacturers can reduce product costs and significantly enhance their competitive position. By optimal arrangement, we mean that no other arrangement can be better with regard to the chosen criteria. Other arrangements may be equally good, but none of them is better. In a manufacturing system, the term *departments* includes machines, workstations, inspection stations, washing stations, locker rooms, rest areas, and other such support facilities. In a service system, departments include offices, lounges, restrooms, and cafeteria. By *physical arrangement*, we mean the assignment of departments to specific locations on the floor. A physical arrangement of departments that minimizes the movement of personnel and material between departments, and thereby decreases material handling costs, increases a system's efficiency and productivity. In practice, many more factors need to be considered in addition to minimizing the cost involved in movement between departments. Some other factors to consider in the physical arrangement are

- Reducing congestion to permit smooth flow of people and material
- Utilizing the available space effectively and efficiently
- Facilitating communication and supervision
- Providing a safe and pleasant environment for personnel

When developing a facility layout, designers should note the following constraints:

- Some department pairs need to be in adjacent sites for safety reasons regardless of the volume of material flow between them. An example is the forging and heat-treatment stations. Due to fire hazards, these two stations must be next to each other even if relatively few parts pass between them.
- Some department pairs need to be located in nonadjacent sites. Sometimes technological reasons dictate that two or more departments cannot be close together even if a large number of parts have to visit them for successive operations. Because the welding station generates sparks that could possibly ignite flammable solvents in the painting station, the two stations should be in nonadjacent sites, as far apart as possible, although there may be much interaction between them.
- Certain departments have to be in specific locations. Consider this real-world example. A manufacturer of industrial mops and buckets has a drill press that is taller than the roof of the building. To install it, a hole had to be made in the roof. If this company plans a layout change, it cannot justify changing the position of this drill press.
- Federal, state, and local governmental regulations and hazard insurance company regulations must be observed. Fire codes may require a certain number of fire exits. The Occupational Safety and Health Administration (OSHA) requires separate restrooms for men and women depending on the number of employees and other factors. Other OSHA regulations include entrance and exits to the manufacturing facility, health and safety norms, requirements that hazardous processes or equipment be located so that employee contact is minimal, and so on. OSHA also dictates how much lighting is required and how much noise an employee can be exposed to without having to wear hearing protection devices.

1.4   TYPES OF LAYOUT PROBLEMS

Designers face the facility layout problem not only when they create a new manufacturing or service system but also when they expand, consolidate, or modify existing systems. Even established manufacturing companies need to change the layout of departments every 2 or 3 years (Nicol and Hollier, 1983). The frequency of layout changes has increased in the last three decades partly because product mixes have changed at more frequent intervals than in the past. Customers demand constant changes in style and functionality. Consider the following examples:

1. A manufacturer is implementing the just-in-time manufacturing concept and must identify machine cells and the position of these cells on the factory floor. In addition, the location of machines within cells must be determined.
2. The production activity in an appliance manufacturing company has doubled in the last 2 years, and no additional space is available for expansion. A consulting industrial engineer suggests that old machines that are rarely used be disposed of and the remaining machines rearranged to consolidate space. The new arrangement will free up space for new high-capacity production equipment that can help meet the increased demand.
3. Due to a recent merger, customer traffic in a regional bank has significantly increased, so the bank wants to expand in its current location.
4. A manufacturer of garden equipment that has two production and assembly sites has purchased 80,000 square feet of space and wants to move operations from the two facilities to the new third site.
5. An insurance company has leased a multistoried office building and wants to move its headquarters there. It must divide the interior space of each floor of the building into offices, lounge, restroom, work area, and conference rooms.
6. The research and development investment made by a defense equipment manufacturer on *dual-use* technology has paid off, and hence, the company is moving gradually into the consumer business. It needs to expand at the existing facility by adding a new automated manufacturing system that consists of numerically controlled machines and automated guided vehicles.
7. Due to a new trade agreement, a foreign car manufacturer has decided to move its assembly operations to the United States and must design a new assembly line.
8. A manufacturer of consumer items does not have the funds or space to expand and hence must use existing floor space in a warehouse to store finished products.
9. A large retailer has discontinued a line of merchandise and is actively promoting a new line of items, so it must redesign merchandise display areas in all its departmental stores.

The examples demonstrate that layout problems occur in manufacturing as well as service organizations and in both the design of new facilities and modification of existing ones. In general, layout problems may be classified into the following four categories:

1. Service systems layout problem
2. Manufacturing layout problem
3. Warehouse layout problem
4. Nontraditional layout problems

1.4.1   Service Systems Layout Problem

Although most of this book focuses on facilities design and layout problems in manufacturing systems, design aspects are just as important in service systems. The layout of the reception area, dining area, kitchen, cocktail lounge in a licensed restaurant; an insurance office; registrar's office at a university; runways at an airport; emergency facilities in a hospital; and county government office and public library are a few examples of facilities design and layout problems in service systems.

To develop a service systems layout, designers must know the number of entities or departments that are to be located, the area that will likely be occupied by each, the interaction between departments, and special layout restrictions for any department or pairs of departments. The departments in a registrar's office at a university might be the receptionist's booth, registrar's office, restrooms, offices of other personnel working in the office, and conference room. In a restaurant, the departments may include the cocktail lounge, kitchen, dining area, cash register desk, order delivery and pickup area, restrooms, coat-checking area, and supplies-loading zone. The layout of a particular service system, a dentist's office, is illustrated in Figure 1.8.

There must be well-justified reasons for developing a new service systems layout. Thus, a thorough operational review including addressing the following questions is necessary (Suskind, 1989):

1. Is the company outgrowing its available space?
2. Is the available space too expensive?
3. Is the building in the proper location?
4. How will a new layout affect the organization and its service?
5. Are office operations too centralized or decentralized?
6. Does the office structure support the strategic plan?
7. Is the layout in tune with the company's image?

As noted before, a layout must meet the following objectives:

1. Minimize unnecessary personnel movement within the building or floor.
2. Facilitate communication and provide privacy where necessary.
3. Conform to building codes.
4. Provide safety and security for people in the building.

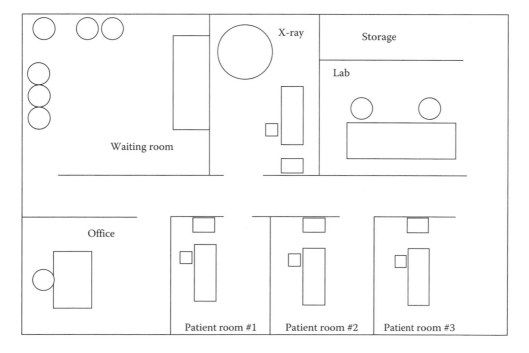

FIGURE 1.8 Layout of a dentist's office.

In addition, consideration must be given to aesthetics in developing service system layouts because there is a more direct customer contact in these systems than in a manufacturing system. The receiver of a service, i.e., a customer, typically participates in the service process, so a pleasant environment is important. Dental offices, for example, have a television, fish tank, or colorful displays in the waiting area. They may also have large mirrors or windows. These features not only improve the office's appearance but also make waiting more tolerable. Similarly, some shopping malls have miniature waterfalls located in strategic locations for visitors and employees. Modern office layouts have departed from traditional designs, with square or rectangular offices placed along a straight hall. Instead, flexibility is emphasized in office layouts. Collapsible partitions not only provide privacy when needed but also can be moved to change the layout. To the extent possible, offices are laid out so that each has at least one window; in fact, architects modify building shapes to permit each office in the building to have a separate window. For example, all the rooms in a pyramid-shaped casino hotel in Las Vegas have windows. Also, the trend is to build energy-efficient buildings made of *environment-friendly* material. The final layout developed must be in tune with the organizational image. For example, if a bank's objective is to serve its customer in a pleasant atmosphere, then the layout designer must plan for as much open space and lobby space as possible.

There are four general office structures, which are illustrated in Figures 1.9 through 1.12:

1. Closed structure
2. Semiclosed structure
3. Open structure
4. Semiopen structure

In a *closed structure*, privacy is an important concern, so the entities are separated by opaque partitions that are at least 4 feet high (see Figure 1.9). In fact, the partitions may extend from floor to ceiling. Opaque partitions are found in post offices where the mail collecting and sorting area is separated from the stamp-selling booths.

A *semiclosed structure* has half walls that are typically opaque. These are used to separate customers from servers in banks, dry-cleaning establishments, fast-food outlets, and post offices. As shown in Figure 1.10, a semiclosed structure satisfies the need for both communication and privacy.

Open structures have no partitions between entities. For example, there are no partitions between dining tables in a restaurant, between tellers in a bank, or between sections of a department store. Although the display shelves in a department store may serve as partitions, their primary function is to display and store merchandise. The open structure shown in Figure 1.11 shows an arrangement of study tables in a student union.

FIGURE 1.9   Closed structure.

FIGURE 1.10 Semiclosed structure.

FIGURE 1.11 Open structure.

FIGURE 1.12 Semiopen structure.

In *semiopen structures*, the entities are separated by transparent partitions. For example, the office spaces of loan officers in a bank, insurance agents in an insurance company, chefs in a kitchen, and the work area in photo-processing laboratories are usually separated by transparent partitions. These transparent partitions ensure that the conversation in one section of the office does not disturb the functioning of another section. They may also serve other functions such as improving employee morale, providing natural light, or increasing visibility. A prescription eyeglass store uses the semiopen structure for its lens preparation and mounting department. A conference room is an example of a semiopen structure (see Figure 1.12).

1.4.2   MANUFACTURING LAYOUT PROBLEM

Layout design is an important task when a manufacturing system is redesigned, expanded, or designed for the first time. Manufacturing layout is different from office layout because different weights are attached to the factors listed in Section 1.8. For example, designers of an office layout place more emphasis on facilitating communication and less on reducing traffic congestion, because easing congestion is not a major objective in developing office layouts. In the design of manufacturing layouts, minimizing material handling costs and providing a safe workplace for employees are major considerations. The layout problem in a manufacturing system involves determining the location of machines, workstations, and other departments to achieve these objectives:

1. Minimize the cost of transporting raw material, parts, tools, work-in-process, and finished goods between the departments
2. Facilitate the flow of traffic
3. Increase employee morale
4. Minimize the risk of injury to personnel and damage to property
5. Where necessary, provide for supervision and face-to-face communication

The departments in a manufacturing system not only include the machines and workstations but also rest areas, inspection stations, clean rooms, heat-treatment stations, supervisor or manager offices, and tool cribs. Figure 1.13 is a layout of assembly equipment and other departments in the shop floor of a company.

1.4.3   WAREHOUSE LAYOUT PROBLEM

Warehouse layout is an important consideration for a facility designer because the cost to rent, lease, or buy real estate in the United States is increasing. A warehouse layout is shown in Figure 1.14. Like a machine layout, a good warehouse layout should use available space effectively to minimize storage cost and material handling cost. Some factors to be considered in warehouse design are shape and size of aisles, height of the warehouse, location and orientation of the docking area, types of racks to be used for storage, and the level of automation involved in storage and retrieval of commodities.

The shape and size of the aisle depend on the following two factors:

1. The type of material handling device used
2. The type of racks used for storage

If a forklift truck is used for material handling, then a narrow aisle may be considered. On the other hand, if a tractor is used, a wide aisle is necessary because the forklift truck can maneuver in narrow aisles better than a tractor. If two-sided racks (in which material can be accessed from both sides) are used, each rack is sufficiently separated from the other to allow for easy storage and retrieval. This arrangement increases aisle space but reduces storage space.

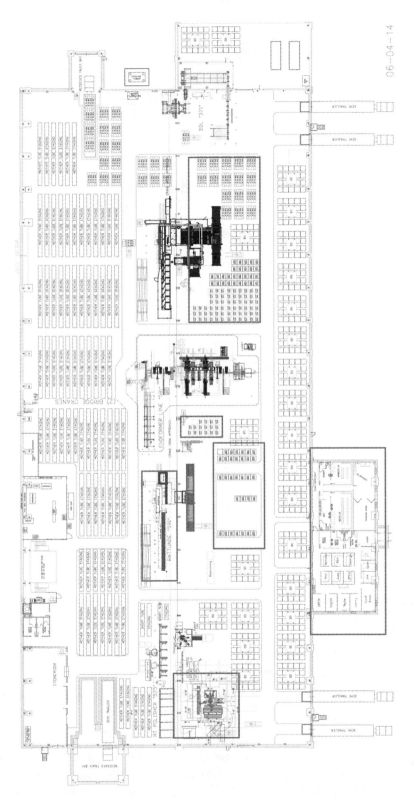

FIGURE 1.13 Factory floor layout of a typical manual assembly facility.

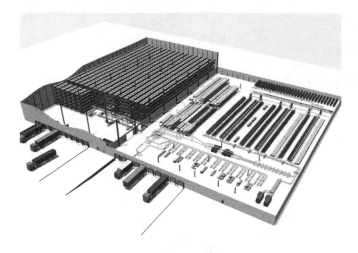

FIGURE 1.14   Sample warehouse layout. (Courtesy of Savoye Logistics.)

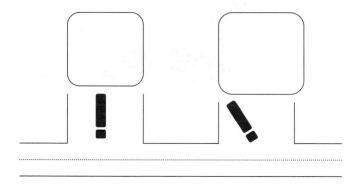

FIGURE 1.15   Wider drive-in roads allow easy access but decrease available space for building.

The location of the docking area depends on factors such as the location of the surrounding roads, entry and exit to nearby access roads, total warehouse space available, size of parking area, and width of drive-in roads. In general, companies tend to prefer narrower drive-in roads, simply because they permit more storage space. As illustrated in the warehouse on the right of Figure 1.15, the docking area layout utilizes less drive-in space (and therefore allows for more storage) than the layout shown on the left.

1.4.4   NONTRADITIONAL LAYOUT PROBLEMS

Layout problems occur in many situations other than those already discussed. Consider the following four examples:

- *Integrated circuit*: Billions of tiny semiconductor components (transistors, diodes, resistors, capacitors, and electrical pins) must be placed on an integrated circuit and connected via metal routings so that the resulting chip can perform specified functions. Because millions of such chips will be manufactured, it is desirable to minimize the area required for the components, optimize their placement for superior circuit performance, and minimize the length of these connections so that they can be manufactured compactly and cost effectively.

- *Keyboard layout*: The layout of keys on a typewriter or computer keyboard is another layout problem. When keys are optimally configured, studies have shown that keyboarding efficiencies are significantly improved. Not only does the keyboarding speed increase, but also the long-term effect of repetitive use of the keyboard (e.g., carpal tunnel syndrome) is minimized.
- *Linear placement*: Another class of layout problems is known as the linear placement problem or the one dimensional space allocation problem, which involves determining an optimal linear ordering of entities to minimize the total traffic cost between each pair of entities. Some applications of this class of problems are single-row machine layout (Heragu and Kusiak, 1991); arrangement of books on a shelf; assignment of files to the cylinders of a disk (Picard and Queyranne, 1981); assignment of aircraft to gates in an airport (Suryanarayanan et al., 1991); and U, L, straight line, or loop layout of machines within a manufacturing cell. These applications are discussed in detail and illustrated in Chapters 5 to 10.

1.5 ENGINEERING DESIGN PROBLEM APPROACH

Solving engineering design problems is complex and time consuming. The approach discussed in this section is based on a time-tested engineering problem-solving approach. It is discussed in the context of the layout problem, but any engineering problem can be solved using these seven steps (also see Figure 1.16):

1. Identify the problem.
2. Gather the required data.
3. Formulate a model for the problem.
4. Develop an algorithm for the model and solve it.
5. Generate alternative solutions, evaluate, and select.
6. Implement the solution.
7. Continuously review after implementation.

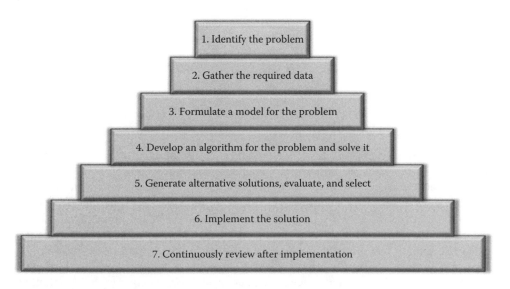

FIGURE 1.16 Seven steps of the engineering design problem approach.

1.5.1 Identify the Problem

The very first step in attempting to solve a problem is to identify it correctly. Inexperienced analysts think that the symptom of a problem is the problem itself. To understand this, let us address this step in the context of layout design. A plant manager notices an unusual level of parts backtracking their flow paths. This has resulted in traffic congestion, which in turn has decreased employee morale eventually resulting in poor-quality parts. The manager also noticed this problem occurred only after the company had a significant change in its product mix and volume. Three suggestions were made: (1) Fire the current machine operators, hire, and train new ones. (2) Buy newer highly automated manufacturing equipment and lay off some junior production personnel. (3) Redesign the layout so that the pieces of manufacturing equipment with the greatest traffic between them are placed close together. Only the last suggestion addresses the problem; the others address its symptoms. Thus, the problem we have is to design a new layout in which parts flow efficiently through the factory floor, minimizing congestion and thereby improving morale and product quality. Notice how easy it is to focus our efforts on the wrong problem and thereby waste precious time and money.

1.5.2 Gather the Required Data

In today's industrial environment, it is common practice to generate and maintain vast amounts of data. Each function (e.g., scheduling, quality control, plant layout, forecasting, accounting) requires its own set of data, and so, in a short period, a company accumulates a large volume of data. Rapid progress in computer technology (hardware and software) has made this accumulation possible and even encouraged it. In undertaking a major project such as layout design, analysts must request and generate only the required information. (The typical information required for layout design is described in detail in Chapters 2 and 3.) Examples of the required information are the type and quantities of production, service, and auxiliary equipment; some measure of interaction between these departments; space requirement for departments; total space available; and special layout considerations. Analysts get this information from production routing data for each part, work center data for each machine, the existing layout, and so on. These data contain information essential for plant layout, but they may also contain other information not required for this problem. For instance, a typical production routing sheet tells the sequence in which parts visit machines; it may also have the setup time, processing time, and the load and unload time for each operation. This additional information is useful for making scheduling decisions, but not necessary for the plant layout problem. Thus, the facility analyst has to exercise care in gathering data. Too often, analysts request excessive amounts of information, and then they are overwhelmed by the data provided. Under such circumstances, key pieces of data may be overlooked.

1.5.3 Formulate a Model for the Problem

The layout problem is an optimization problem and a design problem. (This aspect is discussed in more detail in Section 4.1.) This suggests that at some stage of the problem-solving process, we must develop a model for the problem. Then we can use suitable solution techniques or algorithms to develop alternate preliminary solutions to the problem. Of course, the solutions generated cannot be directly implemented because the models do not represent the actual problem perfectly. This is why we refer to the model solutions as *preliminary* solutions. Much work needs to be done to this preliminary solution before it can be implemented. However, it provides a sound basis on which to build layout designs. If we begin with a randomly generated preliminary solution and use this to develop a layout design, the original problems will persist. Thus, modeling is an important intermediate step in the layout design process.

Prior to the development of mathematical approaches, the layout problem was typically solved using 2D and 3D physical models—that is, icons of departments and material handling devices. An analyst moved the icons around a miniature model of the factory floor until a *satisfactory* layout was obtained (Francis et al., 1992). (Perhaps, this prompted some early researchers to classify the layout problem as a design problem.) Because this method of layout development is very subjective and other models are available, it is rarely used. In practice, the importance of mathematical models for layout design problems has not been fully recognized, and too often there is a tendency to ignore the modeling step altogether.

1.5.4 DEVELOP AN ALGORITHM FOR THE MODEL AND SOLVE IT

The layout problem is very hard to solve optimally. We must therefore develop heuristic algorithms that can generate near-optimal solutions quickly for medium- to large-scale problems. We are interested in near-optimal solutions for three main reasons. First, because the layout problem is complex, efficient optimal solution techniques capable of solving large-scale problems do not exist. Second, the model developed in the previous step is an imperfect formulation of the real problem, so there is no point in focusing on optimal solutions for an imperfect model. Third, the data generated are likely to change. Because any solution generated will undergo some change, we seek near-optimal solutions only.

The solution technique should require very little computational effort because computer time (especially on mainframes and workstations) in most companies is not easily available. Furthermore, we typically generate a large number of solutions before accepting a selected few. If generating each solution takes much computation time, the total time spent analyzing and selecting a few alternative preliminary solutions becomes even greater and expensive.

Throughout this book, we refer to small-scale problems as those problems with fewer than 12 machines; problems with more than 12 but fewer than 30 are considered medium-sized problems. Problems with more than 30 machines are treated as large-scale problems. As discussed in Chapter 11, branch-and-bound, dynamic programming and other optimal solution strategies are capable of solving small-scale problems only. For medium- and large-scale problems, we typically use efficient heuristics, techniques that do not guarantee generation of the best solution for the problem but yet provide good-quality, near-optimal solutions relatively quickly.

There are several solution techniques for the layout problem. Some of these are available as software packages, whereas others are available in the published literature. These techniques are discussed in detail in Chapters 5, 6, and 11. Because these algorithms are easily available, companies do not need to spend time developing their own. Most existing solution techniques provide optimal or near-optimal solution to the model in the form of a block plan, which shows the relative positioning of departments and is drawn to scale (see Figure 1.17).

1.5.5 GENERATE ALTERNATE SOLUTIONS, EVALUATE, AND SELECT

The techniques for solving layout models are extremely useful and important in providing preliminary solutions (block plans) for which detailed layouts can be designed. Because the models cannot capture all the elements of a layout problem, it is almost always necessary to modify the preliminary solution. For presentation and approval purposes, and also to facilitate the move to the new layout, it is necessary to construct a detailed layout that shows detailed shapes of departments, aisles, pickup and drop-off locations for each department, and possibly flow paths. A mathematical program that can develop a detailed layout from any given block plan is discussed in Chapter 10. Most companies are reluctant to employ sophisticated optimization techniques for developing detailed layouts from block plans, so typically, this function is done manually.

In this step, detailed layouts are developed for each of the few block plans selected. Then each is presented to top management and other concerned employees for their feedback and approval. To create a display, any of the tools discussed in Chapter 3—manual drawings, templates, 3D, and

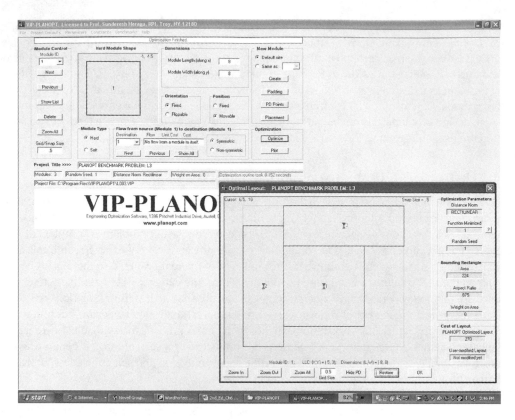

FIGURE 1.17   Layout provided by VIP-PLANOPT.

computer software–generated models—may be used, but the most common layout presentations are templates and computer-aided drawings (CAD). The other options (namely, manual drawings and 3D models) are not preferred at this stage because the layout designs presented by the analyst are likely to change. Incorporating changes and revising with these tools are time consuming and expensive.

The analyst discusses each selected layout with appropriate personnel and points out the advantages and disadvantages. In these discussions, other employees may point out errors, important aspects not considered by the analyst, and their own preferences. They may ask questions about the new layout designs. After detailed consultation with department supervisors, top management, and other employees affected by the layout changes, the most satisfactory design is selected. Of course, there is subjectivity in this decision-making process. For example, a layout with greater material handling costs may be preferred to another that promises lower costs because of the resistance of employees to the latter design. Subjective and objective factors must be carefully considered before a layout design is finalized.

1.5.6   Implement the Solution

After the layout is selected, the next task is to implement it. As Muther (1973) points out, it is a good idea to have the analysts responsible for the layout development oversee its implementation. It is well known that employees are generally reluctant to change. If we wish to have a new layout design work as intended, it must have commitment and support from the employees. A first step toward gaining this support is to communicate the changes clearly and in a timely manner. Employees should be told why changes are necessary, when the changes are expected to take place, how they will take place, what steps will be taken during the transition period, and the advantages and disadvantages of the new layout. Other steps such as planning the move to the new layout and assigning responsibilities are also important and must be considered.

1.5.7 CONTINUOUSLY REVIEW AFTER IMPLEMENTATION

Sometimes, a new layout design may not bring about the changes as planned. Despite careful analysis, proper communication, and careful planning, the new layout may create problems. Problems that crop up must be studied and measures taken to alleviate them. Thus, there is a need to continuously monitor the changes implemented. Whenever there is a deviation (favorable or unfavorable) between the actual results and the expected results, we must thoroughly investigate the reasons. If the deviation is not favorable, we must take steps immediately to correct the situation. If the deviation is favorable, we must analyze it so that it can be duplicated in other areas of the plant.

1.6 SUMMARY

This book is devoted to the study of design and planning problems. We examine how and where facilities are located, designed, grouped, and laid out. We also consider how the products are to be handled and stored. To the extent necessary, we use mathematical models. We wish to caution the reader that although there is heavy emphasis on models in some of the chapters, mathematical models alone cannot be used to design, locate, or lay out facilities. Often, these models are simplified in order to make them solvable, and as a result, subjective factors are typically not considered. However, the latter have a significant impact on the proper functioning of a facility. If we completely ignore the objective factors and arrive at a design decision based solely on subjective factors, we again run the risk of generating an inferior design. At the outset, the reader should understand that neither of the two factors can be ignored and thus a purely quantitative approach or a purely qualitative approach should not be used in making design or planning decisions. Rather, the two must be used in conjunction. For example, we may obtain a preliminary solution via a mathematical modeling (quantitative) approach and refine it via a qualitative approach.

1.7 REVIEW QUESTIONS AND EXERCISES

1. Compare and contrast design and planning problems. Explain why it is desirable to solve the two problems simultaneously rather than sequentially.
2. List the criteria generally used for evaluating a facility layout. Discuss each.
3. List and discuss some real-world constraints that analysts encounter while developing a facility layout.
4. List various types of facility layout problems and discuss an application for each.
5. How does the term facility layout differ from machine layout, plant layout, and office layout?
6. Visit the following *offices*:
   (a) Your local bank
   (b) A grocery store
   (c) Your favorite department store
   (d) An automobile dealer's showroom
   (e) The service department of an automobile dealer or your car service garage
   (f) The Department of Motor Vehicles office
   (g) A doctor's office
   (h) A real estate office
   (i) The Social Security Administration office
   (j) An airport
   (k) The Post Office

Conduct a thorough office operations review. Consult additional sources if necessary (e.g., Suskind, 1989).

7. List at least three examples for each of the following:
    (a) Closed structure office layout
    (b) Semiclosed structure office layout
    (c) Open structure office layout
    (d) Semiopen structure office layout

2 Product and Equipment Analysis

2.1 INTRODUCTION

Even when companies realize the importance of developing efficient layouts and designing efficient departments, many of them, especially the smaller ones, find that it is too time consuming and difficult to collect and generate data in the format required for problem solving. This raises two questions: What are the required pieces of information? Why is it difficult to obtain them? We answer the first question in detail in this and in the next chapters. The second question is answered in less detail, but from the answer to the first question, the reader will be able to answer the second.

Two kinds of analysis—product analysis and process analysis—are important for developing the data required to solve the facility design and layout problems. Product analysis is determining the sequence of machines visited by a part and production volume. It is discussed in this chapter. Equipment and personnel requirement analysis (which covers make-or-buy decisions), production capacity decisions, production space analysis, and employee requirement analysis are also discussed in this chapter. In Chapter 3, we discuss process analysis by focusing on material flow analysis.

2.2 PRODUCT ANALYSIS

The products manufactured by a company provide important information about processes or equipment required, material handling methods and systems needed, and the arrangement of production, auxiliary equipment, and service departments. Before beginning the facility design process, we must know not only the *types* of products manufactured but also their production *volume*. Both have an impact on the final layout of departments. As we discuss in Chapter 3, the production volume (along with other factors) determines the type of layout selected. We now describe how information about products is used to generate data that are essential in developing layouts.

2.2.1 Bill of Materials

As an illustration, we consider a single product manufactured by a medical equipment company—the 3.5 volt halogen pneumatic otoscope, a product used by physicians to examine a patient's ears. The multilevel exploded bill of materials for this product is shown in Figure 2.1. Notice that the top-level assemblies are indicated by the digit 1 in the first column of the figure, the second-level subassemblies by 0.2, third level by ..3, and so on. Also shown are the part numbers, their description, and quantity used per unit of the final product—the 3.5 volt halogen otoscope. Purchased parts have a part number with M or 9 as the first character.

2.2.2 Assembly Charts

To understand the flow of material in a factory, layout analysts use a variety of charts and diagrams. The assembly chart is a companion chart to the indented bill of materials. It tells us how subassemblies and components manufactured separately at various places and stages of manufacturing are assembled to make the end product. An assembly chart for a valve actuator is shown in Figure 2.2 and its bill of materials in Figure 2.3.

9/14/93 7:33:02		PRODUCT_STRUCTURE_INQUIRY C/N 100 WAMS		PKPSTQ02		
		MULTILEVEL EXPLOSION				
Part Number 21701		U/M EA Select Date 9/14/93 Select Alternate				
3. 5V HAL OPER OTOS W/O SPACES Planner · · · 401 Select Display..						
S LEVEL_____I0		Part Number	Description	Qty/Per____UM		A
	1	A01786	21700 ALIGNMENT	0.000	EA	Y
	1	900202	#2 BOX-5 X 2 ¾ X 1 7/8	1.000	EA	Y
	1	900242-6	LABEL W/ADHESIVE	1.000	EA	Y
	1	217001-501	BODY ASSY	1.000	EA	Y
	0.2	M30445	HAR-TECH HTL-15 COMPOUND	0.000	DR	Y
	0.2	M30452	3/8 F-50 MASS FINISHING MEDIA	0.000	LB	Y
	0.2	M30454	3/4 F-50 MASS FINISHING MEDIA	0.000	LB	Y
	0.2	M30458	3/4 F-50 MASS FINISHING MEDIA	0.000	LB	Y
	0.2	M31412	EASY FLOW 45	0.000	TO	Y
	0.2	M31418	ULTRA-FLUX	0.000	LB	Y
	0.2	217002-2	BASE	1.000	EA	Y
	..3	M01052	BRASS ROD ROUND	0.276	LD	Y
	0.2	217003	POST	1.000	EA	Y
	..3	M01062	BRASS ROD ROUND	0.049	LB	Y
	0.2	217016	LENS HOLDER STUD	1.000	EA	+
	..3	M01032	BRASS ROD ROUND	0.004	LB	Y
	0.2	217017-1	SOLDE PREFORM	1.000	EA	Y
	1	217024-1	LIGHT PIPE ASSY	1.000	EA	Y
	0.2	M30445	HAR-TECH HTL-15 COMPOUND	0.000	DR	Y
	0.2	M30446	HAR-TECH HTL-980 COMPOUND	0.000	DR	Y
	0.2	M30454	3/4 F-50 MASS FINISHING MEDIA	0.000	LB	Y
	0.2	217024-501N	LIGHT PIPE ASSY	1.000	EA	Y
	..3	A00296	USEABLE LIGHT OUTPUT TEST	0.000	EA	Y
	..3	M30241	3469 HARDENER	0.000	GL	Y
	..3	M30244	RE-2038 RESIN	0.000	GL	Y
	..3	202017-1N	FLARED TUBE	0.500	EA	Y
	···4	M04424-1	STAINLESS STEEL TUBE ROUND	0.350	FT	Y
	..3	31200958-119	FIBER BUNDLE	0.023	EA	Y
	1	209026-501	LENS HOLDER ASSY-BLACK	1.000	EA	Y
	0.2	209024-1	LENS HOLDER-BLACK	1.000	EA	+
	0.2	209009	OTOSCOPE LENS	1.000	EA	Y
	1	217005	TOP COVER	1.000	EA	Y
	1	203010	LENS HOLDER WASHER	2.000	EA	Y
	0.2	203010N	LENS HOLDER WASHER	0.010	EA	Y
	..3	M04467	STAINLESS STEEL STRIP	0.020	LB	Y
	1	203009	LENS HOLDER SPRING WASHER	1.000	EA	Y
	0.2	M02032	BERYLLIUM COPPER STRIP	0.002	LB	Y
	1	209012	LENS HOLDER SCREW	1.000	EA	Y
	1	217009	LOCKNUT	1.000	EA	Y
	0.2	M01049	BRASS ROD ROUND	0.067	LB	Y

FIGURE 2.1   Indented bill of materials.   (*Continued*)

0.2	M30438	HIGH DENSITY POLY MEDIA-ROTO	0.000	LB	Y
1	217010-1	LOCK RING	2.000	EA	Y
1	209025-1	SPEC HOLDER	1.000	EA	Y
0.2	209025	SPEC HODER	1.000	EA	Y
1	106008	PRISM SCREW	1.000	EA	+
0.2	106008N	PRISM SCREW	0.010	EA	Y
..3	M01023	BRASS ROD ROUND	0.040	LB	Y
1	250004	LAMP COLLAR	1.000	EA	Y
1	031100-503	031 LAMP ASSY	1.000	EA	Y
1	M30322	GE RTV-103 BLACK	0.000	TB	Y
1	217013	2-56 X 3/16 LG SET SCREW	1.000	EA	Y
1	M30338	LOCTITE 242	0.000	BT	Y
0.2	A01880	SPEC F/HANDLING LOCTITE MATL	0.000	EA	Y
1	M11044	LANOLIN	0.000	JR	Y

FIGURE 2.1 (*Continued*) Indented bill of materials. (Courtesy of Welch Allyn Medical Division.)

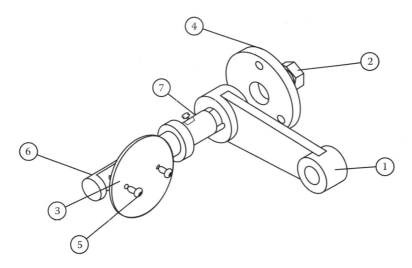

FIGURE 2.2 Assembly chart for a valve actuator.

Item no.	Part number	Description	Quantity
1	WPI-50	VALVE ACTUATOR ARM	1
2	WPI-22	.375-24UNF HEX HEAD NUT	1
3	WPI-30	VALVE PLATE	1
4	WPI-40	SHAFT RETAINER PLATE	1
5	WPI-31	#4-48UNF X .250 ROUND HEAD MACHINE SCREW	2
6	WPI-20	VALVE SHAFT	1
7	WPI-21	WOODRUFF KEY	1

FIGURE 2.3 Bill of materials for a valve actuator.

Assembly charts are useful in scheduling production activities. If the due date for an end product and the time required to procure or manufacture a component or subassembly are known, analysts can work backward from the due date to determine when orders for procurement or production are to be released for each component. The time required to procure or manufacture a component is the elapsed time from when the purchase or production order is released until the component is received.

2.2.3 ENGINEERING DRAWING

In this section, we examine some details of the body assembly of the 3.5 volt halogen otoscope introduced in Section 2.3.1. The engineering drawing of one of the otoscope's assemblies (the light pipe assembly with a part code of 217024-1) is shown in Figure 2.4. This engineering drawing assists the manufacturing or process engineer in determining the processes necessary for manufacturing the product. The drawing also provides some additional notes (not shown) for various operations that are indicated by circled numbers. These are notes on special toler-ances, material used, minimum functional capabilities of the finished product (e.g., the light pipe assembly must transmit at least 80% of usable light output); filling material to be used, if any; minimum dimensions before certain operations such as bending; and what to do in the event of deformation or cracks (e.g., reject, rework, correct process, but allow current batch to proceed). The drawing also gives information on where, when, and by whom the part was done, as well as its number.

In addition to the engineering drawing and bill of materials, a layout analyst may have an exploded view of the drawing of the end product. This diagram provides the same information as the bill of materials but presented pictorially. Such a drawing is most useful for parts with fewer subassemblies. As the number of subassemblies and assemblies increases, the exploded diagram showing all the individual component drawings may be difficult to read.

2.2.4 ROUTE SHEET

A route sheet specifies the operations required for a part and the sequence of the machines visited by the part for these operations. It may also provide the setup time for each operation at each machine, processing times, and labor time. A route sheet for the body assembly mentioned in Section 2.2.3 is shown in Figure 2.5. The information contained in a route sheet may vary slightly from company to company, but it includes two key pieces of information: the machines visited and the sequence

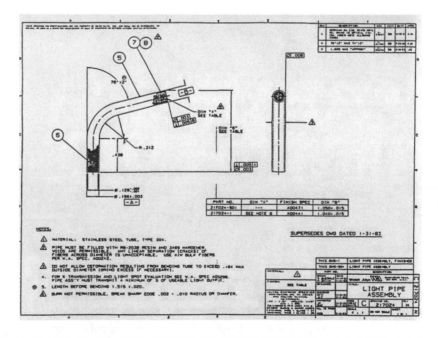

FIGURE 2.4   Part drawing. (Courtesy of Welch Allyn Medical Division.)

```
                              MASTER ROUTING LIST
--PART NUMBER--  -------DESCRIPTION-------    DATE     ALT CODE  BUYER/PLANNER    DRAWING REVISION
H6709           HANDLE, DENSPLY PROBE         6/25/92  B         239                      G

OPER   WORK    OPER   SETUP   CREW  MACH   --TOOLINGREF--  --------STANDARD-------           TIME         MOVE
& ALT  CENTER  CODE   FACTCR  CODE  GROUP    NUMBER        --SETUP-  -LABOR-  -MACHINE--  I/O  BASIS       TIME    -----EFFECTIVE----
       CODE                                                 HOURS    HOURS     HOURS           CDE/-QTY-  -DAYS-    FRCM       TO

10     01226          1.0     01226 T9330                  12.000   336.880   336.880         4  1000     .000      0/00/00    99/99/99
       ---------ROUTING  DESCRIPTIONS-------------
       MAKE @ AUTO 804843P
       C 804843P B
PARTIAL AHEAD QTY
       A A02247 A
15     02053          1.0     02053 T9712                   1.500    41.670    83.330         4  1000     .000      0/00/00    99/99/99
       --------ROUTING  DESCRIPTIONS-------------
       DRILL & TAP 804843P1
PARTIAL AHEAD QTY
       B 804843P1 A
20     02053          1.0     0.2053 T9713                  1.500     8.330    16.670         4  1000     .000      0/0C/00    99/99/99
       --------ROUTING  DESCRIPTIONS-------------
       KNURL OD
PARTIAL AHEAD QTY
30     03029          1.0                                    .000    33.330    33.330         4  1000     .000      0/0C/00    99/99/99
       ---------ROUTING  DESCRIPTIONS-------------
       SCOTCHBRITE/BELT
PARTIAL AHEAD QTY
40     03105          1.0     03105                          .000     3.000     3.000         4  1000     .000      0/0C/00    99/99/99
       ---------ROUTING  DESCRIPTIONS-------------
       PASSIVATE
PARTIAL AHEAD QTY
50     03005          1.0                                    .000    54.170    54.170         4  1000     .000      0/0C/00    99/99/99
       ---------ROUTING  DESCRIPTIONS-------------
       BUFF
PARTIAL AHEAD QTY
60     03007          1.0     03007                          .000    50.000    50.000         4  1000     .000      0/0C/00    99/99/99
       ---------ROUTING  DESCRIPTIONS-------------
       GLASSBEAD KNURL/  SHIP
PARTIAL AHEAD QTY
```

FIGURE 2.5 Part-routing sheet. (Courtesy of Welch Allyn Medical Division.)

in which the machine visits take place. For example, the route sheet in Figure 2.5 shows that the Densply Probe Handle (Part # H6709) has seven operations in which machines 1226, 2053, 3029, 3105, 3005, and 3007 are visited in that sequence for some preliminary operation, drilling and tapering, knurling, scotchbriting, passivating, buffing, and glassbead knurling operations. The route sheet also shows the setup, machining and labor times, and crew factors for each operation.

2.2.5 OPERATION PROCESS CHART

The operation process chart is like an assembly chart except that it is more detailed. It provides information on the processes and time required for each component. This chart uses symbols from the American Society of Mechanical Engineers (ASME) standards to represent each production activity. The five basic activities in manufacturing are listed with their symbols in Figure 2.6.

An *operation* is defined as an activity in which one of the characteristics of an item is changed. Examples are drilling a hole in a part and filling out an order form. *Inspection*, as the name indicates, is an activity in which the characteristics of an item are compared with an established standard. Checking the diameter of a hole in a part to see whether it is within the allowed tolerance and verifying an order form are examples of inspection activities. *Transportation* involves movement of material from one location to another. *Storage* refers to an activity in which an item is kept at a designated place until authorization is received for it to be moved. For example, tools are stored in a tool crib and raw materials are stored in the receiving area until authorization for their release to the shop floor is received. *Delay* is an activity in which the item is waiting for its next planned action to take place. Items received from a supplier may be waiting at the staging area for someone to inspect the packing slip to ensure that they were received in the right quantities and without damage. A suitcase at an airport baggage claim area circling on a conveyor waiting to be picked up by its owner is another example of a delay activity.

An operation process chart for the 3.5 volt halogen pneumatic otoscope (its bill of materials appears in Figure 2.1) is shown in Figure 2.7. Some of the purchased parts and assemblies in Figure 2.1 are not included in the operation process chart in Figure 2.7. Though not shown, other information—such as time for each operation, number of drilling, turning, or other operations—may be included in the operation process chart.

Symbol	Name
●	Operation
■	Inspection
▶	Transportation
▼	Storage
◗	Delay

FIGURE 2.6 Symbols for five basic manufacturing activities.

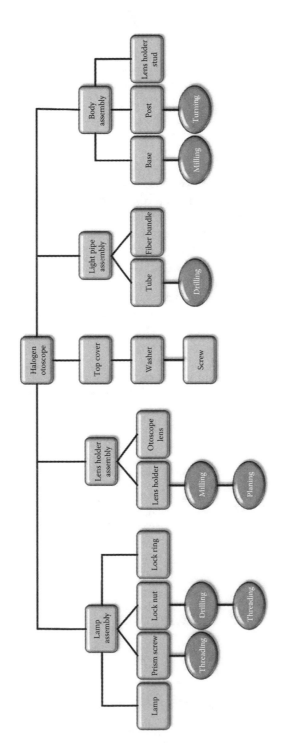

FIGURE 2.7 Operation process chart for a 3.5 volt halogen otoscope.

2.3   EQUIPMENT SELECTION[*]

The key to success of automated manufacturing systems is proper selection and effective use of production, service, and auxiliary equipment. The proper use of these resources has significantly increased productivity. In recent years, the manufacturing industry has witnessed significant developments. The increase in the number and types of automated systems in the industry is a testament to the developments taking place. However, these developments have taken place at the expense of system design problems. Design problems have become even more complex, and designers and users of automated manufacturing systems have developed new tools to cope with these problems. Manufacturing system design, a hierarchical combination of these problems, is therefore a complex activity. It involves solving a number of problems arranged in a hierarchy. At times, it may be necessary to backtrack to higher level problems. The two-level hierarchical approach, illustrated in Figure 2.8, involves solution of the following two problems: production and material handling equipment selection and facility layout (Kusiak and Heragu, 1988). For cellular manufacturing systems (discussed further in Chapter 6), another approach is suggested (Figure 2.9). Group technology (GT) provides the basis for the four-level hierarchical approach to design cellular manufacturing systems. It should be emphasized that the four-level hierarchical approach is generally suitable for only manufacturing systems that produce a large number of components and for which manufacturing activities can be decomposed into almost mutually independent cells. For mass production or continuous production systems, such an approach may be unsuitable.

Before making layout decisions, analysts must know the required *types* and *quantities* of production and support equipment. This problem is sometimes referred to as the machine requirements problem or equipment selection problem as illustrated in Figures 2.8 and 2.9. To determine the required types of equipment, one must first know what types of basic production processes are required (e.g., forging, casting, forming, metal removal, fabricating, drilling holes, planing surfaces, finishing surfaces, and so on) and then match the available equipment with these processes. For example, a forge is used for forging operations; cold rolling or extruding machines are used for forming processes; hobbing machines are suitable for gear cutting operations; mechanical shears and numerically controlled punch presses are useful for metal forming operations; radial drill presses or boring machines are suitable for drilling and enlarging holes; a planer is required for

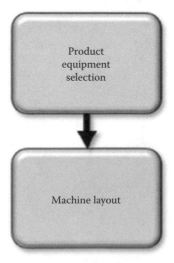

FIGURE 2.8   Two-level hierarchical approach to manufacturing design.

[*] Parts of this section have been reproduced with permission from Heragu and Lucarelli (1996), which appears in *The Industrial Electronics Handbook*, J.D. Irwin, ed., CRC Press, Boca Raton, FL, 1996.

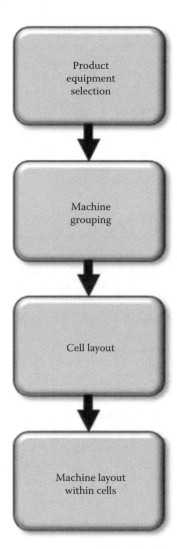

FIGURE 2.9 Four-level hierarchical approach to manufacturing design.

planing surfaces; a sanding machine, honer, centerless grinder, and rotary surface grinder are used for surface finishing; and milling machines are required for certain types of metal removal such as facing and slotting. Production equipment is typically classified into various types based on its function. Some equipment types are lathes, horizontal milling machines, vertical milling machines, planers, shaping machines, and vertical turret lathes.

By *quantity* of production equipment, we mean the number of pieces of each type of equipment available for production purposes. A company may have available three lathes, four horizontal milling machines, two vertical milling machines, one planer, one shaping machine, and two vertical turret lathes. Each type of equipment may require some support facilities as well. For instance, forging equipment will require heat treatment stations and painting stations will require drying equipment. The type and quantity of support and auxiliary equipment are determined by the type of production equipment. In fact, some types of production equipment are sold with the auxiliary equipment.

While exact numbers are not necessary, design analysts must at least have a rough idea of what types of machines are capable of meeting processing needs, what support facilities are required, and approximately how many of these are needed. Automated production equipment, though capable of performing a variety of operations, is typically expensive. Thus, the production equipment selection problem is critical in the design of a system. By determining the right quantities and type of equipment, one can achieve the following benefits (Heragu and Kusiak, 1987):

- Make efficient use of capital equipment purchase budget.
- Make efficient use of maintenance and operating budgets.
- Increase machine utilization.
- Make efficient use of available space, as fewer equipment sufficient to meet production now and in the future are purchased.

2.3.1 TRADITIONAL MODEL

There are several models for determining the type and quantity of required production equipment. In this section, we discuss traditional and queuing-based approaches.

Traditional approaches are very simple. Based on the number of products, desired production rate, production efficiency of the equipment required to process the products, standard processing times for the operations required on the products, and time for which machines are available, we can develop a simple formula to determine the number of production equipment required. One such formula originally from Shubin and Madeheim (1951) is presented below. It uses this notation:

P Desired production rate in units per day
η Efficiency of the machine
τ Time for which machine is available per day, in hours
t Time required to process one unit of product at the machine, in hours
NM Number of units of the machine required

The following formula determines the number of units of machine required, assuming that only one product is processed on this machine:

$$NM = \left\lceil \frac{tP}{\tau\eta} \right\rceil, \quad \text{where} \quad \left\lceil \bullet \right\rceil \text{ is the smallest integer} \geq \bullet \tag{2.1}$$

The preceding analysis can be easily extended when we have more than one stage of production (Miller and Schmidt, 1984). A backward analysis is used to determine the desired production rate at each stage. Based on the number of good units required at the output of the last production stage, we can determine the number of units that must enter this last stage. This depends on the percentage of scrap at the last stage of operation. For example, if S_l is the scrap rate at stage l, expressed as a fraction, N_{ol} is the number of good units required at the output of stage l, and N_{il} is the number of units required at the input of stage l, then the number of good units required at the output of stage l is equal to the number of input units minus scrap; that is, $N_{ol} = N_{il} - S_l N_{il}$. Therefore,

$$N_{il} = \left\lceil \frac{N_{ol}}{1 - S_l} \right\rceil \tag{2.2}$$

Performing a backward analysis for each stage of operation, we can determine the number of units of raw material required to produce the desired number of good finished units. Formula (2.2) assumes that an item not processed within allowed tolerances cannot be used further and must be scrapped. In many situations, units that do not conform to product specifications may be reworked. In that case, formula (2.2) can be modified to account for rework. This is left as Exercise 12 to the reader at the end of the chapter.

Although traditional approaches are simple, they have certain drawbacks. For example, it is difficult to handle the case when more than one product is processed at a given machine or workstation. Also, traditional approaches do not take into consideration budget, overtime, floor space, and other constraints. When such constraints are imposed, mathematical programming approaches are useful.

Another traditional approach uses time standards to determine the quantities of machines of each type. This approach is illustrated using Example 2.1.

Example 2.1

Consider a simple job shop manufacturing system that makes three *class A* products requiring three types of machines. The three products include seven parts shown in Table 2.1. Table 2.1 also shows the time standards in units per hour. Assuming an hour has only 55 minutes of productive time (5 minutes is lost due to operator or machine unavailability and machine downtime), 12,000 units of each part are to be made per day, and the job shop operates only one 8-hour shift per day, determine the number of units of machine types A, B, and C that would be required to achieve the desired production rate.

Solution

Dividing the values in Table 2.1 by 55, we get the time standards expressed in minutes per unit in Table 2.2. Ignoring the setup time, we add up the time taken by all the seven parts on each machine type as shown in the last column of Table 2.2. Multiplying these values by 12,000 and dividing by 440, we find that we need 4.9 units of machine A, 5.85 units of machine B, and 4.3 units of machine C. Rounding up these numbers gives us 5, 6, and 5 units of machine types A, B, and C, respectively.

2.3.2 Queuing Model

The traditional approach is a static model in the sense that it assumes all parameters are known with 100% certainty. It assumes that each part takes the same amount of time to produce, unit after unit. Machine breakdowns are taken into consideration rather crudely by reducing the time available on a machine by a certain percentage. A better method to take into consideration the problem dynamics is to use queuing models to determine the number of units of each machine type required. In Example 2.2, we see how a myopic approach such as the traditional approach provides erroneous results especially when there is significant variability in one or more of the parameters.

TABLE 2.1
Time Standards for Seven Parts in Units per Hour

Part Machine	1	2	3	4	5	6	7
A	2000	—	1200	1500	—	2300	1200
B	1200	1800	1200	—	1600	2000	1000
C	—	—	1200	2000	1200	—	1400

TABLE 2.2
Time Standards for Seven Parts in Minutes per Unit

Part Machine	1	2	3	4	5	6	7	Total Time
A	0.0275	—	0.0459	0.0367	—	0.0239	0.0459	0.1799
B	0.0459	0.3056	0.0459	—	0.0344	0.0275	0.0550	0.2143
C	—	—	0.0459	0.0275	0.0459	—	0.0393	0.1586

Example 2.2

Manufacturing engineers at the Widget Manufacturing Company recently convinced their manager to purchase a more expensive but flexible machine that can do multiple operations simultaneously. The rate at which parts arrived at the machine that was replaced by the flexible machines follows a Poisson process with a mean of 10 parts per hour. The service rate of the flexible machine is 15 units per hour compared with the 11 units per hour service rate of the machine it replaced. (All service times follow an exponential distribution.) The engineers and manager were convinced that the company would have sufficient capacity to meet higher levels of demand, but just after 2 months of purchasing the machines it turned out that the input queue to the flexible machine was excessively long. Part flow times at this station were so long that the flexible machine became a severe bottleneck. The engineers noticed that more parts were routed through this machine, and that the parts' arrival rate to the flexible machines had increased from 10 units per hour to about 14 units per hour, but were puzzled why the part flow time at this station doubled from 30 minutes to 1 hour and the work-in-process (WIP) inventory increased nearly threefold from 5 to 14 when the arrival rate only increased 40%. Use a queuing model to justify the results observed at Widget Manufacturing Company.

Solution

The problem can be modeled as an M/M/1 queuing model. Using results from the M/M/1 model in Chapter 13, we get the results shown in Table 2.3.

The queuing model explains very well why the flow time has doubled and the WIP has increased threefold. The traditional approach neglects to consider variability, and if the variability in the part interarrival or service times is high, as is the case when these random variables follow an exponential distribution, it has a dramatic impact on the WIP and flow times.

2.4   PERSONNEL REQUIREMENT ANALYSIS

Along with the production and material handling equipment selection problem, another problem needs to be addressed: employee requirements. Simple models provide a basis for determining the actual number of employees needed by considering factors such as whether the labor force is unionized or nonunionized, level of automation, production rate, management policies on subcontracting and overtimes, salary rates in the area, health insurance rates, and rules set by the OSHA.

The number of employees required in a new facility is typically proportional to the volume and variety of production. For example, if

n Number of types of operations
O_i Aggregate number of operations of type i required on all the pseudo (or real) products
    manufactured per day

TABLE 2.3
Results for Example 2.2

Parameters	Replaced Machine	Flexible Machine
Arrival rate (λ)	10	14
Service rate (μ)	12	15
Machine utilization (λ)	0.83	0.93
WIP [$L = \lambda/(\mu - \lambda)$]	5	14
Flow time [$W = 1/(\mu - \lambda)$]	0.5	1

T_i Standard time required for an average operation O_i
H Total production time available per day
η Assumed production efficiency of the plant

then the number of production employees required is given by

$$N = \sum_{i=1}^{n} \frac{T_i O_i}{\eta H} \tag{2.3}$$

The factor η in Equation 2.3 adjusts the available production time to account for time lost to maintenance, machine failure or repair, and worker/material/machine/material handling system unavailability. Many assumptions made in Equation 2.3 may not hold in practice. For example, the model implicitly assumes that the same operations are performed day after day and that machine setup times are dependent on the volume of production. Furthermore, the model determines only the required number of manufacturing employees and ignores the number of nonmanufacturing workers. Generally, as production activity increases, the number of nonmanufacturing workers required also increases because companies that have a large workforce must provide additional services—security, janitorial, food, physical recreation, secretarial, and so on. To provide these services, separate departments such as security department, visitor information center, cafeteria, and janitorial department must be formed and adequately staffed. Simple rules of thumb are used to determine the number of support personnel, such as 1 secretary for every 20 production employees and 1 secretary for each executive.

Personnel decisions impact design of support departments such as parking lot, restrooms, locker rooms, and cafeteria. Governmental regulations and company policy specify these support departments. For example, local, state, and federal government regulations may dictate the location of parking lots, fire exits, location, and number of restrooms. Productivity and cost considerations may dictate whether or not cafeteria service is provided. Although a company may incur more fixed costs as a result of providing cafeteria service due to increased heating and maintenance, it may increase productivity because workers dine at the same place and have a chance to talk about work-related problems. A company cafeteria may reduce lunch break times and minimize the problem of workers returning late.

As in the equipment selection problem, let us now use a stochastic (queuing) model to capture the impact of variability and thus help in determining the number of employees required.

Example 2.3

The American Automobile Drivers' Association (AADA) is the only office serving customers in New York's greater capital district area. Ahead of the busy summer season, the office manager wants to hire additional staff members to help provide these services to members effectively— summer travel planning, membership renewal, airline, hotel, cruise booking, and other travel-related services. It is anticipated that each customer typically requires 10 minutes of service time and customers arrive at the rate of one customer every 3 minutes. The arrival process is Poisson and the service times are exponentially distributed. Determine how many staff members are required if the average wages and benefits per staff member are $20 per hour and the *cost* to AADA for every hour that a customer waits to be served is $40.

Solution

We use an M/M/m model to formulate the problem. Using the shortcut formula due to Sakasegawa (1977) for determining the time spent in queue, we find the results shown in Table 2.4 with four,

TABLE 2.4
Results for Example 2.3

Number of staff members (m)	4	5	6
Arrival rate (λ)	20	20	20
Service rate (μ)	6	6	6
Machine utilization ($\rho = \lambda/m\mu$)	0.83	0.67	0.56
Time in queue	0.1447	0.0261	0.0075
Hourly cost ($\$$)	195.74	120.85	126.01

five, and six staff members. Based on the hourly costs (calculated and shown in Table 2.4), it is clear that AADA must employ five staff members. Values in the hourly cost row were obtained by adding the labor cost and the waiting cost per hour. For example, with five staff members, the labor cost is 5∗$20 = $100 and the waiting cost is 0.0261∗20∗$40 = $20.85.

2.5   SPACE REQUIREMENT AND AVAILABILITY

Other factors that need to be considered in developing a layout are the requirement and availability of space. By understanding the flow between machines or the interaction between departments, a designer can determine the space necessary to locate machines. Generally, the shape, size, and space requirement for each department are found by actually measuring the dimensions of the equipment within the department and obtaining space requirement information from experienced personnel on the shop floor. To obtain machine footprint dimensions, one could also use a layout generated using computer-aided design (CAD) software, if one is available.

When determining the space required for laying out departments, the analyst must allow room for operator movement, loading and unloading of parts, a buffer storage area for incoming material and WIP, and shelves for storage of machine accessories and tools. Again, simple rules of thumb are used to determine the extra space that is required. In some companies, 3 or 4 feet is added to the length and width of each machine or workstation. In others, the additional space is calculated as a percentage of the actual area occupied by a workstation, typically 200%–300%.

The preferred method of determining extra space is to calculate the space required for the workstation, auxiliary equipment, operator space, incoming material and WIP space, and other additional space (e.g., load and unload access, material handling carrier clearance) and add the separate quantities to determine the total space required. This method is illustrated for a manufacturing cell in Figure 2.10. Basically, for each department (department names are listed in column 1), we enter the production equipment name and code in columns 2 and 3.* We also enter the length and width of each production equipment (columns 4 and 5) and use this information to calculate the area required by the equipment (column 6). We then determine the space required for the corresponding auxiliary equipment, operators, and material (columns 7–9) and add these quantities in column 10. The space for auxiliary equipment is usually determined by calculating actual space required or by estimating it as a percentage of space required by production equipment. We then multiply the total space by a factor (entered in column 11), typically 150%, to allow for material handling equipment tracks and aisles, storage of WIP during peak operating periods, space required for maintenance operations, and anticipated increase in production volumes in the near future. The result is the total area required for each machine type (column 12). Multiplying this value by the number of units of machines of the corresponding type (entered in column 13) provides the space required for each machine type.

---

* If the preliminary layout is not developed yet, we do not know the allocation of specific machine types to departments, and hence column 1 is not needed.

Department Name	Work Center Name	Work Center Code	Length (Feet)	Width (Feet)	Area (Square Feet)	Auxiliary Area (Square Feet)	Operator Space (Square Feet)	Material Space (Square Feet)	Sub-Total (Square Feet)	Allowance (Square Feet)	Total Space per Machine (Square Feet)	Number of Machines	Total Space Machine Type (Square Feet)
General machining	Vertical milling	1202	15	15	225	70	30	50	375	150%	565	2	1130
	Planer	2005L	25	5	125	40	20	40	225	125%	290	1	290
	Punch press	3058	10	10	100	30	20	20	170	140%	240	2	480
	Injection molding	6078	20	10	200	60	50	100	410	150%	615	3	1845
Otoscope cell	NC-machine	9087	20	8	160	50	30	30	270	125%	340	2	680
	Lathe	1212	15	8	120	40	20	30	210	150%	315	1	315
	Auto-chucker	2056	5	5	25	10	5	5	45	125%	60	1	60

FIGURE 2.10 Production space requirement sheet.

Designers must also allow space for future expansion and growth. In general, gauging whether and when expansion and growth will take place is rather difficult. In fact, in the present industrial climate, many manufacturing companies in the United States are downsizing production activities, facilities, and personnel. Not providing room for expansion while designing departments may prove costly, however, if future growth in the production activity calls for employing more equipment and people. Of course, providing too much room for expansion also has its disadvantages such as higher investment and leasing costs as well as heating and cooling costs. Thus, a facility designer has to find the right balance between providing too much and too little space for future expansion.

Although much of the discussion in this section has focused on horizontal space, determining the vertical space requirement is also important, especially if overhead material handling devices are to be used or commodities are to be stored, as in a warehouse.

2.6 SUMMARY

In this chapter, we discussed the importance of product analysis and its use in developing some of the basic data required for facility design. (The remaining data are discussed in the next chapter.) Although some of the discussion may not be applicable to office layout or other service organization design, the discussion nevertheless points out the factors that need to be considered before a complete design is developed. Engineering drawings and bill of materials of the products manufactured tell us what processes are needed. Using analysis tools such as the assembly chart, operation process chart, and route sheets, we get a rough idea of how the material flow takes place. Knowing what general manufacturing processes are required helps us determine the types and quantities of production equipment. Mathematical models are used for this purpose. We also discussed methods of determining space required for production and auxiliary equipment. Simple models for determining the number of employees were also provided.

In addition to the traditional approach to determine the required number of employees and equipment, queuing models may also be used. For example, a queuing model may be used to determine the number of employees to hire in a shipping department to fill out orders. Sometimes data required for queuing models may be too time consuming to collect or facility designers may be unaware of analytical models. For these reasons, simple formulae such as the one in Section 2.6 or rules of thumb tend to be used in practice. More and more companies are recognizing the value of analytical models, however. They are gradually gaining acceptance in industry.

2.7 REVIEW QUESTIONS AND EXERCISES

1. Get the engineering drawing and bill of materials for a single product from a local manufacturing company. Using the engineering drawing and bill of materials, prepare a list of assemblies and subassemblies that are
   (a) Manufactured within the plant
   (b) Purchased from an outside vendor or obtained from a supplier company
   For each assembly and subassembly manufactured within the plant, obtain a route sheet. Delete all the information not relevant for layout planning purposes. In other words, retain only the machines visited (and the sequence) for each assembly/subassembly.
2. Draw the assembly chart for the following products:
   (a) Pedestal fan
   (b) Garden hose hanger
   (c) Bicycle
   (d) Stereo rack
   (e) Computer desk
   (f) Pedestal lamp
3. Show the indented bill of materials for the products in Exercise 2.

4. Consult additional sources in your library (for example, Groover, 1996) and list the various production processes and the type of available equipment for each. (Examples: gear cutting process, hobbing machine; hole drilling, drill press; and so on.)

5. In what ways are the equipment selection and facility layout problems related? Explain.

6. Hosreel, Inc., a manufacturer of a plastic garden hose reel, is contemplating the purchase of a new type of injection molding equipment that can produce the end product by itself. In other words, no additional operations or equipment is needed. The company wants to produce 1000 units per day and the machine efficiency is 99%. Assuming that each reel can be manufactured in 57 seconds, determine how many pieces of the new injection molding equipment Hosreel must purchase if it operates on only one 8-hour shift per day.

7. If the injection molding equipment in Exercise 6 has a scrap rate of 10%, would the answer to Exercise 6 change? Why or why not? Show calculations to support your answer.

8. Assume that 30% of the *scrap* coming from the injection molding equipment in Exercise 6 can be reworked to yield usable end products, i.e., hose reel. Appropriately modify formula (2.1) and use it to determine how many pieces of the injection molding equipment are needed. Is the new answer different from that obtained in Exercise 7? Explain.

9. The route sheet for a handle used in medical diagnostic equipment is given in Table 2.5. Also given are the desired hourly output and scrap rate for each machine. Assuming that each machine has 90% manufacturing efficiency, determine (1) the required input and (2) how many machines of each type Mediquip Company, Inc., must purchase if the production rate at the lathe, drill, knurl, and buff is 6, 10, 15, and 18 units per hour respectively.

10. A manufacturer is contemplating the purchase of a punch press. Approximately 10,000 units are processed on the press each day and the machine efficiency is 95%. Assuming that each punching operation takes 10 seconds, determine how many pieces of the press must be purchased if the company operates two 8-hour shifts per day.

11. If the press in Exercise 10 has a scrap rate of 10%, would the answer to Exercise 10 change? Why or why not? Show calculations to support your answer.

12. Assume that 70% of the *scrap* coming from the press in Exercise 10 can be reworked. Appropriately modify formula (2.1) and use it to determine the quantities of punch presses needed. Is the new answer different from that obtained in Exercise 11? Explain.

13. Parries Confectionery Company is considering the introduction of a new toffee. It requires four processes: A, B, C, and D in sequence. The equipment required for each process is available in the company now and is dedicated to the production of another type of candy. Although some capacity is available on all the equipment types, management is not sure whether they need to buy additional pieces. To aid management in decision making, the following information is available:

(a) The four pieces of equipment—one for each type—are available 50% of the time.

(b) The company operates one shift per day (8 hours) and 250 days per year.

TABLE 2.5

Production Data for Mediquip Co., Inc., in Exercise 9

Part Name: Handle		Prepared by SSH		
Oper Code	Machine Code	Machine Name	Required Output	Scrap Rate (%)
10	2053	Lathe	103	5
20	1226	Drill	100	8
30	3029	Knurl	100	12
40	3007	Buff	100	10

TABLE 2.6

Production Data for Parries Confectionery in Exercise 13

Time (in Seconds)	Process/Equipment			
	A	B	C	D
Setup/maintenance/cleaning per setup	1800	2700	600	1000
Processing time per unit	25	4	2	4
Scrap rate (%)	10	5	1	1

(c) If the company decides to produce toffee, then it must allocate half a day for candy production and the remaining half day for toffee.

(d) The time required for equipment setup, cleaning, and maintenance and *scrap rate* (i.e., spillage, production items that do not meet shape and other quality standards and hence must be discarded, etc.) for each process is given in Table 2.6.

(e) The demand for toffee is expected to be 1,000,000 per year. If 1,000,000 toffees are required to be made per year, determine how many input and output units are required for each process.

(f) Determine whether the company can produce the new toffee with the available equipment. If not, how many pieces of each equipment must Parries purchase? Show your calculations.

(g) The company can operate half a second shift by using 15 employees who must be paid an overtime salary of $20 per hour. The overhead expenses are estimated to be $10,000. It is known that equipment for processes A, B, C, and D cost $50,000, $200,000, $100,000, and $100,000, respectively. Is it better to operate the (half) second shift or purchase necessary additional equipment. Explain your answer.

14. Repeat Exercise 13 assuming the four types of equipment are available 70% of the time.

15. Hosreel, Inc., a manufacturer of a plastic garden hose reel, is contemplating the purchase of a new type of injection molding equipment that can produce the end product by itself. In other words, no additional operations or equipment is needed. The parts arrive at the injection molding machine at the rate of 125 units per hour. Assuming that each reel can be manufactured in 57 seconds, determine how many pieces of the new injection molding equipment Hosreel must purchase if it operates on only one 8-hour shift per day. Assume that the arrival process is Poisson and that the service times are exponentially distributed.

16. How would your answer to Exercise 15 change if the service times followed some known but general distribution with a mean of 57 seconds and standard deviation equal to 5 seconds. Explain.

17. An employee working for a state office processes incorporation applications from individual companies. Until last year, the employee used to receive an average of 2.5 forms per week and she could process these at the rate of 5 forms per week. This year, the average number of forms received nearly doubled to 4.9 forms a week. The employee's manager noticed that the number of applications at the employee's desk increased from about one (on average during the last year) to about 25 (this year). He was puzzled why the backlog increased 25 times when the workload had only doubled. Use queuing theory to show that the employee was diligently performing her duties. What should be done to bring back the backlog to an average of one form?*

---

* Based on a real-world problem in Winston (1994).

TABLE 2.7
Data for Exercise 18

Machine Type	Length	Width	No. of Machines
Hobbing	10	5	1
Punch press	10	10	1
Extruding equipment	20	10	2
Injection molding	10	10	3
Surface finishing	5	5	4
Deburring	10	10	2

18. CellMan, Inc., has three manufacturing cells, each of which houses a variety of manufacturing equipment. One of the cells has six machines. Data pertaining to these six machines are provided in Table 2.7.

(a) Add 10 feet to each side of the first two machine types and 8 feet to the remaining to allow for operator access, material handling carrier access, in-process material storage, etc. Determine the total space required.

(b) Prepare a table similar to the one shown in Table 2.5, assuming that for each equipment type
  (i) Auxiliary equipment dimensions are 10% of those of the primary equipment.
  (ii) Operator area is 8% of that required for auxiliary and primary equipment.
  (iii) Material storage area is 20% of the actual area required for the primary equipment.
  (iv) Additional allowance is 80% of the areas determined in (i), (ii), and (iii).

3 Process and Material Flow Analysis

3.1 FAST-FOOD INDUSTRY

Fast-food restaurants have been in the United States since at least the 1920s when the first White Castle opened in Wichita they are expanding in much of the world, notably in Asian countries. In 2015, this industry employed more than 4 million people in the United States and is considered a $250 billion industry worldwide. These restaurants use process, material flow, layout, and technology effectively to ensure more people are served as quickly as possible at the lowest possible cost. For example, McDonald's has a layout that allows unskilled workers to do just one thing repetitively—like an assembly line. A worker receives orders and payments from customers, another grills burgers at a grill station, a third fries potatoes at the frying station, and so on. The orders are displayed after payment is received on monitors at each station so the worker at that station knows exactly what to do for each order.

The 43,000 Subway stores also have an assembly line layout with bread, meat, topping, and dressing stations laid out in a sequence. This layout allows customers to customize their orders and yet be served in a minimum amount of time. Whereas Subway does not deliver sandwiches, Jimmy John's is a fast-food chain that specializes in *freaky fast* service—for drive-through, dine-in, or delivery orders. A layout of the Jimmy Johns store is shown in Figure 3.1. Note that the simple, open layout allows for efficient customer service.

3.2 INTRODUCTION

Process and material flow analyses tell us what processes and equipment are needed and how the general material flow will take place. We discussed some elements of process analysis—determining the type and quantities of processing equipment and calculating their space requirements—in Chapter 2. We also discussed product analysis and its importance in development of some of the data required for facility design in Chapter 2. In this chapter, we focus on material flow analysis and how to develop flow and distance data, which are essential in creating layouts. We also list the tools for presenting layout designs and discuss a case study that shows how to generate layout data.

3.3 DATA REQUIREMENT FOR LAYOUT DECISIONS

Given that a good layout minimizes the cost of moving material and people, the following data can assist a designer in layout planning:

1. Frequency of trips or flow of material or some other measure of interaction between departments
2. Shape and size of departments
3. Floor space available
4. Location restrictions for departments, if any
5. Adjacency requirements between pairs of departments, if any

Some of these data are not essential (for example, location restrictions for departments), but knowing the frequency of trips between departments, and the shape and size of departments, is necessary

FIGURE 3.1   Layout of a typical Jimmy John's restaurant. (Courtesy of Certified General Contractors, Inc.)

to make a preliminary layout. The frequency of trips between departments is used to determine the interaction between department pairs. If frequency data are not available, the facility designer must at least have subjective information about the traffic or flow intensities between departments (e.g., heavy, medium, low). In fact, as discussed later in this chapter, the relationship chart uses qualitative measures to determine the desired closeness between departments. Designers must know the dimensions of each department, along with the space requirement for each. For machines and workstations, space must be allowed for clearance between them and aisle space for material handling systems that serve the machines.

3.3.1   Flow Patterns

The first step in the design of a manufacturing facility is to determine the general flow pattern for material, parts and work-in-process inventory, through the system. *Flow pattern* refers to the overall pattern in which the product flows from the beginning to the end—that is, while it is being transformed from raw material (at the receiving stage) through semifinished product (at the fabrication stage) to the finished product (at the assembly stage). A flow pattern established at the nearly 800-acre Nissan Motor Manufacturing Company close to Smyrna, TN, is illustrated in Figure 3.2 (Dilworth, 1989). Although the flow pattern can be changed at a later time if necessary, it is the basis for layout development. If the facility has limited area, then the number of flow patterns from which to choose is also limited. For instance, if we are not unduly restricted by available space, we may choose from a U-shaped flow pattern to an S-shaped pattern to a Z-shaped pattern and so on. As pointed out in Francis et al. (1992), a number of possible patterns exist—one for each letter of the alphabet! Notice that the Nissan plant (which has dimensions of approximately 1.25 miles by 0.75 miles) has a large E-shaped flow pattern.

3.3.1.1   Common Flow Patterns

Some common flow patterns are illustrated in Figures 3.3 through 3.5. In these figures, square or rectangular boxes represent workstations and lines between workstations represent the flow of material between them. The arrows indicate where the flow begins and ends. The pattern in Figure 3.4, known as the dendrite pattern, is suitable for assembly operations. Each vertical line can be thought of as a subassembly line, whereas the horizontal line as a main assembly line. The subassembly lines are arranged so that they feed subassemblies to the main assembly line in the order the subassemblies are required. One can also imagine a variation of the dendrite arrangement in which there

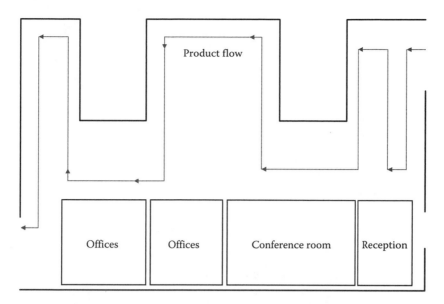

FIGURE 3.2 Flow pattern at Nissan plant in Tennessee. (Adapted from Dilworth, J.B., *Production and Operations Management: Manufacturing and Non-Manufacturing*, Random House Publishing Company, New York, 1989.)

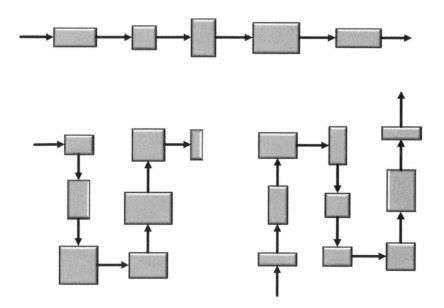

FIGURE 3.3 Various possible flow patterns.

are two sets of subassembly lines, one on each side of the main assembly line (see Figure 3.5). This arrangement, sometimes called a spine arrangement, is useful where the flow of traffic involving people, material, utilities, and information must be consolidated along the central (main assembly) line (Askin and Standridge, 1993). For example, the main assembly line serves as the primary aisle for material movement, while the secondary aisles will be those (subassembly lines) used to traverse the concerned departments. Although the flow of material within departments is shown in one direction only, some situations may call for flow in both directions.

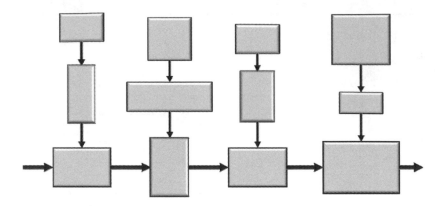

FIGURE 3.4   Dendrite flow pattern.

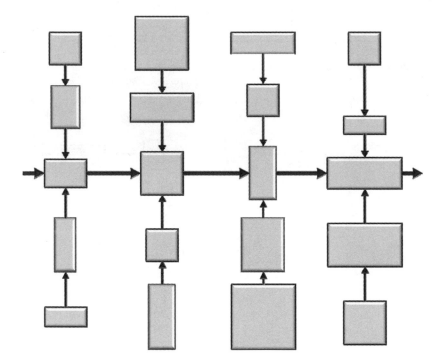

FIGURE 3.5   Spine flow pattern.

3.3.1.2   Five Types of Layouts

Determining the *pattern* is the first step in developing layouts. Designers must then determine the *type* of layout to be used. Five general types of layouts are discussed here. Although much of our discussion is focused on manufacturing systems, the following five layout types are also seen in nonmanufacturing (service) applications:

1. Product layout
2. Process layout
3. Fixed-position layout
4. Group technology (GT)-based layout
5. Hybrid layout

3.3.1.2.1 Product Layout

The product layout is known by names such as flow-line layout, production-line layout, assembly line layout, and layout by product. In a product layout, the machines and workstations are arranged along the product route in a sequence that corresponds to the sequence of operations the product undergoes. If a product undergoes milling, drilling, assembly, and packing operations in sequence, the vertical milling machines, drilling machines, assembly equipment, and packing equipment must be arranged one after another in a line. Typically, the product layout is used by companies that manufacture a single or few items in large quantities, such as a minivan assembly plant. Benefits of the product layout include reduced material handling time, reduced processing time, and easier planning and control. Its main disadvantage is the lack of flexibility. Once a particular product layout is adopted, the cost to make changes is significant, so a product layout is not suitable for a company that makes frequent product changes. The reader is encouraged to identify other disadvantages in Exercise 3 at the end of the chapter. Figure 3.6 is an example of a product layout.

3.3.1.2.2 Process Layout

Machines and workstations in a process layout are arranged based on the operations they perform. Thus, all milling machines are placed together in one department, all turning machines are placed together in another, and so on. In Figure 3.7, a typical process layout, all the lathes (indicated as L) are in the leftmost department. All the milling, drilling, grinding, painting, and assembly machines (identified as M, D, G, P, and A, respectively) are in their respective departments. Other names for process layout include layout by process and job-shop layout. The process layout is useful for companies that manufacture a variety of products or jobs in small quantities, where each job is usually different from any other. While the process layout offers flexibility and allows personnel to become experts in a particular process or function, it has some major disadvantages—increased material handling costs, traffic congestion, long product cycle times and queues, complexity in planning and control, and decreased productivity.

3.3.1.2.3 Fixed-Position Layout

In a fixed-position layout, the product does not move from one location to another. Instead the processes and equipment for making the product are brought to where the product is located. This layout is usually adopted when the manufactured or processed product is bulky and cannot be

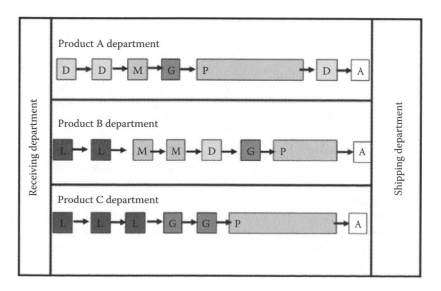

FIGURE 3.6 A product layout. (Courtesy of John S. Usher.)

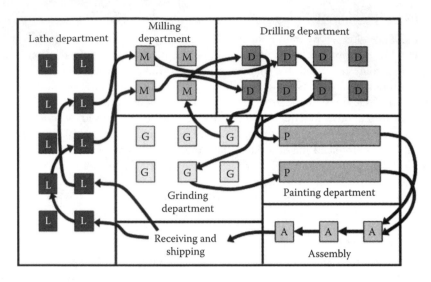

FIGURE 3.7   A process layout. (Courtesy of John S. Usher.)

easily transported or transported at all, such as in shipbuilding and repair; aircraft manufacturing and servicing; and dam, road, and house construction industries. Automobile manufacturing companies (e.g., Nissan) that manufacture high-end models (e.g., Infiniti) also have adopted the fixed-position layout for some of their operations. Due to stringent quality and manufacturing requirements, minimal product movement is desired in some processes. The advantages of fixed-position layout are that the product, which is usually bulky and expensive, is not moved from place to place. Hence, there is lesser chance for damage to the product and the cost of moving it is reduced. On the other hand, there is a significant increase in the cost of moving equipment to and from the work area, so the utilization of equipment is low. Once the equipment is brought to the work site, it must remain there until the entire job is done. For instance, we often see bulky road-laying equipment sitting idle at a construction site. Although the equipment is used intermittently only for a few days, it must be kept at the work site until the entire road-laying job is completed. A fixed-position layout is shown in Figure 3.8.

3.3.1.2.4   Group Technology–Based Layout

The principle of GT and how it is used to develop a cellular manufacturing system are discussed in detail in Chapter 6. In this section, we provide a basic understanding of the GT-based layout. Since the late 1960s, it has been recognized that many medium to large manufacturing systems can exert better control over operation and planning by dividing a large system into two or more, much smaller, independent, subsystems.* In such companies, a large number of parts (usually thousands) are manufactured on a large number of machines (typically hundreds). To divide such a large system into smaller subsystems, planners must first determine the sets of machines and the corresponding parts (which are entirely processed by these sets of machines). Each machine set and the corresponding part set form a manufacturing *cell* and a part *family*. By identifying machine cells and part families that are independent of the others, planners form a number of smaller, mutually independent subsystems that are easier to plan and control. More specific advantages are provided in Chapter 6. The basic principle of GT in the manufacturing context is to identify machines dedicated to the manufacture of a set of parts and group the machines into a machine cell and the corresponding parts into a part family.

---

* This is sometimes referred to as the divide and conquer manufacturing strategy.

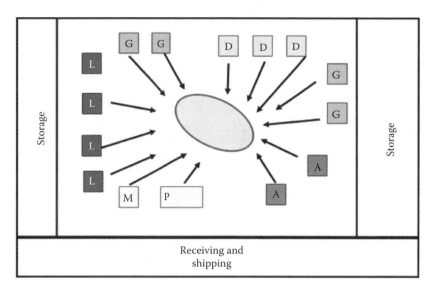

FIGURE 3.8   A fixed-position layout. (Courtesy of John S. Usher.)

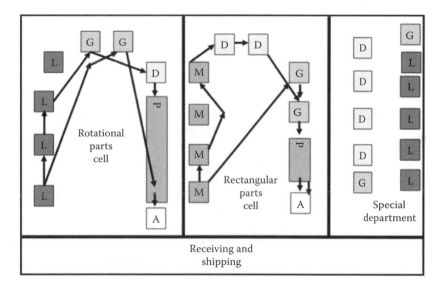

FIGURE 3.9   A group technology–based layout. (Courtesy of John S. Usher.)

A GT-based layout for the machines in Figure 3.6 is illustrated in Figure 3.9. Notice the difference between the two. Machines in a cell are not necessarily similar in their processing capabilities. Further, it will be shown in Chapter 6 that traffic congestion, material handling costs, work-in-process inventory, production lead times, and other forms of waste are significantly reduced in a GT-based layout, because the parts in a family are processed almost entirely in their respective cell.

3.3.1.2.5   Hybrid Layout

Not all companies can adopt a single type of layout. As a company expands by increasing its product lines and volume, it may find that none of the layout types discussed in this section meets its needs entirely. Whereas some production items may require a product layout, others may call for a

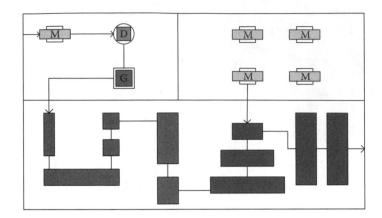

FIGURE 3.10   Hybrid layout.

fixed-position layout. Hence, a number of companies use hybrid layouts, a combination of the layouts discussed. A sample hybrid layout that has characteristics of a GT, process, and assembly line layout is shown in Figure 3.10 (see the upper left, right, and lower departments). The company has a manufacturing cell and a process department dedicated to making two major subassemblies. These two subassemblies are introduced to an assembly line at different stages of assembly of the end product. This company uses a combination of GT-based layout in its manufacturing cells, product-based layout in its assembly area, and process-based layout in the milling section.

3.3.2   Flow Process Chart

The flow process chart, sometimes also known as the layout planning chart (Reed, 1961), graphically illustrates the various steps the product undergoes on its way from receiving to shipping. It lists the operations and the corresponding standardized American Society of Mechanical Engineers (ASME) symbols and includes the standard time for each operation as well as the number of pieces and how they are moved (if appropriate). The chart could be expanded to include the operator and department where the operation takes place, the type and quantities of processing equipment required, type and quantities of material handling devices (MHDs) required, remarks on how process improvements can be made, and other information. Figure 3.11 shows a flow process chart prepared for a press department. It includes many of the aforementioned pieces of information and a partially completed summary table that lists the total number of each operation type as well as the time required and distance moved. This summary table allows planners to find ways of improving the process efficiency. For example, some of the non-value-added activities, such as inspection, delay, and transportation, may be reduced; or two or more operations may be combined into one. Significant process improvements can then be documented in a new flow process chart that includes a column in the summary table comparing the number of each type of operation (delay, inspection, transportation, etc.) in the new chart with the old values. A variation of the flow process chart that lists more than one product is called a multiproduct flow chart. Where appropriate, a multiproduct flow process must be developed only for the major product lines; otherwise, there are too many columns and a visual analysis becomes difficult.

3.3.3   Flow Diagram

The flow diagram shows the operation symbols and sequence on a proposed or current layout. A sample flow diagram is illustrated in Figure 3.12. C, D, L, M, R, S, VT, in the figure denote Conveyor, Drill, Lathe, Mill, Receiving, Shipping, and Vertical Turret Lathe, respectively. Using the

Subject Charted <u>TB03100 Face Panel</u>							Chart No. <u>112XAG</u>	
Drawing No. _____ Part No._____							Chart of Method <u>Present</u>	
Chart Begins <u>Receiving</u>							Charted By <u>N.L.</u>	
Chart Ends <u>Steel Dept.</u>							Date Feb. 5/90	
Sheet__1__ of __1__								

Distance (feet)	Time	Chart Symbol	Oper ID	Dept ID	M/C ID	# of pieces	How moved	Process Description
	.02	1 ▼	A1	S&R		100	Truck	Received material 0.022 wcs (51″ × 102″)
220	.02	1 ♦	H2	MF1		100	Forklift	To crane bay area
		2 ▼		WIP1		100		Stored temporarily
20	.02	2 ♦	H2	MF1		100	Conveyor	To hydra shear
	.01	1 ●	M1	MF1	HS1	100		Cut to length (front panel)
50	.02	3 ♦	H3	MF2	13G 142	95	Forklift	To machine #13G (100 ton press) or Komatsu (machine 142) or to HYMAC 101 (7″ & over neck)
	.01	2 ●	M2	MF2	101	95		Necking operation (punch hole)
160	.03	4 ♦	H3	MF2		95	Forklift	To machine # 136
	.03	3 ●	M2	MF2	136	95		Punch holes
240	.01	5 ♦	H1	MF3		95	Forklift	To machine #155 or machine #104 or machine #111
	.05	4 ●	M2	MF3	155	95		Braking operation
260	.01	6 ♦	H1	WIP2		94	AGV	To marshalling area
		3 ▼		S&R		94		Temporary storage
150	.01	7 ♦	H3	WIP3		94	AGV	To steel dept.

Summary

Event	Total	Time	Distance
● Operations	4		
■ Inspections	—	—	—
♦ Transportations	8		(min) 1100 feet
▼ Storages	3		
♦ Delays			

FIGURE 3.11   Flow process chart.

flow process chart and the layout diagram, the analyst can determine whether the existing process has any unnecessary material movements and operations.

In addition to the flow process chart and flow diagram, there are two key pieces of data required by facility designers—flow and distance data.

3.3.4   FLOW DATA

We now discuss how the interaction between departments can be measured. The interaction may be determined using objective measures or subjective criteria.

3.3.4.1   Qualitative Flow Data

It is rather difficult for some small companies to generate all of the quantitative data required for layout design. This may be because information is not adequately computerized, the company does not have significant changes in product mix (so there is no need to record routing, bill of materials,

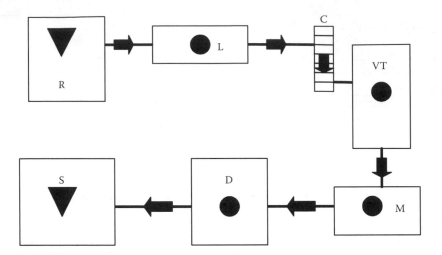

FIGURE 3.12   Flow diagram.

or other information), or the company does not have enough personnel or management commitment to collect, store and process the required data. In such cases, a facility designer must often work with surrogate data that specify the interaction between each pair of machines. Muther (1973) designed a method that allows experienced facility analysts to capture this information subjectively. It is called systematic layout planning (SLP) and is discussed in Chapter 4. We now describe how qualitative information necessary for designing layouts can be developed.

3.3.4.2   Relationship Chart

Muther (1973) used a systematic qualitative approach to solve the facility layout problem. His approach is based on first defining the adjacency relationships for each department pair. The following classes of adjacencies or closeness rating are used and assigned a specific letter code:

- A Absolutely necessary
- E Especially important
- I   Important
- O Ordinary
- U Unimportant
- X Undesirable

The closeness ratings indicate the importance of locating department pairs next to each other. For example, it is absolutely important that department pairs with an A rating be placed adjacent to each other. Similarly, it is especially important that department pairs with an E rating be placed adjacent, provided such a placement does not cause other department pairs with an A rating to be nonadjacent. Also, if a department *i* has A relationships with so many others that some of these cannot be adjacent, then the departments are placed as close as possible to department *i*. This is why it is important to keep the number of A relationships in a chart within 5% of the total relationships, as explained in the next section. The listing of ratings is in order of their importance, except X, which receives top priority like A. Just as department pairs with an A rating *must* always be adjacent, department pairs with an X (undesirable) rating *must not* be adjacent under any circumstance.

The letter codes for the closeness ratings may easily be converted into numerical values. Some layout software packages allow the user to set numerical values for closeness ratings.

3.3.4.2.1 Constructing the Relationship Chart

Generally, the relationship chart is needed only when there are factors other than flow influencing the layout decision. Most practical layout problems are likely to have some factors other than flow, so it is almost always necessary to construct a relationship chart. Figure 3.13 is a sample relationship chart. Notice that the departments or processes are entered in the left column, one per line, and the activity-pair relationships are on the right. To find the relationship code, we locate the box formed at the intersection of the downward sloping and upward sloping lines corresponding to the chosen activities. For example, the relationship between activities 1 (form) and 4 (assembly) is O. The letter in the box indicates the relationship between departments and the number indicates the reason the relationship code is assigned. The box to the right of the relationship chart lists the reasons. For example, activities 3 and 4 are given an X relationship for safety reasons; it is unsafe to locate the treating area adjacent to the assembly area, where a number of employees are working in an open space. As mentioned in Muther (1973), the letters could also be given color codes for easy identification (see Table 3.1).

In the relationship chart, it is desirable to have ratings occur in increasing frequency from A through U—2% to 5% for A, 3% to 10% for E, 5% to 15% for I, 10% to 25% for O, and 25% to 60% for U. The frequency of X ratings depends on the layout problem.

The second step in constructing the relationship chart is to list all the activities on the chart, and the third step is to determine the relationship (letter code) for each department pair. Analysts must carefully complete the relationship chart by using their own knowledge, by actually visiting the factory to observe and discuss operations with floor personnel, and by holding group discussions and consultation with concerned personnel, supervisors, and managers. It is important that each relationship established for a department pair have a reason. A careful review may indicate that minor revisions must be made to the relationship chart. For example, during the initial stages, analysts may have assigned too many A relationships. It is important to share the chart with supervisors and managers and obtain their feedback and approval.

To construct the relationship chart as shown in Figure 3.13, it is first of all necessary to identify all the activities involved in the layout project—that is, departments, work areas, offices, locker

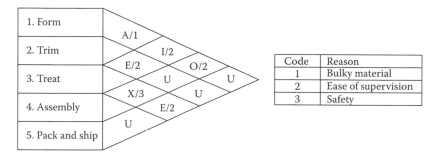

FIGURE 3.13 A sample relationship chart.

TABLE 3.1
Color Codes for Closeness Ratings

Relationship Code	A	E	I	O	U	X
Color	Red	Orange or yellow	Green	Blue	Uncolored	Brown

Source: Muther, R., *Systematic Layout Planning*, Van Nostrand Reinhold Company, New York, 1973.

rooms, and other facilities. This is then shared with managers and supervisors to ensure that no activities have been left out and that appropriate terminology is used. For a large number of activities, there are so many relationships that a computer would be needed to monitor them. For example, 45 activities lead to 45[45 − 1]/2 entries in a relationship chart. Hence, analysts must often consolidate appropriate activities into groups. A practical upper bound on the number of activities that can be considered is 20, and even 20 activities lead to 190 relationships.

3.3.4.3 Quantitative Flow Data

Measures of flow indicate the level or extent of interaction between department pairs. It is obvious that pairs with greater interaction between them should be placed closer than those with limited interaction. This consideration must be given to all pairs of departments. Of course, while using the interaction as a guide in determining the relative locations of departments, we must also keep in mind the area, shape, and space requirement of the departments.

One quantitative flow measure often used by analysts is the *frequency of trips* between departments. If material handling systems are used for transporting material, then the trips made by these systems are considered. If people move material or travel between departments, as in an office setting, then the trips made by people are taken into consideration. To capture the frequency of trips between departments, the following two types of matrices are generally used:

1. From–to frequency of trips matrix
2. Frequency of trips between departments matrix

As the name indicates, the latter matrix shows the number of trips made by the material handling carrier or people *between* departments, whereas the former shows the number of trips *from* one machine *to* another. Thus, the from–to matrix shows the number of trips made in each direction for every department pair, and the between matrix combines the two directions and shows the total number of trips between each pair. The between matrix is symmetric, whereas the from–to is not. (See Figures 3.14 and 3.15, and note that the data in Figure 3.14 were used to construct the matrix in Figure 3.15.)

3.3.5 Distance Measures

The following are some distance measures that can be used in practice:

- Euclidean
- Squared Euclidean
- Rectilinear
- Tchebychev

	Department 1	Department 2	Department 3	Department 4	Department 5	Department 6
Department 1	—	12	3	3		
Department 2	21	—				
Department 3	4	5	—	8	8	8
Department 4				—	4	4
Department 5	1	2	2	15	—	19
Department 6	4	9		7	19	—

FIGURE 3.14 From–to frequency of trips matrix.

	Department 1	Department 2	Department 3	Department 4	Department 5	Department 6
Department 1	—	33	7	3	1	4
Department 2	33	—	5		2	9
Department 3	7	5	—	8	10	8
Department 4	3		8	—	19	11
Department 5	1	2	10	19	—	38
Department 6	4	9	8	11	38	—

FIGURE 3.15 Frequency of trips between departments matrix.

- Aisle distance
- Adjacency
- Shortest path

The actual measure used depends on the availability of qualified personnel, time to gather data, and type of material handling systems used. For example, for an overhead material handling carrier that moves along perpendicular rails, the rectilinear metric may be appropriate, whereas, if the material is moved via automated guided vehicles (AGVs), the aisle distance may be more appropriate. Distance is typically measured from the center of one department to another, even though it may be more accurate to measure the distance between the pick-up and drop-off points of department pairs. Measurement between centers enables analysts to use simpler mathematical models.

3.3.5.1 Euclidean

The Euclidean metric measures the straight-line distance between the centers of departments. Although it may not be realistic in some instances, it is the most commonly used measure because it is useful (shortest distance between two points is a lower bound on the distance) and easy to understand and model. The Euclidean metric has applications for certain conveyor models and transportation and distribution networks. In general, this metric is used more in facility location problems than in layout problems. Consider the problem of locating a new warehouse in an existing network of other warehouses and customers (retail stores). Using a Euclidean metric for the distance between departments may be appropriate. On the other hand, using this metric to calculate the distance traveled by an overhead crane may not be appropriate, because the overhead crane moves along perpendicular tracks.

To develop an equation for the Euclidean metric, consider the following notation and Figure 3.16:

x_i coordinate of the centroid of department i
y_i coordinate of the centroid of department i
d_{ij} distance between the centroid of departments i and j (under various metrics)

Under the Euclidean distance metric, $d_{ij} = ((x_i - x_j)^2 + (y_i - y_j)^2)^{0.5}$ is represented by the straight line that joins the centroids of departments i and j in Figure 3.16.

3.3.5.2 Squared Euclidean

As the name implies, this metric is square of the Euclidean. The squaring assigns greater weights to distant department pairs than to those that are nearby. There are relatively few applications for the

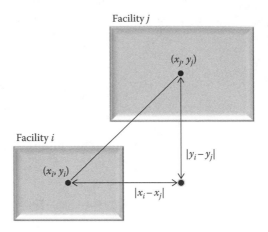

FIGURE 3.16 Distance calculations under various metrics.

squared Euclidean distance metric; however, it does provide some insight into problems, especially some location problems. The squared Euclidean distance metric $d_{ij} = ((x_i - x_j)^2 + (y_i - y_j)^2)$.

3.3.5.2.1 Rectilinear

The rectilinear metric is also called the Manhattan, right-angle, or rectangular metric. It is commonly used because it is easily computed, easy to understand, and appropriate for many practical problems (e.g., distance between points in a city, distance between departments served by MHDs that can move only in a rectilinear fashion). The rectilinear distance metric $d_{ij} = |x_i - x_j| + |y_i - y_j|$ is represented by the horizontal and vertical lines between the centroid of departments i and j in Figure 3.16.

3.3.5.2.2 Tchebychev

Consider the problem of material movement in a heavy machinery factory via overhead cranes powered by two independent motors, one permitting movement in the x direction and the other in the y direction. The time to reach the center of department j from center of department i depends on the larger of the x and y distances. Thus, the Tchebychev distance is the larger of the two values: $d_{ij} = \max(|x_i - x_j|, |y_i - y_j|)$. If we assume that the vertical component of the distance between the centroids of departments i and j is greater than the horizontal component, then the vertical line in Figure 3.16 represents the Tchebychev distance metric. This metric has applications in order-picking problems, in which a person on board a crane in an automated storage and retrieval system picks and deposits items from their respective storage spaces. If movement in the third dimension is also considered—for example, if the crane is powered by three independent motors that allow it to move in the x, y, and z dimensions—then the Tchebychev distance metric is $d_{ij} = \max(|x_i - x_j|, |y_i - y_j|, |z_i - z_j|)$.

3.3.5.2.3 Aisle Distance

The aisle distance is different from all other metrics because it is the actual distance traveled along the aisles by the material handling carrier. Consider the aisle distance between departments i and j in Figure 3.17; it is the sum of a, b, c, and d. The rectilinear metric would have underestimated the distance traveled in this instance. The aisle distance metric has its primary applications in manufacturing layout problems. Because the path of the material handling carrier is not known in the initial design stages, the aisle distance metric is used only in the planning or evaluation stages.

3.3.5.2.4 Adjacency

Yet another metric attempts to determine whether or not departments are adjacent; it is called the adjacency metric. The drawback is that this metric does not differentiate between nonadjacent departments. For example, if departments i and j are 5 feet apart and nonadjacent and departments i and k are 100 feet apart and also nonadjacent, the adjacency metric assigns a value of 0 to both d_{ij} and d_{ik}. In Figure 3.17, $d_{ik}=d_{jk}=1$ and $d_{ij}=0$ because department pairs i, k and j, k are adjacent and i, j is not. This metric is used by the SLP technique in calculating the score for a layout.

3.3.5.2.5 Shortest Path

In network location problems, the shortest path metric is used to determine the distance between two nodes. A network consists of nodes and arcs, with the nodes representing departments and arcs between a pair of nodes representing a path between the two. Usually, a weight is attached to each arc representing the distance or time or cost to travel between the nodes connected by the arc. Because there is typically more than one path between any pair of nodes, the shortest path is an important consideration. Location and distribution problems can be represented on networks, and the shortest path metric is often used in such problems.

3.3.6 Layout Evaluation Criteria

To evaluate a given set of layouts, analysts must first determine the criteria on which to judge the layouts. Several qualitative and quantitative criteria were mentioned in Chapter 2. In this section, we focus on the most commonly used quantitative criteria for evaluating layouts, given by

$$\sum_i \sum_j c_{ij} f_{ij} d_{ij} \qquad (3.1)$$

where
c_{ij} is the cost of moving a unit load of material a unit distance between departments i and j
f_{ij} is the number of loads or trips required between departments i and j
d_{ij} is the distance between departments i and j

It is easy to see that f_{ij} and d_{ij} vary among department pairs, but it is not immediately obvious that c_{ij} can also vary. Generally, the cost to move a unit load between departments depends on the type of material handling carrier used. For example, conveyors cost less than AGVs to move a load of the same size through the same distance. It is assumed that c_{ij} includes apportioned fixed and variable

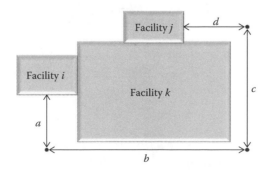

FIGURE 3.17 Distance calculations for the aisle distance and adjacency metrics.

cost of material handling equipment, operating cost, associated labor cost, and inventory cost for time in transit. In practice, it is too time consuming and expensive to determine c_{ij} because more than one type of MHD could be used to transport material between workstations. For example, cartons of finished products are transported via a conveyor to a central staging area and then sent to their respective temporary storage areas in the shipping department via forklift trucks. Another reason that c_{ij} is hard to determine exactly is that it may be difficult to determine the size or volume of the *average* unit load between machines. Also, the variable maintenance and operating costs of each type of MHD may not be readily available. Because it is difficult to determine exact c_{ij} values for each pair of workstations, analysts attempt to determine relative costs as opposed to actual costs. Thus, the c_{ij}'s are all equal if the same handling equipment is used for material transfer (e.g., parts are loaded on carts and moved manually). When different handling devices are used, the relative unit load unit distance handling costs are often guesstimated.*

The formula we provide to determine c_{ij} was developed by Tamashunas et al. (1990) and used in layout software packages such as Flow Planner and Workplace Planner. (The software packages are discussed in detail in Chapter 5.) The formula assumes that no more than one type of MHD is used between any given pair of machines, and the load and unload time is dependent on the MHD and not on the material handled.

Notation

r	Number of MHD *types*
m	Number of *machines*
f_{ijk}	Number of trips required to transport material from machine i to j via MHD k
d_{ij}	Distance from machine i to j, for a given layout (in feet)
TL_k	Average percentage of time MHD k travels loaded
$LULT_k$	Average loading and unloading time per move with MHD k (in minutes)
S_k	Average speed of MHD k (feet per minute)

$$Y_{ijk} = \begin{cases} 1 & \text{if MHD } k \text{ is used to transport material from machine } i \text{ to machine } j \\ 0 & \text{otherwise} \end{cases}$$

N_k	Number of units of MHD k
C_k	Investment/leasing cost for MHD k per year
OP_k	Labor plus nonlabor (fuel, power, and maintenance) cost to operate MHD k per minute
T_{ijk}	Time to transport material from machine i to j via MHD k per trip (in minutes)
MHC_{ijk}	Total material handling cost to transport material from machine i to j via MHD k
MHR_{ijk}	Material handling cost per unit distance per trip to transport material from machine i to j via MHD k

$$T_{ijk} = LULT_k Y_{ijk} + \frac{d_{ij} Y_{ijk}}{S_k TL_k} \tag{3.2}$$

$$MHC_{ijk} = N_k C_k \left(\frac{T_{ijk} f_{ijk}}{\sum_{i=1}^{m} \sum_{j=1}^{n} T_{ijk} f_{ijk}} \right) + T_{ijk} f_{ijk} OP_k \tag{3.3}$$

$$MHR_{ijk} = \frac{MHC_{ijk}}{f_{ijk} d_{ij}} \tag{3.4}$$

---

* Guesstimate means guessing and estimating.

Using this notation, T_{ijk} can be obtained by adding the load and unload time for MHD k and the travel time as shown in Equation 3.2. The second part of the right-hand side of Equation 3.2 determines the travel time between a pair of machines. We divide the distance traveled by the speed of the MHD and adjust it for the time the MHD travels empty to get to workstation i to make the loaded trip. Once again, TL_k is difficult to determine exactly and must be guesstimated. Equation 3.3 determines the fraction of time MHD k serves machines i and j (given by the bracketed terms in the first part of the right-hand side) and multiplies it by the annualized investment or leasing cost to determine the fixed cost of MHD k that can be apportioned to the material handling effort between workstations i and j. The second part of the right-hand side of Equation 3.3 yields the variable cost of operating MHD k between workstations i and j, and OP_k includes labor as well as fuel, power, and maintenance costs. Equation 3.4 calculates the cost of transporting material per trip per unit distance by dividing the total fixed and variable costs of transporting material between workstations i and j by the product of distance and the number of trips between the two workstations. Because it was assumed that only one type of MHD is used between each pair of machines, c_{ij}—the cost of moving a unit load of material through a unit distance between departments i and j—is equal to MHR_{ijk}.

3.4 TOOLS FOR PRESENTING LAYOUT DESIGNS

Thus far, we have examined the basic data required for making layout decisions. Models and solution techniques that use these data to determine the positioning of departments are discussed in Chapters 5 and 10. After the department positions have been determined, the next step is to develop a layout design. A number of tools are available to develop and present layout designs.

1. Drawings
2. Templates
3. Three-dimensional physical models
4. Computer models

Before the proliferation of computers and related computer-aided design (CAD) software, analysts generated entire layout designs using only tools such as templates and manual drawings. By moving and repositioning templates, for example, they created different layout designs and then selected one of them based on personal preferences. They did not consider material flow or any other data. It is astonishing that some companies still use this approach. See Section 4.1 for more details on this topic.

3.4.1 DRAWINGS

Drawings can be manually generated or generated from a CAD file on a plotter or printer. Figure 1.13 is a CAD drawing of a facility. Today, with the increased use of computers and computer software, manual drawings are becoming obsolete because it is too time consuming to make them and they must be redrawn whenever changes are made to the layout. Typically, many changes are made before the final design is reached, so manual drawings of layout designs are not preferred.

3.4.2 TEMPLATES

Like drawings, templates can be generated manually (from cardboard, stiff plastic material, sheet metal, wood, and paper) or from a computer. Commercial templates of machines are also available in various sizes and shapes (see Figure 3.18). Templates are typically placed on a base board (also made of cardboard or other lightweight material) to indicate positions of machines, workstations, and other auxiliary equipment. Self-adhesive tapes indicate the flow of material and show aisles

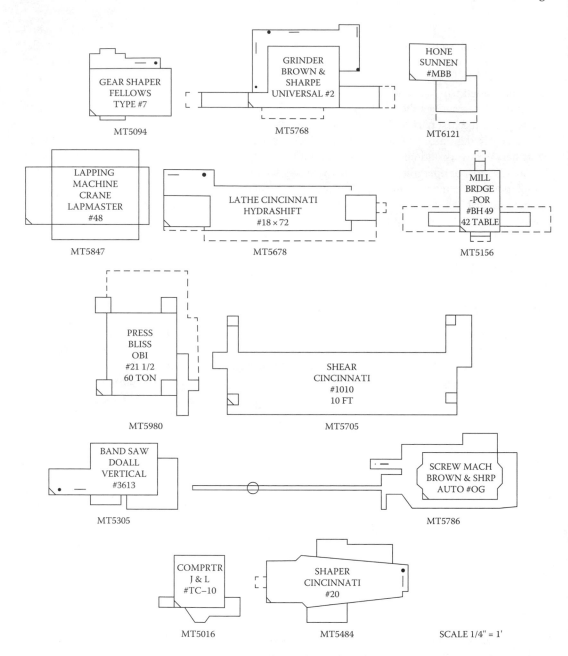

FIGURE 3.18 Illustration of typical machine templates. (Courtesy of Planprint Company.)

or paths of MHDs. Tapes, which can be purchased in a variety of colors, widths, and thicknesses, greatly improve the presentation of a layout design.

3.4.3 THREE-DIMENSIONAL PHYSICAL MODELS

These models are 3D versions of templates. Also available commercially in various shapes, sizes, and materials, physical models provide additional visual information that is helpful in certain circumstances. For example, when transporting parts and material via an overhead crane, models help analysts visualize the paths for such transportation between machines. Because it is generally not

FIGURE 3.19   Three-dimensional model of an automobile assembly plant. (Courtesy of Opel GmbH.)

advisable to transport parts over tall equipment such as vertical punch presses and vertical mills, analysts need to know which paths are feasible and which ones are not. Figure 3.19 is a 3D model of one of the OPEL car manufacturing plants near Dortmund, Germany.

3.4.4   Computer Software Tools

Computer software tools are the most effective for preparing and presenting layout designs. CAD systems are computer systems made up of an operating system (including application software, graphics utility, and device drivers), a model database, and input/output devices (Chang et al., 1991). The user interacts with the CAD system via an interface to develop computer drawings or models of various objects, large and small. Examples include manufactured parts; printed circuit board layout and design; architectural layout of a building, theater, dam, or other large constructions; and artistic designs.

CAD systems allow users to create 2D and 3D drawings. They have gained popularity because of the improved speed and memory capabilities of computers and the availability of 2D and 3D CAD software. Their main advantage is the speed with which we can add, delete, and change drawings and generate new ones. They also allow us to *zoom in* on selected areas and get a detailed top view, side view, or cross-sectional view of a facility. Although most CAD tools are used to design and draw parts, many can also be used to develop the layout of facilities. In Chapter 5, we discuss several CAD-based software packages, many of which allow us to develop facility designs. A 3D view of a temple and cultural center drawn using a CAD software is shown in Figure 3.20.

3.5   GUIDELINES FOR DATA DEVELOPMENT AND GENERATION

In this section, we present a step-by-step approach for gathering the relevant data and constructing them in the required format for developing a layout. The approach is discussed in the context of a manufacturing layout problem. The steps are as follows:

1. Obtaining basic data
2. Extracting relevant data
3. Generate data in the required format
4. Sharing preliminary data and obtaining feedback
5. Documenting data

Southeast view

FIGURE 3.20   Three-dimensional drawing of a temple and cultural center. (Courtesy of the Hindu Temple Society of the Capital District, N.Y., Inc.)

3.5.1   OBTAINING BASIC DATA

The basic data needed to solve most manufacturing layout problems are usually available from personnel on the factory floor—production managers, supervisors, and process engineers. In some companies, data requests must be directed to the information system department. Because the new layout is for the *future*, the data collected should correspond to manufacturing activities expected to take place in the specified future period. For example, part-routing and machine information should be for parts that will be manufactured and machines that will be used for production during the years for which the layout is planned. This means that analysts must be able to forecast the production mix and volume as well as the processes and equipment that will be used. It is well known that the reliability and accuracy of forecasts diminish as they go out into the future. If the forecast is done for a very short period, however, the layouts may have to be redesigned frequently, which is time consuming and expensive. Thus, a trade-off decision has to be made about how far into the future to forecast. The decision depends to a large extent on the nature of the industry. High-technology industries, for example, cannot afford to make forecasts for long periods, whereas manufacturing companies typically are able to make reasonable predictions about product mix and volume for 2–3 years. We now list and discuss key pieces of data required for the layout design problem.

Name and code for each part: Companies typically have a name for each part and a corresponding unique numeric or alphanumeric code. This information is needed to refer to the parts that will be manufactured.

Name and code for each machine: Like parts, machines also have names and unique codes. In smaller companies, machines of the same type may all be given the same code. For example, all lathes may have a code of MATSURA011. In that case, there must be a unique code for each machine, MATSURA011A, MATSURA011B, and so on.

Routing information for each part: Each part has a routing sheet that specifies the sequence in which operations will be performed on the part. This sheet can then be used to determine the sequence of machines visited by a part. If the routing sheet specifies only the machines visited but not the sequence, it is not helpful for the layout problem. Routing sheets may also contain operation cost, time, and other information. Although some of this information may be valuable in planning decisions, it is not needed for developing layouts, which is a design decision.

Production volume for each part: The number of units of each part type expected to be manufactured may or may not be available in the routing sheet. If it is not, other company personnel—perhaps production planning personnel—can supply these data.

Machine dimensions and space requirement: If the machines expected to be used in future manufacturing activities are known, their footprint dimensions may be obtained in one of the following four ways:

1. From equipment (machine) catalogs
2. From hard copies of drawings (if the drawing is to scale)
3. From CAD drawings
4. From actual measurements of the machines

Among the four methods, the first and second require the least time. In addition to the machine footprint, analysts must determine the additional space required for each machine by estimating the space required for operator, MHD, and material storage as illustrated in Figure 2.10.

Cost of handling unit load of material between each department pair: As mentioned earlier, this cost typically depends on the material handling carrier likely to be used to transport the part. For example, if an AGV is used to transport parts, the cost of handling a unit load will be more than the cost with a conveyor. The cost for each machine pair can be found from Equation 3.4. If it is difficult to obtain some of the parameters in Equation 3.4, however, the unit load, unit distance handling cost may be guesstimated by experienced personnel for each pair of machines. In many layout applications, this is how the unit load, unit distance handling cost is obtained.

Special location restrictions or adjacency information, if any: Companies typically have some machines or departments that cannot be moved from their existing location. Three examples are as follows:

1. The expense or the level of difficulty in reassigning a machine to a new location is too high.
2. Special auxiliary support services (e.g., battery-charging facilities) are available only at the existing location, so the department cannot be relocated.
3. The punch press height is greater than that of the building. If the press is to be moved to a new location, another hole in the roof will have to be made at the new location.

Information about such constraints or reasons that certain department pairs must or must not be adjacent may be obtained from personnel who have special knowledge of products, processes, and equipment.

Other qualitative information: Other qualitative information that should be considered while designing the facility layout includes building and construction constraints and location of power and coolant lines. For example, the location of large support beams may dictate the placement of some departments; in some cases, certain departments should not be near the reception or office or cafeteria because of noise, odor, appearance, or other aesthetic reasons. Generally, analysts can get this kind of information only by actually visiting the site a few times. Visiting the company frequently and sharing data collected with concerned company personnel will often reveal much of the qualitative information that cannot be captured in models but needs to be considered while designing the final layout.

3.5.2 Product Data Analysis

At this point, the quantitative information has been obtained and we are ready to do preliminary processing. With the increased computing power of computers and the availability of software

```
--PART NUMBER--    -------DESCRIPTION-------      DATE        ALT CODE    BUYER/PLANNER    DRAWING REVISION
999396-1            LAMP BASE                   0/00/00                       305
```

OPER & ALT	WORK CENTER	OPER CODE	SETUP CODE	CREW FACTOR	MACH GROUP	--TOOLING REF-- NUMBER	--STANDARD-- SETUP HOURS	-LABOR- HOURS	-MACHINE- HOURS	TIME I/O CDE	MOVE BASIS /-QTY-	TIME -DAYS-	EFFECTIVE FROM	TO
11	01205			1.0	01205	T4231	2.000	3.640	7.280	4	1000	.000	0/00/00	99/99/99

```
        -------ROUTING DESCRIPTIONS-------
        MAKE/AUTO 999396-1P
PARTIAL AHEAD QTY
```

| 22 | 03038 | | | 1.0 | 03038 | | .000 | 2.420 | 5.260 | 4 | 1000 | .000 | 0/00/00 | 99/99/99 |

```
        -------ROUTING DESCRIPTIONS-------
        ROTO FINISH
PARTIAL AHEAD QTY
```

| 41 A | 03105 | | | .0 | 03105 | | .000 | .350 | .350 | 4 | 1000 | .000 | 0/00/00 | 99/99/99 |

```
        -------ROUTING DESCRIPTIONS-------
        IRRIDITH/SHP
PARTIAL AHEAD QTY
```

| 43 B | 08003 | | | 1.0 | | | .000 | .540 | .000 | 4 | 1000 | .000 | 0/00/00 | 99/99/99 |

```
        -------ROUTING DESCRIPTIONS-------
        PLATING INSPE /SHP
PARTIAL AHEAD QTY
```

FIGURE 3.21 Sample routing sheet.

999396-1	01205
	03038
	03105
	08003

FIGURE 3.22   Sequence of machines visited by part #999396-1.

tools such large volumes of data files can be processed on a computer or server. Before analysts can generate data in the desired format, they must examine the initial information provided and remove data that are not relevant to the layout problem. A sample routing sheet obtained from a company is shown in Figure 3.21. Notice that much of the information in the figure is textual and not relevant to the layout problem. We do not need information concerning setup, labor, or machine hours, nor do we need to know the names of machines and operations. In fact, the only information needed is an ordered listing of the part numbers and the corresponding machine numbers visited in sequence. Figure 3.22 shows this information extracted for one part (part #999396-1) from the routing sheet in Figure 3.21. Data relevant to the problem must be extracted for each part as shown in Figure 3.23.

3.5.3   GENERATE DATA IN REQUIRED FORMAT

This section discusses the generation of data as various matrices. To construct a frequency of trips matrix, analysts must first construct a part-routing information matrix, which indicates the machines required to perform operations on each part as well as the sequence of these operations. The matrix consists of as many rows as the number of parts. Throughout this chapter, we refer to a part type as simply a part. The number of columns corresponds to the number of machines.

We assign a unique number to all machines of a given type and thus treat each individual machine as a unique type for data-generation purposes. If two or more part types require processing on a machine type of which there are multiple copies, then it may be reasonable to assume that *all* the units of a part type are routed to one of the copies. For example, if part types 2 and 4 require processing on machine type 3 (of which there are two copies), then we create two columns for machine type 3 (say, 3a and 3b) and assume that each unit of part type 2 is processed on the first machine of type 3 (i.e., 3a) and each unit of part type 4 is processed on the second machine (3b). In some cases, this assumption may not be justified. For example, 20% of the units of part type 2 may visit machine 3a with 80% visiting 3b. If this is the case, analysts must know what percentage of the units of a given part type is routed to each copy of a machine type, define additional variables, and then attempt to calculate the frequency of trips between machines. We leave this as Exercise 10, part g.

Each entry in the part-routing information matrix consists of integers (Figure 3.24). An entry k in row i and column j of the matrix indicates that part i undergoes operation k on machine j. If a machine j performs two or more consecutive operations (say operations 3, 4, and 5) on a part i, then there will be three entries in row i and column j—namely, 3, 4, and 5. If a machine performs nonconsecutive operations on a part, then this information is captured by additional dummy columns corresponding to this machine. For more information, see Heragu and Kakuturi (1997). Generally, it is not desirable to have the same machine perform nonconsecutive operations on a part because it increases the setup time. We suggest that the part be redesigned or the process plan (routing) changed so that the multiple operations done by a machine on a part are consecutive.

The part-routing information matrix can be generated from the ordered list of machines visited by each part by a small program in any software programming language such as C++, VisualBASIC, Java, Python or even the macro functions in a spreadsheet software such as Excel. Two columns

999396-1	LAMP BASE A	(Part # 1)	
01204			(Machine # 1)
03029			(Machine # 8)
03105			(Machine # 10)
08003			(Machine # 11)
999441	WEDGE TERMINAL A	(Part # 2)	
01204			(Machine # 1)
03001			(Machine # 7)
03105			(Machine # 10)
999441-3	WEDGE TERMINAL B	(Part #3)	
01204			(Machine # 1)
03029			(Machine # 8)
03105			(Machine # 10)
999441-4	WEDGE TERMINAL C	(Part # 4)	
01204			(Machine # 1)
03105			(Machine # 10)
08003			(Machine # 11)
999442-2	LAMP INSULATOR	(Part # 5)	
01204			(Machine # 1)
03001			(Machine # 7)
999459	LAMP BASE B	(Part # 6)	
01204			(Machine # 1)
01204			(Machine # 1)
03029			(Machine # 8)
03105			(Machine # 10)
H6709	HANDLE, DENSPLY PROBE	(Part # 7)	
01217			(Machine # 2)
02053			(Machine # 6)
02053			(Machine # 6)
03029			(Machine # 8)
03105			(Machine # 10)
03001			(Machine # 7)
03001			(Machine # 7)
H3342-1	STAND OFF	(Part # 8)	
01204			(Machine # 1)
02053			(Machine # 6)
H3888	KNOB	(Part # 9)	
01204			(Machine # 1)
02053			(Machine # 6)
02053			(Machine # 6)
03029			(Machine # 8)
H255	GUIDE, BLOCK	(Part # 10)	
02048			(Machine # 5)
02053			(Machine # 6)
03037			(Machine # 9)
H2883-2	BENDING NECK WASHER A	(Part # 11)	
02035			(Machine # 4)
02053			(Machine # 6)
03001			(Machine # 7)
H2912-1	BENDING NECK WASHER B	(Part # 12)	
02035			(Machine # 4)
02053			(Machine # 6)
03001			(Machine # 7)
H3076-1	BENDING NECK WASHER C	(Part # 13)	
02035			(Machine # 4)
02053			(Machine # 6)
03001			(Machine # 7)
H3397-2	BENDING NECK WASHER D	(Part # 14)	
02035			(Machine # 4)
02053			(Machine # 6)
03001			(Machine # 7)
02028			(Machine # 3)
H3513	HANDLE	(Part # 15)	
02048			(Machine # 5)
02053			(Machine # 6)
02053			(Machine # 6)
03105			(Machine # 10)
03001			(Machine # 7)

FIGURE 3.23   Ordered list of parts and the machines visited by these parts.

		Machine										Part Demand	Batch Size
		1	2	3	4	5	6	7	8	9	10	11		
Part	1	1	0	0	0	0	0	0	2	0	3	4	2000	10
	2	1	0	0	0	0	0	2	0	0	3	0	3000	10
	3	1	0	0	0	0	0	0	2	0	3	0	1000	10
	4	1	0	0	0	0	0	0	0	0	2	3	3000	10
	5	1	0	0	0	0	0	2	0	0	0	0	2000	10
	6	1,2	0	0	0	0	0	0	3	0	4	0	1000	10
	7	0	1	0	0	0	2,3	6,7	4	0	5	0	2000	10
	8	1	0	0	0	0	2	0	0	0	0	0	1000	10
	9	1	0	0	0	0	2,3	0	4	0	0	0	1000	10
	10	0	0	0	0	1	2	0	0	3	0	0	2000	10
	11	0	0	0	1	0	2	3	0	0	0	0	3000	10
	12	0	0	0	1	0	2	3	0	0	0	0	1000	10
	13	0	0	0	1	0	2	3	0	0	0	0	1000	10
	14	0	0	4	1	0	2	3	0	0	0	0	3000	10
	15	0	0	0	0	1	2,3	5	0	0	4	0	1000	10

FIGURE 3.24 Part-routing information matrix.

may be added to the matrix to indicate the production volume and transfer batch size for each part. The transfer batch size indicates how many units of the part can be transported in one trip by the material handling carrier. The production volume and batch size information for each part is used to determine the number of trips made between each pair of machines for transporting that part. The numbers are aggregated to determine the total number of material handling trips between each pair of machines and to quantify the level of interaction between machine pairs.

Although there may be more than one type of material handling carrier, this process implicitly assumes that each part is transported using a specific MHD. Generally, this is true. If it is not, then determine the number of trips between each pair of machines required for each part, and the analyst must

1. Append as many more columns as the number of different types of MHDs used
2. Keep track of which device is used at what stage of transportation

This is left as Exercise 16. From the part-routing information matrix, the flow matrix or frequency of trips matrix is constructed by writing another code that uses the following formula:

$$f_{ij} = \sum_{k=1}^{m} F_{ij}^{k} \tag{3.5}$$

$$F_{ij}^{k} = \begin{cases} (D_k/B_k) & \text{if } |R_{ki} - R_{kj}| = 1 \text{ and } R_{ki}, R_{kj} > 0 \\ 0 & \text{otherwise} \end{cases} \tag{3.6}$$

f_{ij} is the number of trips between machines i and j
m is the number of parts
D_k is the expected demand for part k
B_k is the batch size of part k
R_{ki} is the operation number for the operation done on part k by machine i

Note that F_{ij}^k is defined only for consecutive operations. Because we are interested in determining the number of trips that will be made between machines i and j, B_k is included in the denominator in Equation 3.6. From the flow matrix, the unit load handling cost matrix is constructed. This matrix shows the cost of handling a unit load of material through a unit distance between each pair of machines. As discussed earlier, the formula in Equation 3.6 may be used to generate the elements of this matrix. Finally, a relationship matrix indicating whether or not machine pairs must be adjacent is to be constructed. This matrix has three indicators: A, X, and O. A and X correspond to the adjacency relationships defined by Muther (1973) and discussed in Section 3.3. O in the matrix indicates that there is no special adjacency requirement for the corresponding machine pair and hence their relative positioning will be determined by the layout algorithm.

The last piece of information is the dimension of each machine, which can be obtained in several ways as discussed earlier. One source is the footprint dimensions from a hard copy of a layout drawing. If the layout has been generated with a CAD software package, the coordinates of the corner points give the dimensions of the machine footprint. These dimensions are used to estimate the space required by each machine, and appropriate allowances for operator, material storage, handling carrier access, and other factors are also included.

3.5.4 SHARING PRELIMINARY DATA AND OBTAINING FEEDBACK

Once all the preliminary data set has been collected, analysts should share the generated data with concerned personnel (middle managers, supervisors, and others) and get their feedback. This exercise may reveal oversights in the data collection and generation steps as well as incorrect assumptions. This is also the time to check whether any other special information not considered in the previous steps is needed. Such information is usually of qualitative nature.

3.5.5 DOCUMENTING DATA

The last step involves collating all the generated data and placing them in an electronic file. Because there may be a variety of uses for these data, it is better to store this file in a form compatible with most software programs, e.g., ASCII format.

3.6 CASE STUDY: APPLICATION OF METHODOLOGY AT A MANUFACTURING COMPANY

We now illustrate the data collection steps discussed in the preceding section with a real-world problem at a manufacturing company. The company is Welch Allyn Medical Division, located in New York State. Welch Allyn is a manufacturer of medical diagnostic equipment such as otoscopes, endoscopes, and other probes used by physicians to examine ear, nose, and throat passages. The manufacturing activities are divided into three main sections: (1) a general machining section in which much of the manufacturing is done, (2) a finishing station where the components manufactured in the general machining section are taken for finishing operations, and (3) an assembly station where the end products are assembled. Our objective is to generate a layout for the machines in the general machining and finishing sections only. The production manager at the company has specifically asked that material movement between general machining and finishing stations and the assembly area be ignored because it is difficult to reconfigure equipment in the assembly area.

Companies typically use dedicated computer servers for processing bill of material data, part-routing, process plan, and work center information. For our purposes, let us use the master routing list data available at http://sundere.okstate.edu/Book, which are routinely generated by the Information Systems department at Welch Allyn. These raw data can be easily obtained in ASCII format and transferred to a computer. Information pertaining to the assembly area has not been provided in the master routing list. Notice that some of the data in it (e.g., part routing, sequence of work centers visited) are necessary for generating the part-routing information matrix, whereas much of the remaining information (e.g., setup time, machine and labor hours required at each operation, and time basis) is not necessary for solving the layout problem. Thus, we discard much of the information in the master routing sheets. The data are reduced via spreadsheet and word processing software packages to the list shown in Figure 3.23, which contains only the desired information—namely, the ordered list of machines visited by each part. We have assigned a code for each machine and part; they are shown in parentheses. To solve the layout problem, we must also know part annual production volume and transfer batch size of each part (both are necessary to determine the number of trips required annually between each pair of machines). All the materials are placed in carts and moved manually. Also, we know that the parts are roughly the same size and that 10 units of each part can be transported in one trip. The part demand information is obtained from supervisors who have production planning data, and these data along with the transfer batch size are shown in the last two columns of Figure 3.24.

Using formulas (3.5) and (3.6), we can easily construct a flow matrix as shown in Figure 3.25. For example, the flow between machines 1 and 7 in the aforementioned matrix is calculated as follows: Expressed in the notation used in formulas (3.5) and (3.6), $F_{17}^2 = 300$, $F_{17}^5 = 200$, and all other $F_{17}^k = 0$. Thus, the flow between the machines 1 and 7, f_{17}, is equal to 500. Because all the parts are transported manually between machines, the cost of moving a unit load of material through a unit distance is the same between any pair of machines. For this problem, the only location restrictions are for the following two machine pairs: 1204 (Machine 1), 2035 (Machine 4) and 2035, 2048 (Machine 5). The first pair must be adjacent for safety reasons, but the second pair must not be together for technological reasons. Assume that this information is obtained separately from personnel in the production department. These data are captured in the form of a machine relationship matrix (see Figure 3.26).

From the CAD drawing of the existing layout, we can find the machine dimensions. To allow work space for loading and unloading parts, storing kanbans and tools, we include 1 foot clearance on each side of every machine. The machine dimensions with this clearance added to the length and width of each machine are listed in Table 3.2. A data file can be generated for this problem by collating the data in Figures 3.24 through 3.26 and Table 3.2. In Chapter 6, we show how the data file is used in developing layouts.

Machine

	1	2	3	4	5	6	7	8	9	10	11
1	0	0	0	0	0	200	500	400	0	300	0
2	0	0	0	0	0	200	0	0	0	0	0
3	0	0	0	0	0	0	300	0	0	0	0
4	0	0	0	0	0	800	0	0	0	0	0
5	0	0	0	0	0	300	0	0	0	0	0
6	200	200	0	800	300	0	800	300	200	100	0
7	500	0	300	0	0	800	0	0	0	600	0
8	400	0	0	0	0	300	0	0	0	600	0
9	0	0	0	0	0	200	0	0	0	0	0
10	300	0	0	0	0	100	600	600	0	0	500
11	0	0	0	0	0	0	0	0	0	500	0

FIGURE 3.25 Calculated flow matrix.

Machine

	1	2	3	4	5	6	7	8	9	10	11
1	0	0	0	A	0	0	0	0	0	0	0
2	0	0	0	0	0	0	0	0	0	0	0
3	0	0	0	0	0	0	0	0	0	0	0
4	A	0	0	0	X	0	0	0	0	0	0
5	0	0	0	X	0	0	0	0	0	0	0
6	0	0	0	0	0	0	0	0	0	0	0
7	0	0	0	0	0	0	0	0	0	0	0
8	0	0	0	0	0	0	0	0	0	0	0
9	0	0	0	0	0	0	0	0	0	0	0
10	0	0	0	0	0	0	0	0	0	0	0
11	0	0	0	0	0	0	0	0	0	0	0

(Machine — left axis)

FIGURE 3.26 Machine relationship matrix.

TABLE 3.2
Machine Sizes with One Foot of Clearance Added to Each Side

Machine Number	1	2	3	4	5	6	7	8	9	10	11
Length	10	10	20	20	10	20	10	10	10	10	10
Width	20	10	20	10	10	20	10	10	10	10	10

3.7 SUMMARY

In this chapter, we discussed material flow analysis and its usefulness in developing flow and distance, which are so essential for solving the facility layout design problem. We reviewed several common material flow patterns and five types of layouts. We also listed and discussed some basic pieces of information that enable us to develop a facility design as well as to compare and select alternative designs. This information includes the flow process chart, flow diagram, quantitative and qualitative flow data, and distance data. The relationship chart, which captures qualitative flow data, is used in the SLP layout design technique discussed in Chapter 4. The quantitative flow and distance data as well the quantitative layout evaluation criteria are used in the layout models and algorithms discussed in later chapters.

Four tools for presenting layout designs were briefly discussed with illustrations and guidelines for collecting data for the layout design problem. We then presented a case study applying these guidelines to an industrial problem.

3.8 REVIEW QUESTIONS AND EXERCISES

1. List the types of data useful in facility design. Discuss in one paragraph how each data type can be obtained.
2. List various possible flow patterns and discuss the situations under which each would be suitable.
3. List all the advantages and disadvantages of the following types of manufacturing layouts. Consult additional sources of information if necessary.
   (a) Product layout
   (b) Process layout

    (c) Group technology layout

    (d) Fixed-position layout

    (e) Hybrid layout

4. Visit a local machine shop. Develop a layout planning chart for one of the products processed in the shop.

5. Construct a flow diagram for the product in Exercise 4.

6. Visit a local manufacturing facility. Develop a flow process chart (similar to the one in Figure 3.9). Summarize the number of operations, inspections, transportations, storages, and delays. If possible, eliminate or minimize some of the activities by simplifying or combining some of them. Develop a new flow process chart and compare it with the previous one.

7. Visit the following places in your college or university, list all the departments in each, and draw a relationship diagram. Based on your observation, assign a relationship code for each pair of departments and state the reason for assigning such a code.

    (a) Bookstore

    (b) Cafeteria

    (c) Registrar's office

    (d) Ground floor of library

    (e) Main computer center

    (f) Visitor information center

    (g) Physical recreation center or gymnasium

    (h) Infirmary

8. A real estate agent's office is located in a busy downtown area in a single-floor building that has the dimensions of 20 feet × 60 feet. There is a parking lot behind the building with dimensions 20 feet × 60 feet. The office has four real estate agents and one secretary and must have an office, reception area with space for seating customers, restrooms, photocopy room, conference room, and a multipurpose room with the space requirements shown in Table 3.3.

    (a) Construct a relationship chart for the departments.

    (b) Using the relationship chart and space requirement shown in Table 3.3, draw a block plan (similar to the dentist's office layout in Figure 1.8).

9. An insurance office has leased a small building along a state highway with the following dimensions: 40 feet × 50 feet. There is a parking lot behind the building with dimensions 20 feet × 50 feet. The office has three insurance agents and one secretary and must have an office, reception area with space for seating customers, restrooms, photocopy room, conference room, and a multipurpose room with the space requirements shown in Table 3.4.

    (a) Construct a relationship chart for the departments.

    (b) Using the relationship chart and space requirement shown in Table 3.4, draw a block plan (similar to the dentist's office layout in Figure 1.8).

TABLE 3.3

Space Information for Exercise 8

Room	Space Required in Square Feet
Real estate agent office	80
Office-cum-reception area	200
Men's room	36
Ladies' room	36
Photocopying room	80
Conference room	200
Multipurpose room	150

TABLE 3.4

Space Information for Exercise 9

Room	Space Required in Square Feet
Insurance agent office	120
Office-cum-reception area	400
Men's room	40
Ladies' room	40
Photocopying room	60
Conference room	400
Multipurpose room	380

10. In Figure 3.27, you are given the routing sheet for three types of widgets manufactured by Weedjets Manufacturing International.
    (a) Make a list of the product types and their codes.
    (b) Make a list of the equipment types and their codes.
    (c) Assign a numeric code for each widget type—1, 2, 3. Assign an alphabetic code for each equipment type—A, B, C, etc.
    (d) Write a spreadsheet software code which will
        (i) Remove all the information contained in the route sheets except the equipment types on which the widgets are processed and the sequence in which the equipment types are visited by the widget for processing.
        (ii) Write the retained information in another spreadsheet file.
    (e) Write another spreadsheet code that will use the information stored in the spreadsheet file created in Equation 3.4 to create a from–to frequency of trips matrix. Store this matrix in another spreadsheet file. For this question, use the demand and batch size information shown in Table 3.5.
    (f) Using the spreadsheet file created in Equation 3.5 and another appropriate spreadsheet code, develop a frequency of trips between departments matrix. Store this information in a third spreadsheet file.
    (g) Assume there are two units of machine type 01204 and that 30% of all part types that require machine type 01204 are processed on the first copy and the remaining 70% on the other. Determine the new frequency of trips matrix.
    (h) Using your answer to the machine coordinate information in Exercise 11, determine the length and width of each machine. Concatenate all of the data generated in aforementioned questions (a) through (f) and prepare a data input file.
11. Weedjets Manufacturing International has currently arranged the five equipment types mentioned in Exercise 10 as shown in Figure 3.28. Coordinates of the five machines A, B, C, D, and E are (10, 7.5), (20, 20), (60, 12.5), (35, 7.5), and (5, 20), respectively.
    (a) Determine the dimensions of each equipment type.
    (b) Using a spreadsheet software, determine the following distance measures for each equipment pair:
        (i) Euclidean
        (ii) Squared Euclidean
        (iii) Rectilinear
        (iv) Tchebychev
    (c) Using the distances calculated in part (b) and frequency of trips between matrix calculated in Exercise 10, part (f), determine the sum of flow times distance for the equipment pairs for each distance metric.

6/25/92   10:10:01   WEEDJETS MANUFACTURING INTERNATIONAL C/N 100   PAGE   1   HRIBICK   RPRTEP01

MASTER ROUTING LIST

--PART NUMBER--	--------DESCRIPTION--------	DATE	ALT CODE	BUYER/PLANNER	DRAWING REVISION
999441	WEEDJET 1	0/00/00		400	

OPER & ALT	MACH CODE	OPER CODE	SETUP CODE	CREW FACTOR	--TOOLING REF-- NUMBER	--SETUP- HOURS	--LABOR- HOURS	--MACHINE- HOURS	I/O CDE/	BASIS -QTY-	TIME -DAYS-	MOVE ----EFFECTIVE---- FROM	TO
10	01204			1.0	T10190	4.000	2.325	4.650	4	1000	.000	0/00/00	99/99/99

----ROUTING DESCRIPTIONS----
FORM & CUT OFF BROWN & SHARPE 00

| 30 | 03001 | | | .0 | | .000 | 1.000 | .000 | 4 | 1000 | .000 | 0/00/00 | 99/99/99 |

----ROUTING DESCRIPTIONS----
TUMBLE/RAMPE

| 40 | 03101 | | | .0 | | .000 | 1.500 | .000 | 4 | 1000 | .000 | 0/00/00 | 99/99/99 |

----ROUTING DESCRIPTIONS----
FLASH BARREL NI/ SHIP

6/25/92   10:10:01   WEEDJETS MANUFACTURING INTERNATIONAL C/N 100   PAGE 1   HRIBICK   RPRTEP01

MASTER ROUTING LIST

--PART NUMBER--	--------DESCRIPTION--------	DATE	ALT CODE	BUYER/PLANNER	DRAWING REVISION
999441-3	WEEDJET 2	0/00/00	400		A

OPER & ALT	MACH CODE	OPER CODE	SETUP CODE	CREW FACTOR	--TOOLING REF-- NUMBER	--SETUP- HOURS	--LABOR- HOURS	--MACHINE- HOURS	I/O CDE/	BASIS -QTY-	TIME -DAYS-	MOVE ----EFFECTIVE---- FROM	TO
1.0	T2909			3.000	.910	1.820	1.000	.000	4	0/00/00	.000	0/00/00	99/99/99

----ROUTING DESCRIPTIONS----
MK @ AUTO 999441-3P

| 15 | 03022 | | | 1.0 | .000 | .000 | .250 | .250 | 4 | 1000 | .000 | 0/00/00 | 99/99/99 |

----ROUTING DESCRIPTIONS----
DEGREASE

| 31 | 03101 | | | 1.0 | .000 | .000 | .030 | .030 | 4 | 1000 | .000 | 0/00/00 | 99/99/99 |

----ROUTING DESCRIPTIONS----
BARREL NICKEL /SHP

(Continued)

FIGURE 3.27 Routing sheet for three types of widgets manufactured by Weedjets Manufacturing International.

```
6/25/92   10:10:01              WEEDJETS  MANUFACTURING INTERNATIONAL  C/N 100              PAGE 1      HRIBICK      RPRTEP01

--PART NUMBER--   ------------DESCRIPTION------------  DATE      ALT CODE      BUYER/PLANNER          DRAWING REVISION
999441-4          WEEDJET 3                            0/00/00                     400                       A
                                          MASTER ROUTING LIST
```

OPER & ALT	MACH CODE	OPER CODE	SETUP CODE	CREW FACTOR	--TOOLING REF-- NUMBER	--SETUP- HOURS	-LABOR- HOURS	-MACHINE- HOURS	I/O CDE	BASIS /-QTY-	TIME -DAYS-	MOVE TIME	EFFECTIVE FROM	TO
10	01204			1.0	T6624	5.000	.520	1.010	4	1000	.000		0/00/00	99/99/99

```
              ----------------ROUTING DESCRIPTIONS----------------
              MAKE/AUTO  999441-4P
```

| 30A | 03001 | | | .0 | | .000 | .750 | 1.250 | 4 | 1000 | .000 | | 0/00/00 | 99/99/99 |

```
              ----------------ROUTING DESCRIPTIONS----------------
```

| 40B | 08003 | | | 1.0 | T6625 | .000 | .500 | .000 | 4 | 1000 | .000 | | 0/00/00 | 99/99/99 |

```
              PLATING INSP/SHP
```

FIGURE 3.27 (Continued) Routing sheet for three types of widgets manufactured by Weedjets Manufacturing International.

TABLE 3.5
Demand and Batch Size Information for Exercise 10

Widget Type	Annual Demand	Number of Units That Can Be Transported in One Trip
1	10,000	10
2	14,500	5
3	8,000	20

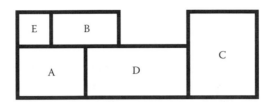

FIGURE 3.28 Current layout for Weedjets Manufacturing International in Exercise 11.

  (d) By trial and error alter the block plan shown and generate a new one that reduces the overall flow times distance by at least 10% when compared to the values obtained in (c). Note that you may change the positions of equipment and their orientation, but not their length, breadth, or area.

12. Here are the x, y coordinates of the centers of six departments 1, 2, 3, 4, 5, and 6.
    $(x_1, y_1) = (26, 40)$
    $(x_2, y_2) = (0, 23)$
    $(x_3, y_3) = (26, 12)$
    $(x_4, y_4) = (0, 0)$
    $(x_5, y_5) = (15, 20)$
    $(x_6, y_6) = (41, 0)$
    Using a spreadsheet software package, determine each of the following distances between the centers of each pair of departments:
    (a) Euclidean
    (b) Squared Euclidean
    (c) Rectilinear
    (d) Tchebychev

13. Given the x, y coordinates in Exercise 12, draw a block diagram so that
    (a) The horizontal boundaries of departments 2 and 4 touch each other
    (b) The vertical boundaries of departments 2 and 5, departments 5 and 3, and departments 3 and 6 touch each other
    Assume that the departments are square or rectangular in shape. Their lengths and breadths are to be chosen so that the departments do not overlap and the conditions (a) and (b) are satisfied. If there is more than one feasible configuration, you are required to identify only one.

14. Consider the "frequency of trips between departments" matrix shown in Figure 3.14. For each distance metric calculated in Exercise 12, determine the flow times distance for each pair of departments. Add these quantities and determine which metric yields the least sum of flow times distance. Perform your calculations using a spreadsheet software package. Alter the configuration you obtained in Exercise 11 so that the departments are closer to one another and the new block plan is feasible. While altering the positions of departments, try to reduce the overall flow times distance at least 15% over the previous configuration. Note that you must use the

same shapes, lengths, and breadths you assumed in Exercise 11. Also note that you may have to generate a different block plan for each distance metric. Perform your calculations using a spreadsheet.

15. Collect product literature on templates and 3D physical models. List specific situations where the following tools would be useful for layout presentation. Justify your answer by providing sufficient explanation.
    (a) Manual drawing
    (b) Templates
    (c) Three-dimensional physical models
    (d) CAD drawings

16. Consider the part-routing information matrix in Figure 3.24. Assume that two material handling devices—a forklift truck and an AGV—are used to transport the material. Further assume that the AGV can transport 20 units of each part type, whereas the forklift truck can handle only 10 units. The AGV is used to transport all the parts to and from machine 6. The forklift truck is used for the remaining material transfers. Calculate the new frequency of trips matrix for the aforementioned problem. How does it compare to the one in Figure 3.24?

17. *Project*: Part-routing data similar to the one in Figure 3.21 is available in http://sundere.okstate.edu/Book. Using a spreadsheet software, generate
    (a) An ordered list of parts and machines visited (like the one shown in Figure 3.22)
    (b) A part-routing information matrix (assume reasonable values for part demand and batch size)
    (c) A flow matrix

4 Traditional Approaches to Facility Layout

4.1 INTRODUCTION

There has been some debate over the characterization of the layout and location problems as design and optimization problems. Simon (1975) suggests that the layout problem is a design problem but the location problem is an optimization problem. Design problems are those in which there is no well-defined optimum solution or solutions. A solution is optimal if every other possible solution to the problem is worse or as good in terms of the chosen criteria. Designers are not interested in finding the best solution, but rather a satisficing solution. Design problems are often solved by synthesizing component decisions that are selective, cumulative, and tentative (Simon, 1975). Thus, not only are different designers likely to come up with different designs, but the procedures or methods they adopt in arriving at the final design are also significantly different. Our discussion of the commonly used facility design strategies in this chapter will illustrate these differences. Optimization problems, on the other hand, have a well-defined optimum solution or solutions. For such problems—or more precisely, at least for small instances of such problems—well-defined solution procedures or techniques that can identify at least one optimum solution are available.

Francis et al. (1992) argue that the location problem should be treated as a design problem because it has some of the characteristics of a design problem. For facility layout problems, we seek the *best* layout solution, and likewise, for facility location problems, we seek the *best* facility locations. According to Francis et al. (1992), even though the location problem may be formulated more precisely than the layout problem, the location problem should not be characterized as an optimization problem. In fact, many qualitative factors cannot be captured in mathematical formulations, so different designers arrive at different designs.

In our opinion, the layout and location problems have elements of both design and optimization problems. As a result, they can be classified neither as a design problem nor as an optimization problem. To illustrate this, consider a real-world example. Facility analysts were responsible for redesigning a new automated manufacturing facility. They classified machines and workstations into three categories based on size—*large*, *medium*, and *small*. In the final design approved by top management, large machines were placed in a certain part of the factory floor; medium-sized machines were placed in an adjacent location; and machines in the small category were placed beside those in the medium group. No consideration was given to material flow or any of the other criteria listed in Section 1.3! Now, if we classify the layout problem as a design problem only, then this design is *acceptable*, but common sense tells us that it will lead to problems such as traffic congestion and excessive material handling delays. On the other hand, if we classify the layout problem as an optimization problem only in which we seek an optimal location of machines and workstations to minimize material handling costs, then the design is also incorrect because it ignores other factors that cannot be easily captured. To formulate the layout problem mathematically (and to solve it), analysts make limited assumptions in the mathematical model and ignore factors that are difficult to formulate mathematically. For instance, the quadratic assignment problem (QAP), which is often been used to formulate the layout problem, implicitly assumes the departments are squares with equal area.

Layout and location problems therefore have elements of optimization and design problems. To see this, let us examine the manner in which the layout and location problems are solved.

In general, we use two phases to find the final design. In the first phase, we formulate the problem as an optimization problem (after making assumptions so that a model can be formulated and, more important, can be solved using an appropriate algorithm) and generate a number of preliminary solutions using optimization techniques. In the second phase, we include factors ignored by the model and modify the preliminary solutions suitably. Then, we select a *satisfactory* design. Because this phase is satisficing rather than optimizing, it has characteristics of the design problem. Thus, the layout and location problems have characteristics of design and optimization problems.

In practice, analysts have a tendency to think of the layout problem as a pure design problem. To a large extent, three of the contemporary approaches discussed in this chapter also treat the layout problem as a design problem. It is our hope that after reading this book, the reader will fully understand that the layout and location problems have elements of both optimization and design and treat the two problems accordingly. This chapter is devoted to study the layout problem as a design problem and Chapters 5 and 10 study it as an optimization problem.

We discuss two approaches to facility design in detail. The first called systematic layout planning (SLP) is by far the most popular approach used in practice. Although this technique was developed in the 1960s, it is widely used even today! In fact, there are software packages that support elements of this approach (PROPLANNER, for example). SLP is based on the two approaches developed by Reed (1961) and Apple (1977), respectively.

4.2   SYSTEMATIC LAYOUT PLANNING

A primary reason the SLP technique has remained popular for more than 30 years is its simple step-by-step approach to facility design. It consists of the four phases as illustrated in Figure 4.1:

> *Phase I—Determination of the location of the area where departments are to be laid out*: Phase I involves identifying the locations for the departments. For example, this area may be on the north side of a building or in a new building adjacent to the existing manufacturing building. This phase is the easiest of the four phases.
>
> *Phase II—Establishing the general overall layout*: Phase II involves determining the flow of materials between departments, examining special adjacency requirements, determining the space required for each department, balancing it with the space available, incorporating practical constraints (e.g., budget, safety), and generating up to five alternate layout plans. The plans are then evaluated based on cost and other noncost considerations and a layout is selected for departments and general work areas.
>
> *Phase III—Establishing detailed layout plans*: The relative positions of departments found in Phase II do not provide details about the layout and location of each specific machine, auxiliary equipment, and support services such as restrooms, cleaning rooms, inspection station, and battery-charging rooms. This detailed layout of departments and support services is done in Phase III. The procedures for generating layouts in Phase III are the same as in Phase II, except that Phase II deals with the layout of departments, whereas Phase III deals with the layout of machines and other auxiliary equipment in each department.
>
> *Phase IV—Installing the selected layout*: The detailed layout must be approved by all concerned people: affected employees, supervisors, and managers. Then the final layout is prepared. The drawings must show much more detail because they are used to plan the move to the new facility. In Phase IV, funds and time are appropriated for the move, and the actual relocation of machinery and services takes place.

In our discussion of the SLP technique, we describe how Phases II and III are carried out. As mentioned, Phase I involves determining the area where the departments will be laid out and is rather simple. Phase IV actually deals with moving departments to the new site and is not discussed.

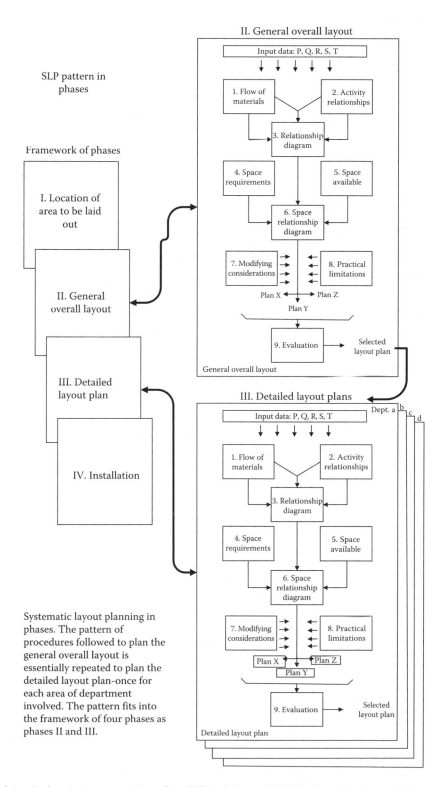

SLP pattern in phases

Framework of phases

I. Location of area to be laid out

II. General overall layout

III. Detailed layout plan

IV. Installation

Systematic layout planning in phases. The pattern of procedures followed to plan the general overall layout is essentially repeated to plan the detailed layout plan-once for each area of department involved. The pattern fits into the framework of four phases as phases II and III.

II. General overall layout

Input data: P, Q, R, S, T

1. Flow of materials
2. Activity relationships
3. Relationship diagram
4. Space requirements
5. Space available
6. Space relationship diagram
7. Modifying considerations
8. Practical limitations

Plan X   Plan Z
Plan Y

9. Evaluation   Selected layout plan

General overall layout

III. Detailed layout plans

Dept. a b c d

Input data: P, Q, R, S, T

1. Flow of materials
2. Activity relationships
3. Relationship diagram
4. Space requirements
5. Space available
6. Space relationship diagram
7. Modifying considerations
8. Practical limitations

Plan X   Plan Z
Plan Y

9. Evaluation   Selected layout plan

Detailed layout plan

FIGURE 4.1 A pictorial representation of the SLP technique. (Redrawn from Muther, R., *Systematic Layout Planning*, Van Nostrand Reinhold Company, New York, 1973. With permission.)

The procedure used by SLP to generate the overall and detailed layouts in Phases II and II is illustrated in Figure 4.2. The input data required by SLP are classified into the following five categories:

P   Product: Types of products to be produced.
Q   Quantity: Volume of each part type.
R   Routing: Operation sequence for each part type.
S   Services: Support services, locker rooms, inspection stations, and so on.
T   Timing: When are the part types to be produced? What machines will be used during this time period?

The SLP technique is explained in Figure 4.1. With P–Q–R data, a from–to material flow matrix is constructed (box 1). This matrix indicates the intensity of flow between each machine pair. Similarly, using the P–Q–S data, the relationship chart is constructed (box 2). Next, the relationship diagram is drawn with information from the flow matrix and relationship chart (box 3). As mentioned in Chapter 3, this diagram enables us to capture nonflow relationships between machine pairs. From the quantitative flow data and qualitative relationships, machines are connected by lines with intensities that correspond to the desired closeness of the machine pairs. For example, four lines are used to connect machine pairs that either have an A relationship or flow values within 10% of the largest value in the flow matrix. Three lines are drawn between machine pairs that have an E relationship or flow values between 10% and 30% of the largest flow matrix value, and so on. Of course, the percentages may vary depending on the application. Also, instead of using two, three, and four lines to connect department pairs with I, E, and A relationships, we may use thin, thick, and extra-thick lines, respectively. Whatever line drawing convention or percentages are used, the important task now is to combine data in the flow matrix and relationship chart and to draw the relationship diagram. Although this diagram does not provide the overall layout, it does provide some preliminary information about which department pairs must be adjacent, which ones must not, and how the machines should be positioned relative to one another, thus enabling analysts to construct an initial, but rough layout.

Although we have suggested that the relationship diagram be constructed after combining information from the from–to material flow matrix and relationship chart, in the following discussion, we use only the relationship chart in Figure 3.13 to construct the relationship diagram shown in Figure 4.3. Notice that lines in the relationship diagram indicate the level of desired closeness between departments. If a department pair has an A relationship, these two are connected by four lines. Similarly, department pairs having an E, I, and O relationship are connected by three lines, two lines, and one line, respectively. Department pairs with a U relationship (departments 4 and 5, for example) are not connected at all. Those with an X (undesirable) relationship are connected with a zigzag line. Also, the American Society of Mechanical Engineers (ASME) symbols in Figure 2.6 are used to denote activities (departments) in the relationship diagram.

Knowledge of space requirements and availability is essential to construct a meaningful layout, so spatial analysis is the next step (box 4). We create a list of machines and departments and calculate the area required by each. Of course, as discussed in Chapter 2, we must include the space required for operators, material handling access, and auxiliary equipment. Figure 4.4 is a sample space requirement worksheet.

The next step in the SLP technique is to determine the available space; special spatial restrictions, if any (e.g., beams or other structures that do not allow full utilization of the space); and other such factors (box 5). Boxes 4 and 5 and the relationship diagram are then used to construct what is called a space relationship diagram (box 6). Figure 4.5 is a sample space relationship diagram. The corresponding relationship diagram is shown in Figure 4.6. Notice that the diagram shows the required space in addition to the adjacencies between activity pairs.

After other factors (e.g., material handling methods, storage equipment, utility distribution) and practical limitations (cost, building codes, power availability, existing structures or beams,

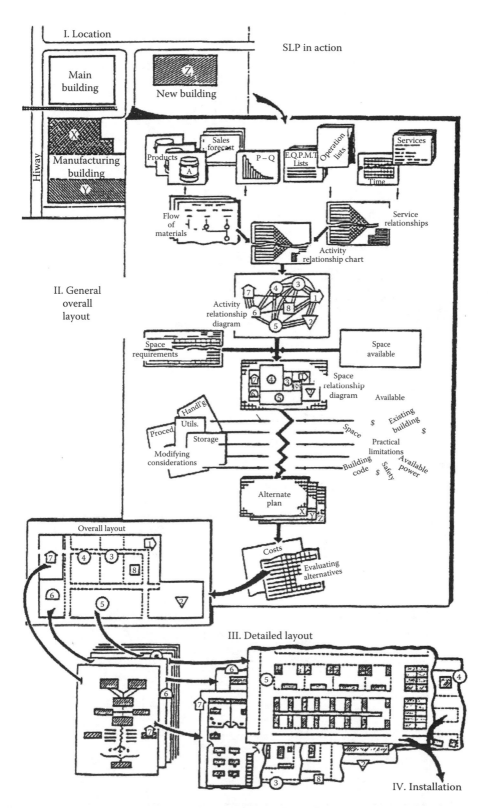

FIGURE 4.2 Four phases of the SLP technique. (Reprinted from Muther, R., *Systematic Layout Planning*, Van Nostrand Reinhold Company, New York, 1973. With permission.)

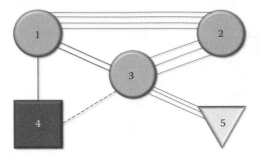

FIGURE 4.3   Relationship diagram for five activities from the relationship chart in Figure 3.13.

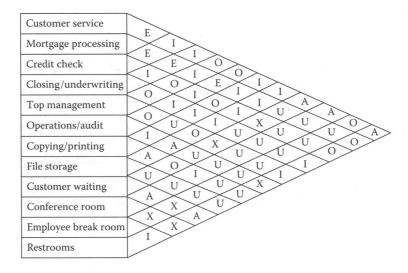

FIGURE 4.4   Relationship chart for MortAmerica, Inc.

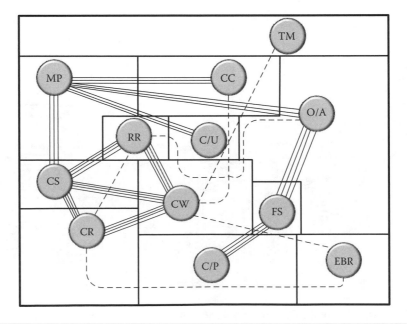

FIGURE 4.5   Space relationship diagram constructed using activity relationship diagram in Figure 4.6.

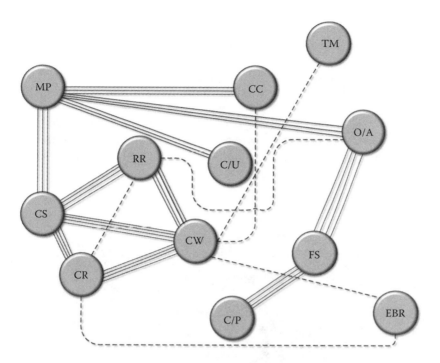

FIGURE 4.6 Activity relationship diagram for MortAmerica, Inc.

personnel and equipment safety) are considered, the space relationship diagram previously constructed is modified to create as many as five alternative layout plans. Typically, two or three alternative layouts are generated at this stage. Generating more is time-consuming and does not provide significant new information. Each alternative layout is then evaluated based on cost and other intangible factors, and the best plan is selected.

In Phase III, we use the same except that we are now dealing with detailed layouts of departments as opposed to an overall layout.

Thus, at the end of Phase III, we have an overall layout that indicates the relative positions of departments and a detailed layout for each. This information is used in developing a detailed drawing that can be used in the actual relocation of departments.

4.3 SPECIAL CONSIDERATIONS IN OFFICE LAYOUT

Some of the main objectives in designing an office layout are as follows:

1. Minimizing distance traveled by employees
2. Permitting flexibility so that the current layout can be changed, expanded, or downsized easily
3. Providing a safe and pleasant atmosphere for people to work in
4. Minimizing capital and operational costs of the facility

Before embarking on an office layout or office redesign project, a thorough operational review must be conducted. Typical questions raised during an office operations review were discussed in Chapter 1 and are repeated here:

1. Is the company outgrowing available space?
2. Is the available space too expensive?

3. Is the current building not in the proper location?
4. How will a new office layout affect the organization?
5. Are office operations too centralized or decentralized?
6. Does the office structure support the strategic plan?
7. Is the office layout in tune with the company's image?

In Chapter 2, we saw several types of office structures. We will now discuss some of the interior office designs. Figure 4.7 show the lobby space of a research building at the University of Louisville. High ceilings and open space combined with mirrors, plants, and other displays (e.g., fish tanks) provide a pleasing and welcoming environment to visitors and employees alike. The student union lobby at Oklahoma State University is three stories high (see Figure 4.8).

FIGURE 4.7   High ceilings and open spaces enhance building design.

FIGURE 4.8   Student union at Oklahoma State University.

Some organizations, for example, a high-end interior decorator or a five-star hotel, may wish to provide a luxurious waiting area with lavish appointments.

Many office layouts have cubicles for employees. Cubicles facilitate better communication between employees, permit supervision, and allow layouts to be easily modified, expanded, or shrunk. They do not provide privacy, nor do they help in isolating noise. Several cubicles are shown in Figure 4.9.

The layout of a state department of transportation office and the Denrer International Airport terminal are shown in Figures 4.10 and 4.11, respectively.

FIGURE 4.9   Cubicles layout.

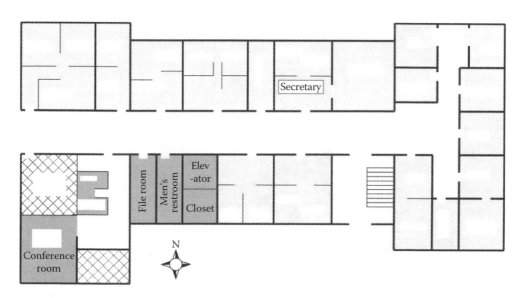

FIGURE 4.10   Layout of an Iowa state department of transportation office. (From http://www.sysplan.dot. state.ia.us.office_layout.htm.)

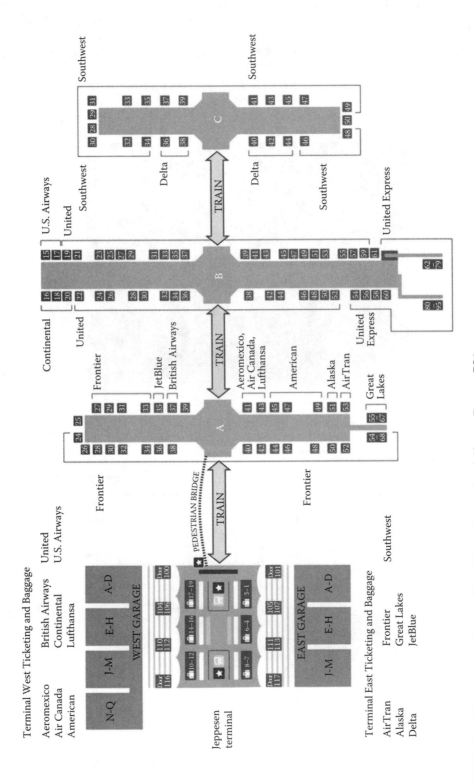

FIGURE 4.11 Airport terminal layout. (From Denver International Airport, Denver, CO.)

4.4 OFFICE PLANNING PROJECT FOR A MORTGAGE COMPANY

In this section, we discuss an office layout project that employs many of the ideas in the SLP technique. The plan is for the office of a medium-sized mortgage company (MortAmerica, Inc.). Four major phases are typically involved in office planning:

1. Evaluation
2. Planning
3. Site selection
4. Design and layout

4.4.1 EVALUATION

This phase essentially involves determining the cost of building, renting, or leasing along with interior decoration and furnishing costs. Customer service, marketing, operational aspects, and strategic plans are thoroughly evaluated. An office operations review is a necessary first step because it tells whether the new office plan is justified.

For the mortgage company, four office operations review questions are relevant:

1. Is there a significant increase in mortgage lending operations of MortAmerica, Inc.?
2. Are the costs of leasing and refurbishing interior space too high?
3. Is there a problem with the current location? For example,
    a. There is not enough space for expansion.
    b. Major attorneys' offices, other related financial institutions and restaurants, are not located within a reasonable distance of MortAmerica, Inc.
    c. Adequate parking space is not available.
    d. Traffic is too congested.
4. Will a change in office location improve business?

4.4.2 PLANNING

Planning is the most important phase of the office layout project. It is also a time-consuming step. The time is well spent during this phase, however, because the quality of design and layout decisions made in Phase IV are directly dependent on careful and thorough planning in Phase II. Phase II consists of four distinct tasks:

1. *Review current space utilization*: This task is to conduct an inventory of current space by examining each department within the organization. In MortAmerica, Inc., the departments are
    a. Customer service
    b. Mortgage processing/marketing
    c. Credit check
    d. Closing/underwriting
    e. Operations/audit
    f. Top management
    In addition to the departments, support services such as lounge, customer waiting areas, and conference, mail, and copy rooms are considered. The actual square footage of usable space (excluding aisle and wall space) for each department and support facility is then calculated.
2. *Determine space projections*: This is also a critical task. The corporate strategic plan should be considered in projecting space requirement because it provides a good indication

of how the organization will look in the future. Specifically, it tells us what sort of organizational structure is likely to exist. It also tells us (a) areas of the organization that will receive an increased emphasis and those that will receive a decreased emphasis and (b) whether some departments will be expanded, scaled down, or even merged with other departments. From the projected organizational structure, staff and support services are projected. Participation of top management is essential at this stage because they are the ones who make strategic decisions concerning future organizational structure.

For projecting staff requirements, we must clearly list all the personnel categories and the number of people in each. The next step is to determine the space required for the various categories of personnel projected. A spreadsheet is a useful tool for estimating projected staff space requirements. Table 4.1 is the employee space projection spreadsheet for MortAmerica, Inc. This table also lists current space requirements for employees.

While determining space requirements (current and future), the analyst must also keep track of special considerations for each type of office. For example, executives and senior staff typically prefer a semiclosed structure, whereas an open structure is typical for secretaries.

Just as space requirements are projected for employees, a similar analysis is done for support services (see Table 4.2). For MortAmerica, Inc., the support services are as follows:
  i.   Copying/printing area
  ii. Files storage rooms
 iii. Customer waiting lounge
  iv. Conference rooms
  v.   Employee break rooms
  vi. Restrooms

3. *Determine level of interaction between departments*: This critical step attempts to quantify or qualify the level of interaction between departments and support areas. Because it is time-consuming and expensive to develop quantitative data for MortAmerica (and indeed for office planning in general), qualitative relationships have been developed and placed in a relationship chart as shown in Figure 4.4.

4. *Identifying special consideration*: This step considers other special factors such as (a) the need for being within a reasonable distance from attorneys' offices and restaurants, (b) ground floor location, and (c) ample space for display of company logo.

4.4.3 Site Selection

Site selection is a separate topic and is discussed in Chapters 9 and 11.

4.4.4 Design and Layout

Using the relationship chart and space requirement tables, we draw an activity relationship diagram, a space relationship diagram, and a prearchitectural layout. These are shown in Figures 4.5, 4.6, and 4.12, respectively.

The activity relationship diagram in Figure 4.6 highlights the relationship between departments using lines. Department pairs with an A relationship have four lines and those with an E relationship have three lines. Department pairs with a U relationship have no lines and those with an X relationship have a dashed line. To maintain clarity in the figure, the O and I relationships (normally indicated by one and two lines, respectively) are not shown. The activity relationship diagram does not consider activity areas, so the next step is to draw a space relationship diagram in Figure 4.5, which shows not only the relationships between departments (using lines in Figure 4.6) but also their areas.

Using information from the space relationship diagram, we determine the relative positioning of departments so that department pairs with A relationship are adjacent, those with an X relationship

TABLE 4.1

Current and Projected Total Employee Space Requirements at MortAmerica, Inc.

Department Name	Current/Future Requirements	Categories of Employees and Number in Each Category					
		Senior Executive	Senior Staff	Staff	Clerical/ Secretary	Net Space Required	Gross Space, 150% of Net Space
Customer service (CS)	Current space/employee		150	100	75		
	Number of employees		1	4	1		
	Current total space/category		150	400	75	625	938
	Future space/employee			120	75		
	Number of employees			6	1		
	Future space/category			720	75	795	1,193
Mortgage processing/marketing (MP/M)	Current space/employee		200	100	75		
	Number of employees		2	10	2		
	Current total space/category		400	1000	150	1550	2,325
	Future space/employee	250	200	100	75		
	Number of employees	1	1	15	1		
	Future space/category	250	200	1500	75	2025	3,038
Credit check (CC)	Current space/employee			100	75		
	Number of employees			10	1		
	Current total space/category			1000	75	1075	1,613
	Future space/employee			80			
	Number of employees			5			
	Future space/category			400		400	600
Operations audit (O/A)	Current space/employee	200	100	90	75		
	Number of employees	2	4	15	5		
	Current total space/category	400	400	1350	375	2525	3,788
	Future space/employee	250	100	100	75		
	Number of employees	3	4	20	2		
	Future space/category	750	400	2000	150	3300	4,950

(Continued)

TABLE 4.1 (*Continued*)
Current and Projected Total Employee Space Requirements at MortAmerica, Inc.

Department Name	Current/Future Requirements	Categories of Employees and Number in Each Category					
		Senior Executive	Senior Staff	Staff	Clerical/ Secretary	Net Space Required	Gross Space, 150% of Net Space
Top management (TM)	Current space/employee	250	200		100		
	Number of employees	5	2		5		
	Current total space/category	1250	400		500	2150	3,225
	Future space/employee	250	200		100		
	Number of employees	5	4		8		
	Future space/category	1250	800		800	2850	4,275
Closing/underwriting (C/U)	Current space/employee		185	100	100		
	Number of employees		1	3	1		
	Current total space/category		185	300	100	585	878
	Future space/employee		185	100	100		
	Number of employees		1	3	1		
	Future space/category		185	300	100	585	878
Total current space		1650	1535	4050	1275	8510	12,767
Total future space		2250	1585	4920	1200	9955	14,934

TABLE 4.2
Current and Future Support Services Space Projections

Support Service Area	Current Net Space	Current Gross Space 150% of Net Space	Future Net Space	Future Gross Space 150% of Net Space
Copying/printing area (C/P)	300	450	465	700
File storage room (FS)	300	450	80	120
Customer waiting lounge (CW)	300	450	800	1200
Conference rooms (CR)	500	750	1000	1500
Employee break room (EBR)	200	300	850	1275
Restrooms (RR)	200	300	500	750
Total	1800	2700	3695	5545

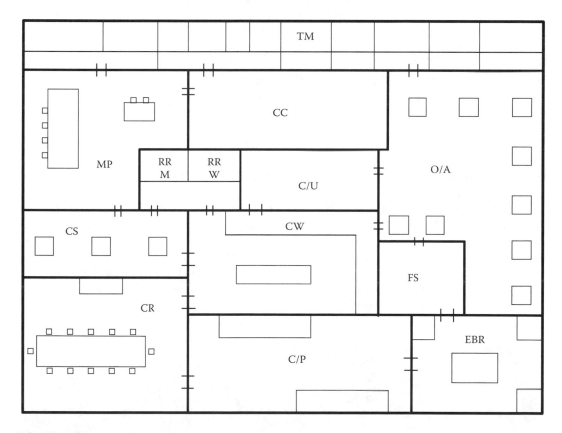

FIGURE 4.12 Prearchitectural layout.

are nonadjacent (preferably in distant locations), those with an E relationship are placed as close as possible, and so on. The relative positioning of departments is typically determined by trial and error.

A reasonably *good* arrangement for the MortAmerica office is shown in Figure 4.5. While developing the prearchitectural layout (see Figure 4.12), we take into consideration walls and other aspects of the layout design ignored thus far. Department shapes that were square or rectangular may be altered to a suitable shape so that all the departments will fit within the building dimensions. This step requires the facility designer to be creative and generate alternative designs, listing their advantages and disadvantages.

Note that the layout in Figure 4.12 for MortAmerica, Inc., takes into account the space requirements for staff and support services from Tables 4.1 and 4.2. It also shows aisles and individual offices for the employees. If the designer is creative, a number of alternative configurations can be drawn for the individual offices, rooms, and aisles. Computer software packages are very useful in the final step because they enable designers to generate several alternative prearchitectural layouts and modify each as needed.

4.5   CODE COMPLIANCE, OSHA, ADA REGULATIONS, AND OTHER CONSIDERATIONS IN FACILITY DESIGN

After developing detailed layout designs for each department within a facility and the entire facility itself, a number of regulatory, safety, and ergonomic issues must be considered prior to the construction of a facility. We will begin this section with the Americans with Disabilities Act (ADA) and then discuss Occupational Health and Safety Administration (OSHA)-related topics. More details on these topics are available at these websites: www.ada.org, www.osha.gov, and www.nfpa.org.

4.5.1   SUPPORT FACILITIES

Employees working in a facility have needs for various services, equipment, and facilities that do not directly relate to their job. Some examples include locker rooms, exercise equipment, break areas, cafeteria, parking space, restrooms, and other facilities. This section describes many of these facilities.

4.5.1.1   Cafeteria

The cafeteria is empty for much of the day, but during peak hours, a number of customers are served in a very short period. This is an important part of an organization that must not be overlooked. A cafeteria is the place where employees gather during breaks to discuss topics that are both related and unrelated to work. The cafeteria should be designed for effective noise dissipation. High ceilings; well spread out tables; multiple cash registers; good ventilation and climate control; multiple places to pick up trays; silverware and paperware; courteous, efficient, and fast service; and a clean environment will make the cafeteria a place where employees can quickly get their food and have a relaxing lunch in a comfortable setting. Photographs of the indoor and outdoor student union food courts at the Oklahoma State University are shown in Figure 4.13a and b.

(a)   (b)

FIGURE 4.13   Cafeteria layout. (a) Indoor dining and (b) outdoor dining.

In addition to a large general cafeteria where there is a larger choice of items, an organization serving lunch to a large number of people must consider decentralizing the cafeteria service by providing food stalls with limited, ready-to-go food, located strategically throughout the organization and keeping these locations open for extended hours. It may also be desirable to have eating areas for executive lunches, lunches with distinguished visitors, and serving meals for special occasions—for example, retirement celebrations, welcoming new employees, celebrating achievement of individuals, and so on. Smaller companies that cannot justify the expense involved in running a cafeteria may opt for well-stocked and well-subsidized vending machines.

4.5.1.2 Locker Rooms

Employees would like a room to change into and out of work clothes, a place to store their personal belongings including lunch and paperwork, and if the facility has an exercise room, a shower stall, washbasins, and toilets to refresh after a workout. Separate locker rooms for men and women are required. As a rule of thumb, the size of a well-equipped locker room is five times the number of employees. Thus, it is desirable to designate 2500 square feet for locker rooms in a facility with 500 employees. The architectural floor plan of the Hindu Temple Cultural Center showing a detailed layout of restrooms, kitchen, dining, auditorium seating, greenrooms, stage, and classrooms is shown in Figure 4.14.

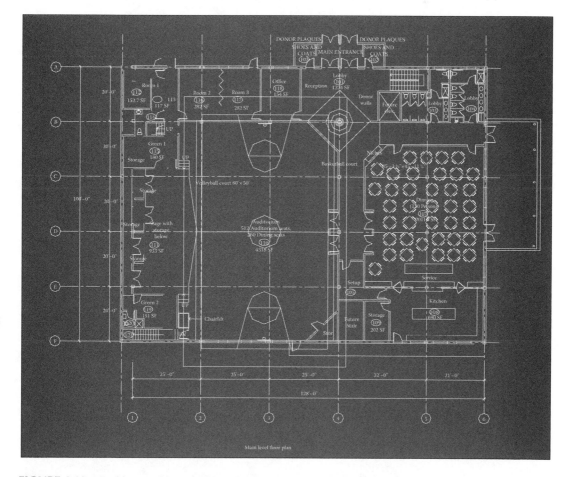

FIGURE 4.14   Architectural layout of a multipurpose cultural center.

TABLE 4.3
Sample Building Code for Plumbing Fixtures for Various Types of Facilities

Organization	Showers	Lavatories	Water Closets	Water Fountain	Others
		Service and Manufacturing Facilities			
Restaurants	—	1 per 200	1 per 75	1 per 500	1 service sink
Arenas (capacity more than 3000)	—	1 per 200 (male); 1 per 150 (female)	1 per 120 (male); 1 per 60 (female)	1 per 1000	1 service sink
Churches	—	1 per 200	1 per 150 (male); 1 per 75 (female)	1 per 1000	1 service sink
Schools	—	1 per 50	1 per 50	1 per 100	1 service sink
Airports	—	1 per 750	1 per 500	1 per 1000	1 service sink
Factories	Section 411	1 per 100	1 per 100	1 per 1000	1 service sink
Hospitals	1 per 15	1 per room	1 per room	1 per 100	1 service sink
Prisons	1 per 15	1 per cell	1 per cell	1 per 100	1 service sink
Hotels	1 per room	1 per room	1 per room	—	1 service sink
Dormitories	1 per 8	1 per 10	1 per 10	1 per 100	1 service sink

Source: www.metroped.org.

4.5.1.3 Water Closets, Sinks, Showers, and Drinking Fountains

In addition to complying with the minimum number of plumbing facilities required by building codes, it is important that a facility be designed with the following principles in mind:

1. Maintain clean, safe, and sanitary restroom facilities.
2. Install many, smaller toilet room facilities instead of one large facility with the same number of closets, urinals, and sinks. They should be easily identifiable, but not very conspicuous.
3. Install toilet facilities so that no restroom is more than 150 feet from any employee's workspace.
4. Provide more women's facilities per employee than men's facilities.
5. Where appropriate, provide baby changing stations in men's and women's restrooms.
6. Design doors and interior partitions of the restrooms so that total privacy is afforded to users of these facilities when others open the door to enter or exit the restroom.

Table 4.3 lists the minimum number of water closets, lavatories, showers, drinking fountains, and other facilities according to the Florida Building Code.

4.5.1.4 Parking Lots

The desire to provide adequate spaces to employees and visitors close to buildings must be traded off against environmental and aesthetic considerations. The Center for Watershed Protection (www.cwp.org) recommends that the designer keep these goals in mind when designing parking lots:

- Minimize paved surface
- Reduce stall dimensions
- Reduce water runoff and promote other environmental benefits by incorporating landscape into the design.

Parking lots that meet local codes can be designed in many ways. A rough estimate of the total parking space required can be calculated by assuming each employee requires a total of 250 square feet

TABLE 4.4

Sample Code for Design of Parking Lots

Service and Manufacturing Facilities	
Organization	**Parking Spaces**
Restaurants (with drive-through facilities)	One space per 75 square feet of floor area or 1.5 persons (whichever is greater)
Theaters, arenas, and assembly areas	One space per 8 feet of bench length or 4 seats (whichever is greater)
Secondary schools and colleges	One space per 8 students, one-and-a-half spaces per classroom, and number of spaces for gymnasium/assembly hall seating
Factories	One space per 1000 square feet of area plus number of spaces for offices
Hospitals	Two spaces per bed
Churches	One space per three persons
Hotels	One space per guest room plus number of spaces for accessory uses
Warehouses	One space per 2000 square feet of floor area

Source: www.ioniacounty.org.

TABLE 4.5

Minimum Dimensions for Parking Stalls

Parking Angle (°)	Aisle Width (Two-Way) (Feet)	Aisle Width (One-Way) (Feet)	Stall Width (Feet)	Stall Length (Feet)
76–90	25	15	9	20
30–75	25	12	9	22
0–29	18	12	9	25

TABLE 4.6

Accessible Spaces for Persons with Disability

Total Spaces in Parking Lot	1–25	26–50	51–75	76–100	101–150	151–200	201–300	301–400	401–500	501–1000
Minimum accessible spaces	1	2	3	4	5	6	7	8	9	2%

of space including aisles and stalls. The minimum stall width, length, aisle width, and number of spaces are specified by local codes and by the ADA. A sample code from Ionia County in Michigan is provided in Tables 4.4 through 4.6.

4.5.1.5   Exercise Area

Providing a good exercise facility is nowadays thought to be an important part of facility design. In addition to providing standard exercise equipment such as treadmills, stair climbers, elliptical cross-trainers, steppers, rowing machines, exercise bicycles, stretching mats, and weights, larger organizations provide a swimming pool, personal trainers, and special breathing, meditation, and physical exercise classes. Large mirrors, towel service, ceiling or wall-mounted flat-screen television sets, locker rooms, massage, and other services will ensure more employees use the exercise

FIGURE 4.15   An exercise and fitness center.

facilities. After all, one of the reasons for providing such facilities is to promote a healthier lifestyle for employees so that their productivity increases and they remain in good health. Fortune 100 corporations such as Google provide a fully equipped exercise facility with trainers, massage parlor, cafes with menus that promote a healthy lifestyle, on-campus daycare center, babysitting service, dry-cleaning drop-off service, and many other perks for their employees. An exercise facility is shown in Figure 4.15.

In addition to the aforementioned facilities, other facilities that may need to be provided include medical facilities, employee and visitor lounges, and security services.

4.5.2   ADA

The ADA, enacted in 1990, prohibits discrimination on the basis of disability in employment. The act lists five titles covering businesses and religious activities employing 15 or more people, state and local government, public accommodations, commercial facilities, transportation, and telecommunications. It also covers the U.S. Congress. A person is considered disabled if he or she has (1) a physical or mental impairment that substantially limits his or her major life activities, or (2) has a history or record of such an impairment, or (3) is perceived by others as having such an impairment. The goal of ADA is to ensure that people with disabilities have an equal opportunity to benefit from the full range of employment-related opportunities that are available to other individuals. The ADA prohibits discrimination in recruiting, hiring, promoting, training, and other benefits of employment and requires employers (unless it results in undue hardship) to make reasonable accommodation to the known physical or mental limitations of people with disabilities, who are otherwise qualified. State and local governments are required to provide individuals with disabilities an equal opportunity to benefit from all of their programs, services, and activities including, but not limited to, public education, employment, transportation, recreation, health care, social services, courts, voting, and town meetings. State and local governments are required to follow specific architectural standards in the new construction and alteration of their buildings, unless they can demonstrate that complying with the regulations will impose undue financial and administrative burden. Public transportation services covering local and long-distance buses, trains, and commuter rails cannot discriminate against individuals with disabilities in providing their services. The ADA covers facilities such as stores, restaurants, hotels, recreational facilities, theaters, private schools, sports stadiums, convention centers, dentists' offices, facilities operated by nonprofit organizations, transportation depots, zoos, funeral homes, and child care centers, and these facilities must not exclude, segregate, or offer unequal treatment for people with disabilities. Factories and warehouses must comply with the ADA's architectural standards for new construction and alterations. The ADA also addresses telephone and television access for people with hearing and speech impairments. Telecommunications devices for the deaf (TDD) and teletypewriters (TTYs) allow callers with speech and hearing disabilities to communicate. We cite next a few requirements of the ADA to give the reader a rough idea on what this law means to facility designers. Readers are strongly encouraged to get additional details from www.ada.org.

- Table 4.4 shows some of the parking requirements. If the total number of parking spaces is greater than 1000, then 20 spaces plus 1 for each 100 over 1000 must be provided. Thus, a facility with 1250 spaces must allot 22 spaces for access by individuals with disabilities.
- Minimum width for a passage allowing a single wheelchair access is 32 inches at any point, for example, doorways, but 36 inches continuously. The minimum width for two wheelchairs to pass is 60 inches (Figure 4.16).
- The maximum slope of a ramp from a walkway to the street cannot exceed a ratio of 1:20 (Figure 4.17).
- The minimum dimensions of elevator cars must be as shown in Figure 4.18.
- The minimum dimensions for toilet stalls must be as shown in Figure 4.19.

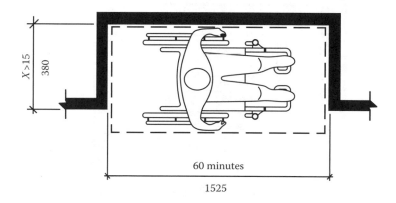

FIGURE 4.16 Passage widths for wheelchair access. (Courtesy of Occupational Safety and Health Administration.)

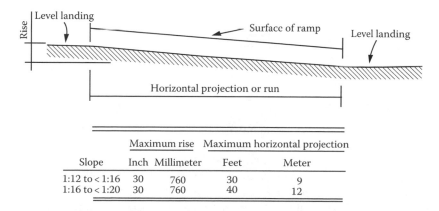

Slope	Maximum rise		Maximum horizontal projection	
	Inch	Millimeter	Feet	Meter
1:12 to < 1:16	30	760	30	9
1:16 to < 1:20	30	760	40	12

FIGURE 4.17 Minimum slope for ramps. (Courtesy of Occupational Health and Safety Administration.)

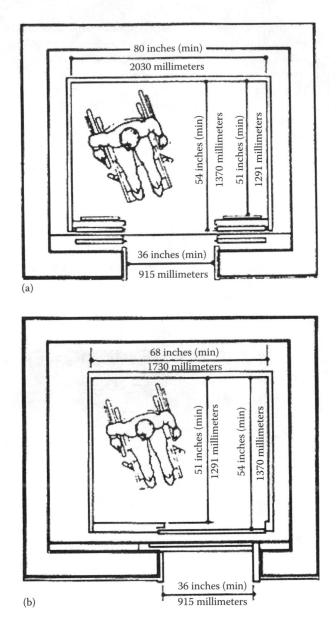

FIGURE 4.18 Minimum elevator car access requirements. (a) Door in the middle and (b) door on one side. (Courtesy of Occupational Health and Safety Administration.)

4.5.3 OSHA REGULATIONS

The primary mission of the Occupational Safety and Health Act, enacted in 1970, is to "assure the safety and health of America's workers by setting and enforcing standards" and "encouraging continual improvement in workplace safety and health." OSHA's jurisdiction is vast and covers almost every working individual in the United States. It is not practical to cover all the rules and regulations that apply to facility design and materials handling. We recommend the reader to peruse the OSHA website for specific information about the act pertaining to specific facilities (see Exercise 3) but briefly describe three aspects covered by OSHA.

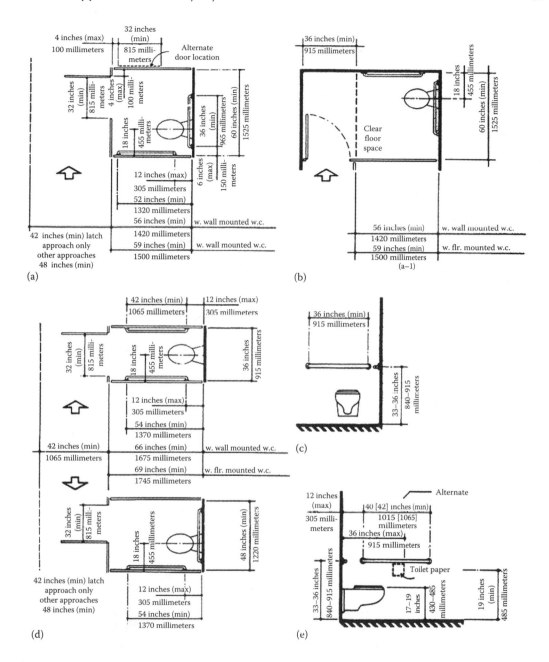

FIGURE 4.19 Minimum access dimensions for toilet stalls. (a) Standard stall, (b) standard stall (end of row), (c) rear wall of standard stall, (d) alternate stalls, and (e) side walls. (Courtesy of Occupational Health and Safety Administration.)

4.5.3.1 Noise

More than 30 million people in the United States are exposed to noise that is considered hazardous in their occupation. By utilizing hearing conservation programs, loss of hearing due to noise can be reduced or eliminated. The noise and hearing conservation programs consists of five steps:

1. Recognition
2. Evaluation

3. Control
4. Compliance
5. Training

It is important that companies recognize common health hazards including air contaminants and noise, evaluate hazards via sampling methods, implement sound control methods (for example, providing proper ventilation in the work area and providing personal protective equipment to employees exposed to hazards), and comply with all OSHA regulations. Depending on the industry and length of exposure to noise, there are different minimum requirements. The list of topics to be covered on this subject are exhaustive and beyond the score of this book. We recommend the reader to OSHA's website at www.osha.gov.

4.5.3.2   Lighting

Given that 80% of the U.S. economy is service related (rather than manufacturing or agriculture), many people work at desks or in work areas for much of the workday. If not properly controlled, lighting can become a hazard to employees who are working in relatively poor conditions for 6 hours or more. The following are two general principles in work desk arrangement:

- Minimize glare from overhead lights, desk lamps, windows, and other sources.
- Maintain appropriate air circulation. Avoid sitting directly under air-conditioning vents that *dump* air right on top of you.

Eye fatigue may be significantly reduced by eliminating glare or targeting bright lights away from the screen. It can also be reduced by placing rows of lights parallel to the line of sight, eliminating bright sources of light directly behind the monitor, and using light diffusers (e.g., blinds) and supplemental lighting for other computer unrelated tasks such as reading, writing, phone conversation (e.g., desk lamps specifically for reading or writing). The basic idea is to target light away from the computer monitor (see Figure 4.20). In general, the lighting for reading and writing tasks at a desk should be between 20 and 50 foot-candles. With liquid-crystal display (LCD) monitors, an additional 20–30 foot-candles is required. (Fluorescent lights with 150 Watt bulbs on an 8 foot high ceiling will produce 60 candles of light on a desk.) It is desirable to have natural light in the room, but not in the field of view of the employee. The OSHA website suggests horizontal blinds be used

FIGURE 4.20   Computer monitor must be placed perpendicular to window blinds to minimize eye fatigue.

for windows facing the north or east direction and vertical blinds for the other two directions. It also suggests that light, matte colors, and finishes be used on walls and ceilings so that contrast and dark shadows are reduced and indirect lighting is reflected in a way that it does not cause glare.

4.5.3.3 Ventilation

In general, there are three strategies for ensuring clean air quality in work areas—ventilation, dilution, and air cleaning. Ventilation can be thought of a passive way to maintain indoor air quality. If the level of air contaminants is not too high, proper ventilation ensures that fresh air minimizes the hazards posed by contaminated air. Dilution is a semi-active way of maintaining clean air in a work area. Fresh air is pumped in at a desired rate so that the steady-state percentage of uncontaminated air is at an acceptable level. The third strategy is considered an active one because it ensures the air contaminants are trapped at the source before they can pollute the indoor air and are treated and directed outside.

Poorly designed or improperly functioning ventilation systems cause discomfort to employees especially when exposed for a long period. Placing a vent directly above a work area is never a good idea. Hot air can cause irritation, air that is not properly circulated can make the room stuffy, and room temperatures that are too cold or too hot affect the productivity of employees. OSHA recommends that the air flow rates should range from 3 to 6 inches per second, or 7.5 centimeters and 15 centimeters per second, respectively, relative humidity must be between 30% and 60%, and the indoor temperatures must be between 68°F and 74°F (20°C and 23.5°C) during the winter season and between 73°F and 78°F (23°C and 26°C) during the summer season.

It is better to isolate printers, copiers, and fax machines by placing them in a separate room with proper ventilation and abundant fresh air because these release undesirable chemicals and pollutants into the air.

In addition, the reader must consider other aspects in facility design including environmental, recycling, using fire-rated and fireproof materials (where possible), general safety principles in designing facilities, designing utility networks, proper materials and design of the ceiling, flooring, and building siding for various industries and application, and so on. We recommend the reader to the following websites and reference—www.dol.gov, www.osha.gov, www.nfda.org, and Hanna and Konz (2004).

4.6 SUMMARY

The layout problem has elements of both design and optimization problems. In this chapter, we treated it as primarily a design problem and discussed two conventional approaches to facility design. Traditional approaches translate data into qualitative measures that reflect the relationships between layout entities and departments. They then use these qualitative measures to construct a usable layout. While doing so, they attempt to satisfy as many of relationships between the layout entities as possible. To construct a layout, traditional approaches rely on visual search techniques, which are useful only for a small number of entities and departments. For larger problems, we need a more systematic and computerized method of searching for layouts. The engineering design problem approach suggests the use of mathematical models and algorithms for this purpose. (The various mathematical models and corresponding heuristic and optimal algorithms that can be used for the layout problem are discussed in Chapters 5 through 10.) Suitable models and algorithms should be used because the layout problem is an optimization problem as well as a design problem. When the models are developed and solved, they provide one or more preliminary solutions to the layout problem. We must then attempt to develop a full-fledged layout design using these.

In this chapter, we also discussed regulations governed by OSHA and ADA and how these affect facility design. Support services such as parking areas, cafeterias, locker rooms, exercise rooms, and others are important considerations in facility design.

4.7   REVIEW QUESTIONS AND EXERCISES

1. What do the terms "optimization" problem and "design" problem mean? Is facility design an optimization problem or a design problem? Discuss.
2. What do you understand by the terms "facility design" and "facility layout?" Compare and contrast the two, discussing the scope of each.
3. Visit the Department of Labor, Americans with Disability Act, Occupational Health and Safety Act, and the National Fire Prevention Association websites located at www.dol.gov, www.ada.gov, www.osha.gov, and www.nfpa.org. Learn the regulations and recommendations that pertain to facility design for the following facilities:
   (a) Warehouse
   (b) Automobile assembly facility
   (c) University
   (d) Bank
   (e) Retail store
4. Consider the space requirement sheet in Figure 2.10. Add 5 units to the length and width of each work center in the figure. Assuming that the space required for auxiliary equipment, operator, and material is 35%, 10%, and 20%, respectively, determine the total space required for each work center. Using the allowance and number of machines information provided in Figure 2.10, develop a new production space requirement sheet. Use spreadsheet software for this purpose.
5. Consider the relationship charts you had developed in Chapter 3, Exercise 7 for (a) bookstore, (b) cafeteria, (c) registrar's office, (d) ground floor of library, (e) main computer center, (f) visitor information center, (g) physical recreation center or gymnasium, and (h) infirmary. Develop a corresponding space relationship diagram for each chart.
6. Consider Exercise 8 in Chapter 3. Develop a layout using the SLP technique.
7. Consider Exercises 9 and 10 in Chapter 3.
   (a) Convert the from–to frequency of trips matrix to a relationship chart so that the percentage of A, E, I, O, and U occurring in the chart is approximately 5%, 10%, 15%, 25%, and 45%, respectively. Assume there are no two departments with an X relationship. Of course, the A, E, I, O, and U ratings should correspond to the from–to frequency of trips.
   (b) Using the relationship chart from part (a) and the areas obtained from Exercise 10, Chapter 3, develop a layout using the SLP technique.
   (c) Explain how the layout you obtained in (b) compares with the one obtained in Chapter 3?
8. The relationship codes and areas (in square feet) for six departments in a manufacturing facility are provided in Figure 4.21. Develop a layout using the SLP technique.

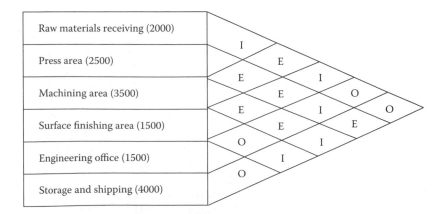

FIGURE 4.21   Relationship chart for departments in a small manufacturing facility.

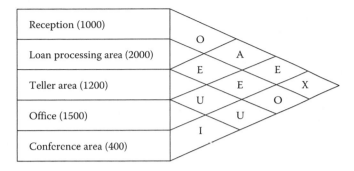

FIGURE 4.22   Relationship chart for facilities in a regional bank.

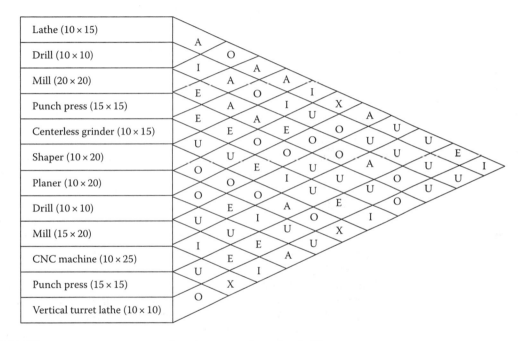

FIGURE 4.23   Relationship chart for a large manufacturing facility.

9. The relationship codes and areas (in square feet) for five departments in a regional bank are provided in Figure 4.22. Using the information provided and the SLP technique, develop a layout.

10. The relationship codes and dimensions of 12 machines in a manufacturing facility are provided in Figure 4.23. Using the information provided and the SLP technique, develop a layout. The dimensions of each machine include allowance for operator access, material and tool storage, and material loading/unloading. You cannot alter the machine shapes while developing a layout.

11. The relationship codes and areas of 12 sections in a regional lending institution are provided in Figure 4.24. Using the information provided and the SLP technique, develop a layout.

12. Compare and contrast the SLP and engineering design approaches to the facility design problem.

13. *Project*: Visit a local grocery store. Make a list of all the *departments* in the store. Review current space utilization, determine space projections and level of interaction between departments, and identify special considerations, if any. Following the approach outlined in Section 4.4, design a new layout for the grocery store.

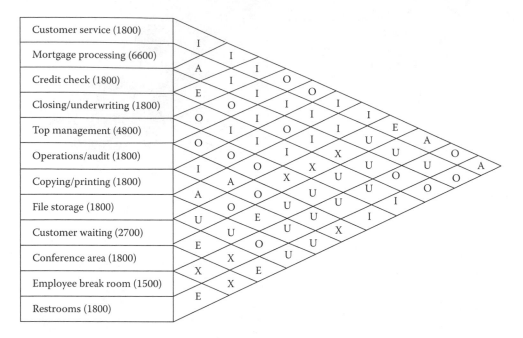

FIGURE 4.24   Relationship chart for a regional lending institution.

14. *Project*: Visit a large departmental store. Make a list of all the departments in the store. Review current space utilization, determine space projections and level of interaction between departments, and identify special considerations, if any. Following the approach outlined in Section 4.4, design a new layout for the departmental store.

15. *Project*: Visit a large home improvement store. Make a list of all the departments in the store. Review current space utilization, determine space projections and level of interaction between departments, and identify special considerations, if any. Following the approach outlined in Section 4.4, design a new layout for the departmental store.

16. *Project*: Visit a large public facility such as courthouse, county government office, library, and department of motor vehicle office. Make a list of all the departments in the office. Review current space utilization, determine space projections and level of interaction between departments, and identify special considerations, if any. Following the approach outlined in Section 4.4, design a new layout for the departmental store.

5 Basic Algorithms and Software for the Layout Problem

5.1 ALGORITHMS FOR THE LAYOUT PROBLEM

An algorithm is a step-by-step procedure that finds a solution to a model, and hence to the problem, in a finite number of steps. In Chapters 5 through 10, we introduce algorithms and software that have been developed for the layout problem. Some have been in use for the last 50 years, whereas others are relatively new and have been used successfully to solve many problems, including the layout problem. To make the models amenable to solution by means of algorithms, we make a number of assumptions—some realistic and others unrealistic. For instance, the quadratic assignment problem (QAP) discussed in Chapter 10 assumes the departments are square. Similarly, all but one of the models presented in Chapter 10 assume that material flow occurs between the centers of departments and that the cost of moving a unit load is implicitly assumed to be proportional to the distance. These models are useful even though many of these assumptions may not be realistic. The layout analyst has to be aware of the assumptions made in the models and use the solution generated by a corresponding algorithm with caution. It must be used only as a basis for generating solutions that can be applied in the real world (Apple, 1977). Thus, despite their assumptions, models provide the layout analyst with a mechanism to generate meaningful solutions that can be implemented in practice.

The layout problem has been studied formally since the late 1940s. Apple (1977) provides a chronological list of the research efforts in the 1940s and 1950s. Early efforts to solve the layout problem involved the use of flow charts, process charts, and the experience and knowledge of the facility analyst. Other methods used the relationship chart to determine the layout. As discussed in Chapter 4, the relationship chart formed the basis for the development of the systematic layout planning (SLP) technique (Muther, 1973). SLP was one of the first systematic attempts to solve the layout problem. Wimmert (1958) presented a mathematical method for the facility layout problem that used the criterion of minimizing the product of flow values and distances between all combinations of departments. Flaws in the theorem on which Wimmert's method was based were pointed out and proved using a counter example by Conway and Maxwell (1961b). Buffa (1955) proposed a method called sequence analysis based on the study of the operation sequence of parts processed in a plant. Additional efforts were made in the late 1950s and early 1960s. Many of these did not provide good-quality solutions, so they are not discussed here. However, a reader interested in the history of development of techniques for the layout problem may consult Apple (1977).

Since the formulation of the layout and location problems as a QAP by Koopmans and Beckman (1957), substantial progress has been made in solving the layout problem. A number of algorithms have been proposed and these may be classified as follows:

- Optimal algorithms
- Heuristic algorithms

Optimal algorithms and several heuristic algorithms are discussed in Chapter 11. Several popular layout software packages used in industry are discussed in this chapter. In the next three sections, we will discuss a few heuristics.

All the optimal algorithms developed for the layout problem require extremely high memory and computational time, and they increase exponentially as the problem size increases. It is therefore, not surprising that there are many more heuristic algorithms for solving the facility layout problem than optimal algorithms. Heuristic algorithms can be divided into three classes:

1. Construction algorithms
2. Improvement algorithms
3. Hybrid algorithms

5.2 CONSTRUCTION ALGORITHMS

Construction algorithms generate a facility layout from scratch. Starting with an empty layout, they add one department (or a set of departments) after another until all the departments are included in the layout. The main differences among the various construction algorithms relate to the criteria used to determine the following:

1. First department to enter the layout
2. Subsequent department or departments added to the layout
3. Location of the first (and subsequent) departments in the layout

Many of the construction algorithms developed in the last 30 years are listed and reviewed in Kusiak and Heragu (1987b). Tompkins and Moore (1984) provide detailed descriptions of five construction and improvement algorithms, specifying how data are to be entered, various procedures used in the algorithms, and sample data input and output. The construction algorithm discussed in the next section is easy to understand, use, and explain. It provides reasonably good solutions for small-sized problems.

5.2.1 MODIFIED SPANNING TREE ALGORITHM FOR THE SINGLE-ROW LAYOUT PROBLEM

A single-row layout problem involves determining the optimal placement of departments along a linear row so that the total cost to transfer material between each pair of departments is minimized. Applications and further discussions are presented in Chapter 10.

We now discuss a simple algorithm called the modified spanning tree (MST) algorithm, which is very similar to the spanning tree algorithm (Heragu and Kusiak, 1988). First, we present some definitions that will clarify the terminology used. A graph $G = (V, E)$ consists of a set of vertices V and edges E with each edge connecting two vertices. Vertices are sometimes referred to as nodes and edges as arcs. In a complete graph, there is an edge between each pair of distinct vertices. Given a complete graph with undirected edges and a weight for each, the spanning tree algorithm attempts to find a set of edges such that

1. Each vertex is connected to every other through these edges.
2. The sum of the weights of these edges is minimized or maximized depending on the criterion selected.
3. The edges do not form a cycle—that is, there is a unique path between each pair of vertices.

In other words, the spanning tree algorithm attempts to find a tree (set of edges connecting each vertex to every other without forming cycles) that spans all vertices and is such that the sum of the weights of the edges is minimized or maximized.

If machines are thought of as vertices and each pair of adjacent machines is connected by an edge with the weight of the edge is equal to the product of flow and distance between the corresponding adjacent machines, then the MST algorithm attempts to develop a layout by placing in adjacent locations those machine pairs that have a large contribution to the objective function when placed next to each other. This is because doing otherwise will increase this machine pair's contribution to the objective function value (OFV) even more. Like the spanning tree algorithm, the MST algorithm also generates a tree. However, it is done such that each vertex, except the first and last in the sequence, has exactly two edges emanating from it. The first and last vertices have only one. The MST algorithm cannot be optimal because it does not consider nonadjacent weights. Also, because it determines only the sequence of machines, additional factors must be considered in generating a layout, such as the clearance between machines and whether the layout is linear or semicircular. For example, when an automated guided vehicle (AGV) is used, the layout is typically linear to maximize the operating efficiency of the AGV. On the other hand, when a robot is used as the material handling device, the layout is semicircular because in one circular motion, the robot can access all the machines. The steps of the MST algorithm are as follows:

Step 1: Given the flow matrix $[f_{ij}]$, clearance matrix $[d_{ij}]$, and machine lengths l_i, compute an adjacency weight matrix $[f'_{ij}]$ where $f'_{ij} = (f_{ij})(d_{ij} + 0.5(l_i + l_j))$.

Step 2: Find the largest element in $[f'_{ij}]$ and the corresponding i, j. Denote this pair of i, j as i^*, j^*. Connect machines i^*, j^*. Set $f'_{i^*j^*} = f'_{j^*i^*} = -\infty$.

Step 3: Find the largest element $f'_{i^*j^*}, f'_{j^*i^*}$ in row i^*, j^* of matrix $[f'_{ij}]$. If $f'_{i^*k} \geq f'_{j^*l}$, connect k to i^*; remove row i^*, column i^* from matrix $[f'_{ij}]$; and set $i^* = k$. Otherwise, connect l to j^*; remove row j^*, column j^* from matrix $[f'_{ij}]$; and set $j^* = l$. Set $f'_{i^*j^*} = f'_{j^*i^*} = -\infty$.

Step 4: Repeat step 3 until all machines are connected. The sequence of machines obtained determines the arrangement of machines.

Note that we are implicitly assuming that the machine orientations are known *a priori*. By assuming the distance between adjacent machines to be $(d_{ij} + 0.5(l_i + l_j))$, we assume that the machines are positioned so that their lengthwise sides are parallel. In step 2, the MST algorithm finds the first pair of machines (i^*, j^*) that must be adjacent in the layout using the adjacency weights. Then, it determines the machine k or l that must be placed adjacent to either i^* or j^*, respectively. If the adjacency weights indicate that machine k be placed adjacent to machine i^*, the latter is removed from further consideration because it already has two adjacent machines (k, j^*). Note that any machine can have at most two adjacent machines in a single-row layout. At this stage, we have a partial sequence in which machine k is on one end, j^* on the other end, and i^* in the middle. The algorithm removes machine i^* from consideration, sets $i^* = k$, and attempts to find the next machine k or l that must be placed adjacent to either j^* or the new i^*. The algorithm determines adjacent pairs such that one of them is already in the partial sequence and the other is not. If the current i^*, j^* pair has the largest adjacent weight, this pair cannot be selected because it is already in the sequence. Hence, $[f'_{i^*j^*}]$ and $[f'_{j^*i^*}]$ are set to $-\infty$.

If the adjacency weights indicate that machine l be placed adjacent to machine j^*, then we do this: connect machine l to j^*, remove machine j^* from consideration, set $j^* = l$, and attempt to find the next machine k or l that must be placed adjacent to either i^* or the new j^*. This is continued until we reach a sequence that includes all the machines. As mentioned previously, this sequence is used in generating a layout. One may attempt to improve this solution if possible using an improvement algorithm such as 2-opt discussed in Section 5.3.1. The computer program SINROW.EXE is available at the following website: http://sundere.okstate.edu/downloadable-software-programs-and-data-files allows the user to improve the solution produced by the MST algorithm using 2-opt.

Example 5.1

Using the following flow data for six machines and their lengths in Figure 5.1, develop a layout using the MST algorithm. Assume that the clearance between each pair of machines must be at least 2 meters.

Solution

Step 1: First, compute the adjacency weight matrix $[f'_{ij}]$ as in Figure 5.2, where $f'_{ij} = (f_{ij})(d_{ij} + 0.5(l_i + l_j))$.

Step 2: Find the largest element $[f'_{16}]$ and connect machines 1 and 6. Set $f'_{16} = f'_{61} = -\infty$.

Step 3: Select the largest elements f'_{12} and f'_{64} in rows 1 and 6, respectively. Because $f'_{12} = f'_{64}$, select f'_{12} arbitrarily. Connect 2 to 1 and remove row 1 and column 1 from matrix $[f'_{ij}]$. Set $f'_{26} = f'_{62} = -\infty$.

Step 4: Because all machines are not connected yet, repeat step 3.

Repeating the aforementioned steps, we see that machine 4 is connected to machine 6, machine 3 is connected to machine 4, and machine 5 is connected to machine 2 in subsequent steps. A layout generated for this sequence is shown in Figure 5.3. Notice that the clearance between machines

	1	2	3	4	5	6	Machine lengths (meter)
1	—	12	3	6	0	20	20
2	12	—	5	5	5	0	10
3	3	5	—	10	4	2	16
4	6	5	10	—	2	12	20
5	0	5	4	2	—	6	10
6	20	0	2	12	6	—	10

FIGURE 5.1   Flow and length data for six machines.

	1	2	3	4	5	6
1	—	204	60	132	0	340
2	204	—	75	85	60	0
3	60	75	—	200	60	30
4	132	85	200	—	34	204
5	0	60	60	34	—	72
6	340	0	30	204	72	—

FIGURE 5.2   Adjacency weight matrix.

5	2	1	6	4	3

FIGURE 5.3 Layout generated using the sequence produced by MST.

is also shown. As mentioned before, this layout may be improved using 2-opt, which is left as an exercise to the reader (see Exercises 21 and 22).

5.2.2 GRAPH THEORETIC APPROACH

Chapter 4 showed how the layout problem can be represented graphically using Muther's SLP technique. In the SLP approach, the department areas as well as the adjacency relationships are indicated pictorially on graphs. Then the relative positioning of departments is determined simply by "eye balling" and a trial-and-error approach. Based on this, a human designer generates a final detailed layout showing aisles and department pickup or drop-off points. It is clear that if the intermediate eye balling and trial-and-error step yields poor-quality suboptimal solutions, then the final layout will not help us achieve the objective of minimizing material handling costs. Hence, another approach based on graph theoretical concepts has been proposed (Foulds, 1983).

To understand the graph theoretic approach, we must first understand the meaning of the terms *planar graph* and *maximal planar graph*. A graph consisting of a nonempty, finite set of nodes and an unordered set of arcs is said to be planar if it can be drawn in two dimensions without any arc crossing any other. The graph in Figure 5.4 is called a complete graph on five nodes because it has five nodes and each node is connected to every other node. It can be shown that this graph is nonplanar, i.e., it is impossible to connect each pair of the five nodes without some of the arcs crossing some others. However, the graph shown in Figure 5.5 is planar. A planar graph is called maximal planar if it is already planar, and the addition of just one more edge makes it nonplanar.

The graph theoretic approach determines the pairs of departments that must be adjacent so that the sum of the profits is maximized. It identifies a maximal planar graph in which the sum of the weights is maximized. The weight of an arc between two nodes corresponds to the level of interaction between the corresponding departments. The weight actually represents the numeric value assigned to the relationship code between the node (department) pairs connected by an arc. If the interactions are provided as a material flow matrix, we can use the material flow values

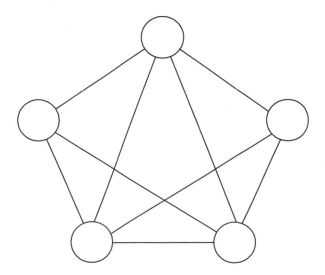

FIGURE 5.4 Complete graph on five nodes.

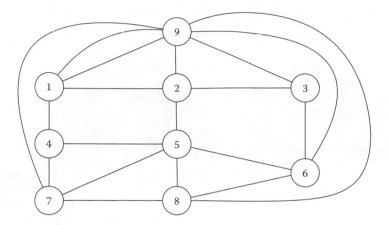

FIGURE 5.5   A planar adjacency graph.

as weights. The arcs in the maximal planar graph specify which pairs of departments are to be adjacent. From these adjacencies, a designer can then come up with a feasible layout in which space and shape requirements are satisfied and department pairs that have large weights are adjacent. Most graph theoretic methods focus on maximizing the sum of the weights of only adjacent departments; the weights of nonadjacent departments are not considered at all. Also, they do not take into consideration areas or distance between departments in determining adjacencies. There are some exceptions and one example is the method by Cimikowski and Mooney (1995), which considers relationships between nonadjacent departments as well. Another example is the approach by Montreuil et al. (1987), which explicitly considers department areas and adjacencies. The graph theoretic method is a tool that provides us with relatively good-quality layouts, which can perhaps be improved by using 2-opt or 3-opt presented in this chapter and simulated annealing presented in Chapter 11.

For every feasible layout or block plan, we can draw a corresponding adjacency graph. To do so, we place a node inside each department and join two nodes by an arc only if the departments corresponding to the two nodes are adjacent in the layout. Two departments are considered to be adjacent only if they have a common boundary of positive length. For example, departments 5 and 8 in Figure 5.6 are adjacent, but not departments 1 and 5. Figure 5.5 is the adjacency graph for the

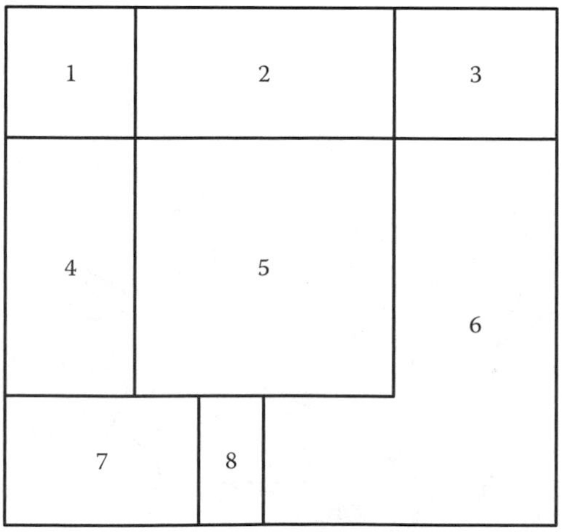

FIGURE 5.6   Layout obtained from the planar adjacency graph in Figure 5.5.

layout in Figure 5.6. It is obtained by placing a node inside each department in Figure 5.6 and drawing arcs between the nodes so that they intersect the common boundaries shared by the adjacent departments. The region exterior to the eight departments (*outside world*) is also considered to be another department. The node labeled 9 corresponds to this external department in Figure 5.5. It is connected to the nodes corresponding to the departments that are on the exterior part of the building, i.e., all the departments except department 5.

If an adjacency graph is planar, it is called the *planar adjacency graph* (PAG). If the PAG is maximally planar, it is called a maximal PAG. From this definition and our description of the graph theoretic method, it is perhaps apparent to the reader that the method is concerned with the identification of a maximal PAG in which the sum of the weights is maximized. If the weights assigned to the relationship codes are nonnegative, at least one optimal solution will be a maximal PAG because any optimal solution that is not a maximal PAG can be made one by adding arcs to it (Foulds, 1983). Moreover, doing so will not decrease the total weight.

Just as every feasible layout has a corresponding PAG, every PAG (including the maximal PAG) has a corresponding feasible layout. To identify the layout from a corresponding PAG, we place a node inside each face and draw arcs between the nodes so that they intersect the common boundaries shared by adjacent faces. A region bounded by arcs is called a face. For example, the region bounded by arcs 1–4, 1–9, and 4–9 in Figure 5.7 is a face. Figure 5.7 shows how the layout in Figure 5.6 is used to draw the PAG in Figure 5.7. Notice that the bounded regions enclosing the labeled nodes in Figure 5.7 indicate a layout that does not consider shape and space requirements. For clarity, these region boundaries are shown using thicker lines. Some of the departments have irregular shapes. A creative designer can incorporate shape and space requirements and come up with a variety of feasible layouts (including the one in Figure 5.6) that satisfy the adjacencies in Figure 5.7. Figure 5.7 is the dual of the layout in Figure 5.6 and the PAG in Figure 5.5 (or Figure 5.7) is the dual of the layout in Figure 5.6.

When the number of nodes (n) is greater than or equal to 3, the maximum number of arcs in a planar graph is $3n - 6$. Thus, any planar graph with three or more nodes can satisfy at most $3n - 6$ arcs; so only $3n - 6$ adjacencies among the $n(n - 1)/2$ possible ones can be satisfied. Note that the number of possible adjacencies in a graph with n nodes is the summation series $n - 1 + n - 2 + \cdots + 3 + 2 + 1$, which is equal to $n(n - 1)/2$. If there are six departments in a layout problem, the total number of nodes is seven because we use one node to represent the external region. Among the $7(7 - 1)/2 = 21$ possible adjacencies, $3(7) - 6 = 15$ will be satisfied. The gap between the maximum

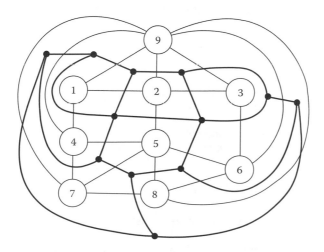

FIGURE 5.7 A sample layout.

possible adjacencies and the maximum number that can be satisfied in a layout widens nonlinearly as the number of departments in the problem increases.

Finding a maximal PAG in which the sum of the weights is maximum is a hard problem because we have to examine all possible $3n - 6$ arc combinations from a total of $n(n - 1)/2$ adjacencies. However, an interesting property of maximal planar graphs may be exploited to devise heuristic methods. This property tells us that the number of faces in a maximal PAG is equal to $2n - 4$ and that each face is triangular. Thus, beginning with a triangular face in which the sum of the weights of the three arcs is maximum, we add one node at a time, in such a way that

1. The sum of the weights of arcs is large.
2. The planarity of the resulting graph is maintained.

When all the nodes are included in the graph, we get a PAG. Using the dual of the resulting PAG, a human designer can construct a detailed layout.

Steps of the graph theoretic algorithm can be summarized as follows. Identify a maximal PAG in which the sum of the weights is maximized, determine its dual, and use the dual to generate a feasible layout. A heuristic algorithm for identifying maximal PAG is as follows:

Step 1: Identify the department pair in the flow matrix with the maximum flow. Place the corresponding nodes in a new PAG and connect them.

Step 2: From the rows corresponding to the connected nodes in the flow matrix, select the node that is not yet in the PAG and has the largest flows with the connected nodes.

Step 3: Update PAG by connecting the selected node to those in step 2. This forms a triangular face in the PAG.

Step 4: For each column of the flow matrix corresponding to a node not present in the PAG, examine the sum of flow entries in the rows corresponding to the nodes of the triangular face selected in step 3. Select the column for which this sum is the largest. Update PAG by placing the corresponding node within the selected face and connect it to nodes of the face. This forms three new triangular faces.

Step 5: Arbitrarily select one of the faces formed and go to step 4. Repeat step 5 until all the nodes have been included in the PAG.

Because we add $(n - 3)$ nodes and $3(n - 3)$ arcs in step 4, the total number of arcs added in all the steps is $3(n - 3) + 3 = 3n - 6$. Furthermore, because we add nodes in such a way that graph planarity is maintained, we always have a maximal PAG. However, the PAG generated may not have the greatest sum of weights among all the available maximal PAGs. Thus, the aforementioned algorithm does not guarantee optimal solutions. Several other heuristics for the maximal PAG are reviewed in Foulds (1983).

Example 5.2

Consider the flow matrix presented in Figure 5.8. Develop a layout for the 12 machines in the example using the graph theoretic heuristic.

Solution

Step 1: Because machine pair {4, 7} has the largest flow, connect the corresponding nodes in the PAG (see Figure 5.9).

Step 2: From rows 4 and 7 of the flow matrix, select node 8 that has the largest flows with nodes 4 and 7.

Step 3: Update PAG by connecting node 8 to nodes 4 and 7 (see Figure 5.10). Select the triangular face formed by arcs 4–7, 4–8, and 7–8.

Flow matrix (number of trips in thousands)

Machine

	1	2	3	4	5	6	7	8	9	10	11	12
1	—	1	0	8	0	2	3	0	0	0	0	0
2	1	—	0	1	1	1	0	0	0	0	0	0
3	0	0	—	0	2	0	0	0	0	0	0	0
4	8	1	0	—	0	4	14	11	0	0	0	0
5	0	1	2	0	—	1	0	0	0	0	0	0
6	2	1	0	4	1	—	3	0	0	3	0	0
7	3	0	0	14	0	3	—	5	5	9	8	2
8	0	0	0	11	0	0	5	—	8	0	0	0
9	0	0	0	0	0	0	5	8	—	0	0	0
10	0	0	0	0	0	3	9	0	0	—	6	0
11	0	0	0	0	0	0	8	0	0	6	—	4
12	0	0	0	0	0	0	2	0	0	0	4	—

(Machine — row labels)

FIGURE 5.8   Flow matrix (in thousands of trips) for Example 5.2.

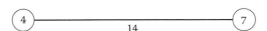

FIGURE 5.9   Arc connecting nodes 4 and 7.

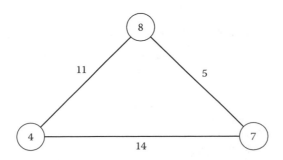

FIGURE 5.10   Triangular face formed by connecting node 8 to nodes 4 and 7.

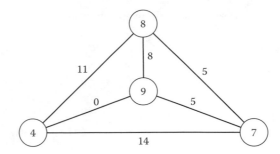

FIGURE 5.11   Three faces formed by placing node 9 in the triangular face of Figure 5.10.

Step 4: For each of the columns 1, 2, 3, 5, 6, 9, 10, 11, and 12, examine the sum of flow entries in rows 4, 7, and 8 of the flow matrix, one column at a time. Select column 9 because it has the largest sum (0 + 5 + 8 = 13). Update PAG by placing node 9 within the triangular face formed by arcs 4–7, 4–8, and 7–8. Connect node 9 to nodes 4, 7, and 8 (see Figure 5.11).

Step 5: Select face formed by arcs 4–8, 4–9, and 8–9. Go to step 4.

Step 4: For each of the columns 1, 2, 3, 5, 6, 10, 11, and 12, examine the sum of flow entries in rows 4, 8, and 9 of the flow matrix, one column at a time. Select column 1 because it has the largest sum (8 + 0 + 0 = 8). Update PAG by placing node 1 within the face formed by arcs 4–8, 4–9, and 8–9.

When step 5 is repeated until all the nodes are included, a maximal PAG is obtained. The intermediate steps are shown in Table 5.1. The final step is to add a node corresponding to the external boundary and connect it to nodes 4, 8, and 7 as shown in Figure 5.12. We now have a maximal PAG.

To obtain a layout corresponding to the maximal PAG, we draw its dual by placing a dot in the middle of each face in Figure 5.12 and connecting the dots to enclose all the nodes (machines) except node 13, which corresponds to the external boundary. A very rough preliminary layout emerges from this dual (see the *layout* formed by the thicker lines in Figure 5.13). Now the facility designer must use information concerning machine areas and construct a detailed layout—a time-consuming and tedious task. We have generated a layout by trial and error and it is shown in Figure 5.14. Because the machine shapes are fixed, we were not able to satisfy all the adjacencies specified in Figure 5.8 but we have developed a feasible layout.

TABLE 5.1

Summary of Intermediate Steps of the PAG Heuristic in Example 5.2

Iteration No.	Arcs of the Selected Face	Nodes Available	Node Selected	Sum of Flows
3	7–8, 7–9, 8–9	2, 3, 5, 6, 10, 11, 12	10	9
4	4–7, 4–9, 7–9	2, 3, 5, 6, 11, 12	11	8
5	1–4, 1–9, 4–9	2, 3, 5, 6, 12	6	6
6	1–8, 1–9, 8–9	2, 3, 5, 12	2	1
7	7–9, 7–10, 9–10	3, 5, 12	12	2
8	8–9, 8–10, 9–10	3, 5	3	0
9	7–8, 7–10, 8–10	5	5	0

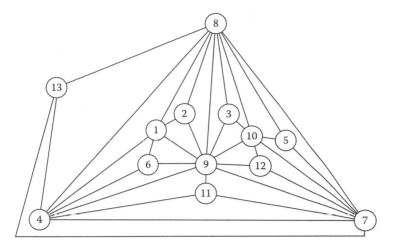

FIGURE 5.12 Maximal planar adjacency graph for Example 5.2.

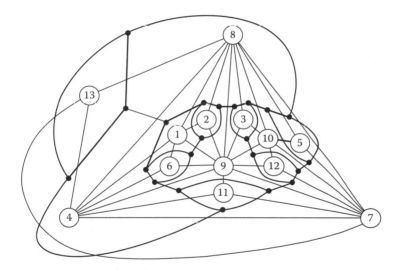

FIGURE 5.13 Dual of maximal planar adjacency graph in Figure 5.12.

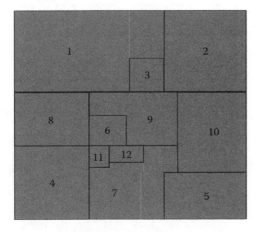

FIGURE 5.14 Final layout obtained from dual in Figure 5.13.

5.3   IMPROVEMENT ALGORITHMS

Improvement algorithms are very simple and are easy to understand and implement. Because they produce reasonably good solutions, these algorithms have also been applied to solve the scheduling problem, graph partitioning problem, and other complex combinatorial problems that are very difficult to solve optimally. In fact, the 2-opt and 3-opt algorithms discussed in this section were first used to solve the traveling salesman problem (Syslo et al., 1983). As their name indicates, improvement algorithms improve a user-provided starting solution, which must be feasible. Improvement algorithms systematically modify the starting solution and evaluate the resulting modified solution. If it is better, the modification is made permanent. If not, the systematic modification is continued until it is no longer possible to produce better solutions.

5.3.1   2-OPT ALGORITHM

Consider the multirow layout problem in which the departments are all squares with equal area. Assume that the number of departments in every row and column is equal to m and n, respectively, as shown in Figure 5.15, so mn departments can be accommodated in this layout. For this problem, the number of locations in which the departments will be located is also equal to mn and is known (see Figure 5.16). We discuss how the 2-opt algorithm is used to solve the equal-area, multirow layout problem heuristically.

The 2-opt algorithm consists of three steps:

Step 1: Let S be the initial solution provided by the user and z its OFV. Set $S^* = S$, $z^* = z$, $i = 1$; $j = i + 1 = 2$.

Step 2: Consider the exchange between the positions of departments i and j in the solution S. If the exchange results in a solution S' that has an OFV $z' < z^*$, set $z^* = z'$ and $S^* = S'$. If $j < mn$, set $j = j + 1$; otherwise, set $i = i + 1$, $j = i + 1$. If $i < mn$, repeat step 2; otherwise, go to step 3.

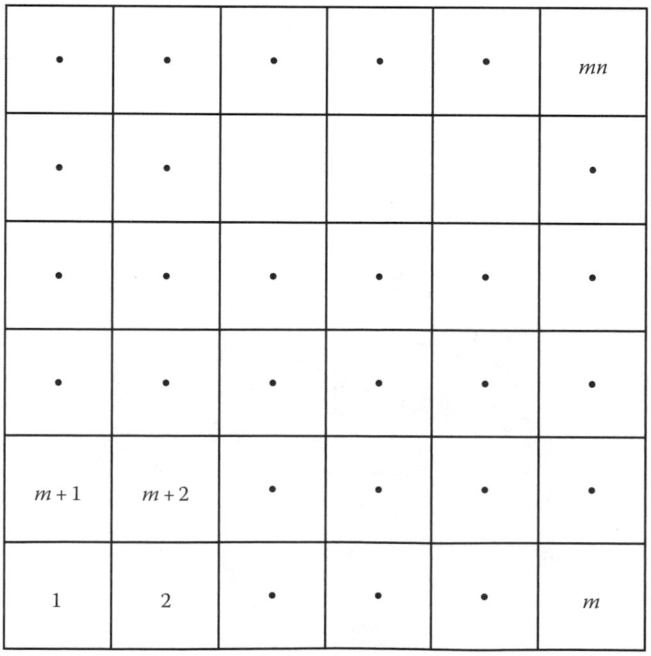

FIGURE 5.15   A layout of mn departments.

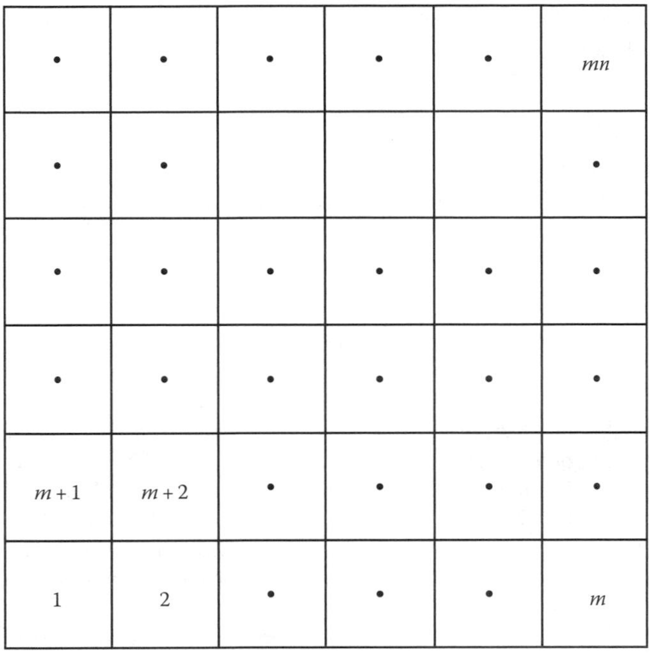

FIGURE 5.16 Locations in which the mn departments in Figure 5.15 are to be placed.

Step 3: If $S \neq S^*$, set $S = S^*$, $z = z^*$, $i = 1$, $j = i + 1 = 2$ and go to step 2. Otherwise, return S^* as the best solution to the user. Stop.

Observe that the 2-opt algorithm considers only pairwise exchange; that is, only two departments at a time are considered for exchange. Initially, the algorithm considers exchanging the positions of departments 1 and 2. If the resulting solution's OFV is greater than that of the initial solution, it is stored as a candidate for future consideration. If not, it is discarded and the algorithm considers an exchange in the positions of departments 1 and 3. If this yields a solution with an OFV greater than what is currently the best, it is stored as a candidate for future consideration; if not, it is discarded. Thus, whenever a better solution is found, the algorithm discards the previous best solution. This procedure continues until all the pairwise exchanges are considered. Because each department can be exchanged with $mn - 1$ other departments, and there are mn departments in total, $mn(mn - 1)$ exchanges are possible. The exchange of department in position i with that in position j is the same as the exchange of department in position j with that in position i, so we need to consider only half of the $mn(mn - 1)/2$ exchanges.

These $mn(mn - 1)/2$ exchanges are considered in step 2. The solution retained at the end of step 2 is the one that provides the most improvement in the OFV. Starting with this as the new solution, the algorithm repeats step 2 to find another better solution. At some stage, no improvement in the current best solution is possible, and then the algorithm terminates. The current best solution is returned to the user.

The following variation of the 2-opt algorithm is commonly used. In step 2, as soon as an exchange in the positions of departments i and j results in an improvement, we can make this permanent. With this solution as the initial one, we repeat the procedure until it is not possible to improve the solution any further. This variation of 2-opt is called greedy 2-opt. As the name indicates, this algorithm, unlike 2-opt, does not wait until all the $mn(mn - 1)/2$ exchanges are evaluated before initializing the procedure—hence the term *greedy*. Because 2-opt typically evaluates more exchanges, it provides slightly better solutions than greedy 2-opt. For a specific layout problem, however, the user must be warned that it is better to try both algorithms and choose the better solution.

Example 5.3

Develop a layout for the matrix in Figure 5.17 using the 2-opt algorithm.

Solution

Assume that an initial solution in which departments 1, 2, 3, and 4 are assigned to locations 1, 2, 3, and 4 is available (see Figure 5.18). In the figure, the locations are identified in parentheses. The OFV of the initial solution is $(17)(1) + (12)(1) + (11)(2) + (12)(2) + (4)(1) + (4)(1) = 83$. Note that m and n are 2 and 2, respectively.

Step 1: Let S be the initial solution in which departments are assigned as shown in Figure 5.18 and $z = 83$; Set $S^* = S$; $z^* = 83$; $i = 1$; $j = i + 1 = 2$.

Step 2: Consider exchanging the positions of departments 1 and 2. The resulting solution is shown in Figure 5.19. The OFV of this solution is $(17)(1) + (12)(2) + (11)(1) + (12)(1) + (4)(2) + (4)(1) = 76$. Because exchanging the positions of departments 1 and 2 results in a solution with OFV = 76 < 83, set $z^* = 76$ and store this solution as S^*. Because $j = 2 < mn = 4$, set j to $j + 1 = 2 + 1 = 3$.

We have $i = 1 < 4$, so repeat the same procedure to get a solution with an OFV of 81. Because this is not better than the current best solution, it is discarded. Because

$$
[f_{ij}] = \underset{\text{Office}}{}
\begin{array}{c|cccc}
 & 1 & 2 & 3 & 4 \\
\hline
1 & - & 17 & 12 & 11 \\
2 & 17 & - & 12 & 4 \\
3 & 12 & 12 & - & 4 \\
4 & 11 & 4 & 4 & - \\
\end{array}
\qquad
[d_{ij}] = \underset{\text{Site}}{}
\begin{array}{c|cccc}
 & 1 & 2 & 3 & 4 \\
\hline
1 & - & 1 & 1 & 2 \\
2 & 1 & - & 2 & 1 \\
3 & 1 & 2 & - & 1 \\
4 & 2 & 1 & 1 & - \\
\end{array}
$$

FIGURE 5.17   Flow and distance matrices for Example 5.2.

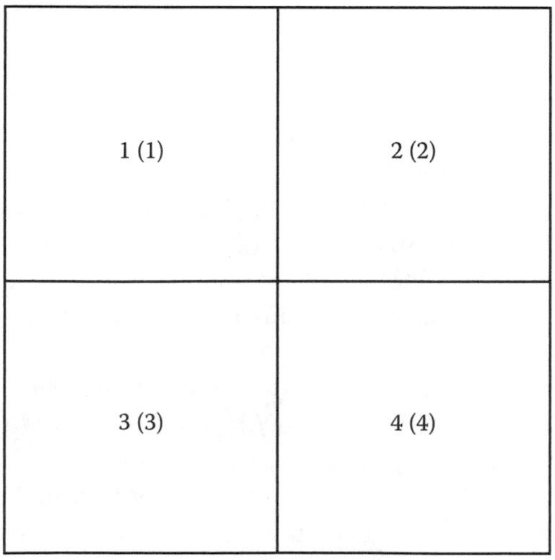

FIGURE 5.18   Initial solution for Example 5.2 using 2-opt.

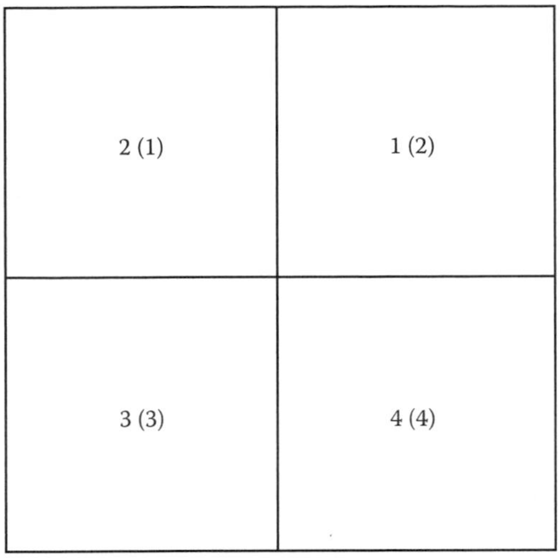

FIGURE 5.19   Solution after exchanging the positions of departments 1 and 2.

$j = 3 < mn = 4$, set j to $j + 1 = 3 + 1 = 4$. We get $i < 4$, so repeat step 2. The next exchange has the same OFV as the initial solution, so it is discarded.

Step 3: Because $S \neq S^*$, set $S = S^*$, $z = 76$, $i = 1$, $j = i + 1 = 2$ and go to step 2.

Step 2: Because none of the exchanges result in a solution with an OFV lower than 76, go to step 3.

Step 3: Because $S = S^*$, return S^* as the best solution to the user.* Stop.

5.3.2   3-OPT ALGORITHM

The 3-opt algorithm is similar to the 2-opt algorithm except that it considers exchanging the position of three departments at a time. There are two possible ways of exchanging the positions of three departments:

$$i \rightarrow k \rightarrow j \rightarrow i \quad \text{and} \quad j \rightarrow k \rightarrow i \rightarrow j$$

In the 3-opt algorithm described here, we consider the second of these exchanges—that is, changing the position of department i to that of department j, j to that of k, and k to that of i. The exchange of three departments at a time allows us to evaluate many more alternate layouts. For example, consider a layout problem with mn departments. For each layout, 2-opt considers $mn(mn - 1)/2$ pairwise exchanges for each layout. The 3-opt algorithm considers $(mn)!/[(mn - 3)!3!]$ exchanges. The difference in the number of layouts evaluated between 2-opt and 3-opt is significant, especially for larger problems (see Exercise 23). Because we are searching and evaluating more layouts in 3-opt than in 2-opt, 3-opt should yield better results, but it also takes significantly more computation time. Thus, in deciding whether to use 2-opt or 3-opt, the user has to trade off between solution quality and computation time. In improvement (search) algorithms, solution quality can be improved only at the expense of computation time. Of course, this trade-off decision is needed only for large problems—those with more than 30 departments. For smaller ones, we can solve the layout problem using both algorithms and keep the best solution. Although 3-opt in general will yield a better solution than

---

* See Example 10.3. Further improvement is not possible and 76 is the optimum OFV.

2-opt, the final solution produced depends on the starting solution used. This is why we suggest that the user try both algorithms for small- and medium-sized problems.

The following are the steps of the 3-opt algorithm:

Step 1: Let S be the initial solution and z its OFV; Set $S^* = S$, $z^* = z$, $i = 1$; $j = i + 1$; $k = j + 1$.

Step 2: Consider changing the position of department i to that of j, j to that of k, and k to that of i, simultaneously. If the resulting solution S' has an OFV $z' < z^*$, set $z^* = z'$ and $S^* = S'$.

Step 3: If $k < mn$, set $k = k + 1$ and repeat step 2. Otherwise, set $j = j + 1$ and check if $j < mn - 1$. If $j < mn - 1$, set $k = j + 1$ and repeat step 2. Otherwise, set $i = i + 1$, $j = i + 1$, $k = j + 1$ and check if $i < mn - 2$. If $i < mn - 2$, repeat step 2. Otherwise, go to step 3.

Step 4: If $S \neq S^*$, set $S = S^*$, $z = z^*$, $i = 1$, $j = i + 1$, $k = j + 1$ and go to step 2. Otherwise, return S^* as the best solution to the user. Stop.

5.4 HYBRID ALGORITHMS

Because all the improvement algorithms require a starting solution and, in general, the better the initial solution, the better the final layout produced by an improvement algorithm, the astute reader may have thought of using a solution produced by a construction algorithm as a starting solution for an improvement algorithm. In fact, algorithms using two or more types of solution techniques do exist and are known as hybrid algorithms or composite algorithms. Some other algorithms which have the characteristics of optimal and heuristic algorithms have also been placed in the category of hybrid algorithms (Kusiak and Heragu, 1987). In Chapter 10, we present a hybrid algorithm that generates an initial solution using a modified penalty algorithm and improves it using the simulated annealing algorithm.

5.5 LAYOUT SOFTWARE

We will cover some layout software packages that have been developed over the past five decades. Some of the newer ones, especially, VIP-PLANOPT and Layout-iQ, are powerful software packages that use sophisticated algorithms and have a very user-friendly interface. Others, such as the Flow Path Calculator, are useful in evaluating a given layout and also have a rich user interface.

There are many layout software packages available. Many are 3-d software and allow walk-through or fly-through visualization. In other words, they allow the user to visualize the factory by *walking through* the plant virtually before it is even built.

Some of the software we will describe include computerized relative allocation of facilities technique (CRAFT), BLOCPLAN, Production Flow Analysis and Simplification Toolkit (PFAST), Factory Design suite included in Autodesk, Proplanner, Layout-iQ, and VIP-PLANOPT.

5.5.1 CRAFT

CRAFT was originally presented in Armour and Buffa (1963) and Buffa et al. (1964). It uses the 2-opt solution strategy in developing a layout but has some differences in its implementation. For example, CRAFT does not examine all possible pairwise exchanges before generating an improved layout.

CRAFT requires the following data as input:

- Dimensions of the building in which the departments are to be housed
- Dimensions of the departments
- Flow of material or frequency of trips between department pairs and cost per unit load per unit distance
- An initial layout
- Restrictions on the location of departments, if applicable

Given the initial layout, CRAFT computes the distance between the centers of each department pair and determines the cost of the initial layout. Note that the cost can be easily computed by determining the following items for each pair of departments:

- Product of the number of trips between departments i and j and the cost to make one trip between the two
- Distance between departments i and j

We refer to the aforementioned first item as the flow component and the second item as the distance component. The product of the flow and distance components provides the cost of moving material between departments i and j. If the calculations are performed for each department pair and the resulting values added, we get the cost of the initial layout.

CRAFT considers exchanging the locations of department pairs that are adjacent or have the same area. Then, it makes the location exchange that results in the greatest *estimated* cost reduction. If all the departments are of equal area or if every nonadjacent pair of departments has the same area, the algorithm examines $n(n - 1)/2$ possible exchanges, where n is the number of departments in the layout problem. While this is the maximum number of exchanges possible, in a general department layout problem, the number of actual exchanges examined are much less than $n(n - 1)/2$ because not all departments will have the same area.

To calculate the estimated cost reduction, CRAFT interchanges the coordinates of departments i and j whose exchange is being considered using the following expression:

$$\sum_{k=1,k\neq i,k\neq j}^{n} f_{ik}d_{ik} + \sum_{k=1,k\neq i,k\neq j}^{n} f_{jk}d_{jk} - \sum_{k=1,k\neq i,k\neq j}^{n} f_{jk}d_{ik} - \sum_{k=1,k\neq i,k\neq j}^{n} f_{ik}d_{jk} \qquad (5.1)$$

where d_{ij} and f_{ij} refer to the distance and flow between departments i and j, respectively. The first two terms in expression (5.1) indicate the preexchange material handling cost contribution of departments i and j. The last two *estimate* the postexchange material handling cost contribution. Note that the product of f_{ij} and d_{ij} is not included in any of the four terms in expression (5.1) because the distance between departments i and j is the same before and after the exchange. To estimate the postexchange contribution, CRAFT multiplies the distance vector of department i with the flow vector of department j and the distance vector of department j with the flow vector of department i. Thus, CRAFT implicitly assumes that the coordinates of departments i and j will be reversed after the exchange. Obviously, this will happen only if the two departments are of the same area and shape. When they are not, the postexchange centroid of departments i and j and hence the distance estimated by expression (5.1) will be different from the actual distance.

To understand this, consider Figure 5.20 that shows the layout of four departments and the flow matrix. The coordinates of their centers are (5, 25), (30, 25), (10, 10), and (10, 35), respectively. The total material handling cost for this layout is 3420. If expression (5.1) is used to determine the estimated cost reduction of exchanging departments 1 and 3, we obtain a value* of 2(3∗25 + 6∗20 + 19∗45 + 6∗20 + 16∗35 + 0∗25 − 16∗25 − 0∗45 − 3∗35 − 19∗25 − 2∗6∗20) = 1020. If departments 1 and 3 are indeed exchanged, the actual reduction will be equal to 810. This can be verified using the new layout and the distance matrix provided in Figure 5.21.

Due to the differences between the estimated and true costs, it is possible that CRAFT may not make an exchange even though it will reduce the material handling cost. This occurs rarely, however, and can occur only when unequal area departments are being exchanged. Besides, the gap between the estimated and true costs is typically very small. In addition, we want to know only whether it is a good idea to exchange the two departments under consideration. Because the actual

---

* The calculations may help the reader understand why the last term is included in expression (5.1).

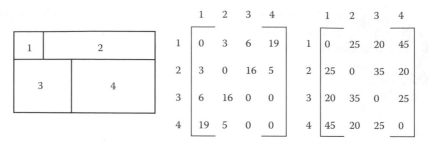

FIGURE 5.20   Layout of four departments and the corresponding flow and distance matrices.

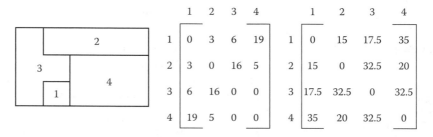

FIGURE 5.21   Layout after exchanging the positions of departments 1 and 3 in Figure 5.20 and the corresponding flow and distance matrices.

cost is calculated after the exchange is made and the savings in computation time outweigh the problems caused by the slightly inaccurate estimation of total cost, CRAFT computes the estimated cost reduction using expression (5.1).

As mentioned earlier, rather than examining all possible exchanges, CRAFT considers the exchange of only adjacent department pairs or pairs that have the same area. For example, to go to the next improved layout from the one in Figure 5.20, CRAFT will not consider the exchange of department pair 1, 4 because they neither have a common boundary nor have the same area.* The exchange that results in the highest estimated cost savings is made. The new layout, the new material handling cost, the cost reduction effected, and the departments involved in the exchange are provided as output. Then, starting with the new layout as the initial one, the same procedure is continued until there is no exchange that results in a cost lower than that of the current layout.

In addition to two-way exchanges, CRAFT can also perform the following:

- Three-way exchanges
- Two-way exchanges followed by three-way exchanges
- Three-way exchanges followed by two-way exchanges

Furthermore, the user can have CRAFT consider the best of 2-opt and 3-opt. The positions of any department may be fixed by simply specifying that it is not a candidate for exchange. This feature is helpful because in practice, there are always some departments that cannot be moved from their current location due to several constraints—cost, physical, and others. CRAFT was one of the first algorithms that used quantitative input data. The original version was capable of handling only 40 departments.

---

* Of course, CRAFT would consider the exchange of departments 1 and 4 if it were to proceed further from the layout in Figure 5.21 because they are adjacent there.

One major disadvantage with all improvement algorithms is that their ability to produce good solutions is highly dependent on the initial solution provided to them. Thus, in general, the better the starting solution, the better the final solution. This is also true with CRAFT also. To overcome this disadvantage, one may provide various starting solutions to the improvement algorithm and develop a final layout for each starting solution and select the best one. Another disadvantage of CRAFT is that it assumes that when a department i is interchanged with another department j, the same material handling carrier will serve the two departments in their new location. Also, flow is assumed to occur between the centers of departments, and the cost of material handling is assumed to be directly proportional to the distance between departments. Although many of these assumptions may not hold in practice, algorithms such as CRAFT are useful because they provide a framework with which to develop a layout. The user has to be aware of the assumptions made by algorithms and use the solution only as a guide in developing an implementable layout.

CRAFT has been modified a number of times. COFAD (Tompkins and Reed, 1976), a four-step algorithm, tackles the selection of material handling system and layout. In the first step, it determines a layout. Next, it selects a material handling system for the layout obtained in step 1 from a candidate list of equipment. In other words, it assigns specific material handling equipment from the available list to each move. While doing so, it attempts to maximize equipment utilization. Then, it calculates the cost of each move based on the material handling equipment assignment to the individual moves. The cost includes fixed and variable costs, and the calculations are done differently for fixed-path equipment (e.g., conveyors or wire-guided AGVs) and for mobile equipment (e.g., trucks). For this revised cost, it determines the new layout. The procedure of determining a layout, selecting a material handling system, and revising the cost of each move is repeated until a satisfactory solution is obtained. COFAD can consider rectilinear or straight-line distance metrics because for some material handling systems, the straight-line distance metric is a more appropriate measure.

To get a good-quality final solution, the user must provide different starting solutions—perhaps generated randomly—evaluate the final solution obtained by CRAFT for each of these and choose the best one. To overcome this problem, the *biased sampling technique* samples solutions in a biased manner so that good solutions have a high probability of being sampled (Nugent et al., 1968). The technique assigns positive probabilities to each exchange that results in a cost reduction. Of course, the greater the estimated cost reduction, the greater the probability of being sampled. Whereas CRAFT selects the interchange resulting in the most reduction, the biased sampling technique allows any interchange that results in a cost reduction to take place. We will discuss a relatively more recent and superior version of this technique, called simulated annealing, in Chapter 10. A spreadsheet-based CRAFT software is available at the following website:

http://www.me.utexas.edu/~jensen/ORMM/omie/computation/unit/lay_add/lay_add.html. This website can be accessed by clicking on the CRAFT link at http://sundere.okstate.edu/ downloadable-software-programs-and-data-files.

Example 5.4

Ten departments are to be laid out on a factory floor in a way that minimizes the total material handling cost. The data input file is provided in Figure 5.22. Develop a layout using CRAFT.

Solution

The spreadsheet-based CRAFT version available at http://www.me.utexas.edu/~jensen/ ORMM/omie/computation/unit/lay_add/lay_add.html was used to solve the layout problem. The output file generated by the program is shown in Figure 5.23. The OFV of the initial layout is 3213. After three iterations, CRAFT produces the final layout shown in the output file. The final cost of the layout is 2740. No other exchange will improve the solution, so the program terminates.

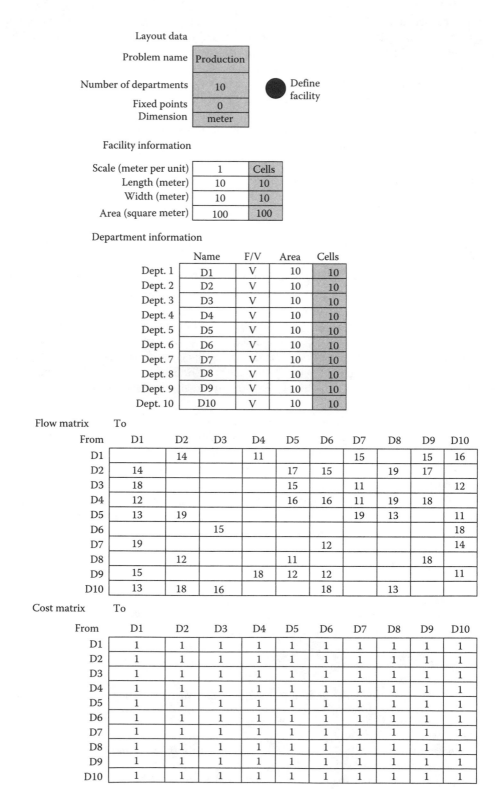

Layout data

Problem name	Production
Number of departments	10
Fixed points	0
Dimension	meter

● Define facility

Facility information

		Cells
Scale (meter per unit)	1	
Length (meter)	10	10
Width (meter)	10	10
Area (square meter)	100	100

Department information

	Name	F/V	Area	Cells
Dept. 1	D1	V	10	10
Dept. 2	D2	V	10	10
Dept. 3	D3	V	10	10
Dept. 4	D4	V	10	10
Dept. 5	D5	V	10	10
Dept. 6	D6	V	10	10
Dept. 7	D7	V	10	10
Dept. 8	D8	V	10	10
Dept. 9	D9	V	10	10
Dept. 10	D10	V	10	10

Flow matrix   To

From	D1	D2	D3	D4	D5	D6	D7	D8	D9	D10
D1		14		11			15		15	16
D2	14				17	15		19	17	
D3	18				15		11			12
D4	12				16	16	11	19	18	
D5	13	19					19	13		11
D6			15							18
D7	19					12				14
D8		12			11				18	
D9	15			18	12	12				11
D10	13	18	16			18		13		

Cost matrix   To

From	D1	D2	D3	D4	D5	D6	D7	D8	D9	D10
D1	1	1	1	1	1	1	1	1	1	1
D2	1	1	1	1	1	1	1	1	1	1
D3	1	1	1	1	1	1	1	1	1	1
D4	1	1	1	1	1	1	1	1	1	1
D5	1	1	1	1	1	1	1	1	1	1
D6	1	1	1	1	1	1	1	1	1	1
D7	1	1	1	1	1	1	1	1	1	1
D8	1	1	1	1	1	1	1	1	1	1
D9	1	1	1	1	1	1	1	1	1	1
D10	1	1	1	1	1	1	1	1	1	1

FIGURE 5.22   Data input file for Example 5.4.

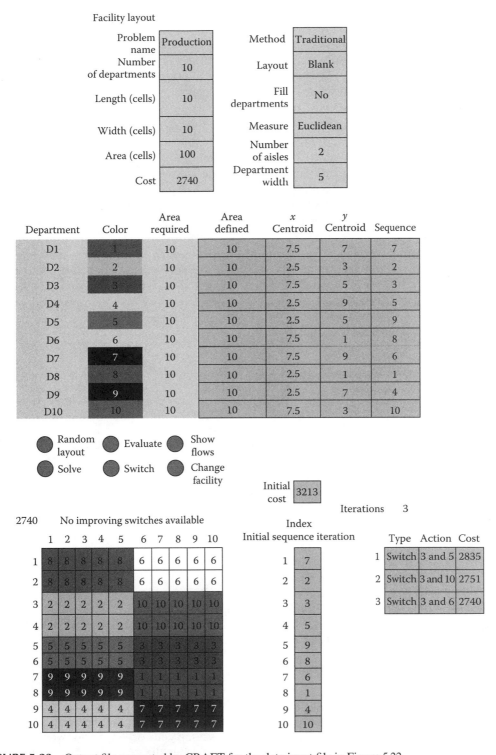

FIGURE 5.23   Output file generated by CRAFT for the data input file in Figure 5.22.

5.5.2 BLOCPLAN

BLOCPLAN is an interactive program developed by Donaghey and Pire (1991) that can develop a single-story or multistory layout and has a number of useful features. It offers a number of heuristic algorithms for solving the layout problem and can handle quantitative as well as qualitative data. It allows the user to enter product routing data. BLOCPLAN provides ample opportunity to view and edit data. The user provides flow data in the form of a relationship chart, which is then used to develop a layout. BLOCPLAN can develop a layout randomly or using an automatic search algorithm. The user can also manually insert departments.

The *random layout algorithm* generates layouts without considering the flow or interaction between departments. The *automatic search algorithm* generates an initial layout randomly using the random layout algorithm and then uses the improvement algorithm to find a better layout. The exchange of departments continues until an improved layout is obtained. Then, the procedure is repeated (a maximum of 20 times) all over again with the improved layout as the initial layout. It calculates the *adjacency score* and *rel-dist score* of the new layout. These terms are explained later. The automatic search algorithm is identical to the greedy 2-opt algorithm discussed in this chapter. To determine the adjacency score of each layout, BLOCPLAN computes the sum of the positive numeric values corresponding to all the relationship codes. Next, it examines the relationship codes of adjacent departments in the layout and adds the corresponding numeric values. The latter is divided by the former to get the adjacency score. To determine the rel-dist score, BLOCPLAN computes the actual distance times the numeric value of the corresponding relationship between each pair of departments. Then it adds these computed values to obtain the rel-dist score. Mathematically, adjacency score is given by

$$\frac{\sum_{i=1}^{n-1}\sum_{j=i+1}^{n} R_{ij}D_{ij}}{\sum_{i=1}^{n-1}\sum_{j=i+1}^{n} R_{ij}} \tag{5.2}$$

and rel-dist score is given by

$$\sum_{i=1}^{n-1}\sum_{j=i+1}^{n} d_{ij}R_{ij} \tag{5.3}$$

where

$$D_{ij} \begin{cases} 1 & \text{if departments } i \text{ and } j \text{ are on the same floor and adjacent} \\ 0 & \text{otherwise} \end{cases}$$

R_{ij} is the numeric value assigned to the relationship code between departments i and j
n is the total number of departments
d_{ij} is the rectilinear distance between the centers of departments i and j

Department pairs with a negative numeric value are not included in the summation in the denominator of the adjacency score formula (5.2). Instead of printing the rel-dist score directly, BLOCPLAN normalizes the score and prints the resulting value. To do so, it computes a simple lower and upper bound on the value of the rel-dist score by multiplying two vectors—a distance vector whose elements are arranged in nondecreasing order and a relationship code or *flow* vector whose elements are arranged in nonincreasing order. The first vector includes the distance, whereas the second includes the *numeric values* corresponding to the relationship code between each pair of departments. Because we are trying to minimize the sum of the product of flow times distance calculated for each pair of departments, it is well known that a lower bound can be obtained by arranging elements of the flow vector in nonincreasing order and the distance vector elements in nondecreasing order. Thus, we are trying to match the largest flow elements with the smallest distance element. Although such a matching may not lead us to the OFV of a feasible solution, it provides us a reasonably tight lower bound to compare the OFV's of other feasible solutions. Obviously, the optimal solution is the one whose OFV is closest to the lower bound (among all possible feasible solutions). To get an upper bound, we may do the reverse of what we did to compute the lower bound—arrange elements of the two vectors so that the largest flow element is matched with the largest distance element and multiply the two vectors. This means that both the vectors are to be arranged so that their elements are in nondecreasing order. BLOCPLAN thus computes the lower and upper bounds and determines the normalized rel-dist score, called R-score, as

$$R\text{-score} = 1 - \frac{(\text{rel-dist score-lower bound})}{(\text{upper bound-lower bound})} \tag{5.4}$$

Thus, an R-score of 1 would mean that the corresponding solution is optimal. Of course, such a value for R-score is highly unlikely because it means that the rel-dist score of the corresponding layout is equal to the lower bound. Similarly, an R-score value of 0 (also highly unlikely) means that the corresponding layout is the worst possible layout because its OFV is equal to the upper bound.

The main advantage of BLOCPLAN is its user-friendliness. It allows the user to edit previously entered data, fix the position of departments, and manually insert them at desired locations. It also prints a table of ranked layouts that shows the raw rel-dist scores as well as the normalized R-score for each layout along with some other information. In addition to single-story layouts, BLOCPLAN can produce multistory layouts, and this is discussed in a later section.

Example 5.5

A manufacturing facility has six major departments for which a single-story layout is to be developed. The department names and areas are shown in Figure 5.24. Using the relationship chart in Figure 5.25, and BLOCKPLAN default values for the relationship indicators, develop a single-story layout using BLOCPLAN.

Solution

After we enter the aforementioned data interactively, BLOCPLAN calculates the score vectors for the six departments (see Figure 5.26). When the layouts are randomly generated via the single-story layout menu, BLOCPLAN produces the output screens shown in Figures 5.27 and 5.28.

5.5.3   PFAST

Production flow analysis simplification toolkit (PFAST) is a program for machine grouping, part family formation, cell layout, and shop design layout (Irani et al., 2000). It emphasizes the importance of examining the material flow and handling aspects in a facility before undertaking any physical design of the layout. PFAST requires the user to provide the following information:

1. Parts to be manufactured
2. Work centers where the parts will be processed
3. The part routing information

Based on these inputs, PFAST analyzes the material flow, determines how the machines are to be grouped, and develops a layout. PFAST analyzes the material flow data, provides descriptive statistics for routing data, groups similar routings, identifies *misfit* (or outlier) routings, and creates

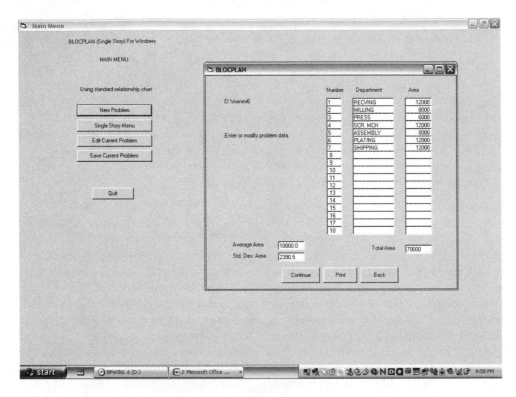

FIGURE 5.24   Area for six departments in Example 5.5.

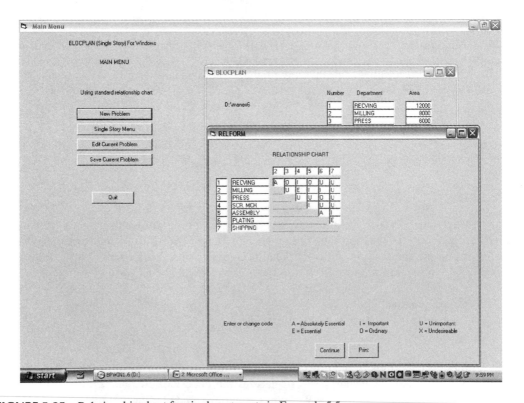

FIGURE 5.25   Relationship chart for six departments in Example 5.5.

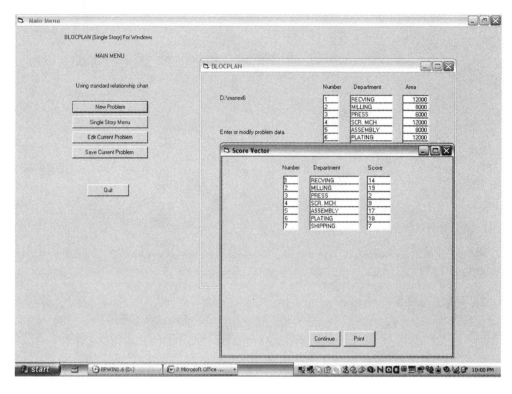

FIGURE 5.26 Score vector for the six departments in Figure 5.24.

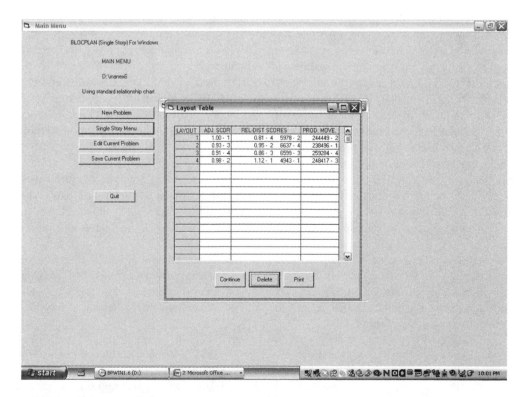

FIGURE 5.27 Layouts randomly generated by BLOCPLAN.

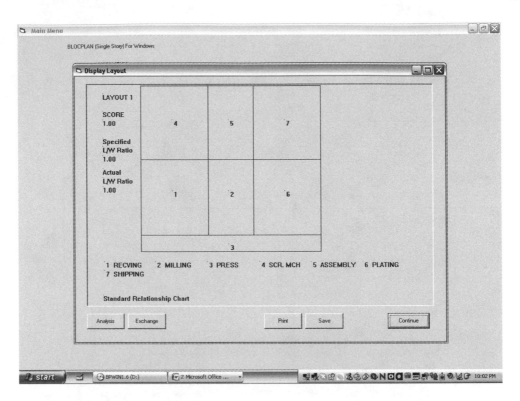

FIGURE 5.28 Graphical display of layout 1.

alternative routings for key products. In designing a facility layout, it is helpful in identifying non-traditional layouts and designing a flexible layout using multiple samples of routings. Similarly in planning for cellular manufacturing, it is helpful in rapidly assessing the feasibility of conversion to manufacturing cells and analyze the stability of cell compositions.

To understand the working methodology of PFAST, consider a 12-machine and 19-part example (see Tables 5.2 through 5.4). PFAST requires input data to be in Microsoft Excel format. Table 5.2 shows data about the parts to be manufactured—part number, part name, quantity required per unit of time, revenue generated by each part for the same time period, number of operations required to manufacture each part, and the production runs required to fulfill the demand.

Table 5.3 provides information about the work centers available in the facility—work center name, description (if available), number of work centers available, floor space required by each work center, and the percentage of time for which each machine is available (uptime). Table 5.4 shows routing information for the parts to be manufactured in the facility. It includes the sequence of work centers visited by each part as well as the corresponding machine setup and processing times.

All the required information in the Excel data file can be imported into PFAST using the "Import External Data" option in the "File" menu. After the data are imported, the following report can be generated using the "Quick Excel Report" option in the "Report" menu.

Product routing map is a quick reference guide to understand the diversity of product and machines in the facility (Table 5.5).

P–Q analysis: The P–Q analysis, also referred to as "ABC" or "Pareto" analysis, sorts the parts in decreasing order of their production quantities (Figure 5.29). A major advantage of this analysis is that it allows the facility designer to focus efforts on reducing material handling costs, work-in-process (WIP) inventory, queuing times, production led times, and operating costs only on high-volume parts.

TABLE 5.2

Parts Data Set

Part No.	Description	Annual Quantity	Revenue	Number of Operations	Number of Production Runs
1	Part 1	20	68	4	2
2	Part 2	50	9	6	5
3	Part 3	10	6	6	5
4	Part 4	20	134	4	5
5	Part 5	50	15	5	4
6	Part 6	20	49	4	8
7	Part 7	40	66	4	4
8	Part 8	90	91	7	5
9	Part 9	90	5	6	4
10	Part 10	70	56	4	3
11	Part 11	50	40	1	5
12	Part 12	70	143	3	2
13	Part 13	80	94	2	4
14	Part 14	60	130	3	5
15	Part 15	80	66	6	3
16	Part 16	40	5	6	6
17	Part 17	40	102	3	5
18	Part 18	30	109	3	2
19	Part 19	70	47	1	6

TABLE 5.3

Work Center Data

Work Center No.	Description	Quantity	Area	Availability
1	1	1	1	0.8
2	2	1	1	0.8
3	3	1	1	0.8
4	4	1	1	0.8
5	5	1	2	0.8
6	6	1	1	0.8
7	7	1	1	0.8
8	8	1	4	0.8
9	9	1	1	0.8
10	10	1	1	0.8
11	11	1	1	0.8
12	12	1	1	0.8

P–Q–$ analysis: In the *P–Q–$ analysis*, the potential revenue generated by the parts is taken into consideration. All parts to be manufactured in the facility are plotted in a 2D plot called the *P–Q–$* plot (see Figure 5.30). The *X* and *Y* axes represent the revenue generated by each product and the corresponding quantities, respectively. By dividing the chart into a few small cells, we can select a small, yet representative sample of parts representing a large proportion of high-value products manufactured.

TABLE 5.4

Part Routing, Setup, and Processing Time Data

Part No.	Work Center No.	Sequence No.	Setup Time	Processing Time
1	1	1	0	11
1	4	2	0	23
1	8	3	0	22
1	9	4	0	29
10	4	1	0	20
10	7	2	0	17
10	4	3	0	29
10	8	4	0	10
11	6	1	0	21
12	11	1	0	15
12	7	2	0	19
12	12	3	0	15
13	11	1	0	21
13	12	2	0	23
14	11	1	0	16
14	7	2	0	28
14	10	3	0	17
15	1	1	0	14
15	7	2	0	22
15	11	3	0	17
15	10	4	0	23
15	11	5	0	24
15	12	6	0	21
16	1	1	0	23
16	7	2	0	19
16	11	3	0	28
16	10	4	0	22
16	11	5	0	19
16	12	6	0	25
17	11	1	0	24
17	7	2	0	20
17	12	3	0	23
18	6	1	0	25
18	7	2	0	11
18	10	3	0	22
19	12	1	0	17
2	1	1	0	14
2	4	2	0	29
2	7	3	0	24
2	4	4	0	10
2	8	5	0	22
2	7	6	0	13
3	1	1	0	16
3	2	2	0	17
3	4	3	0	11
3	7	4	0	11
3	8	5	0	17

(Continued)

TABLE 5.4 (*Continued*)
Part Routing, Setup, and Processing Time Data

Part No.	Work Center No.	Sequence No.	Setup Time	Processing Time
3	9	6	0	13
4	1	1	0	23
4	4	2	0	14
4	7	3	0	10
4	9	4	0	25
5	1	1	0	13
5	6	2	0	16
5	10	3	0	26
5	7	4	0	19
5	9	5	0	16
6	10	2	0	19
6	7	3	0	27
6	8	4	0	22
6	9	5	0	10
7	6	1	0	27
7	4	2	0	22
7	8	3	0	27
7	9	4	0	24
8	3	1	0	25
8	5	2	0	24
8	2	3	0	23
8	6	4	0	13
8	4	5	0	24
8	8	6	0	22
8	9	7	0	20
9	3	1	0	18
9	5	2	0	29
9	6	3	0	12
9	4	4	0	19
9	8	5	0	24
9	9	6	0	22

Flow diagram: The flow diagram shows a near-optimal location for all the work centers in the facility. Workstation pairs with high intensity of traffic between them are located adjacently. The thickness of the arrows between two work centers in Figure 5.31 represents the intensity of the product traffic between the corresponding work center pairs.

From–to chart: The from–to chart (Table 5.6) is a numeric representation of the traffic flow intensity visually depicted in the product flow diagram. It allows the facility analyst to focus on those areas where blocking or queuing is likely to occur.

The *0–1 matrix*: The 0–1 matrix (Table 5.7) is the result of applying the machine-part matrix clustering (MPMC) algorithm available in PFAST. This matrix lists the part numbers along the rows and the machines across the columns. PFAST arranges this table in such a way that parts requiring similar or identical sets of machines are grouped together into blocks along the diagonal of the matrix. Each block—indicated by a set of consecutive rows and columns—represents a family of parts that could potentially be produced in a single manufacturing cell identified by the collection of corresponding rows. We will study more about the 0–1 matrices in Chapter 6.

TABLE 5.5
Product Routing Map

Part No.	Quantity	Revenue	Routing					
1	20	68	1	4	8	9		
2	50	9	1	4	7	4	8	7
3	10	6	1	2	4	7	8	9
4	20	134	1	4	7	9		
5	50	15	1	6	10	7	9	
6	20	49	10	7	8	9		
7	40	66	6	4	8	9		
8	90	91	3	5	2	6	4	8	9
9	90	5	3	5	6	4	8	9
10	70	56	4	7	4	8		
11	50	40	6					
12	70	143	11	7	12			
13	80	94	11	12				
14	60	130	11	7	10			
15	80	66	1	7	11	10	11	12
16	40	5	1	7	11	10	11	12
17	40	102	11	7	12			
18	30	109	6	7	10			
19	70	47	12					

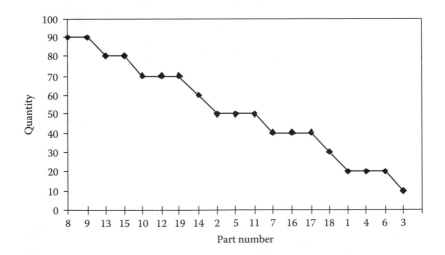

FIGURE 5.29 *P–Q* analysis.

Dendogram: A clustering dendogram is depicted in Figure 5.32. This figure shows the various machine groups that can be identified at various threshold levels. A more detailed discussion of threshold levels and dendograms is provided in Chapter 6.

Common subsequences: Table 5.8 shows that each of the substrings of operations 4–8, 8–9, and 4–8–9 has a high frequency of occurrence in the original routings shown in Table 5.4. Subject to equipment constraints, this string matching result could be used to replace the three single-function machines 4, 8, and 9 by a multifunction machining center. This planned introduction

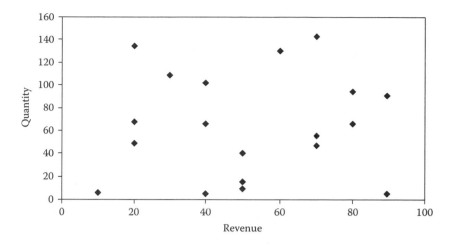

FIGURE 5.30 *P–Q–$* analysis.

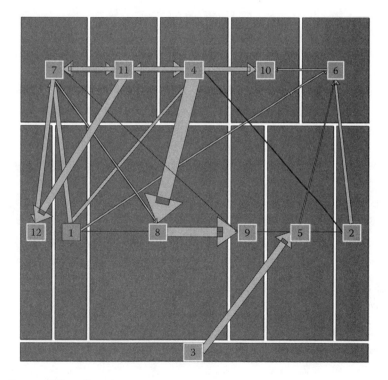

FIGURE 5.31 Proposed flow diagram.

of flexible automation would eliminate interoperation material handling, loading and unloading, and setups for the parts 1, 2, 3, 6, 7, 8, 9, and 10, thereby significantly improving operational efficiency.

Thus, PFAST is capable of providing much of the required information to design and make decisions toward the design of a new facility layout or to modify existing layouts. Upcoming versions of PFAST are expected to be more robust and include algorithms to provide more data-oriented conclusions, which a designer can then implement directly in an existing or new factory.

TABLE 5.6
From–To Chart

	1	4	8	9	7	2	6	10	3	5	11	12
1		90			120	10	50					
4			360		150							
8				270	50							
9												
7		120	30	70				90			120	110
2		10					90					
6		220			30			50				
10					70						120	
3										180		
5						90	90					
11					170			120				200
12												

TABLE 5.7
0–1 Matrix

	2	5	3	6	9	8	4	1	7	10	11	12
11				1								
8	1	1	1	1	1	1	1					
9		1	1	1	1	1	1					
7				1	1	1	1					
1					1	1	1	1				
3	1				1	1	1	1	1			
4				1			1	1	1			
2						1	1	1	1			
10						1	1	1				
6				1		1			1	1		
5				1	1			1	1	1		
18					1				1	1		
14									1	1	1	
15								1	1	1	1	1
16								1	1	1	1	1
12									1		1	1
17									1		1	1
13											1	1
19												1

5.5.4   FACTORY SUITE OF PROGRAMS

Three popular software packages are briefly discussed in this section. AutoCAD or AutoDesk Architectural Desktop must be installed on the computer prior to running any of these three programs.

5.5.4.1   FactoryCAD

FactoryCAD is a software tool that customizes AutoCAD and allows the user to develop new factory layouts or modify existing ones. It does not generate a layout on its own. However, it is an

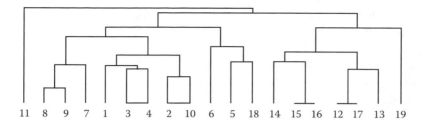

FIGURE 5.32   Clustering of operation sequences into part families.

TABLE 5.8
Common Subsequences

Common Subsequences	Length	Frequency	Quantity (%)	Revenue (%)
1–4	2	3	9.18	7.29
4–8	2	6	36.73	29.15
8–9	2	6	27.55	21.86
4–8–9	3	4	24.49	19.43
4–7	2	4	15.31	12.15
1–4–7	3	2	7.14	5.67
4–7–4–8	4	2	12.24	9.72
7–8–9	3	2	3.06	2.43
7–9	2	2	7.14	5.67
10–7	2	2	7.14	5.67
6–4–8–9	4	3	22.45	17.81
3–5	2	2	18.27	14.57
11–7	2	3	17.35	13.77
11–7–12	3	2	11.22	8.91
11–12	2	3	20.41	16.19
7–10	2	2	9.18	7.29
1–7–11–10–11–12	6	2	12.24	9.72

elegant drafting tool that allows the user to draw building walls, doors, windows, building columns, work center borders, and utility lines and zoom in on selected portions of the drawing. The user can select 3D models of machines, material handling devices, pallet, and other objects contained in a symbol library and thus insert new machines or material handling systems with relative ease into the layout. Object-specific performance factors such as scrap rate, load and unload time, setup, and breakdown can be included in the FactoryCAD model. It is most useful as a drawing tool and should be used only for this purpose.

5.5.4.2   FactoryFLOW

FactoryFLOW is a layout software that requires an existing layout generated using the AutoCAD software. After loading the existing drawing, the user enters three key pieces of information in three separate files—product file, parts routing file, and material handling equipment file. Of course, if the data have already been entered as a spreadsheet or word processor file, they can be converted into the required format and saved as an ASCII file, which may then be imported into the program. The product file—one for each product produced—allows the user to enter the name and number of units of the product produced in one time unit (year, month, week, day, or shift). The user can also assign one of 256 colors to each product, which is helpful when viewing the flow path of the products. After entering product information, information concerning the product's part routings is required to be entered. The parts may be thought of as assemblies or subassemblies that make the end product. For example, a product such as a ladder may have steps, rails, and screws as its parts. FactoryFLOW requires information on each move for each part, quantity moved, material handling device used, and unit load. The last piece of information concerns the material handling equipment. The user selects the type of material handling device used for all the material transfers. Ten types of material handling devices, e.g., conveyor, forklift truck, and AGV, are available in FactoryFLOW. If the handling device used does not correspond to any of the types listed, the user can specify the device as a *special* device. For the device selected, the number of units of that device available, time for which it is available, investment cost per time unit (year), hourly variable nonlabor costs of running the device (e.g. fuel, power, and maintenance costs), hourly variable labor costs, the load and unload time, average speed, and estimated percentage of loaded travel must be entered.

After the data about the product, part routing, and material handling device are entered, the software asks the user to draw the material handling paths for each move. With this information, the data in the three files and the data in the AutoCAD layout file, the software calculates the total move distance and cost. The cost is calculated using a formula similar to the one obtained via Equations 3.2 through 3.4. The user can then generate scaled product flow lines whose thickness (visible on the generated layout) corresponds either to the total cost of that move or the number of trips required for a move. For example, if the cost of a move is twice that of another and the user requests scaled product flow lines with thickness corresponding to move cost, the product flow line corresponding to the more expensive move will appear twice as thick as that of the other. This allows the user to concentrate on the relatively expensive material handling moves. FactoryFLOW is also capable of generating a material handling report that specifies total distance traveled (for alternate distance metrics such as Euclidean and aisle distance), distance and costs between each pair of work centers, material handling device utilizations, and other information helpful to the user in analyzing the layout under consideration.

While FactoryFLOW has a number of useful features, a major drawback is that by itself, it cannot generate layouts. However, the user may examine the product flow diagrams and, based on such an examination, change the locations of certain work centers or material flow path of specific parts and recalculate the material handling costs and distances, as well as utilizations. The aforementioned procedure of examining software output, making location changes for appropriate work centers, changing material flow paths, and recalculating costs may be repeated until the user obtains a satisfactory layout. Because the number of possible layouts is equal to $n!$, where n is the number of departments, the user would have to generate a large number of layouts by trial and error and pick the one that minimizes the objective function measure. Thus, FactoryFLOW is useful as a post processor to evaluate layouts developed previously using another software (e.g., CRAFT). The user could then make minor changes and try to improve the layout under consideration. A Deere & Company plant in Horicon, WI, used FactoryFLOW to determine the material handling costs under a process layout and a group technology–based layout (Tamashunas et al., 1990). Using this software tool, they were able to show that a group technology–based layout would reduce material handling travel distance by about 30% and require fewer material handling carriers compared to a traditional process layout. FactoryFLOW is also useful for performing *what if* analysis. For example, the user can determine the change in material handling costs brought about by a change in the material flow path of a high-volume part.

5.5.4.3 Plant Simulation

Plant Simulation interfaces with FactoryCAD and FactoryFLOW and can be used to develop a detailed simulation model so that a given layout can be evaluated relative to the important operational performance measures such as work center utilization, product flow times, and WIP inventory buildup at specific workstations, departments, or the factory as a whole. FactoryFLOW could be used to develop handful—say three—layouts that minimize material handling costs by trial and error and then a test these layouts for their operational performance via Plant Simulation. Plant Simulation also has a genetic algorithm–based optimizer feature that can help develop near-optimal system parameters such as the number of machines and material handling devices of a given type to include in the layout.

5.5.5 Layout-iQ

Layout-iQ is a powerful tool to quickly develop layouts, draw them, evaluate their traffic costs, and modify them by trial and error. The input data can be manually entered or read from a Microsoft Access data file. Rooms, hallways, machines, and storage areas are modeled as locations. Layout-iQ calculates the traffic cost between each pair of locations and graphically displays them, differentiating the high traffic intensity pairs with thicker lines and the lower intensity locations with thinner lines. Of course, the intensities between each pair of locations depend on the volume of traffic between the corresponding locations. The traffic intensity itself is a function of how many parts, material, or people travel between the location pairs per unit time. The user can provide any combination of the following three pieces of information:

- Product routing data including (which is used by the software to automatically develop a flow matrix)
- The flow matrix
- The relationship indicator data for each pair of locations

The product routing data include product names, quantities manufactured per unit time period, sequence of machines visited by each unit load, and a flow factor. Flow factor is used to model situations where parts can flow to multiple locations and to model assemblies that contain the same part more than once. Flow factor can also be used to place a relatively higher importance on some products than others, for example, more emphasis on class A products than class C products. If necessary, the flow matrix data can be entered directly. In general, this approach is not preferred because it is better to have the program calculate the flow matrix based on product routings, unit load, and flow intensity factors. The third option for indicating the interactions between location pairs utilizes the closeness ratings we studied in Chapters 3 and 4. Layout-iQ allows multiple users (or a single user) to provide the closeness ratings as well as the reason why these ratings were provided. When multiple users provide data, their inputs can be assigned different weights depending on how important the designer considers each person's input to be in the layout design process. For example, a supervisor's input could be considered twice as important as that of machine operator. The closeness rating for a location pair can be based on a single product or a set of user-selected products. With all three flow intensity data options, Layout-iQ allows the user to specify the minimum distance by which to separate location pairs that cannot be adjacent.

The second major piece of information that must be entered pertains to the locations. The user can draw each location or insert previously drawn icons for each location. For example, in designing layout for a hospital, a bed may be considered a location and a suitable drawn icon for a bed may have been previously stored in a database. In this case, the user can simply import that icon for this location. It is rather easy to change the shapes or dimensions of these locations and Layout-iQ provides a user interface to develop simple graphics.

Unlike the factory suite of programs, the layout or the icons need not be developed in AutoCAD. Most graphic file formats will work. If necessary, the layout itself may be overlaid on a previously created floor plan. This feature is useful if a drawing of the building is already available.

In the final step, the user develops a layout by placing each location into the layout—one at a time. As each location is placed in the layout, the software calculates a travel distance or relative score and aisle effectiveness. The primary metric is defined as the total distance traveled when using quantitative input flow data and a relative score when using relationship data and is used to find the best physical location of objects in the workspace. The object of the layout exercise is to minimize travel distance or minimize the relative score depending on the type of data being modeled. Aisle effectiveness is a secondary measure used to evaluate how effective the aisles and pathways are at moving material from one point to another in the layout. Aisle effectiveness is defined as Euclidian travel distance (best possible) divided by the actual travel distance. Aisle effectiveness above 70% is generally acceptable and aisle effectiveness below 70% can usually be improved. As the objects corresponding to the locations are moved, the program dynamically calculates the score and aisle effectiveness thus providing immediate feedback to the user on any changes made in the layout. Layout-iQ also indicates forbidden areas when the user attempts to move a location close to another with which it has an undesirable closeness rating. The user must develop a layout by trial and error. An alternative is to use an efficient algorithm such as the GTLAYPC program provided in the website http://sundere.okstate.edu/downloadable-software-programs-and-data-files and create a detailed final layout using Layout-iQ. In fact, we have done precisely this for the data input file created for the case study in Chapter 3, which is solved in Chapter 6 using the GTLAYPC algorithm (see Figures 5.33 through 5.36).

In addition to those described in this chapter, there are a few other software packages for modeling, designing, and analyzing layout problems. The reader should consult the buyer's guide section of *Industrial Engineering Solutions* magazine for a list of these. Obviously, each software has its advantages as well as disadvantages. Many of them either are simulation tools or do not use the models described in Chapter 10 or the sophisticated solution techniques described in Chapter 11. CRAFT and VIP-PLANOPT are some of the few software packages that use optimization techniques to identify a good layout.

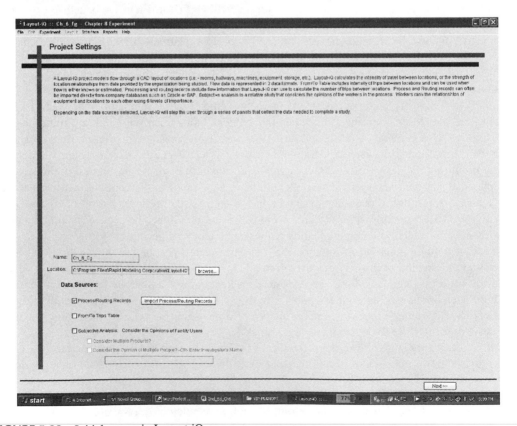

FIGURE 5.33 Initial screen in Layout-iQ.

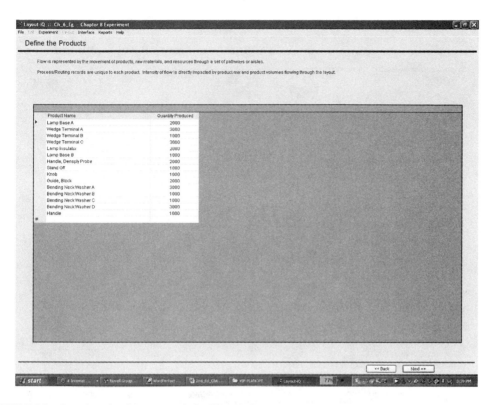

FIGURE 5.34 Part data for the case study in Chapter 3.

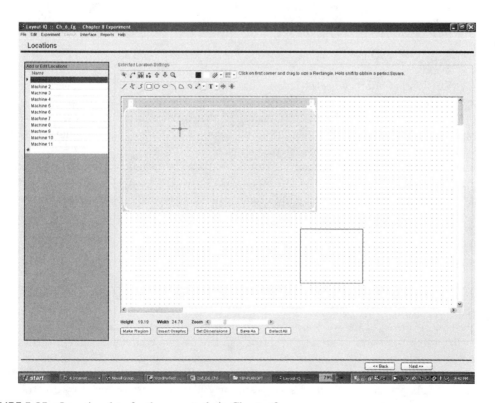

FIGURE 5.35 Location data for the case study in Chapter 3.

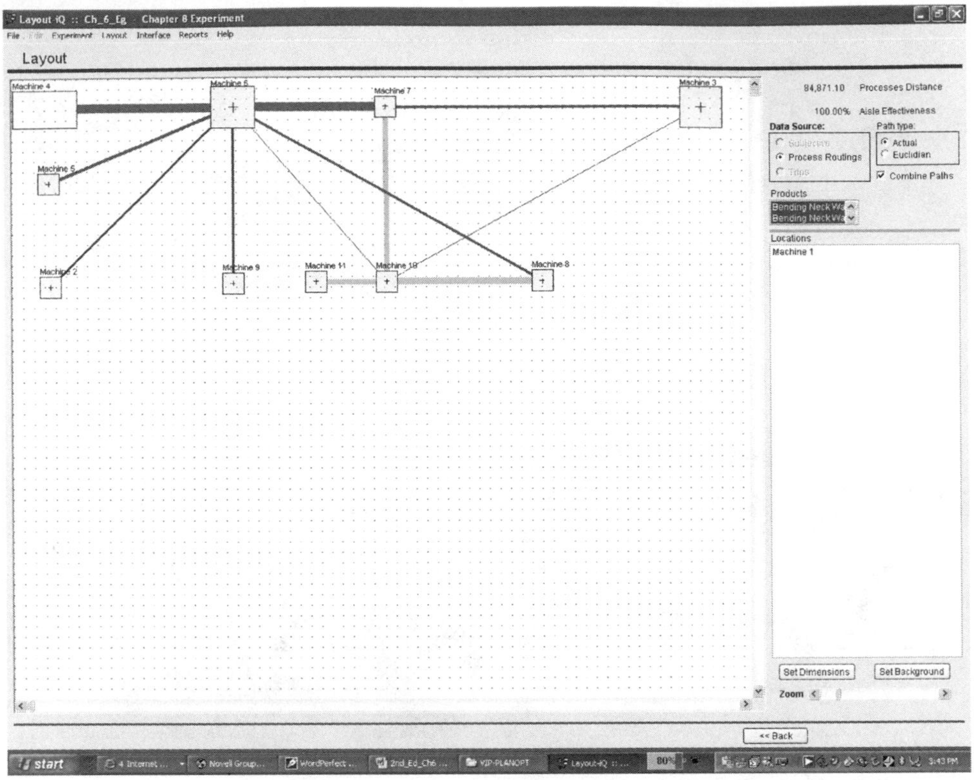

FIGURE 5.36   Final layout developed for the case study in Chapter 3 using Layout-iQ.

5.5.6   VIP-PLANOPT

VIP-PLANOPT is yet another useful layout software package that can also generate near-optimal layouts. Although the software website (www.planopt.com) claims that the software can generate optimal layouts, we will discuss in Chapter 10 that the layout problem is nondeterministic polynomial-time hard, and thus, by definition, no solution technique that can guarantee optimal solutions for medium- to large-scale problems can exist.

Practitioner-oriented layout software has traditionally focused more on the graphics than on solving the layout problem near-optimally. Academic software packages that provide very good solutions exist, but these do not have the high-quality and user-friendly interface that is required by practitioners. VIP-PLANOPT fills this gap well and is a software that places equal importance on

1. Providing a good graphical user interface
2. Generating near-optimal layouts using sophisticated algorithms

A particularly useful feature of VIP-PLANOPT is that as the mouse is dragged over various areas of the screen, appropriate information on the items over which the mouse is being moved is displayed. The professional version of the software can solve up to 500 department problems.

The educational version available at http://sundere.okstate.edu/downloadable-software-programs-and-data-files can solve layout problems with five departments only. Departments are called modules and the user can fix the positions of modules or have the software determine its optimum location. Their shapes can be fixed or allowed to vary within a certain range by controlling the shape ratio parameter that permits the designer to create long and narrow departments or square

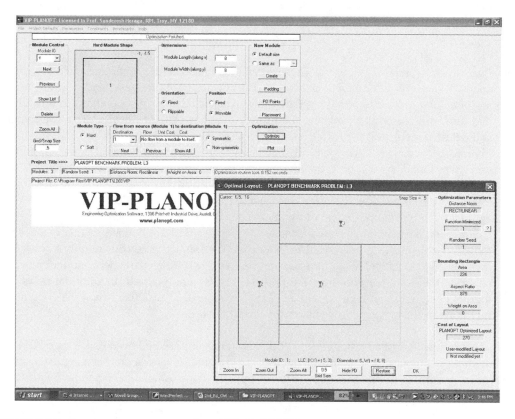

FIGURE 5.37   Input and output screens from VIP-PLANOPT.

departments. The objective function used is similar to the QAP objective discussed in Chapter 10, with an additional term, if invoked, encouraging the user to provide the smallest possible area of the building enclosure. The distance metrics used are rectilinear, Euclidean, or squared Euclidean, and the user can choose to use the distance between pickup and drop-off points or the centroids of departments (modules). The user can *pad* the area of a module to allow for operator access, aisle space, and other reasons. The optimization algorithm uses a simulated annealing–based procedure and uses a default random seed or a user-selected seed. A sample input and output screen is shown in Figure 5.37.

5.5.7   FLOW PATH CALCULATOR*

The Flow Planner (FP) application, developed by Proplanner (www.proplanner.com), is a useful tool to evaluate material flow in manufacturing companies. It is integrated with AutoCAD and calculates detailed material flow information within the layers of the underlying AutoCAD layout drawing. In that sense, FP is a descriptive model rather than a prescriptive model. It does not generate a layout by itself and the user must provide one first. Once a layout has been drawn using AutoCAD, and part, material handling, processing, and location information have been entered, rich information about the material flow is readily available and the graphic and tabular display allow the user to quickly focus on the important material flows and enable the user to improve the initial layout by moving the location of machines and workstations to minimize material flow costs.

---

* Parts of this section were written by David Sly.

We provide a list of the input data required in FP and briefly explain them first. The output provided by the software is then discussed briefly. The input data for FPC include the following:

- *Parts within Product Groupings*: These are the individual components consumed and produced in the facility for which the color-coded flow diagrams are generated.
- *Methods and Method Types*: These are the movement devices (people and/or equipment) that move the material throughout the facility. The flows of each device are color coded and detailed reports on their time and utilization are generated.
- *Locations*: Locations correspond to specific coordinates where material is picked up or set down. Locations can be aggregated such that the user can view flows to hundreds of locations along an assembly line or the flows grouped to a central location on that line. Viewing aggregated flows between lines and departments helps evaluate opportunities for aggregated high-volume delivery (conveyors, AGVs, tugger trains, etc.).
- *Containers*: The containers used to store the parts moved are defined so that quantities moved per trip can be referenced to determine the frequency of trips required between locations. For example, if 150 parts are delivered a day, with 25 parts per container moved using 2 containers per trip, then 3 total trips per day are required. In addition, flows of unique containers can be color coded. This helps the user visualize, when the container might affect the move method chosen for certain routes.

Once the part flow data (see Figure 5.38) have been entered, or imported from Excel, the FP application automatically computes the flow frequencies between all locations and then aggregates these frequencies according to one of many different options. For example, aggregating frequency by Product (default) generates a unique flow line between each location pair for each product in the analysis. As such, if there are four different products with parts moving between location "Dock"

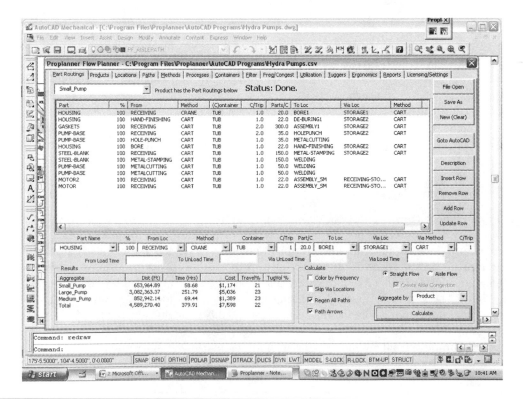

FIGURE 5.38   Data input required for Flow Planner.

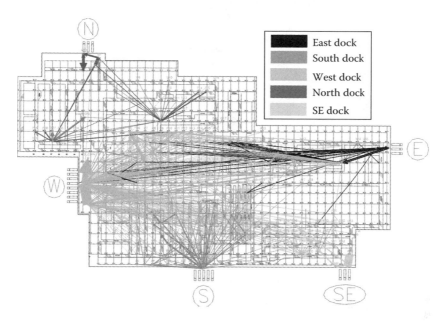

FIGURE 5.39   Straight flow diagram of an automotive plant, aggregated by entry dock.

and location "Store," then four lines will be created between those locations, and each line will be assigned a unique color, and the width of those lines will be scaled according to the frequency of trips occurring for all parts within that product group (thicker widths indicate greater frequencies). Notice the difference between two different aggregates of the same part flows in the same factory. Figure 5.39 is aggregated by receiving Dock and Figure 5.40 is aggregated by part type.

Flow lines, and thus travel distances, are automatically created as a straight line between each location pair, although users can create an aisle network that the application uses to automatically

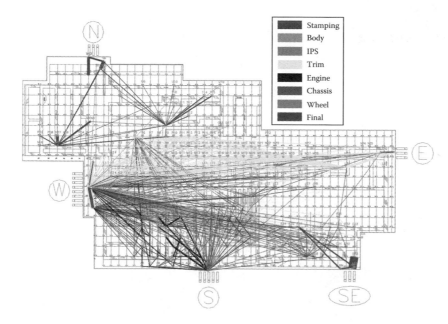

FIGURE 5.40   Straight flow diagram of an automotive plant, aggregated by part type.

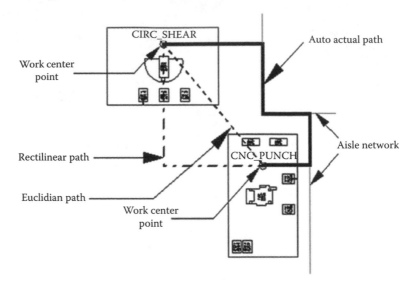

FIGURE 5.41   Aisle diagram of an automotive plant.

generate aisle-path flow lines using an internal shortest path algorithm (Figure 5.41). This algorithm allows for users to specify one-way aisles or even specify multiple aisle networks and map them to specific move method types for situations where some devices are not able to use specific aisle paths.

Aisle flow lines are substantially more accurate in computing travel distances, in method utilization (Figure 5.42), and in determining aisle congestion problems than are straight flow lines;

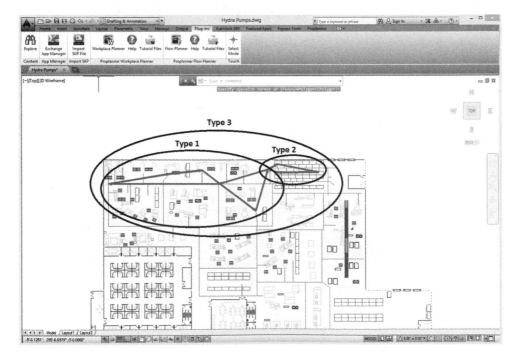

FIGURE 5.42   Automatic flow diagram options in Flow Planner.

AGGREGATE	FROM	TO	FREQUENCY	TOTAL DISTANCE FEET	TRIP DISTANCE FEET	EFF. TRIP DISTANCE FEET	TRAVEL TIME SECONDS	L/UL TIME SECONDS	TOTAL TIME SECONDS	TRIP TRAVEL TIME SECONDS	COST $
GASKETS	RECEIVING	STORAGE2	33.33	5,359.28	160.79	160.75	357.31	1,000.00	1,357.31	10.72	7.54
	STORAGE2	ASSEMBLY1	33.33	4,322.29	129.68	129.67	288.18	1,000.00	1,288.18	8.65	7.16
	RECEIVING	STORAGE2	185.00	29,738.75	160.79	160.75	1,983.05	5,550.00	7,533.05	10.72	41.85
	STORAGE2	ASSEMBLY1	185.00	23,988.95	129.68	129.67	1,599.40	5,550.00	7,149.40	8.65	39.72
SUB TOTAL			436.67	63,409.27			4,227.94	13,100.00	17,327.94		96.27
HOUSING	RECEIVING	STORAGE1	3,950.00	833,134.00	210.96	210.92	55,552.31	118,500.00	174,052.31	14.06	968.96
	STORAGE1	BORE1	500.00	21,500.00	42.97	43.00	1,432.25	15,000.00	16,432.25	2.86	91.29
	HAND-FINISHING	STORAGE2	454.55	69,013.57	151.85	151.83	4,601.37	13,636.36	18,237.73	10.12	101.32
	STORAGE2	DE-BURING1	454.55	80,945.37	178.11	178.08	5,397.38	13,636.36	19,033.74	11.87	105.74
	BORE	STORAGE2	454.55	89,963.55	197.88	197.92	5,996.51	13,636.36	19,632.87	13.19	109.07
	STORAGE2	HAND-FINISHING	454.55	69,013.57	151.85	151.83	4,601.37	13,636.36	18,237.73	10.12	101.32
	STORAGE1	BORE	2,700.00	132,769.00	49.16	49.17	8,848.05	81,000.00	89,848.05	3.28	499.16
	HAND-FINISHING	STORAGE2	3,136.36	476,193.99	151.85	151.83	31,749.46	94,090.91	125,840.37	10.12	699.11
	STORAGE2	DE-BURING	2,454.55	411,553.56	167.70	167.67	27,442.35	73,636.36	101,078.71	11.18	561.55
	BORE	STORAGE2	3,136.36	620,748.96	197.88	197.92	41,375.91	94,090.91	135,466.82	13.19	752.59
	STORAGE2	HAND-FINISHING	3,136.36	476,193.99	151.85	151.83	31,749.46	94,090.91	125,840.37	10.12	699.11
	DE-BURING	DE-GREASING	2,160.00	157,680.00	73.04	73.00	10,517.70	64,800.00	75,317.70	4.87	418.43
	STORAGE1	BORE3	750.00	41,250.00	55.01	55.00	2,750.32	22,500.00	25,250.32	3.67	140.28
	STORAGE2	DE-BURING2	681.82	107,440.88	157.54	157.56	7,161.13	20,454.54	27,615.67	10.50	153.42
SUB TOTAL			24,423.63	3,587,390.45			239,175.57	732,709.07	971,884.64		5,399.35
MOTOR	RECEIVING	RECEIVING-STORAGE	454.55	16,172.71	35.61	35.58	1,079.16	13,636.36	14,715.52	2.37	81.75
	RECEIVING-STORAGE	ASSEMBLY_SM	454.55	35,340.87	77.79	77.75	2,357.17	13,636.36	15,993.53	5.19	88.95

FIGURE 5.43 Tabular part output produced by Flow Planner.

however, they can be difficult to use for evaluating layout improvements, because there are many lines stacked on top of one another. For those situations, the application provides a filter capability where users can selectively turn off and on groups of flow lines based on aggregate, from, and/or to path properties.

In addition to distance, FP also calculates the corresponding material handling time and cost reports using formulae similar to those seen in Chapter 3 (Figure 5.43). With these reports, engineers can identify and classify material movement problems and determine the financial opportunities for improvement. Initially, users will try to find ways to eliminate or remove the material flow by making changes to the factory or the process. Once those options are exhausted, evaluations of alternate move methods and containerization can be studied.

The user has the ability to manipulate the layout by changing part move routes (paths), modifying locations, changing part containerization (frequencies) or selecting different move methods, and recalculating the analysis to generate new reports and diagrams.

A recent addition to the FP application, and flow analysis in general, is the ability to define and evaluate tugger routes. With tuggers, multiple parts in multiple containers are delivered to multiple locations in one trip (route). This creates a unique challenge to evaluate material flow because parts visit machine locations that many not be in their routing. In addition, the entire layout needs to be reevaluated according to this new delivery method, because the objective is now route driver optimization and not necessarily the distance between high-frequency from and to locations. This analysis is substantially more complex due to the addition of the traveling salesman problem whereby the application needs to simultaneously find the shortest path through the aisle network while sequencing the locations to be visited a near optimal manner. FP provides three unique ways that tugger routes can be evaluated (Stage-TO, FROM-Stage, and From-Stage-TO) (Figure 5.44).

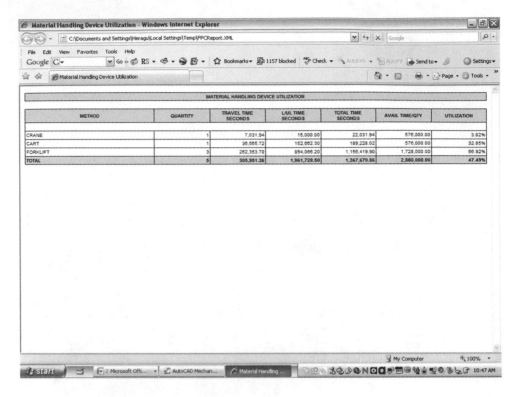

FIGURE 5.44   Tugger route analysis.

5.6   CASE STUDY USING LAYOUT-iQ*

Kirby Risk Service Center (Figure 5.45) produces wiring harnesses and cable assemblies for large off-road mining equipment and other industrial applications. One of its main production areas has experienced over a 100% increase in demand and was also noticing that its production capabilities could not handle the increase while keeping their tight delivery deadlines with their current layout.

Kirby Risk Service Center saw an opportunity to utilize Layout-iQ that would allow them to analyze their current layout and multiple future state layouts. During the analysis of their current state layout, the company confirmed their beliefs that the current layout (Figure 5.46) was not working and that it was running at 60% efficiency.

Utilizing the software's ability to analyze a future state of the area, the company was able to accurately create a layout that improved the efficiency by 25%. This was accomplished very easily with no lost production because the layout changes were made virtually using the software. The final layout is shown in Figure 5.47.

The software helped to develop an efficient layout that provided an estimated $150,000 in savings from better flow through the entire area with an estimated payback of less than 2 weeks on the cost of the software. What would have taken multiple weeks and numerous man-hours to create and test a single layout improvement, only took an afternoon to create several possibilities, with the most

---

* This case study was written by Joel Harrison from Kirby Risk Service Center. If you have any questions concerning this case study, contact Joel at Joel.Harrison@KirbyRisk.com.

Rapid Modeling Corporation is the developer of Layout-iQ. If you have any questions about Layout-iQ or Rapid Modeling's other products, contact Rapid Modeling at www.Rapidmodeling.com.

FIGURE 5.45   Kirby Risk Service Center.

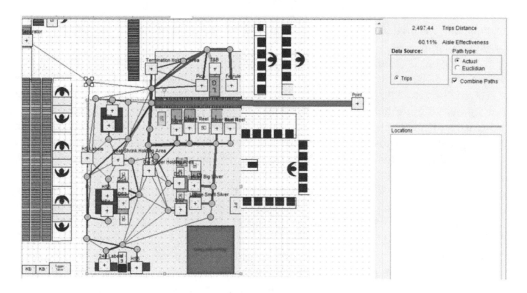

FIGURE 5.46   Current layout analysis using Layout-iQ.

efficient pictured in Figure 5.47. The Layout-iQ software from Rapid Modeling Corporation is a tool that can be used to plan, create, or modify layouts.

5.7   RE-LAYOUT AND MULTIPLE-FLOOR LAYOUT

We believe that re-layout of an existing facility will likely be encountered more often than layout of a new one in future systems. While material handling cost is an important consideration in layout and re-layout, there is another cost that cannot be ignored in a re-layout problem—the cost of moving departments from their current position in the current layout to another in the new layout. Indeed, this cost could be significant and, in the case of some machines, outweigh the material handling costs. In other words, although moving an equipment to a new location may decrease the

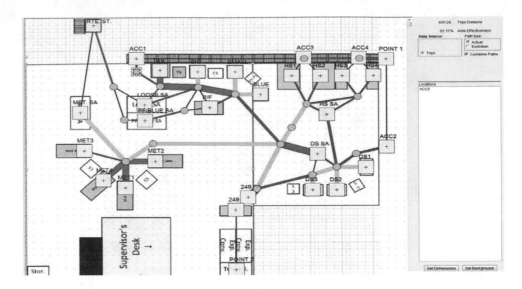

FIGURE 5.47   Final layout produced using Layout-iQ for Kirby Risk Service Center.

cost of handling material between this machine and others, the high cost of physically moving it may prohibit its relocation.

For the re-layout problem, a software called CRAFT-M—an extended version of CRAFT—is available. CRAFT-M takes into consideration the cost of moving a department from its current location to another, in addition to the material handling cost (Hicks and Cowan, 1976). It is a useful program for determining the cost of layout rearrangement as it considers the fixed and variable costs of moving a department. It should be observed that determining the cost of relocating a department is rather difficult. It depends on not only the actual cost of relocation, i.e., labor cost and cost of renting other equipment for relocation, but also the cost due to lost production. In practice, these costs are often guesstimated.

In the literature, the re-layout problem is referred to as the dynamic layout problem. Although many researchers have addressed the dynamic layout problem, most of them make the assumption implicitly or explicitly that the changes in product mix and volume are known at the time the first design decision is made (see, for example, Rosenblatt (1986), Urban (1993), Lacksonen (1994), Montreuil and Venkatadri (1991)). Heragu and Kochhar (1994) mention that this assumption may not be realistic in uncertain and volatile environments where changes in mix and volume may only be known just prior to the revised production run.

The layout problem is often assumed to be 2D. While this assumption generally holds in a manufacturing system, it does not in an office layout problem, where departments are usually located in several floors. Heragu and Kochhar (1994) and Bozer et al. (1994), citing several reasons and examples, argue that even manufacturing systems in the future may have to deal with 3D layout problems. If light machines that do not require elaborate foundations are available, land prices become too high, and the cost of building a new plant is significant more than refurbishing an existing one, it may become attractive for manufacturers to consider expansion in the vertical dimension, thus transforming a 2D plant layout problem into a 3D layout problem. Considering the third (vertical) dimension obviously introduces additional complexity to the already complex layout problem. For example, the distance between two departments situated on different floors may be nonlinear. The number and location of elevators may introduce multiple routes between two departments, requiring each route to be examined and the shortest one retained (Johnson, 1982). There are many algorithms available for the multifloor layout problem (see, for example, BLOCPLAN (Donaghey and Pire, 1991), SPACECRAFT (Johnson, 1982), CRAFT 3D (Cinar, 1975), MULTIPLE (Bozer et al., 1994), and HOPE-M (Kochhar and Heragu, 1998)). We will discuss the multiple-floor layout version of BLOCPLAN briefly in the following.

The basic idea in BLOCPLAN is to treat the 3D layout problem as several 2D layout problems. The number of 2D problems is equal to the number of floors in the layout. In order to develop a multifloor layout, BLOCPLAN first asks the user to provide the number of floors. Next, it requires a list of departments in each. The user can provide this list or let BLOCPLAN automatically generate it. Once generated, the area difference factor (ADF) is calculated for the multistory layout using the following formula:

$$\text{ADF} = \max\left[\frac{(A_{\max} - \overline{A})}{\overline{A}}, \frac{(\overline{A} - A_{\min})}{\overline{A}} \right] \tag{5.5}$$

where
  $\overline{A}$ is the mean area for each floor (obtained by dividing the total space required by all departments in all floors)
  $A_{\max}$ and $A_{\min}$ are the areas of the floors requiring the most and least space

Of course, $A_{\max}$ and $A_{\min}$ depend on the list of departments assigned to each floor and the space required by them. ADF provides a measure of the maximum deviation from the optimal floor area provided by $\overline{A}$. In addition, BLOCPLAN also provides another measure of effectiveness called the partition score. The partition score provides a measure of how well the relationship codes have been satisfied. It assumes that departments on different floors are not adjacent and computes the score using formula (5.5). R_{ij} and D_{ij} in the formula are defined in Section 5.5.2.

$$\text{PS} = \frac{\sum_{i=1}^{n-1} \sum_{j=i+1}^{n} R_{ij} D_{ij}}{\sum_{i=1}^{n-1} \sum_{j=i+1}^{n} R_{ij}} \tag{5.6}$$

Department pairs with negative numeric value are not included in the denominator of formula (5.5). The user has the option of saving the allocation of departments or changing it. After a satisfactory allocation of departments is saved, BLOCPLAN treats each floor as a separate layout problem and saves the relationship code as well as the list of departments on different floors in a separate data file. The user can then run each saved problem as a single-story layout problem as discussed previously and aggregate the resulting layouts to obtain a multistory layout.

There are other software packages in addition to BLOCPLAN for the multistory layout problem and are all extensions of CRAFT. SPACECRAFT is a 3D version of CRAFT and determines the layout of departments in a multifloor building. It has many additional features as the 3D problem requires additional considerations. SPACECRAFT is similar to another 3D version of CRAFT— CRAFT 3D. Details of the latter are not available in Cinar (1975) or any other source according to Bozer et al. (1994).

MULTIPLE (Bozer et al., 1994) and HOPE-M are perhaps the best multiple-floor layout software packages available today. They are not only fast but also produce good-quality solutions. They use space filling curves to represent a layout. The advantage of using space filling curves is that unequal area departments can be easily represented and their shapes can be controlled. These curves allow the positions of departments to be easily exchanged in a manner more powerful than two-way and three-way exchanges. Space filling curves have also been used to solve the order picking and traveling salesman problems, and we refer the reader to Bozer et al. (1994) and Kochhar and Heragu (1998) for details of its application to the multistory layout problem.

5.8   SUMMARY

Various classes of heuristic algorithms and two algorithms belonging to the construction and improvement classes were also presented in this chapter. Many of the heuristic algorithms presented, e.g., 2-opt, greedy 2-opt, and 3-opt exchange algorithms, can be used to solve other complex problems such as the traveling salesman problem and vehicle routing problem. Computer programs for several of the algorithms discussed in this chapter are available at http://sundere.okstate.edu/downloadable-software-programs-and-data-files.

5.9   REVIEW QUESTIONS AND EXERCISES

1. What are the major differences between optimal and heuristic algorithms? Give three examples of each. (Hint: You may want to refer to your favorite Operations Research textbook for a list of algorithms you have previously studied.)
2. If there is an optimal as well as a heuristic algorithm for solving a problem, which would you prefer? Explain your answer.
3. Why is the solution to a layout problem developed using a heuristic or optimal algorithm usually not implementable without some modifications?
4. If the solution produced by an algorithm is not directly implementable, is it worthwhile studying that algorithm? Why or Why not?
5. Explain the advantages and disadvantages of construction, improvement, and hybrid algorithms.
6. Fastfood Inc. prepares and sells five varieties of fast food items—A, B, C, D, and E. Each item has to go through one or more of four stages of preparation (e.g., adding tomato, lettuce, and other ingredients; heating prepared item in microwave; and so on). The four stages must be positioned along a straight line in an assembly-line fashion to maximize space utilization. The demand for each item in a typical week is given in Table 5.9 along with the sequence of stages that each item has to undergo. Develop a flow matrix that shows the number of *transactions* between each pair of stages. Then, determine a single-row layout using the MST algorithm.
7. In Exercise 6, assume that the local building codes require each set of adjacent stages to be separated by a distance equal to half of each stage's length. Solve the problem again under the new restrictions using the MST algorithm.
8. A company makes hundreds of products, but six of them—A, B, C, D, E, and F—are considered class A items because they account for 90% of the company's revenues. Each product undergoes one to five stages of production at five workstations. The five workstations must be positioned along a straight line in an assembly-line fashion to maximize space utilization. The demand for each product in a typical week is given in Table 5.10 along with the sequence of workstations that each product must visit. Develop a flow matrix that shows the number of trips between each pair of workstations. Then, determine a single-row layout using the MST algorithm.

TABLE 5.9

Stages of Preparation and Weekly Demand for Five Varieties of Fast Food

Fast Food	Sequence of Stages	Demand (in Thousands)
A	1–2–4–3	10
B	1–3–4	5
C	1–4–3–2	20
D	1–2–4	8
E	4–1–2–3	15

TABLE 5.10

Sequence of Workstations and Weekly Demand for Six Products

Fast Food	Sequence of Stages	Demand (in Thousands)
A	1–2–5–3	100
B	1–3–5	50
C	1–4–3–2	120
D	1–2–5	18
E	4–1–2	8
F	1–2–3	150

TABLE 5.11

Workstation Lengths

Workstation	Dimensions (in Meters)
1	20×15
2	10×10
3	20×6
4	6×20
5	13×13

9. Consider Exercise 8. Using the workstation dimensions provided in Table 5.11 and assuming that the workstations are oriented so that their longer sides are parallel to the aisle and that the clearance between each pair of adjacent workstations is equal to 3 meters, use the MST algorithm to determine a single-row layout.

10. The check processing department of a regional bank has five square offices each requiring 49 square meters of space. These five offices are to be housed in the third floor of the bank's office at State College, PA. Stairs located in one corner of the building occupy 49 square meters of space (see Figure 5.48). The floor area available for the offices (and stairways) is 14×21 meters. Given the matrix in Figure 5.49, which shows the traffic information between each pair of departments, develop a layout using the greedy and steepest descent versions of the 2-opt and 3-opt algorithms. Which algorithm produces the best solution? You may use any available software for this purpose.

11. In Exercise 10, assume that the third floor where the department is located is the uppermost floor. Further assume that the bank wants to renovate the building and install an elevator instead of stairways. The bank has the freedom to place the elevator at any of the six locations shown in Figure 5.50. (To satisfy local fire codes, the stairways will be located outside the building as shown.) Using the traffic flow information between the lower floors and each office in the third floor given in Table 5.12, develop a layout using the greedy and steepest descent versions of the 2-opt and 3-opt algorithms. Which algorithm produces the best solution? You may use any available software for this purpose.

12. Repeat Exercise 11 using the traffic data in Table 5.13.

13. Are the two graphs in Figure 5.51 planar? Explain your answer.

14. Write the PAG and corresponding dual for the two layouts in Figure 5.52. Be sure to represent the external boundary using a node.

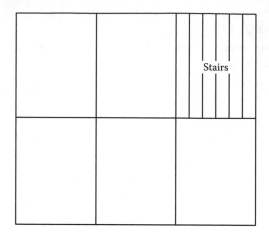

FIGURE 5.48   Layout of check processing department of a regional bank.

	1	2	3	4	5
1	—	5	10	2	0
2	10	—	4	2	4
3	5	6	—	3	6
4	2	3	6	—	4
5	10	2	8	4	—

FIGURE 5.49   Flow matrix for Exercise 10.

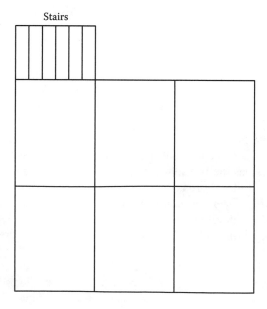

FIGURE 5.50   Third floor layout of a check processing department.

TABLE 5.12
Flow Matrix for Exercise 11

Traffic between Lower Floors and Offices

Office	Traffic
1	10
2	4
3	1
4	0
5	6

TABLE 5.13
Flow Matrix for Exercise 12

Traffic between Lower Floors and Offices

Office	Traffic
1	45
2	20
3	15
4	50
5	30

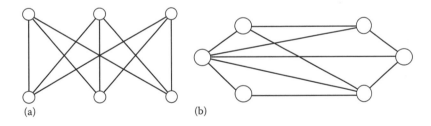

(a)   (b)

FIGURE 5.51 Two graphs. (a) Bi-particle graph and (b) graph with six nodes.

1	2	3	4
5	6	7	8

(a)

(b)

FIGURE 5.52 Two layouts. (a) Two-row layout and (b) multi-row layout.

15. Are the graphs in Figure 5.51 or the PAGs you constructed in Exercise 14 maximally planar? Explain your answer. How many possible adjacencies exist in each?

16. Consider the single-row layout problem and the flow matrix provided in Example 5.1.
    (a) Identify a maximal PAG.
    (b) Draw the dual of the maximal PAG identified in (b).
    (c) Based on the dual and using the department lengths provided in the example, develop a detailed layout by trial and error. Assume that each department has a width of 4 units.
    (d) How does it compare with the solution provided in Example 5.1? Explain.

17. Consider the multirow layout problem and the flow matrix provided in Example 5.3.
    (a) Identify a maximal PAG.
    (b) Draw the dual of the maximal PAG identified in (b).
    (c) Based on the dual and using the department lengths provided in the example, develop a detailed layout by trial and error.
    (d) How does it compare with the 2-opt solution provided in Example 5.3? Explain.

18. Consider the relationship chart in Figure 5.53.
    (a) Convert the relationship codes to numerical flow values by assigning A = 100, E = 80, I = 60, O = 40, U = 20, and X = 0 and generate the *flow between departments* matrix.
    (b) Use the PAG heuristic to identify a maximal PAG.
    (c) Draw the dual of the maximal PAG identified in (b).
    (d) Based on the dual and using the department areas provided in parentheses in the aforementioned relationship chart, develop a detailed layout by trial and error.

19. There are five work centers in one section of a manufacturing company in the northeast heat treatment (HT), painting (P), welding (W), general machining area (MA), and cooling (C). The interaction between the following workstation pairs—(MA, W), (C, W)—is best characterized by an E relationship, between (MA, P) and (C, P) by an I relationship, and between (MA, C) by

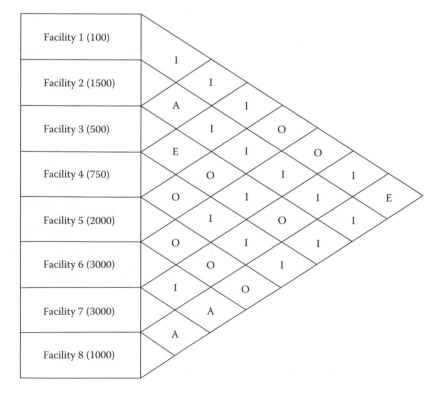

FIGURE 5.53 Relationship chart for eight departments.

TABLE 5.14
Area Information for Five Work Centers in Exercise 19

Work Center	Area (in Square Meters)
HT	6000
P	2000
W	2000
MA	4000
C	4000

an O relationship. All other pairs except (HT, C) and (P, W) have a U relationship. Whereas HT and C work centers must be adjacent due to technological considerations, P and W must not be adjacent due to a relationship chart for the five work centers. Construct a relationship chart for the five work centers. Use the numeric values of 64, 32, 16, 8, 4, and 2 for A, E, I, O, U, and X and the department areas shown in Table 5.14 to find a solution to the problem using the PAG heuristic.

20. Explain the difference between 2-opt and greedy 2-opt algorithms. Why is the latter called "greedy"?

21. In Example 5.1, a solution to the single-row layout problem was generated using MST. Apply 2-opt algorithm to improve the solution (if possible). What is the new OFV?

22. In Example 5.1, a solution to the single-row layout problem was generated using MST. Apply greedy 2-opt algorithm to improve the solution (if possible). What is the new OFV?

23. Calculate the number of layouts evaluated in each pass of step 2 in the 2-opt and 3-opt algorithms for a layout problem with six departments. What is the number of layouts evaluated (again, in each pass of step 2 for both algorithms) for a problem with 10 departments. Explain why there is such a large increase in the number of layouts evaluated for the two problem sizes.

24. Solve Example 5.3 using 3-opt algorithm (by hand).

25. Draw a flow chart to illustrate how CRAFT works.

26. Consider the flow and clearance matrices in Figure 5.54 that show the number of trips and clearance between each pair of five machines, respectively. Assuming the machine lengths are equal, develop a layout using the CRAFT software.

27. Consider the flow and clearance matrices in Figure 5.55 that show the number of trips and clearance between each pair of five machines, respectively. Assuming the machine lengths are equal, develop a layout using the CRAFT software.

Flow matrix

Machine

$[f_{ij}] =$ Machine

	1	2	3	4	5
1	—	16	16	8	12
2	16	—	14	3	1
3	16	14	—	8	9
4	8	3	8	—	20
5	12	1	9	20	—

Clearance matrix

Machine

$[d_{ij}] =$ Machine

	1	2	3	4	5
1	—	1	1	2	1
2	1	—	2	2	1
3	1	2	—	1	2
4	2	2	1	—	1
5	1	1	2	1	—

FIGURE 5.54 Flow and clearance matrices for Exercise 26.

Flow between matrix

Machine

$$[f_{ij}] = \begin{array}{c} \text{Machine} \end{array} \begin{array}{c|ccccc} & 1 & 2 & 3 & 4 & 5 \\ \hline 1 & - & 18 & 16 & 8 & 12 \\ 2 & 16 & - & 1 & 3 & 10 \\ 3 & 16 & 14 & - & 12 & 19 \\ 4 & 8 & 3 & 8 & - & 20 \\ 5 & 12 & 1 & 9 & 12 & - \end{array}$$

Clearance matrix

Machine

$$[d_{ij}] = \begin{array}{c} \text{Machine} \end{array} \begin{array}{c|ccccc} & 1 & 2 & 3 & 4 & 5 \\ \hline 1 & - & 1 & 2 & 2 & 2 \\ 2 & 1 & - & 2 & 2 & 2 \\ 3 & 1 & 2 & - & 2 & 2 \\ 4 & 2 & 2 & 1 & - & 2 \\ 5 & 1 & 1 & 2 & 1 & - \end{array}$$

FIGURE 5.55 Flow and clearance matrices for Exercise 27.

$$\begin{array}{c|ccccc} & B & D & L & M & SG \\ \hline B & - & 20 & 4 & 2 & 16 \\ D & 20 & - & 12 & 12 & 12 \\ L & 4 & 12 & - & 4 & 8 \\ M & 2 & 12 & & - & 5 \\ SG & 16 & 12 & 8 & 5 & - \end{array}$$

FIGURE 5.56 Flow matrix for Exercise 28.

28. Five machines—broach (B), drilling (D), lathe (L) milling (M), and surface grinding (SG)—are to be arranged linearly along two rows in a factory. The factory manufactures various types of parts, each having a different processing sequence. Hence, a process layout is preferred. The number of unit loads (in thousands) transported between each pair of machines in 1 year is given in the matrix in Figure 5.56. For access, safety, space for material handling carrier, and other reasons, each machine must be separated from its immediate, horizontal and vertical neighbors by a minimum of 3 and 15 meters, respectively. The machine dimensions are given in Table 5.15. Assume the clearance between machines in the horizontal and vertical dimensions is zero. Select an initial layout randomly. Develop a solution using CRAFT.

29. In Example 5.1, a solution to the single-row layout problem was generated using MST. Apply the CRAFT algorithm to improve the solution (if possible). What is the new OFV? Compare it to the one obtained in Example 5.1.

30. Evaluate any three layout software packages to which you have access.

TABLE 5.15
Machine Dimensions for Exercise 28

Machine	Length × Width
B	3×1
D	5×2
L	3×2
M	3×2
SG	2×2

31. Develop a layout for the factory layout problem discussed in Example 5.2 using the automatic search algorithm in BLOCPLAN. Is the solution obtained different from the one obtained in Example 5.2. Why or why not?
32. Consider the flow matrix in Figure 5.17. Develop a layout using the improvement algorithm in BLOCPLAN. Keep performing the exchanges manually until you get the optimal solution shown in Figure 5.19.
33. Consider the flow matrix shown in Figure 5.17. Prepare an input file so that the layout problem may be solved via CRAFT. Generate a layout using CRAFT.
34. Use BLOCPLAN to develop a near-optimal layout for the problem described in Exercise 26.
35. Use BLOCPLAN to develop a near-optimal layout for the problem described in Exercise 27.
36. Consider the real estate office layout described in Exercise 8 of Chapter 3. Develop a near-optimal layout using (1) BLOCPLAN, (2) VIP-PLANOPT, and (3) Layout-iQ.
37. Repeat Exercise 36 for the data provided in Exercise 9 in Chapter 3.
38. Repeat Exercise 36 for the data provided in the following Exercises in Chapter 4.
    (a) Exercise 8
    (b) Exercise 9
    (c) Exercise10
    (d) Exercise 11
    Compare the BLOCPLAN, VIP-PLANOPT, and Layout-iQ solution with the SLP solutions you obtained in Chapter 4.
39. Consider the *product* routings in Exercise 6. Assuming the four stages of preparation require the same amount of space, develop a *good* layout (not necessarily a single-row layout) using BLOCPLAN, Layout-iQ, and VIP-PLANOPT. (Do not calculate the flow values. Instead, use the *production routing* feature in BLOCPLAN for this question.)
40. Consider the *product* routings in Exercise 8. Assuming the five workstations require the same amount of space, develop a *good* layout (not necessarily a single-row layout) using BLOCPLAN, Layout-iQ, and VIP-PLANOPT. (Do not calculate the flow values. Instead, use the *production routing* feature in BLOCPLAN for this question.)
41. Repeat Exercise 40 for the data provided in the following Exercises in Chapter 7.
    (a) Exercise 9
    (b) Exercise 10
    (c) Exercise 11
    (d) Exercise 19
    (e) Exercise 26
    (f) Exercise 27

6 Group Technology and Facilities Layout

6.1 INTRODUCTION

Group technology (GT) is a management philosophy that attempts to group products with similar design or manufacturing characteristics or both (Mitrofanov, 1983). Cellular manufacturing (CM) can be defined as an application of GT that involves grouping machines based on the parts manufactured by them. The main objective of CM is to identify machine cells and part families simultaneously and to allocate part families to machine cells in a way that minimizes the intercellular movement of parts. To successfully implement the CM concept, analysts must develop the layout of machines within the cells so as to minimize inter- and intracellular material handling costs. CM is a relatively recent concept and has been applied successfully in many manufacturing environments and can achieve significant benefits (Black, 1983). Companies surveyed in Wemmerlov and Hyer (1989) have witnessed the following results:

- Setup time reduction
- Work-in-process inventory reduction
- Material handling cost reduction
- Direct and indirect labor cost reduction
- Improvement in quality
- Improvement in material flow
- Improvement in machine utilization
- Improvement in space utilization
- Improvement in employee morale

The main difference between a traditional job shop environment and a CM environment is in the grouping and layout of machines (Burbidge, 1992). In a job shop environment, machines are typically grouped on the basis of their functional similarities (see Figure 6.1). On the other hand, in a CM environment, machines are grouped into cells, with each cell dedicated to the manufacture of a specific part family (see Figure 6.2). Typically, the machines in each cell have dissimilar functions. The CM arrangement allows easier control of a cellular manufacturing system (CMS).

In addition to minimizing the intercellular movement of parts and part families, a CMS designer must satisfy a number of other constraints (Singh, 1993). For example, part families must be allocated to machine cells so that the available capacity of machines in each cell is not exceeded, safety and technological requirements pertaining to the location of equipment are met, and the size of a cell and number of cells do not exceed threshold levels set by the analyst.

Most of the research efforts in GT and CMS design ignore some or all of these constraints and focus only on the identification of machine cells and corresponding part families. CMS designers begin with a part–machine processing indicator matrix such as the one in Figure 6.3. Typically, the matrix consists of 0 and 1 entries only. A 1 entry in row i and column j indicates that the part corresponding to the ith row is processed by the machine listed in the jth column. A 0 entry means that the part is not processed by the corresponding machine, but a 0 is usually replaced by a blank in the matrix, as in Figure 6.3.

Analysts attempt to rearrange the rows and columns of the part–machine processing indicator matrix to get a block diagonal form as shown in Figure 6.4. The clusters of 1s around the diagonal

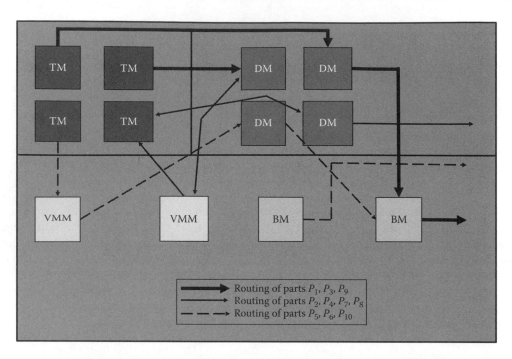

FIGURE 6.1 Arrangement of cells in a job shop environment. (BM, broaching machine; DM, drilling machine; TM, turning machine; VMM, vertical milling machine.)

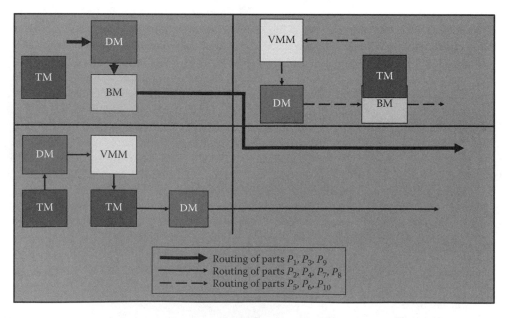

FIGURE 6.2 Arrangement of cells in a cellular manufacturing system. (BM, broaching machine; DM, drilling machine; TM, turning machine; VMM, vertical milling machine.)

$[a_{ij}] =$

Part \ Machine	M_1	M_2	M_3	M_4	M_5	M_6	M_7
P_1	1			1		1	
P_2		1	1		1		
P_3				1		1	
P_4		1	1				
P_5			1				1
P_6		1			1		1

FIGURE 6.3 Sample part–machine processing indicator matrix.

$[a_{ij}] =$

Part \ Machine	M_1	M_4	M_6	M_2	M_3	M_5	M_7
P_1	1	1	1				
P_3		1	1				
P_2				1	1	1	
P_4				1	1		
P_5					1		1
P_6				1		1	1

FIGURE 6.4 Rearranged processing indicator matrix.

Part \ Machine	M_1	M_4	M_6	M_2	M_3	M_5	M_7
P_1	1	1	1				
P_3		1	1				
P_2	1			1	1	1	
P_4				1	1		
P_5						1	1
P_6				1		1	1

FIGURE 6.5 Processing indicator matrix with an exceptional part.

of the matrix indicate two groups of machine cells—$MC_1 = \{M_1, M_4, M_6\}$ and $MC_2 = \{M_2, M_3, M_5, M_7\}$—and the corresponding part families are $PF_1 = \{P_1, P_3\}$ and $PF_2 = \{P_2, P_4, P_5, P_6\}$. Not all part–machine matrices can be rearranged to fit a block diagonal form; in fact, for many matrices, such a block diagonal form may not even exist. For example, assume that part P_2 requires processing on machine M_1 in addition to machines M_2, M_3, and M_5. The corresponding part–machine processing indicator matrix is shown in Figure 6.5. No rearrangement of the rows and columns will produce a block diagonal form such that no 1s lie outside the two blocks.

The rows (parts) corresponding to the 1s that lie outside the diagonal block are usually referred to as exceptional parts because, when they are removed, a block diagonal structure is easily identified. Of course, the remaining rows and columns may still have to be rearranged to uncover the block diagonal structure. In the example shown in Figure 6.5, the exceptional part is P_2. If P_2 is removed from the matrix, two mutually separable part family–machine cell combinations can easily be identified. If we want to have completely independent cells with no intercellular movement of material, exceptional parts must be subcontracted out. If subcontracting is not a viable alternative and the routing of part P_2 is not modified, this part will have to visit cells MC_1 and MC_2. There are methods that rearrange rows and columns (i.e., attempt to identify part families and machine cells) to minimize the intercellular movement of parts for such types of problems.

One could consider machines corresponding to the columns that contain exceptional elements, that is, elements outside the block diagonal structure, and treat these machines as bottleneck machines. Bottleneck machines are so named because two or more part families share them. For example, M_1 is the bottleneck (or exceptional) machine in the matrix shown in Figure 6.5. If the column corresponding to machine M_1 is removed, then two mutually separable clusters of machine cells and part families can be identified. This tells us that if an additional copy of exceptional machine M_1 is purchased and placed in the second cell in Figure 6.5, the intercellular movement of part P_2 can be eliminated. This is done only if the cost of the bottleneck machines is reasonable.

It is sometimes useful to use nonbinary values between 0 and 1 to indicate the degree of match between each part–machine pair in the processing indicator matrix $[a_{ij}]$. Compared with the binary matrix representation, in which the only information available is whether or not a part is processed on a machine, the nonbinary representation is flexible because it allows the user to capture other relationships between each part–machine pair (e.g., cost of processing a part on a machine or processing time).

The part–machine processing indicator matrix may be modified to handle other pieces of valuable information that are used in identifying machine cells and part families. For example, two additional columns can be created to indicate the number of parts to be manufactured and the batch size for each. Similarly, the sequence of machines visited by a part may also be recorded in the matrix (Heragu and Kakuturi, 1997; Nagi et al., 1990). The operations sequence for each part is a critical factor in the identification of machine cells. For example, if a part has one intermediate operation performed in a secondary cell and all the other operations in a primary cell, then two trips are required between the two cells for each batch of parts. On the other hand, if only the last operation is performed in a secondary cell, then only one trip is needed.

Operation sequence information is easily captured by defining x_{ij} as

$$x_{ij} = \begin{cases} k & \text{if part } i \text{ visits machine } j \text{ for the } k\text{th operation} \\ 0 & \text{otherwise} \end{cases}$$

where k is an integer representing the operation for which part i visits machine j. Figure 6.6 is a modification of Figure 6.5 showing the operation sequence. The representation enables us to capture routing sequence information. As long as each machine is visited by a part for any number of consecutive operations, the representation shown in Figure 6.6 is adequate. All the consecutive operations on the same machine are treated as a single operation. If the same machine is visited for two or more nonconsecutive operations, however, the aforementioned representation is not adequate. In that case, each entry in the matrix must be changed to a vector that represents all the operations for which the corresponding machine is visited by the part. For example, if part P_5 visits machine M_2 for the second and fourth operations, then this information will have to be stored in the form of a vector—(2, 4) in the fifth row and second column of the matrix. This obviously increases the memory requirements for data storage. Alternate ways of representing nonconsecutive operations on the same machine with dummy columns have been suggested (Heragu and Kakuturi, 1997). In general, it can be argued that it is advantageous to modify the part design so that the multiple

Machine

	M_1	M_4	M_6	M_2	M_3	M_5	M_7
P_1	2	3	1				
P_3		1	2				
P_2	3			1	4	2	
P_4				2	1		
P_5						1	2
P_6				1		2	3

$[a_{ij}] =$ Part

FIGURE 6.6 Processing indicator matrix showing sequence of operations.

operations required by the part on a given machine are consecutive operations. Otherwise there will be an unnecessary increase in material handling trips and setup time—the very thing the GT approach is trying to minimize!

6.2 CLUSTERING APPROACH

In a broad sense, clustering techniques attempt to uncover and display similar clusters or groups in an object–object or object–attribute data matrix (McCormick et al., 1972). As discussed in Section 6.1, the technique is to rearrange rows and columns of the input matrix—typically a binary matrix that determines whether or not a part is processed on a particular machine. The input matrix looks like the one in Figure 6.3. Clustering techniques have been applied in areas such as biology, data recognition, medicine, pattern recognition, production flow analysis, task selection, control engineering, and expert systems.

Because this chapter deals with CMSs, we define clustering techniques rather narrowly as methods concerned only with the identification of machine cells, corresponding part families, or both. They neither take into consideration the design constraints discussed in Section 6.3 nor the cost factors. Basically, they attempt to rearrange the rows and columns of a binary input matrix until a block diagonal matrix structure can be identified. All the techniques we discuss use process plan or part-routing information in forming machine cells, part families, or both. Such techniques have been classified and reclassified several times (see Heragu, 1994).

We now discuss the following five commonly used clustering algorithms:

- Rank-order clustering
- Bond energy
- Row and column masking (R&CM)
- Similarity coefficient (SC)
- Mathematical programming approach

6.2.1 RANK-ORDER CLUSTERING ALGORITHM

The rank-order clustering (ROC) algorithm determines a binary value for each row and column, rearranges the rows and columns in descending order of their binary values, and then identifies clusters (King, 1980). It is a very simple and easily implemented algorithm to determine clusters in a block diagonal format. Each cluster defines a machine group and corresponding part family. In the following steps of the ROC algorithm, m and n denote the number of machines and parts, respectively.

Step 1: Assign binary weight $BW_j = 2^{m-j}$ to each column j of the part–machine processing indicator matrix.

Step 2: Determine the decimal equivalent DE of the binary value of each row i using the formula

$$DE_i = \sum_{j=1}^{m} 2^{m-j} a_{ij}$$

Step 3: Rank the rows in decreasing order of their DE values. Break ties arbitrarily. Rearrange the rows based on this ranking. If no rearrangement is necessary, stop; otherwise go to step 4.

Step 4: For each rearranged row of the matrix, assign binary weight $BW_i = 2^{n-i}$.

Step 5: Determine the decimal equivalent of the binary value of each column j using the following formula:

$$DE_j = \sum_{i=1}^{n} 2^{n-1} a_{ij}$$

Step 6: Rank the columns in decreasing order of their DE values. Break ties arbitrarily. Rearrange the columns based on this ranking. If no rearrangement is necessary, stop; otherwise go to step 1.

One drawback of the ROC is that the final solution produced depends on the arrangement of rows and columns in the initial matrix. Nevertheless, this algorithm provides good solutions that can be further improved upon or modified to take into consideration practical constraints discussed in the next section. Improved versions of the ROC algorithm attempt to overcome some of its deficiencies. Some examples can be found in King and Nakornchai (1982) and Chandrasekharan and Rajagopalan (1986a,b). Example 6.1 illustrates the use of the original ROC algorithm.

Example 6.1

Consider the part–machine processing indicator matrix in Figure 6.3. Determine a block diagonal form by rearranging the rows and columns of the matrix using the ROC algorithm.

Solution

The binary weights and decimal equivalents of the binary values of each row are shown in Figure 6.7. The binary weights and decimal equivalents of the binary values of each column are shown in Figure 6.8. Proceeding with steps 1–3 of the ROC, we get the matrix shown in Figure 6.9. Proceeding with steps 4–6 of the ROC algorithm, we get the matrix shown in Figure 6.10. Because rearrangement is not necessary, the algorithm stops. The final matrix in Figure 6.10 shows a block diagonal structure.

		M_1	M_2	M_3	M_4	M_5	M_6	M_7		
Binary weight		64	32	16	8	4	2	1	Binary value
	P_1	1			1		1		74
	P_2		1	1		1			52
$[a_{ij}] = $	P_3				1		1		10
	P_4		1	1					48
	P_5		1					1	17
	P_6		1			1		1	37

FIGURE 6.7 Binary values and weights for the rows and columns of a part–machine processing indicator matrix.

$[a_{ij}] =$

	M_1	M_2	M_3	M_4	M_5	M_6	M_7	
Binary value	32	28	26	33	20	33	6	Binary weight
P_1	1			1		1		32
P_2		1	1		1			16
P_4		1	1					8
P_6		1			1		1	4
P_5			1				1	2
P_3				1		1		1

FIGURE 6.8 Rearrangement of the rows of the matrix in Figure 6.7 based on DE values and subsequent determination of binary weights and values.

$[a_{ij}] =$

	M_4	M_6	M_1	M_2	M_3	M_5	M_7	
Binary weight	64	32	16	8	4	2	1	Binary value
P_1	1	1	1					112
P_2				1	1	1		14
P_4				1	1			12
P_6				1		1	1	11
P_5					1		1	5
P_3	1	1						96

FIGURE 6.9 Rearrangement of the columns of the matrix in Figure 6.8 based on DE values and subsequent determination of binary weights and values.

$[a_{ij}] =$

	M_4	M_6	M_1	M_2	M_3	M_5	M_7	
Binary value	48	48	32	14	12	10	3	Binary weight
P_1	1	1	1					32
P_3	1	1						16
P_2				1	1	1		8
P_4				1	1			4
P_6				1		1	1	2
P_5							1	1

FIGURE 6.10 Final rearrangement of rows of the matrix in Figure 6.9 based on DE values.

6.2.2 Bond Energy Algorithm

The bond energy algorithm (BEA) is a heuristic that attempts to maximize the sum of the bond energies for each element $\{i, j\}$ in the part–machine processing indicator matrix $[a_{ij}]$. The bond energy is defined so that a matrix with clusters or diagonal blocks of 1s will have a larger bond energy compared with the same matrix with rows and columns arranged so that the 1s are uniformly distributed throughout the matrix. In other words, the matrix in Figure 6.4 has a higher bond energy than the one in Figure 6.3. The bond energy for element $\{i, j\}$ is given by $a_{ij}[a_{i,j+1} + a_{i,j-1} + a_{i+1,j} + a_{i-1,j}]$. The BEA attempts to maximize the sum of the bond energies over all the row and column permutations of the part–machine processing indicator matrix $[a_{ij}]$. Mathematically, it maximizes the following function:

$$\sum_{i=1}^{n}\sum_{j=1}^{m} a_{ij}[a_{i,j+1} + a_{i,j-1} + a_{i+1,j} + a_{i-1,j}]$$

By definition, $a_{0,j} = a_{n+1,j} = a_{i,0} = a_{i,m+1} = 0$. Rather than maximize the bond energy in one pass, the BEA performs two passes, one for the row and another for the column, and in each pass, it maximizes the row and column bonds, respectively. If a block diagonal structure exists, it is immediately identified. If there are one or more bottleneck elements that prevent the formation of such a block diagonal structure, then like the other techniques in this section, the BEA does not perform well. An advantage of the BEA is that the final clusters identified are relatively insensitive to the initial matrix provided. Improved versions of the BEA exist (see for example, Gongaware and Ham, 1981). The BEA steps are as follows:

Step 1: Set $i = 1$. Arbitrarily select any row and place it.

Step 2: Place each of the remaining $n - i$ rows in each of the $i + 1$ positions (i.e., above and below the previously placed i rows) and determine the row bond energy for each placement using the following formula:

$$\sum_{i=1}^{i+1}\sum_{j=1}^{m} a_{ij}(a_{i-1,j} + a_{i+1,j})$$

Select the row that increases the bond energy the most and place it in the corresponding position.

Step 3: Set $i = i + 1$. If $i < n$, go to step 2; otherwise go to step 4.

Step 4: Set $j = 1$. Arbitrarily select any column and place it.

Step 5: Place each of the remaining $m - j$ rows in each of the $j + 1$ positions (i.e., to the left and right of the previously placed j columns) and determine the column bond energy for each placement using the following formula:

$$\sum_{i=1}^{n}\sum_{j=1}^{j+1} a_{ij}(a_{i,j-1} + a_{i,j+1})$$

Select the column that increases the bond energy the most and place it in the corresponding position.

Step 6: Set $j = j + 1$. If $j < m$, go to step 5; otherwise stop.

If the initial matrix is symmetric, then only one pass for either the column or the row is necessary, because doing the other pass would yield the same solution. Thus, if the input data matrix is rearranged to be a symmetric one, the number of computational steps can be cut in half.

Example 6.2

Consider the matrix in Figure 6.11. Identify clusters of machines and corresponding parts using the BEA. For this example, blank entries in the matrix are represented as zeros.

Solution

Although a solution to the GT problem described in Figure 6.11 can be identified visually, we illustrate the first iteration of the BEA so that the reader understands how the algorithm works.

Step 1: Set $i = 1$. Arbitrarily select row 2 and place it.

Step 2: Place each of the remaining three rows in each of the two positions, i.e., above and below the previously placed row, and determine the row bond energy for each

Column	1	2	3	4
Row				
1	1	0	1	0
2	0	1	0	1
3	0	1	0	1
4	1	0	1	0

FIGURE 6.11   Part–machine processing indicator matrix for Example 6.2.

TABLE 6.1
Placement of Second Row

Row Selected	Where Placed	Row Arrangement	Row Bond Energy	Maximize Energy
1	Above row 2	1 0 1 0 0 1 0 1	0	No
1	Below row 2	0 1 0 1 1 0 1 0	0	No
3	Above row 2	0 1 0 1 0 1 0 1	4	Yes
3	Below row 2	0 1 0 1 0 1 0 1	4	Yes
4	Above row 2	1 0 1 0 0 1 0 1	0	No
4	Below row 2	0 1 0 1 1 0 1 0	0	No

1	0	1	0
1	0	1	0
0	1	0	1
0	1	0	1

FIGURE 6.12   Placement of rows in the first three iterations of BEA.

placement using the formula in step 2 of the BEA. The result is shown in Table 6.1. Placing row 3 above or below row 2 maximizes the row bond energy, so row 3 is placed below row 2 as shown in Figure 6.12. Repeating step 2 for the other rows, the matrix in Figure 6.12 is obtained. Executing the BEA for the columns, it is easy to develop the intermediate Table 6.2 and the final block diagonal matrix shown in Figure 6.13.

TABLE 6.2
Placement of the Second Column

Column Selected	Where Placed	Column Arrangement	Column Bond Energy	Maximize Energy
2	Left of column 1	0 1 0 1 1 0 1 0	4	No
2	Right of column 1	1 0 1 0 0 1 0 1	4	No
3	Left of column 1	1 1 1 1 0 0 0 0	8	Yes
3	Right of column 1	1 1 1 1 0 0 0 0	8	Yes
4	Left of column 1	0 1 0 1 1 0 1 0	4	No
4	Right of column 1	1 0 1 0 0 1 0 1	4	No

FIGURE 6.13 Clusters identified using the bond energy algorithm for the matrix in Figure 6.11.

6.2.3 ROW AND COLUMN MASKING ALGORITHM

This clustering algorithm begins from any arbitrarily selected row and masks all columns that have a 1 entry in that row. It then masks all rows with 1 entries in the masked columns. This procedure continues until it is not possible to go to new unmasked rows or columns and then a cluster of machines and corresponding part families is formed (Iri, 1968). The procedure is repeated to identify other clusters. A major disadvantage of this algorithm is that if there are one or more bottleneck machines or exceptional parts, the algorithm may provide a solution with all the machines in a cell and all the parts in a corresponding part family! For example, try applying the R&CM algorithm to the data in Figure 6.5. The algorithm can be modified to produce reasonable solutions (see Kusiak and Chow, 1988). The R&CM algorithm steps are as follows:

Step 1: Draw a horizontal line through the first row. Select any 1 entry in the matrix through which there is only one line.

Step 2: If the entry has a horizontal line, go to step 2a. If the entry has a vertical line, go to step 2b.

Step 2a: Draw a vertical line through the column in which this 1 entry appears. Go to step 3.

Step 2b: Draw a horizontal line through the row in which this 1 entry appears. Go to step 3.

Step 3: If there are any 1 entries with only one line through them, select any one and go to step 2. Repeat until there are no such entries left. Identify the corresponding machine cell and part family. Go to step 4.

Step 4: Select any row through which there is no line. If there are no such rows, STOP. Otherwise draw a horizontal line through this row, select any 1 entry in the matrix through which there is only one line, and go to step 2.

Example 6.3

Consider the matrix in Example 1. Identify clusters of machines and corresponding parts using the R&CM algorithm.

Solution

Executing steps of the R&CM algorithm, we get the horizontal and vertical lines shown in Figure 6.14. The integer to the right of each horizontal line and below each vertical line indicates the sequence in which the corresponding lines were drawn. After this first cluster is identified, the R&CM algorithm is applied again to the unmarked 1s in the matrix. We then get the horizontal and vertical lines shown in Figure 6.14. This identifies the second cluster—machine cell and part family (Figure 6.15). To maintain clarity, horizontal and vertical lines drawn to identify the first cluster are not shown in Figure 6.14. The two cells and corresponding part families are the same as that shown in Figure 6.4.

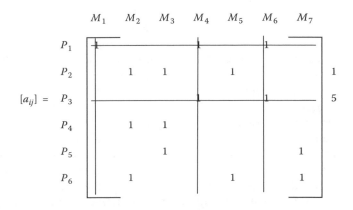

FIGURE 6.14 Identification of the first machine cell and part family.

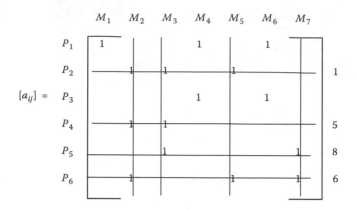

FIGURE 6.15 Identification of the second machine cell and part family.

6.2.4 SIMILARITY COEFFICIENT ALGORITHMS

SC algorithms have been derived from numerical taxonomy and attempt to measure the SC between each pair of machines or parts. Most SC algorithms use the Jaccard SC (Sneath and Sokal, 1973). For a pair of machines, the Jaccard coefficient can be defined as the number of part types that visit both machines divided by the number of part types that visit at least one machine:

$$s_{ij} = \frac{\sum_{k=1}^{n} a_{ki}a_{kj}}{\sum_{k=1}^{n} (a_{ki} + a_{kj} - a_{ki}a_{kj})} \tag{6.1}$$

$$\text{where } a_{ki} = \begin{cases} 1 & \text{if part } k \text{ requires machine } i \\ 0 & \text{otherwise} \end{cases}$$

In the preceding equation, n denotes the number of part types. Instead of a_{ki} and a_{kj}, one may use l_{ki} and l_{kj} in the aforementioned equation, which denote the number of batches of part k that visit machines i and j, respectively, and appropriately modify the denominator in Equation 6.1. This provides a more accurate estimate of the interaction between each pair of machines than Equation 6.1, but most SC algorithms still use Equation 6.1. Examples of algorithms that measure the similarity between machines are the single-linkage clustering algorithm in McAuley (1972), its modification in Rajagopalan and Batra (1975), and the algorithm in Seifoddini and Wolfe (1986).

We now discuss the basic SC algorithm for determining machine groups. Initially, each machine is placed in its own cell. Then we compute the SC values for each machine pair and place a machine pair in a new cell if its SC value is above a user-selected threshold value. Ties are broken arbitrarily. This procedure gives us a new solution with fewer cells such that one or more cells have two machines in them. We then treat each cell as a machine and determine the new set of SC values for the machine pairs, cell pairs, and machine–cell pairs. SC values for machine pairs are calculated using Equation 6.1. For cell pairs, we determine the SC values between each machine in the first cell and every other in the second and use the largest among these values as the SC value for the corresponding cell pair. Similarly, for machine–cell pairs, we determine the SC value between each machine in the cell and the single machine and use the largest of these values as the SC value for the corresponding machine–cell pair. Using a new (lower) threshold value, we decide whether or not to combine two machines or cells into one as before. As a result, we get even fewer cells. This procedure can be repeated until a satisfactory solution is obtained.

Variations of the SC algorithm are possible. For example, the SC values between cell–cell pairs or machine–cell pairs may be calculated using average SC values, rather than maximum values. The threshold values may be predetermined or we can continue the algorithm step by step by setting the threshold value equal to the largest SC value. The algorithm may be terminated when we have a satisfactory number of cells. Instead of using the number of parts to determine the SC values, we may use the number of loads or batches.

The SC algorithm adds a new machine to an existing cell if the SC value between the new machine and any existing machine in the cell exceeds a certain threshold level. An obvious disadvantage of this approach is that a machine with a high SC with just one other machine already in the cell will automatically be included in that cell even if the SC between the new machine and all other machines in the cell is very low. This is called the chaining problem. Along with other deficiencies, it is remedied in a graph theoretic method by Rajagopalan and Batra (1975), which considers some additional design constraints.

SC algorithms that measure the similarity between parts (rather than machines) are presented in Carrie (1973) and Currie (1992). The SC for a part pair $\{i, j\}$ is determined by the ratio of the number of machines visited by parts i and j to the number of machines visited by i or j. Parts are added to a part family provided there is an acceptable level of similarity between the two. Additional constraints may specify, for example, the minimum number of parts allowed per family. The same principles have been applied to develop plant layouts. For more details, see Carrie (1973).

We now illustrate the basic version of the SC algorithm with a numerical example.

Example 6.4

Consider the matrix in Figure 6.3. Identify clusters of machines and corresponding parts using the SC algorithm.

Solution

First place each machine in its own cell and determine the SC values using Equation 6.1 as shown in Table 6.3. Next, using a threshold value of 0.66, combine machines {2, 5} and {4, 6} into two separate cells. Determine the new set of SC values for the machine–cell pairs as shown in Table 6.4. To determine the SC values between machine–cell or cell–cell pairs, we make use of the SC values in Table 6.3 and choose the corresponding largest value. For example, the SC value between machine 3 and the cell consisting of machines 2 and 5 (sixth row in Table 6.4) is obtained by examining the SC values between machines 2 and 3 and between 3 and 5 and choosing the larger value of 1/2. Using a threshold value of 0.5, machines 1, 4, 6 and 2, 3, 5 are combined into two cells, respectively.

Repeating this procedure for a threshold value of 0.33, we get the SC values and machine–cell pairs shown in Table 6.5. Machines 2, 3, 5, and 7 are combined into one cell. When we attempt to repeat the aforementioned procedure for a threshold value of 0.01 (see Table 6.6), we notice that the cells cannot be combined further. Hence, we obtain a solution with two cells: cell 1 with machines 1, 4, and 6 and cell 2 with machines 2, 3, and 5.

The graph shown in Figure 6.16 (called the dendogram) is useful in visually identifying machine groups at various threshold levels. It is constructed using the SC values in Tables 6.3 through 6.6 and tells us the machine cells that can be identified at various threshold levels. For example, at a threshold level of 0.5, three cells—{2, 3, 5}, {7}, {1, 4, 6}—can be formed. The fact that cells with more than one machine—{2, 3, 5, 7} and {1, 4, 6}—can be identified just slightly above zero threshold tells us that the data are such that completely independent cells (with absolutely no interaction among them) can be found as was confirmed by the other methods, for example, the R&CM algorithm.

TABLE 6.3
SC Values for Machine Pairs

Machine Pair	SC Value	Combine into One Cell?
{1, 2}	0/4 = 0	No
{1, 3}	0/4 = 0	No
{1, 4}	1/2	No
{1, 5}	0/3 = 0	No
{1, 6}	1/2	No
{1, 7}	0/3 = 0	No
{2, 3}	2/4 = 1/2	No
{2, 4}	0/5 = 0	No
{2, 5}	2/3	Yes
{2, 6}	0/5 = 0	No
{2, 7}	1/4	No
{3, 4}	0/5 = 0	No
{3, 5}	1/4	No
{3, 6}	0/5 = 0	No
{3, 7}	1/4	No
{4, 5}	0/4 = 0	No
{4, 6}	2/2 = 1	Yes
{4, 7}	0/4 = 0	No
{5, 6}	0/4 = 0	No
{5, 7}	1/3	No
{6, 7}	0/4 = 0	No

TABLE 6.4
SC Values in the Second Iteration

Machine/Cell Pair	SC Value	Combine into One Cell?
{1, (2, 5)}	0	No
{1, (4, 6)}	1/2	Yes
{1, 3}	0	No
{1, 7}	0	No
{(2, 5), (4, 6)}	0	No
{(2, 5), 3}	1/2	Yes
{(2, 5), 7}	1/3	No
{(4, 6), 3}	0	No
{(4, 6), 7}	0	No
{3, 7}	1/4	No

TABLE 6.5
SC Values in the Third Iteration

Machine/Cell Pair	SC Value	Combine into One Cell?
{(1, 4, 6) (2, 3, 5)}	0	No
{(1, 4, 6), 7}	0	No
{(2, 3, 5), 7}	1/3	Yes

TABLE 6.6
SC Values in the Last Iteration

Machine/Cell Pair	SC Value	Combine into One Cell?
{(1, 4, 6) (2, 3, 5, 7)}	0	No

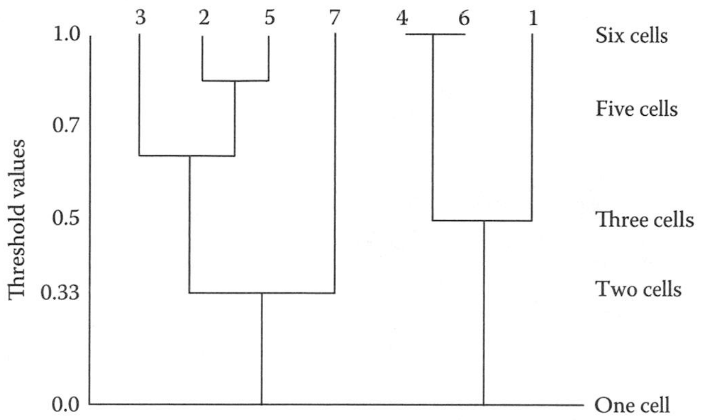

FIGURE 6.16 Dendogram for Example 6.4.

6.2.5 MATHEMATICAL PROGRAMMING APPROACH

In this section, we present a mathematical programming model to identify part families. Just as the SC algorithm can be modified to identify part families, the p-median model can also be modified to identify machine cells. Before we discuss the model, we define the SC for parts by dividing the number of machines visited by parts i and j to the number of machines visited by either i or j:

$$s_{ij} = \frac{\sum_{k=1}^{m} a_{ik} a_{jk}}{\sum_{k=1}^{m} (a_{ik} + a_{jk} - a_{ik} a_{jk})} \tag{6.2}$$

$$\text{where } a_{ik} = \begin{cases} 1 & \text{if part } i \text{ requires machine } k \\ 0 & \text{otherwise} \end{cases}$$

Whereas Equation 6.2 measures the (processing) similarity of parts, the weighted Minkowski metric (Equation 6.3) measures the dissimilarity:

$$d_{ij} = \left[\sum_{k=1}^{n} w_k |a_{ki} - a_{kj}|^r \right]^{1/r} \tag{6.3}$$

where
r is a positive integer
w_k is the weight for part k

We use the d_{ij} notation instead of s_{ij} to indicate that this is a dissimilarity coefficient. The special case where $w_k = 1$, for $k = 1, 2,..., n$, is called the Minkowski metric. It is easy to see that for the

Minkowski metric, Equation 6.3 yields an absolute Minkowski metric when $r = 1$; when $r = 2$, it yields the Euclidean metric. The absolute Minkowski metric measures the dissimilarity between part pairs. Using this metric, we can now set up a p-median model.

p-Median Model:

$$\text{Minimize} \sum_{i=1}^{n}\sum_{j=1}^{n} d_{ij}x_{ij} \tag{6.4}$$

$$\text{Subject to} \sum_{j=1}^{n} x_{ij} = 1 \quad i = 1,2,...,n \tag{6.5}$$

$$\sum_{j=1}^{n} x_{jj} = P \tag{6.6}$$

$$x_{ij} \le x_{jj} \quad i, j = 1,2,...,n \tag{6.7}$$

$$x_{ij} = 0 \text{ or } 1 \quad i, j = 1,2,...,n \tag{6.8}$$

In the model, P is a parameter that represents the number of part families desired, so the user must know it *a priori*. Of course, the user can solve the model for different values of P and use the P for which the total dissimilarity coefficient is minimized. x_{ij} is a decision variable that takes on 0 or 1 integer values only due to constraint (6.8). A 1 indicates that part i belongs to part family j and 0 indicates that it does not. Constraint (6.5) ensures that each part belongs to one part family only, and constraint (6.6) specifies the desired number of families. Constraint (6.7) guarantees that a part i is assigned to part family j only when this family is formed. In other words, if x_{jj} is 0 for some j, then this part family j is not formed, so due to constraint (6.7), no other part i can be assigned to it. The objective function (6.4) minimizes the overall dissimilarities of parts. It is illustrated in the example.

Example 6.5

Consider the machine–part matrix in Figure 6.3. Identify two part families using the p-median model.

Solution

First, the dissimilarity coefficient matrix shown in Figure 6.17 is obtained for the part–machine processing indicator matrix in Figure 6.3 using the absolute Minkowski metric.

		1	2	3	4	5	6
	1	0	6	1	5	5	6
	2	6	0	5	1	3	2
	3	1	5	0	4	4	5
$[d_{ij}] =$	4	5	1	4	0	2	3
	5	5	3	4	2	0	3
	6	6	2	5	3	3	0

FIGURE 6.17 Dissimilarity coefficient matrix.

For the matrix in Figure 6.17, we set up the following p-median model. LINGO model input and solution output are shown next.

```
Sets:
      Part/1..6/;
      ObjFn(Part,Part): D, X;
Endsets
Data:
      D = 0 6 1 5 5 6
          6 0 5 1 3 2
          1 5 0 4 4 5
          5 1 4 0 2 3
          5 3 4 2 0 3
          6 2 5 3 3 0;
      P = 2;
Enddata
      ! Objective function;
      Min= @SUM(ObjFn(i,j): D*X);
      ! Constraints;
      @FOR (Part(i): @SUM(ObjFn(i,j): X(i,j))=1);
      @SUM(ObjFn(j,j): X(j,j))=P;
      @FOR (Part(i): @FOR(Part(j): X(i,j)<=X(j,j)));
      @FOR(ObjFn(i,j): @BIN(X));
```

```
Global optimal solution found.
      Objective value:                    7.000000
      Extended solver steps:                     0
      Total solver iterations:                   7


                  Variable          Value      Reduced Cost
                  X( 1, 3)       1.000000          1.000000
                  X( 2, 4)       1.000000          1.000000
                  X( 3, 3)       1.000000          0.000000
                  X( 4, 4)       1.000000          0.000000
                  X( 5, 4)       1.000000          2.000000
                  X( 6, 4)       1.000000          3.000000
```

The aforementioned solution tells us that the two part families are PF_3 with parts 1 and 3 and PF_4 with parts 2, 4, 5, and 6. Using the matrix in Figure 6.3, it is easy to identify the corresponding machine cells MC_1 with machines 1, 4, and 6 and MC_2 with machines 2, 3, 5, and 7.

6.3 IMPLEMENTATION OF GT PRINCIPLES

In the previous sections, we introduced techniques that had the following three main objectives: (1) to identify parts with similar processing requirements, (2) to identify the set of machines that can process these parts, and (3) to dedicate the set of machines to the manufacture of the parts. The set of machines formed a machine cell and the corresponding set of parts formed a part family. The identification of machine groups and the corresponding part families was based solely on whether a machine was required for processing one or more operations on a part. This is the basis for the cluster identification problem: given a part–machine processing indicator matrix, rearrange the rows and columns to create a block diagonal form. In this process, the techniques in the previous sections did not consider some real-world design constraints that must be incorporated while designing CMSs. For example, a company may wish to arrange machines into cells and parts into families to minimize the number of trips made by the material handling carriers *between* cells. Or a company may be interested in forming machine cells so that machine utilization in each cell is above a minimum threshold and below an upper limit. None of the techniques in the previous sections directly address such issues, so researchers have modified those basic techniques to incorporate other constraints. For a survey of these revised techniques, refer to Heragu (1994).

6.4 DESIGN AND PLANNING ISSUES IN CELLULAR MANUFACTURING SYSTEMS

We now list some real-world design and planning constraints and discuss them in order of their importance. In the next section, we discuss an industrial project that takes into consideration these constraints and outline a solution approach that allows us to form machine cells (and part families) in a way that satisfies these constraints to the extent possible (Heragu and Gupta, 1994). Some important design constraints are listed as follows:

- Available machine capacity must not be exceeded.
- Safety and technological requirements must be met.
- The number of machines in a cell and number of cells must not exceed an upper bound.
- Intercellular and intracellular costs of handling material between machines must be minimized.
- Machines must be utilized effectively.
- Machine purchase and operating costs and work-in-process inventory costs as well as product cycle times must be minimized.

6.4.1 MACHINE CAPACITY

Machine capacity is more important than the other constraints, so design analysts must first ensure that adequate capacity is available to process all the parts. The number of each type of machine must be known *a priori*, or else it must be determined using analytical models. When analysts allocate machines to cells, they must verify that there is adequate capacity in each cell to completely process all the part families assigned to it.

Consider an example with these conditions: (1) Two units of machine M_5 are available and these two machines only process parts P_1, P_4, and P_5. (There is just enough capacity on these two machines to process the three parts.) (2) High-volume parts P_1 and P_4 are in part family PF_1, whereas low-volume part P_5 is in part family PF_2. (3) One unit of machine M_5 is in each machine cell MC_1 and MC_2. Given that there is just enough capacity in the two units of machine M_5 to process all the three parts, the capacity constraint in cell MC_1 may be violated because the two high-volume parts P_1 and P_4 are processed on only one unit of machine M_5. Because of the capacity limitation, it may be necessary to process parts P_1 and P_4 on the other machine M_5 in cell MC_2. This will obviously require some units of part P_1 or P_4 to visit the two cells. This example illustrates that even though the number of pieces of each type of equipment may be sufficient to process all the parts, the machine grouping may be such that there is not enough capacity to process one or more parts entirely within their corresponding cells. Thus, it is necessary to ensure that the capacity constraint in each cell is not violated when machines are allocated to cells and parts to part families.

6.4.2 SAFETY AND TECHNOLOGICAL CONSIDERATIONS

Safety or technological considerations may dictate two or more machines to be placed in the same cell, regardless of the number of parts visiting the two machines. For example, due to fire hazards, the forging and heat-treatment stations must be placed in the same cell even if relatively few parts visit both stations. On the other hand, the painting and welding stations cannot be placed in the same cell, even if a large number of parts are processed by both workstations, because sparks generated in the welding station could ignite flammable solvents in the painting station. Such factors must be considered while allocating each machine to a nonempty cell.

6.4.3   Upper Bound on the Number of Machine Cells and Size of a Cell

In many CMSs, multifunctional workers may be assigned to oversee the operations of several cells in order to improve the utilization of employees. Because the availability of such cross-trained workers is limited, it is necessary to impose an upper bound on the number of cells. There may also be an upper bound on the number of machines that can be included in a cell. In practice, managers use their experience to determine such upper bounds. There are a number of reasons why such upper limits exist. For example, for control purposes, it may be desirable to assign one operator to a machine cell. Because an operator can attend to a limited number of machines, it may be desirable to put an upper bound on the number of machines in a cell. Furthermore, floor plan dimensions may dictate the size of a cell in some GT problems. In addition to upper bounds on the number of machines in a cell, lower bounds may be imposed to ensure each operator is assigned a minimum workload.

6.4.4   Minimization of Intercellular and Intracellular Material Handling Cost

In an ideal CMS, no material flows between cells, but in the real world, it is difficult to form machine groups with no flow of material between them. Given this practical limitation, analysts try to form groups in a way that minimizes intercellular movement of parts. In fact, as discussed before, one of the main objective of most GT techniques is to do just that. An equally important goal is to minimize intracellular cost of transporting parts, which requires that a layout design be done for each cell. One possible approach is to first identify machine cells and corresponding part families and then determine the layout of machines within each cell and the layout of cells. As discussed in Chapter 11, if the maximum number of machines in a cell and maximum number of cells is 15, then optimal techniques enable us to solve the two layout problems. Instead of a sequential approach, a simultaneous solution of the grouping and layout problems would be desirable, but such problems are difficult to solve.

6.4.5   Machine Utilization

Machine utilization is an important factor that must be considered especially when equipment and process selection decisions are made. At later stages, this factor becomes less important because costs have already been incurred in purchasing equipment, and so there is not much to gain by ensuring high equipment utilization when making planning decisions. In fact, queuing models tell us that the product cycle times and work-in-process inventory increase significantly when machine utilizations are over 90%. They become much worse as they begin to exceed 95%.

6.4.6   Cost Minimization

Figure 6.5 illustrated how a bottleneck machine M_1 can prevent identification of a block diagonal form. For such problems, bottleneck machines may be duplicated in appropriate cells so that parts that belong to a part family are processed entirely within a given cell. For the data in Figure 6.5, if an additional unit of machine M_1 is placed in cell MC_1, which already consists of machines M_2, M_3, M_5, and M_7, then part P_2 may be entirely processed within cell MC_2. This is illustrated in Figure 6.18. However, the benefit of forming mutually separable clusters may be more than offset by the expense incurred as a result of purchasing additional units of bottleneck machines, or the addition of duplicate machines may violate the cell size constraint. Under such circumstances, a

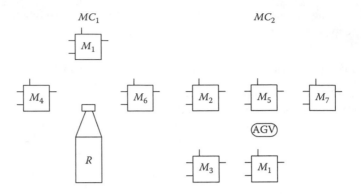

FIGURE 6.18   Duplication of machine M_1 results in the formation of two smaller cells.

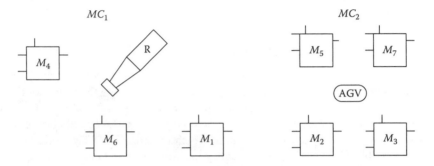

FIGURE 6.19   Physical placement of machine M_1 allows two part families to share the machine.

good alternative solution may be a physical arrangement in which bottleneck machine(s) are placed between machine cells so that the corresponding part families can share the resource. Figure 6.19 shows a physical arrangement in which an expensive bottleneck machine M_1 is shared by two part families. For illustration purposes, it is assumed in Figure 6.18 and 6.19 that the first set of machines is served by a robot and the other set by an automated guided vehicle (AGV).

Our example underscores the importance of considering machine procurement costs while designing CMSs. Others factors that need to be considered are

- Work-in-process inventory
- Machine depreciation
- Machine setup

If these costs are considered while identifying machine cells and allocating part families to machine cells, the resulting solution will not only enable better control of the manufacturing system but also minimize operating costs significantly.

6.4.7   Scheduling of Jobs in Individual Cells

Scheduling jobs in individual cells is an operational problem that should be addressed at the design stage. Because the design problems themselves are very complex, however, integrating scheduling problems is often seen as just a further complication. When treated separately, the scheduling of jobs in a CMS is relatively easier than scheduling in a traditional system because there are fewer jobs in each cell.

6.4.8 THROUGHPUT RATE MAXIMIZATION

A primary goal in any manufacturing system is to maximize production throughput rate. This goal should typically be given more importance than maximizing machine utilization, especially if the machines have already been selected. One way to maximize throughput rate is by reducing the setup time for each operation. In fact, setup times are naturally reduced as a result of implementing GT because similar parts are manufactured within a cell. If setup times can be further reduced by redesigning parts or redesigning fixtures, this will increase the throughput rate even more.

6.5 PROJECT ON MACHINE GROUPING AND LAYOUT

Mediquip Manufacturer Inc. has a process layout in which eight machines belonging to five machine types are arranged as shown in Figure 6.20. For several reasons, including recent drastic changes in product mix and volume, the current layout has led to problems such as high traffic congestion, lack of control in the system, low employee morale, and increased material handling costs. The industrial engineering (IE) team at Mediquip has correctly attributed the source of these problems to the current layout and strongly feels regrouping the machines into cells (so that each is dedicated to the manufacture of a set of parts with similar processing characteristics) will alleviate many of the problems. Because the IE team is busy with the study of a pilot run and process planning pertaining to a new product, it cannot undertake this grouping and layout project. You have been hired to do a thorough analysis of the parts, machines, processes, and material handling methods used and recommend a new machine grouping and corresponding layout. You have the following information:

- Current layout drawn to scale (see Figure 6.20)
- Routing information for 20 parts, including setup times, processing times, batch size, and annual demand (see Table 6.7)
- Material handling devices used between machines and cost of moving a unit load through unit distance on each (see Table 6.8)

The production manager at Mediquip likes to have no more than three machines in each cell because of a lack of skilled personnel. There are no special adjacency restrictions for any of the machine

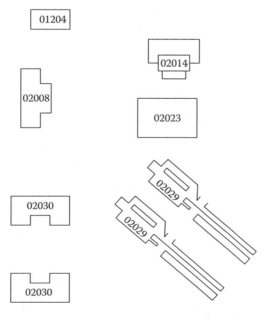

FIGURE 6.20 Current process layout for eight machines at Mediquip.

TABLE 6.7
Part Information

Part No.	Part Name	Demand (No. of Units)	Batch Size (No. of Units)	Machine Name	Setup Time per Batch	Processing Time per Batch
1	115110	220	10	01204	3.00	3.64
				02008	3.00	5.98
2	176022	500	10	01204	5.00	2.00
				02008	2.25	1.00
				02014	0.50	1.00
3	176023	500	10	01204	4.00	2.00
				02008	1.00	3.00
4	177007	500	10	02008	1.50	3.00
				02023	0.50	8.27
				02008	0.00	2.00
				02014	2.00	2.00
5	201052	220	10	02008	1.00	2.00
				02014	1.00	6.86
				02023	1.00	10.00
6	202102-501	120	10	02030	0.50	84.58
				02023	1.00	26.78
7	202103	700	10	02008	1.00	1.00
				02023	1.00	1.00
8	203009	900	10	02030	0.75	1.25
				02014	0.50	1.00
				02029	0.00	0.31
9	209001	200	10	02029	8.00	65.00
				02014	0.00	2.23
10	210001	250	10	02030	0.50	5.00
				02008	1.00	7.00
11	217002-2	400	10	02029	8.00	55.00
				02014	1.00	2.67
12	236235	300	10	01204	2.89	8.68
				02008	1.00	8.00
13	270071-501	250	10	02029	0.00	8.55
				02030	0.00	19.37
				02014	1.50	2.80
14	270080-501	600	10	02029	0.00	7.53
				02030	1.00	3.00
				02014	0.75	1.00
15	280011	450	10	02030	1.00	2.91
				02030	1.50	3.83
16	280024	500	10	01204	4.00	3.64
				02008	0.75	3.00
17	30130	400	10	02030	0.00	30.57
				02023	0.50	2.00
18	324008	300	10	02029	0.00	12.01
				02030	0.00	7.59
				02023	0.75	14.50

(Continued)

TABLE 6.7 (*Continued*)
Part Information

Part No.	Part Name	Demand (No. of Units)	Batch Size (No. of Units)	Machine Name	Setup Time per Batch	Processing Time per Batch
19	325003	10	10	02029	0.00	23.25
				02030	0.00	23.63
				02008	0.00	42.67
20	328008	250	10	02029	0.00	7.28
				02030	0.00	23.00
				02023	1.00	2.00

TABLE 6.8
Material Handling Devices Used[a]

From Machine(s)	To Machine(s)	Device Used
1	All others	Truck
2	All others	Manually by cart
3	All others	AGV
4	1, 2	Manually by cart
4	5	Truck
4	3, 4, 6	AGV
5	All others	Manually by cart
6	All others	Manually by cart
7	1, 3, 5, 6	AGV
7	2	Manually by cart
7	4	Truck
8	1, 2, 6	Manually by cart
8	3, 4, 5	Truck
9	1, 2	Manually by cart
9	3, 5, 6	Truck
9	4	AGV

[a] Although the exact cost of handling a unit load through unit distance on each type of material handling device is difficult to obtain, it is known that the AGV is three times as expensive (per load per unit distance) as transporting manually via carts and transporting via trucks is twice as expensive as transporting manually via carts, for this example.

pairs. If more than one unit of a machine type is available for processing, each part that requires this machine type is assumed to visit each of the units in equal proportion. (For example, if four units of machine type 2 are available and part types 1 and 4 are the only ones that require processing on machine type 2, then the first, second, third, and fourth units of machine type 2 will each process 25% of part types 1 and 4.) Determine a machine grouping and layout using the GTLAYPC.EXE program available at http://sundere.okstate.edu/downloadable-software-programs-and-data-files.

The first step in solving the GT and layout problem is to determine whether there is sufficient capacity to process the projected part mix and volume. From the given annual demand and batch size, we can easily determine the number of batches of each part that are to be processed annually by dividing the demand by batch size (see the fifth column in Table 6.9). Then using the setup and processing times per batch (given in Tables 6.7 and 6.9), we can calculate the time required on each machine type for processing each of the 20 part types by multiplying the number of batches by the

TABLE 6.9
Machine Capacity Calculations

Part No.	Machine Name	Setup Time/Batch	Processing Time/Batch	No. of Batches	01204	02008	02014	02023	02029	02030
1	01204	3.00	3.64	22	146					
	02008	3.00	5.98			198				
2	01204	5.00	2.00	50	350					
	02008	2.25	1.00			163				
	02014	0.50	1.00				350			
3	01204	4.00	2.00	50	300					
	02008	1.00	3.00			200				
4	02008	1.50	3.00	50		225				
	02023	0.50	8.27					439		
	02008	0.00	2.00			100				
	02014	2.00	2.00				200			
5	02008	1.00	2.00	22		66				
	02014	1.00	6.86				173			
	02023	1.00	10.00					242		
6	02030	0.50	84.58	12						1021
	02023	1.00	26.78					333		
7	02008	1.00	1.00	70		140				
	02023	1.00	1.00					140		
8	02030	0.75	1.25	90						180
	02014	0.50	1.00				135			
	02029	0.00	0.31						28	
9	02029	8.00	65.00	20					1460	
	02014	0.00	2.23				45			
10	02030	0.50	5.00	25						138
	02008	1.00	7.00			200				
11	02029	8.00	55.00	40					2520	
	02014	1.00	2.67				147			

(Continued)

TABLE 6.9 (Continued)
Machine Capacity Calculations

Part No.	Machine Name	Setup Time/Batch	Processing Time/Batch	No. of Batches	01204	02008	02014	02023	02029	02030
12	01204	2.89	8.68	30	347					
	02008	1.00	8.00			270				
13	02029	0.00	8.55	25	214				214	
	02030	0.00	19.37							484
14	02014	1.50	2.80				108			
	02029	0.00	7.53	60					452	240
	2030	1.00	3.00							
	02014	0.75	1.00				105			
15	02030	1.00	2.91	45						176
	02030	1.50	3.83							240
16	01204	4.00	3.64	50	382					
	02008	0.75	3.00			188				
17	02030	0.00	30.57	40						100
	02023	0.50	2.00					100		
18	02029	0.00	12.01	30					360	
	02030	0.00	7.59							228
	02023	0.75	14.50					458		
19	02029	0.00	23.25	1					23	
	02030	0.00	23.63							24
	02008	0.00	42.67			43				
20	02029	0.00	7.28	25					182	
	02030	0.00	23.00							575
	02023	1.00	2.00					75		
	Necessary machine hours				1739	1791	1262	1786	5239	3405
	Number of machines required				1.00	1.00	1.00	1.00	3.00	2.00

sum of the setup and processing times per batch (see the last six columns of Table 6.9). Summing the values in each of the last six columns gives the hours required on each machine type. Then, dividing these figures by the time available (assuming 40 hours per week operation for 50 weeks at 90% efficiency, or 1800 hours) provides an estimate of the number of machines of each type required. If it is anticipated that machine utilization will be less than 90% (due to part and material handling equipment or operator unavailability, machine downtime, etc.), then the appropriate figure must be used. For the 90% assumption, we find that three units of machine type 02029, two units of machine type 02030, and one unit of the remaining are required. We know from Figure 6.20, however, that the total number of machines currently available is eight (two units of machine types 02029 and 02030 and one unit of the other types). Hence, an additional copy of machine type 02029 should be purchased.

The next step is to create a data file as discussed in Section 3.5.* For machine types 02029 and 02030, because there is more than one unit available for processing, each part that requires this machine type is assumed to visit each of the units in equal proportion. The columns corresponding to machines 6, 7, and 9 in the input data file in Figure 6.21 are indeed the columns corresponding to the extra units of machine types 02029 and 02030. Moreover, columns 6 and 7 are identical to column 5 and column 9 is identical to 8. Columns 1, 2, 3, 4, 5, and 8 correspond to machines 01204, 02008, 02014, 02023, 02029, and 02030, respectively. Thus, each machine has its own column.

In Section 6.1, we mentioned that whenever a part visits a machine for multiple, nonconsecutive operations, the corresponding entry in the part–machine processing indicator matrix has to be modified as vector. This increases storage and memory requirements, however, because we are storing a vector inside a matrix. Hence, we use the alternate approach suggested in Heragu and Kakuturi (1997). Because part type 4 visits machine type 2 for the first and third operations, we add a dummy column corresponding to machine type 2 (see the third to last column in the part–machine processing indicator matrix in the input data file in Figure 6.21) and indicate in that column that the third operation is done on machine type 2. This dummy column is used only to calculate the flow between machines and not anywhere else. Not only does this approach allow us to capture nonconsecutive operation information, but it does so with minimal changes to the matrix and memory requirement. (Note that adding a column to a matrix is more efficient than storing a vector inside a matrix.) Also, as mentioned in Section 6.1, it is better to modify the design or routing so that parts do not visit the same machine(s) for nonconsecutive operations because that significantly decreases setup times. Although part type 15 requires two operations on machine type 8, we treat the two as a single operation. This does not affect the flow matrix as there is no material movement; the part stays on the same machine for the two operations.

The second matrix in Figure 6.21 is called the fractional processing matrix and it shows the fraction of each part type visiting each unit of each machine type. A 0 (or blank) entry indicates that the part does not visit the machine and a 1 indicates that 100% of this part type is processed on the corresponding machine. The third matrix, the material handling cost matrix, shows the cost of handling a unit load through unit distance between each machine pair. These data are obtained from Table 6.8 and show relative costs because actual costs are not available. The fourth matrix is the relationship indicator matrix and tells us whether or not there are special adjacency restrictions for machine pairs. A 1 (3) in this matrix indicates that the corresponding must (must not) be placed in the same cell. A 0 indicates that there is no special adjacency relationship, and the machine pair placement is to be determined by the layout algorithm. Notice that there is no special adjacency relationship in this problem.

The maximum machines allowed in a cell are assumed to be three. The machine dimensions obtained from a CAD drawing of the existing layout were modified to include clearances of 2 meters

---

* The input data file is modified in order to maintain clarity. The actual way the input data file is to be prepared is explained in the manual accompanying the GTLAYPC.EXE program.

on each side of every machine to allow for work space, storage of kanbans, tools, and so on. The revised machine dimensions are provided in the last lines of the input data file in Figure 6.21.

Before we discuss a solution to the problem, we must briefly describe the GTLAYPC.EXE program, which identifies not only machine cells and corresponding part families but also their layout and the layout of machines within each cell. To an extent, it attempts to integrate the machine grouping and layout problems. The program is based on a three-stage approach and is illustrated in Figure 6.22.

```
Number of machines

9

Number of parts

20

Number of dummy machines

1

Machine names

    1       01204
    2       02008
    3       02014
    4       02023
    5       02029A
    6       02029B
    7       02029C
    8       02030A
    9       02030B

Part names
```

Part-machine processing indicator matrix

Part	1	2	3	4	5	6	7	8	9	2		
1	1	2									220	10
2	1	2	3								500	10
3	1	2									500	10
4		1	4	2						3	500	10
5		1	2	3							220	10
6				2				1	1		120	10
7		1		2							700	10
8			2		3	3	3	1	1		900	10
9			2		1	1	1				200	10
10		2						1	1		250	10
11			2		1	1	1				400	10
12	1	2									300	10
13			3		1	1	1	2	2		250	10
14			3		1	1	1	2	2		600	10
15								1	1		450	10
16	1	2									500	10
17				2				1	1		400	10
18				3	1	1	1	2	2		300	10
19		2			1	1	1	2	2		10	10
20				3	1	1	1	2	2		250	10

	Part names
1	115110
2	176022
3	176023
4	177007
5	201052
6	202102-501
7	202103
8	203009
9	209001
10	210001
11	217002-3
12	236235
13	270071-501
14	270080-501
15	280011
16	280024
17	30130
18	324008
19	325003
20	328008

FIGURE 6.21   Input data file for Mediquip Manufacturer Inc.   *(Continued)*

Fractional processing matrix

	1	2	3	4	5	6	7	8	9	2
1	1	1								
2	1	1	1							
3	1	1								
4		1	1	1						1
5		1	1	1						
6			1					0.5	0.5	
7		1		1						
8			1		0.33	0.33	0.33	0.5	0.5	
9			1		0.33	0.33	0.33			
10		1						0.5	0.5	
11			1		0.33	0.33	0.33			
12	1	1								
13			1		0.33	0.33	0.33	0.5	0.5	
14			1		0.33	0.33	0.33	0.5	0.5	
15								0.5	0.5	
16	1	1								
17				1				0.5	0.5	
18			1		0.33	0.33	0.33	0.5	0.5	
19		1			0.33	0.33	0.33	0.5	0.5	
20			1		0.33	0.33	0.33	0.5	0.5	

Material handling cost matrix

	1	2	3	4	5	6	7	8	9
1	0	2	2	2	2	2	2	2	2
2	1	0	1	1	1	1	1	1	1
3	3	3	0	3	3	3	3	3	3
4	1	1	3	0	2	2	2	3	3
5	1	1	1	1	0	1	1	1	1
6	1	1	1	1	1	0	1	1	1
7	3	1	3	2	3	3	0	3	3
8	1	1	2	2	2	2	2	0	1
9	1	1	2	3	2	2	2	2	0

Machine relationship matrix

	1	2	3	4	5	6	7	8	9
1	0	0	0	0	0	0	0	0	0
2	0	0	0	0	0	0	0	0	0
3	0	0	0	0	0	0	0	0	0
4	0	0	0	0	0	0	0	0	0
5	0	0	0	0	0	0	0	0	0
6	0	0	0	0	0	0	0	0	0
7	0	0	0	0	0	0	0	0	0
8	0	0	0	0	0	0	0	0	0
9	0	0	0	0	0	0	0	0	0

Maximum machines in any cell

3

Machine lengths

20   20   15   25   45   45   45   20   20

Machine widths

10   25   10   20   10   10   10   15   15

FIGURE 6.21 (*Continued*)   Input data file for Mediquip Manufacturer Inc.

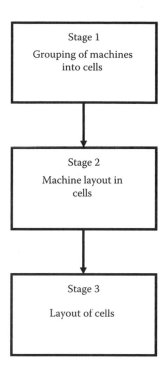

FIGURE 6.22   Flow chart for three-stage approach.

In the first stage, machines are grouped into cells and the corresponding parts into part families. In the second stage, the layout of machines within the cells is determined. In the third stage, the layout of cells is determined. GTLAYPC, an interactive program, printed the scaled layout, calculated dimensions of the cell, and unutilized space in the layout. It prompts the user to perform sensitivity analysis. More details are provided in Heragu and Kakuturi (1997). In the following discussion, we give a brief outline of the algorithms used in GTLAYPC.EXE:

Stage 1 (Grouping of machines into cells): The algorithm used in this stage considers some of the constraints listed in Section 6.4. Machine adjacency requirements must be met, the number of machines in each cell cannot exceed a user-specified value, and the number of cells can also not exceed a user-specified value. Additional features are also incorporated, including the identification of bottleneck machines, which when duplicated in appropriate cells, will minimize or eliminate intercellular movement of parts, and the identification of exceptional parts, which may be subcontracted or processed in a job shop. Exceptional parts are those that prevent the formation of mutually separable clusters of machine cells and part families. The algorithm also considers material flow at every appropriate step to achieve the objective of minimizing intercellular and intracellular material handling costs.

This algorithm first determines the flow between each pair of machines using Equations 3.5 and 3.6. It then identifies a preliminary set of mutually separable machine groups and part families simultaneously using the R&CM algorithm in Section 6.2.3. However, while doing so, it makes sure that machine adjacency constraints are satisfied. Then the algorithm ascertains whether the number of machines in the cell formed is less than the upper limit specified by the user. If this constraint is violated, it attempts to identify bottleneck machines and exceptional parts so that the large cell(s) may be partitioned into two or more smaller cells of the required size. If such a bottleneck machine or exceptional part cannot be identified, machines in the large cell are partitioned based on material flow considerations. In other words, the algorithm identifies the pair of machines that have the greatest flow and places them together. It then adds the remaining machines one by one to the two machines already placed so that the machine added has the greatest interaction with the machines being placed together, until a cell of the required size is formed. This procedure is repeated for the remaining parts and machines.

Stage 2 (Machines layout in cells): Once the machines are grouped into cells, the layout of machines within the cells must be determined. For this purpose, we use the simulated annealing algorithm discussed in Chapter 11.

To develop the final block plan, the program requires the user to provide values for two shape parameters (p and q), which specify the number of rows and columns of machines desired by the user in the layout. By varying these values, the user can generate a variety of layouts ranging from a long and narrow block plan (i.e., a straight line layout) to a square one. The p and q values are supplied by the user interactively and should depend on the number of machines in each cell. For example, if there are six machines in a cell, the maximum number of machines in any row and column must be such that their product is at least 6. Thus, $p = 3$, $q = 2$ is feasible, but $p = 1$, $q = 5$ is not. The block plan of each cell obtained is appropriately scaled and printed. Its objective function value (OFV), dimensions of the cell, and clearance between the machines within the cell are also calculated and printed. The clearance between machines may also be provided interactively, and the computer program provides an option for the user to do so just before generating a layout. The user can also zoom the layout.

If the block plan determined does not meet the user's requirements, for example, the material handling cost is too high, space is not properly utilized, or machines requiring the same operator not placed adjacently, it can be changed by varying the shape parameters for the cell. In other words, p and q values can be changed to generate new block plans. Thus, the designer can try different combinations and retain the one that best satisfies the specific requirements.

Stage 3 (Layout of cells): After the block plans are obtained, the dimensions of each cell and the flow between the cells are calculated. The hybrid simulated annealing algorithm used in stage 2 is applied again to determine the layout of cells. Because the number of cells is relatively small, a near-optimal layout of cells can be generated. Again, the final shape of the cell layout can be changed to satisfy the user requirements by changing the shape parameters for the cells. Even after the final layout of cells is determined, the user has the flexibility of changing the layout of machines within any cell. The final layout retained by the user is printed with appropriate scaling. The output file generated by the program is provided in Figure 6.23.

From the output file, the layout shown in Figure 6.24 can be constructed. Although some parts do visit multiple cells (mainly due to the cell size restriction), the overall intercellular movement is minimized.

6.6 MACHINE GROUPING AND LAYOUT CASE STUDY*

6.6.1 BACKGROUND

Job shops may be ill-advised to undertake a complete reorganization into flexible and lean (FLEAN) cells. A FLEAN cell is flexible enough to produce any and all orders for parts that belong in a specific part family and utilize lean techniques to eliminate waste. For example, FLEAN cells that are implemented in job shops may not be conducive for one-piece flow that is feasible in assembly cells. Still, due to the proximity between consecutively used machines, small batches of parts can be easily moved by hand, on wheeled carts, on short roller conveyors or using jib cranes. In fact, it may be that the production volumes and demand stability for many part families simply cannot justify dedicating equipment, tooling, and personnel to produce any of those families in a stand-alone cell.

---

* Dr. Shahrukh Irani, President of Lean and Flexible, LLC, is the author of this section. His work with Hoerbiger Corporation of America is used with permission.

1. GROUP TECHNOLOGY PROGRAM WILL START NOW

FINAL MACHINE CELLS

MACHINES IN CELL 1

MACH. NO.	MACH. NAME
1	01204
2	02008
3	02014

MACHINES IN CELL 2

MACH. NO.	MACH. NAME
4	02023
9	02030B
8	02030A

MACHINES IN CELL 3

MACH. NO.	MACH. NAME
5	02029A
6	02029B
7	02029C

FINAL PART FAMILIES

PARTS THAT GO INTO CELL 1 ARE

PART NO.	PART NAME
1	115110
2	176022
3	176023
12	236235
16	280024
4	177007
5	201052
7	202103
10	210001
19	325003
8	203009
9	209001
11	217002-3
13	270071-501
14	270080-501

PARTS THAT GO INTO CELL 2 ARE

PART NO.	PART NAME
4	177007
5	201052
6	202102-501
7	202103
17	30130
18	324008
20	328008
8	203009
10	210001
13	270071-501
14	270080-501
15	280011
19	325003

PARTS THAT GO INTO CELL 3 ARE

PART NO.	PART NAME
8	203009
9	210001
11	217002-3
13	270071-501
14	270080-501
18	324008
19	325003
20	328008

GROUPING OF MACHINES AND PARTS IN MATRIX FORM

	1	2	3	4	9	8	5	6	7
1	1	2	0	0	0	0	0	0	0
2	1	2	3	0	0	0	0	0	0
3	1	2	0	0	0	0	0	0	0
12	1	2	0	0	0	0	0	0	0
16	1	2	0	0	0	0	0	0	0

FIGURE 6.23 Output file generated by GTLAYPC.EXE for input shown in Figure 6.21.   *(Continued)*

```
4    0    1    4    2    0    0    0    0    0
5    0    1    2    3    0    0    0    0    0
7    0    1    0    2    0    0    0    0    0
10   0    2    0    0    1    1    0    0    0
19   0    2    0    0    2    2    1    1    1
8    0    0    2    0    1    1    3    3    3
9    0    0    2    0    0    0    1    1    1
11   0    0    2    0    0    0    1    1    1
13   0    0    3    0    2    2    1    1    1
14   0    0    3    0    2    2    1    1    1
6    0    0    0    2    1    1    0    0    0
17   0    0    0    2    1    1    0    0    0
18   0    0    0    3    2    2    1    1    1
20   0    0    0    3    2    2    1    1    1
15   0    0    0    0    1    1    0    0    0
```

GROUP TECHNOLOGY PART OVER

LAYOUT OF MACHINES WITHIN CELLS WILL START NOW

DETAILS AND LAYOUT OF CELL 1

  LAYOUT SCALE 1: 3.5000
  TOTAL CELL LENGTH = 35.00
  TOTAL CELL WIDTH = 35.00
  OPTIMUM VALUE OF FLOW TIMES DISTANCE
  FOR THE CELL IS 12620.000

POSITION	MACH. NO.	MACH. NAME	LENGTH	WIDTH	LAYOUT
1	2	20080	20.00	25.00	2 2 2 2 2 2 3 3 3 3
2	3	01402	15.00	10.00	2 2 2 2 2 2 3 3 3 3
3	1	01204	20.00	10.00	2 2 2 2 2 2 3 3 3 3
					2 2 2 2 2 2 0 0 0 0
					2 2 2 2 2 2 0 0 0 0
					2 2 2 2 2 2 0 0 0 0
					2 2 2 2 2 2 0 0 0 0
					1 1 1 1 1 1 0 0 0 0
					1 1 1 1 1 1 0 0 0 0
					1 1 1 1 1 1 0 0 0 0

  DETAILS AND LAYOUT OF CELL 2

  LAYOUT SCALE 1: 4.5000
  TOTAL CELL LENGTH = 45.00
  TOTAL CELL WIDTH = 35.00
  OPTIMUM VALUE OF FLOW TIMES DISTANCE
  FOR THE CELL IS 7222.500

POSITION	MACH. NO.	MACH. NAME	LENGTH	WIDTH	LAYOUT
1	4	23020	25.00	20.00	4 4 4 4 4 4 9 9 9 9
2	9	0B	20.00	15.00	4 4 4 4 4 4 9 9 9 9
3	8	0A020	20.00	15.00	4 4 4 4 4 4 9 9 9 9
					4 4 4 4 4 4 0 0 0 0
					8 8 8 8 0 0 0 0 0 0
					8 8 8 8 0 0 0 0 0 0
					8 8 8 8 0 0 0 0 0 0

FIGURE 6.23 (*Continued*) Output file generated by GTLAYPC.EXE for input shown in Figure 6.21.
(*Continued*)

```
DETAILS AND LAYOUT OF CELL 3

   LAYOUT SCALE 1: 9.0000
   TOTAL CELL LENGTH = 90.00
   TOTAL CELL WIDTH = 20.00
   OPTIMUM VALUE OF FLOW TIMES DISTANCE
   FOR THE CELL IS 0.000

POSITION    MACH.      MACH.     LENGTH   WIDTH      LAYOUT
            NO.        NAME
    1          5       9A 020    45.00    10.00    5 5 5 5 5 6 6 6 6 6
    2          6       9B 020    45.00    10.00    7 7 7 7 7 0 0 0 0 0
    3          7       9C 020    45.00    10.00
LAYOUT OF MACHINES WITHIN THE CELLS OVER
LAYOUT OF FINAL CELLS WILL START NOW

DETAILS AND FINAL LAYOUT OF ALL CELLS

   FINAL LAYOUT SCALE 1: 9.0000
   TOTAL FINAL LAYOUT LENGTH = 90.00
   TOTAL FINAL LAYOUT WIDTH = 55.00
   OPTIMUM VALUE OF FLOW TIMES DISTANCE
   FOR THE FINAL LAYOUT IS63164.55

POSITION    CELL. NO.  LENGTH   WIDTH      FINAL LAYOUT
    1          2        35.00    35.00    2 2 2 2 2 1 1 1 1 0
    2          1        45.00    35.00    2 2 2 2 2 1 1 1 1 0
    3          3        90.00    20.00    2 2 2 2 2 1 1 1 1 0
                                          2 2 2 2 2 1 1 1 1 0
                                          3 3 3 3 3 3 3 3 3 3
                                          3 3 3 3 3 3 3 3 3 3
```

FIGURE 6.23 (*Continued*) Output file generated by GTLAYPC.EXE for input shown in Figure 6.21.

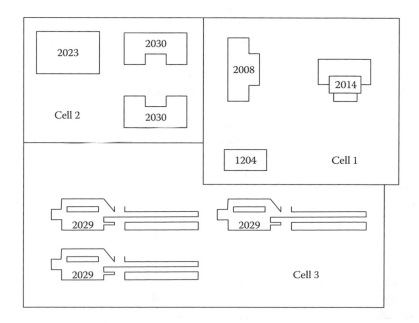

FIGURE 6.24 Layout developed using output shown in Figure 6.23.

6.6.2 FLEAN Cells: Starting Point for Implementing JobshopLean

The starting point for implementing lean principles in a high-mix low-volume facility (Jobshop) is to implement as many FLEAN cells as possible. Management should consider supporting continuous improvement (CI) projects to help each cell become an autonomous business unit (ABU). They must empower employees in each cell to manage their day-to-day operations and make decisions about allocation of orders to operators, cross-training schedule, and other aspects of their work.

6.6.3 Design of a FLEAN Cell at HCA-TX

One of the existing five machining cells at HCA-TX, the MP Cell (MPC), was selected to test a computer-aided methodology for implementing JobshopLean. Routings of all the parts processed in the MPC during a 5-day week were used to create a from–to flow chart, which was then used to construct the flow diagram shown in Figure 6.25a using the PFAST software discussed in Chapter 5. The flows shown in thick lines represent relatively large values in the from–to chart and the flows shown in thin lines represent low values.

A new layout for the MPC (see Figure 6.25b) was designed and implemented by blending:

- Outputs produced by the PFAST software
- Outputs produced by a linear programming software
- Work done by an industrial engineering (IE) intern who was dedicated full time on the project and engaged daily with the employees in the cell
- Work done by a team that partnered with the employees in the cell to implement 5S, housekeeping, and ergonomics-related improvements
- Time study data provided by an in-house IE

6.6.4 Application of the JobshopLean Methodology at HCA-TX

Figure 6.26a through c shows the routings, parts, and work center spreadsheets for the MPC. Together, these three spreadsheets constitute the input file for the PFAST software. Figure 6.27a shows the product–process matrix analysis produced by PFAST using the data for the sample of parts produced in the MPC. The two part families displayed in Figure 6.27a correspond to parts in the MPC part family and parts from the family produced in another cell, the PRR cell (PRRC). The machines required by the PRRC part family could not be placed in the room that housed that cell. Hence, they were intermingled in the same area with the machines that constitute the MPC. Figure 6.27b shows the sequence similarity analysis of the routings for the same sample of parts. Like the product–process matrix analysis, Figure 6.27b is an alternative visualization of a large number of routings of parts produced by machines in a single cell.

Now we are ready to design an actual layout for the MPC. This requires arranging the group of machines required to produce the part family into a U-shape. Figure 6.28a shows the from–to flow chart produced by PFAST using the data in the input file. Figure 6.28b shows the flow diagram that PFAST produces to help contrast the high-volume and the low-volume flows between various machines in the cell. Essentially, Figure 6.28b is a visualization of Figure 6.28a to assist a manual design of the cell layout.

The from–to chart is input to another linear programming software. Figure 6.29a through c shows examples of the different arrangements (layout skeletons) of the machines in the cell that could be produced simply by changing layout settings permitted by the algorithm programmed in this software. Based on consultation with the team members and supervisors and PFAST output, it was evident that a generic routing for the MPC part family was as follows: Turn → Grind → Mill → Rebore → Drill → Insert Pins → Attach Spring. While this became the *backbone* of the cell layout, several adjustments were made to accommodate the differences among the routings that were highlighted by the Sequence Similarity Analysis of Routings (STORM) shown in Figure 6.29b.

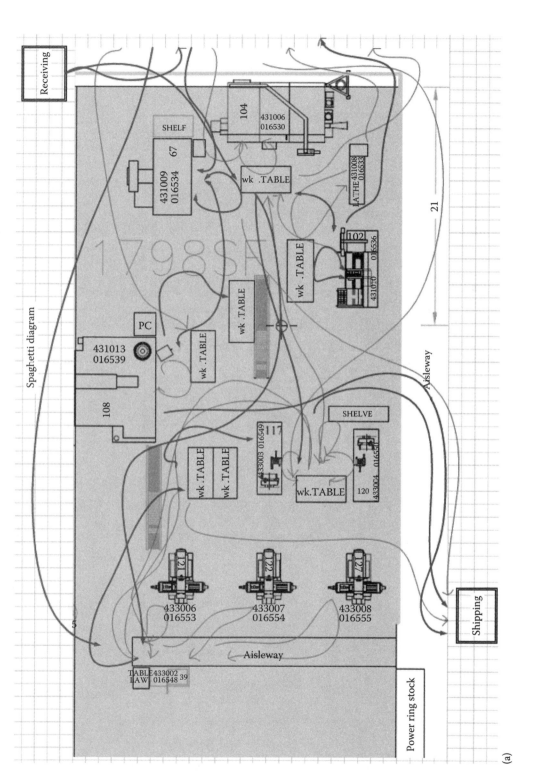

FIGURE 6.25 (a) Material flows in the current layout for the MP cell (MPC).

(Continued)

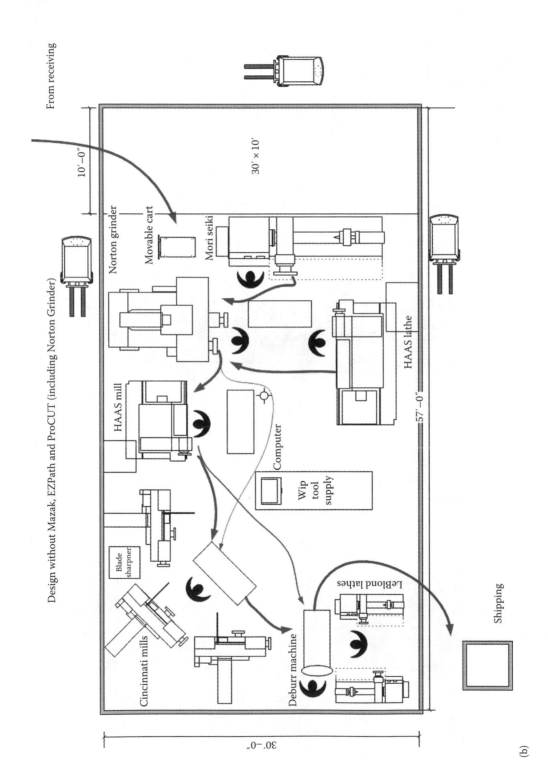

Design without Mazak, EZPath and ProCUT (including Norton Grinder)

FIGURE 6.25 (Continued)   (b) Material flows in the proposed layout for the MPC.

(b)

Part No.	Work Center No.	Sequence No.
1210954	705	1
1210954	240	2
1210954	210	3
1210954	205	4
1210954	255	5
1210954	710	6
1872434	705	1
1872434	240	2
1872434	210	3
1872434	205	4
1872434	255	5
1872434	710	6
1867043	705	1
1867043	240	2
1867043	210	3
1867043	215	4
1867043	205	5
1867043	255	6
1867043	710	7

(a)

Part No.	Description		Annual Quantity	Revenue
1210954	PISTON RING	HY113	25	25
1872434	PISTON RING	0804-00	12	12
1867043	RIDER RING	0872-00	5	5
1206113	RIDER RING	HY112	26	26
1203600	PISTON RING	HY112	4038	4038
1205489	RIDER RING	HY112	178	178
1205702	PISTON RING	HY112	110	110
1230010	PRESSURE BREAKER	HY112	4	4
1203361	PACKING RING	HY112	2,273	2273
1204876	PISTON RING	HY112	100	100
1205529	PACKING RING	HY112	1022	1022
1233569	PISTON RING	HY103	83	83
1233281	PISTON RING	HY112	247	247
1385526	SEALING RING	0309-01	845	845
1875646	PISTON RING	0703-00	8	8
1210542	PISTON RING	CL40CI	6	6

(b)

Work Center No	Description	Area
105	PACKING DOUBLE DISC	1
110	PACKING CNC MILL	1
115	PACKING SPRINGS	1
120	PACKING MANUAL LATHE	1
125	PACKING MISCELLANEOUS	1
126	PACKING SPRINGS AND MISCELLANEOUS IN CELL	1
130	PACKING DRILL & PIN	1
135	PACKING CNC LATHE	1
145	PACKING REBORE	1
150	PACKING SLITTER	1
155	PACKING DEBURR	1
170	PACKING SEGMENT CNC LATHE	1
180	PACKING MELCHIORRE LAPPING	1
205	P/R RING SAW	1
210	P/R RING GRINDER	1
215	P/R RING MILL	1
227	P/R RING HEAT TENSION	1
240	P/R RING MANUAL LATHE	1
245	P/R RING SANDBLAST	1
305	BLANCHARD GRINDER	1
255	TACLOC/EXPANDER BENCH GRIND	1
410	POWER RING GRINDER	1
705	MATERIAL ISSUE	1
710	STOCK & STAGE	1
915	TINNIZE	1
250	P/R RING CNC LATHE	1

(c)

FIGURE 6.26 (a) Routings spreadsheet in the PFAST input file. (b) Parts spreadsheet in the PFAST input file. (c) Work center spreadsheet in the PFAST input file.

(Continued)

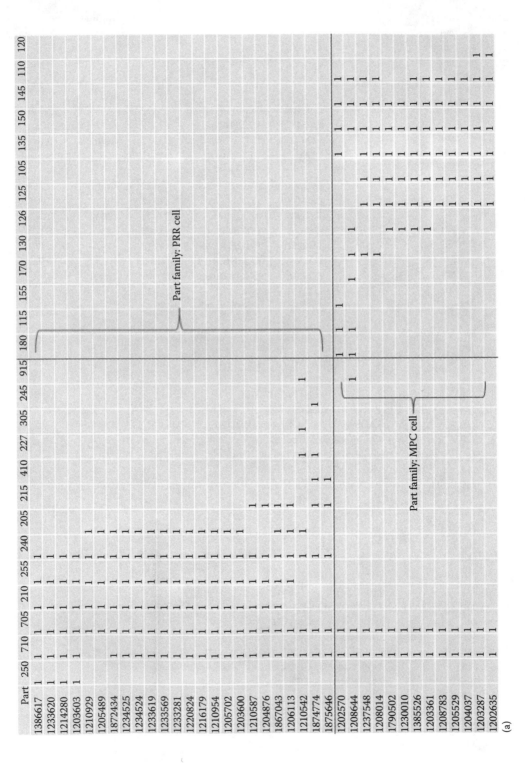

FIGURE 6.27 (a) Product–process matrix analysis.

(a)

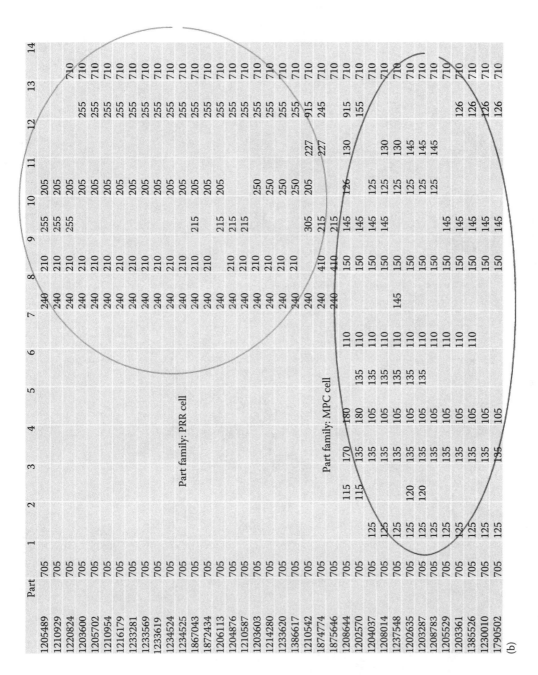

FIGURE 6.27 (*Continued*) (b) Sequence similarity analysis of routings.

(b)

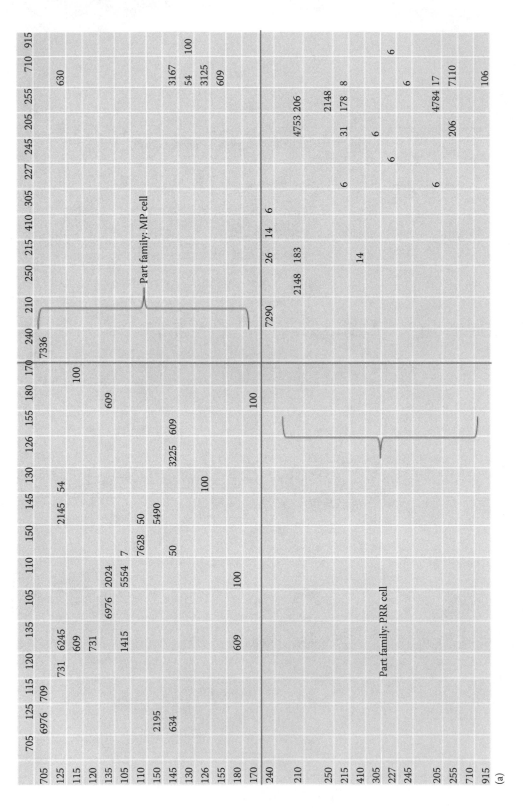

(Continued)

FIGURE 6.28   (a) From–to chart for the MPC part family.

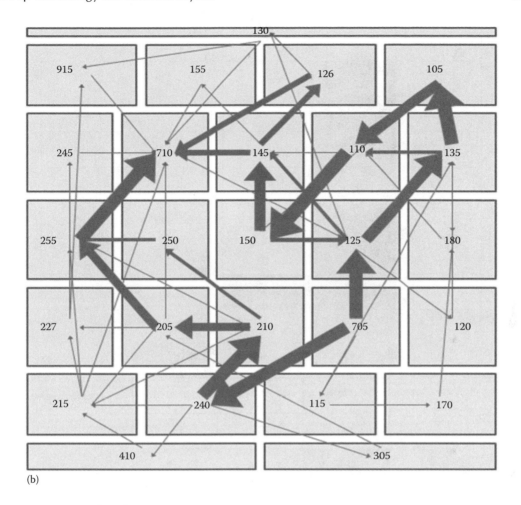

(b)

FIGURE 6.28 (*Continued*)   (b) Flow diagram to visualize the from–to chart.

The proposed layout in Figure 6.25b is a good starting point on which a detailed practical layout can be built. Justifying the investments in relocating the machines already in the area, as well as moving machines currently located elsewhere into the area, was a major challenge. A manager estimated that the following expenses would be incurred:

- Capital investment
  - Purchase a Norton Grinder.
  - Purchase a jib crane for loading/unloading both the Mori Seiki and Haas lathes.
  - Purchase new worktables, toolboxes, and cabinets for all machines.
- Equipment relocation
  - Move Mori Seiki from PRRC into MPC.
  - Move Mazak VTC/Mill, ProCut Lathe, and EZPath Lathe out of the MPC into the PRRC.
- Facility upgrades
  - Relocate and rewire all other machines already in the area occupied by the MPC based on the new floor plan for the cell.
  - Resurface the floor.

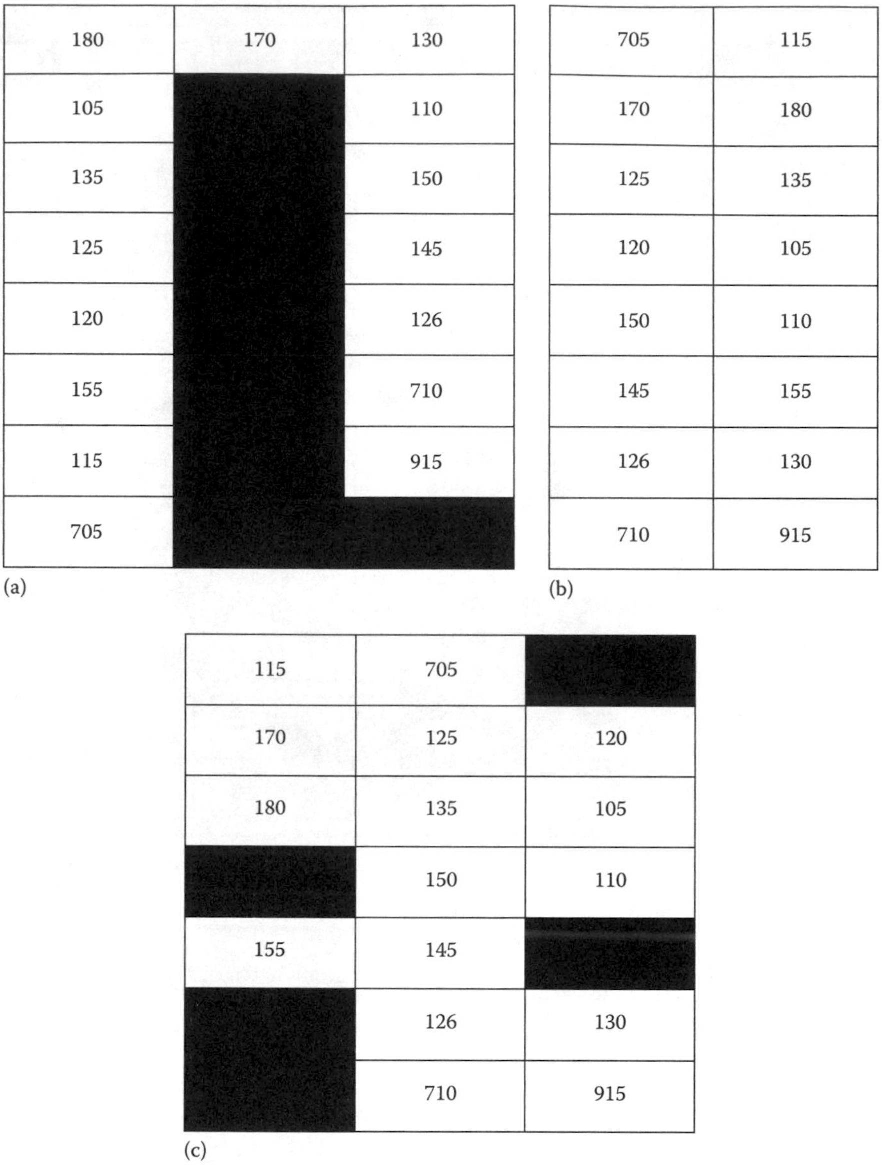

180	170	130
105	■	110
135	■	150
125	■	145
120	■	126
155	■	710
115	■	915
705	■	

(a)

705	115
170	180
125	135
120	105
150	110
145	155
126	130
710	915

(b)

115	705	■
170	125	120
180	135	105
■	150	110
155	145	■
■	126	130
■	710	915

(c)

FIGURE 6.29 (a) U-shaped layout produced by STORM. (b) Parallel-line layout produced by STORM. (c) Block layout produced by STORM.

The earlier list was used in preparing a detailed cost-benefit analysis to justify investments in the implementation of the first FLEAN cell in our facility. Lean principles encourage making improvements that require minimal or no expense. Figures 6.30 through 6.32 present examples of employee-initiated improvements that required negligible investments. Figure 6.30 shows improvements undertaken by each employee with assistance provided to them on an as-needed basis by a graduate intern. Figure 6.31 is an example of in-house benchmarking that implemented ideas successfully implemented in other cells. A tool storage cart fabricated by a senior multitalented employee in another cell designed and fabricated a similar fixture for the tools used on the LeBlond lathes in the MPC. Figure 6.32 displays his solution.

FIGURE 6.30   5S improvements by cell employees to reduce cell footprint.

(a)   (b)

FIGURE 6.31   Milling work center (a) before and (b) after Tiger Team Kaizen.

FIGURE 6.32   Tool storage fixture built over a weekend by an employee.

6.7   SUMMARY

In this chapter, we presented the GT approach to CMS design. The CMS design algorithms discussed in Section 6.2 are very important in developing preliminary groups of machines and corresponding part families. Because they do not take into consideration several practical CMS design constraints, much work needs to be done to develop a CMS design that can actually be implemented.

The design considerations and more complex models that develop more complete CMS designs were also discussed in this chapter. In Section 6.5, we treated the CMS design problem as a series of three subproblems in which cell formation problem is solved first, the layout of machines within each cell is determined next, and the layout of cells is determined at the third stage (see the three-stage approach depicted in Figure 6.22). This approach only deals with design aspects. Using efficient solution techniques such as those described in Chapter 11, it appears that we can develop an approach that solves design and planning problems in a sequential manner. Consider the following four-phase approach. Phase 1 is product and process information gathering and data preparation. Phase 2 is preliminary design using the Benders' decomposition approach and development of machine layouts as well as cell layouts using tools discussed in Chapters 5 and 11. Phase 3 involves examining the operation of the system designed in phase 2 using simulation or other tools. Phase 4 is the final design phase. The output of phase 2 is a preliminary design that tells us the number of machines of each type required, corresponding to part families, optimal number of cells, layout of cells, and layout of machines within each cell. Phase 3 considers planning and scheduling constraints and other downstream information that were not taken into account in the earlier phases. The operational performance is then examined using an appropriate tool. Phase 4 provides a user interface for decision makers to adjust the intermediate design, if necessary, to improve the operational performance. This phase allows the user to perform recursive CMS design changes and operational performance evaluation until a design that achieves a satisfactory level of operational performance is obtained.

Research on the GT and CMS design problem has typically focused primarily on identifying mutually separable part family and machine cell combinations without considering practical design constraints that are frequently encountered. The heuristic approaches presented in this chapter not only incorporate design constraints but are capable of providing optimal or near-optimal CMS design solutions. In this chapter, some of the more important ones were listed and discussed. In addition, we presented a project involving GT and machine layout. The three-stage approach used to solve the problem has produced satisfactory results for some example problems and a real-world problem (see Heragu and Kakuturi, 1997 for details). An important feature of this approach is the flexibility to change the design constraints, perform sensitivity analysis, and consider alternative process plans while grouping machines. The additional advantage is the interactive program, which is relatively easy to use.

6.8 REVIEW QUESTIONS AND EXERCISES

1. Define group technology and cellular manufacturing.
2. Discuss the advantages of implementing CMSs.
3. Compare and contrast traditional job shop systems with CMSs.
4. What is a part–machine processing indicator matrix? List and explain all the relevant information required for machine grouping and layout problems that can be stored in this matrix.
5. Discuss the advantages and disadvantages of the clustering algorithms discussed in this chapter.
6. In your own words, discuss the design and planning problems that arise in CMSs in practice.
7. Table 6.10 lists the sequence of machine types required for nine part types and also shows the annual volume of production for each.
   (a) Set up a part machine matrix for the aforementioned data.
   (b) Assuming the transfer batch size for part types 1–4 is 30, part types 5–8 is 15, and part type 9 is 10, determine the frequency of trips matrix using the matrix you set up in (a).
8. Table 6.11 lists the sequence of machine types required for 12 part types and also shows the annual volume of production for each.
   (a) Set up a part machine matrix for the aforementioned data.
   (b) Assuming the transfer batch size for part types 1, 4, and 5 is 60; part types 2, 3, 6, 10, 11, and 12 is 20; and part types 7–9 is 50, determine the frequency of trips matrix using the matrix you set up in (a).
9. Determine machine groups and part families for the data in Exercise 7 using
   (a) ROC algorithm
   (b) BE algorithm
   (c) R&CM algorithm
   (d) SC algorithm
   (e) Mathematical programming approach
       (For (e), try fixing the number of part families to 2 and 3).

TABLE 6.10
Part Routing and Volume Information for Exercise 7

Part Type #	Sequence of Machine Types Visited	Production Volume
1	3, 5, 2	3000
2	3, 5	1000
3	4, 6, 1	8550
4	1, 4	2750
5	6, 4	9000
6	1, 6	500
7	2, 5, 3	300
8	5, 3	3600
9	6, 1, 4	4110

TABLE 6.11

Part Routing and Volume Information for Exercise 8

Part Type #	Sequence of Machine Types Visited	Production Volume
1	2, 5, 6	13,000
2	1, 4	4,000
3	4, 1, 3	2,500
4	5, 6	6,700
5	6, 2	1,000
6	1, 3, 4	5,000
7	2, 5, 3	9,000
8	5, 2	13,000
9	2, 5	13,000
10	1, 3	13,000
11	4, 3	13,000
12	3, 1, 4	15,000

---

10. Suppose part type 5 in Exercise 7 requires another operation that is to be done on machine type 3. Determine the new machine groups and part families using the following algorithms:
    (a) ROC
    (b) BE
    (c) R&CM
    (d) SC
    (e) Mathematical programming approach
    What do the solutions tell us about the effectiveness of the algorithms? Explain.

11. Determine machine groups and part families for the data in Exercise 8 using
    (a) ROC algorithm
    (b) BE algorithm
    (c) R&CM algorithm
    (d) SC algorithm
    (e) Mathematical programming approach (Hint: Try fixing the number of part families to 2 or 3.)

12. Suppose the third operation for part type 7 in Exercise 8 can be done on machine type 5. Determine the new machine groups and part families using the following algorithms:
    (a) ROC
    (b) BE
    (c) R&CM
    (d) SC
    (e) Mathematical programming approach
    Comparing the results obtained for Exercise 8, what can you say about the effectiveness of the algorithms? Explain.

13. A manufacturer of underground transformers wants to change its existing job shop layout to a CMS layout. The system design team has examined past year's production data and has found that the 20 part types shown in Figure 6.33 account for 90% of the production. Determine how machine groups and part families are to be formed using the following algorithms:
    (a) ROC
    (b) BE
    (c) R&CM

	1	2	3	4	5	6	7	8	9	10
1		1					1			
2		1					1			
3										1
4								1	1	
5			1							
6			1					1		
7			1						1	
8			1					1	1	
9				1		1				
10				1	1	1				
11					1	1				
12	1									1
13								1	1	
14	1									
15				1	1					
16					1					
17			1						1	
18			1							
19			1							
20			1							

FIGURE 6.33 Part–machine processing indicator matrix for Exercise 13.

(d) SC
(e) Mathematical programming approach
    You may assume that production volumes for each part type are roughly the same. For (v), assume that P, the number of part families, equals 5.
14. Suppose that the company in Exercise 13 wants to replace part types 10 and 11 so that neither of them requires processing on machine type 6, but instead both are to be processed on machine type 7. Determine the new machine groups and part families using
    (a) ROC
    (b) BE
    (c) R&CM
    (d) SC
    (e) Mathematical programming approaches
15. A manufacturer of discrete machined parts wants to change its existing job shop layout to a CMS layout. The 25 part types shown in Figure 6.34 are considered class A products and accounted for 90% of last year's production. The system design team does not anticipate

	1	2	3	4	5	6	7	8	9	10	11	12
1					1			1	1			
2	1					1						1
3												
4		1			1			1				
5		1						1	1			
6						1						1
7	1											1
8												
9		1						1	1			
10						1						1
11			1				1			1	1	
12				1			1			1		
13							1					
14			1	1						1	1	
15								1	1			
16			1	1								
17						1			1			
18							1			1		
19										1	1	
20							1			1		
21			1	1			1			1	1	
22			1	1			1			1	1	
23		1										
24							1					
25		1							1			

FIGURE 6.34 Part–machine processing indicator matrix for Exercise 15.

significant changes to its product mix and volume. Determine how machine groups and part families are to be formed using the following algorithms:

(a) ROC

(b) BE

(c) R&CM

(d) SC

(e) Mathematical programming approach

   You may assume that production volumes for each part type are roughly the same. For (v), assume that P, the number of part families, equals 5.

16. Suppose that the company in Exercise 15 wants to replace part types 13 and 24 so that neither of them requires processing on machine type 7, but instead both are to be processed on machine type 3. Determine the new machine groups and part families using

(a) ROC

(b) BE

	1	2	3	4
1		1		
2	1			
3		1	1	
4		1		
5			1	
6			1	
7	1			1
8				1

FIGURE 6.35 Part–machine processing indicator matrix for Exercise 17.

	1	2	3	4
1		1		
2	1	1		
3		1	1	
4		1		
5		1	1	
6		1	1	
7	1			1
8	1			1

FIGURE 6.36 Part–machine processing indicator matrix for Exercise 19.

    (c) R&CM

    (d) SC

    (e) Mathematical programming approaches

17. Consider the binary part–machine processing indicator matrix shown in Figure 6.35. Determine machine groups and corresponding part families using the five algorithms ROC, BE, R&CM, SC, and mathematical programming approach.

18. Assuming that part type 4 in Exercise 17 requires machine type 1 for processing in addition to machine type 2, determine the new machine groups and part families using the five algorithms. ROC, BE, R&CM, SC, and mathematical programming approach. Compare the solutions you get from each algorithm. Are they different? Explain.

19. Consider the binary part–machine processing indicator matrix shown in Figure 6.36. Determine machine groups and corresponding part families using the five algorithms ROC, BE, R&CM, SC, and mathematical programming approach.

20. Assuming that the second operation on part type 2 in Exercise 19 did not require machine type 2 but could be done on machine type 1, determine the new machine groups and part families using the five algorithms. Compare the solutions you get from each algorithm. Are they different? Explain.

21. Identify machine groups and corresponding part families for the binary part–machine process-ing indicator matrix in Example 6.1 using the BE algorithm.
22. Repeat Exercise 21 for the data provided in Example 6.2 using all the algorithms discussed in Section 6.2 except the BE algorithm.
23. Construct a part–machine processing indicator matrix for Exercise 20. Interchange column 1 with column 2 and row 4 with row 8. Determine the machine groups and part families using the ROC algorithm. Is the final ordering of the rows and columns identical to that obtained in Exercise 20 for the ROC algorithm? Why or why not?
24. Show that steps of the ROC algorithm converge to a final solution. Does it converge to a unique final solution? Explain.
25. *Project*: Collect literature on recent algorithms that address two or more of the design and plan-ning constraints listed in Section 6.4, from your library. Explain in your own words how each algorithm works (a maximum of two double-spaced typed pages per algorithm). In a separate section, compare and contrast the algorithms.
26. *Project*: Visit a nearby discrete parts manufacturing company and get the data listed in Tables 6.7 and 6.8 for a maximum of 20 part types. Perform machine capacity calculations as shown in Table 6.9 and then prepare a data input file as required by the GTLAYPC.EXE pro-gram available at http://sundere.okstate.edu/Book. Provide the input data file to the GTLAYPC. EXE program, obtain the results and present it to your class.

7 Material Handling

7.1 MATERIAL-HANDLING SYSTEM IN ACTION

The European Combined Terminals (ECT) in Rotterdam, the Netherlands, is one of the largest container terminals in the world. Goods to and from Europe are transported to the outside world primarily via two types of containers—large and small. Many of the newer docks have been built on reclaimed land in the North Sea (Figure 7.1a). Trucks arriving from Belgium, Germany, France, the Netherlands, and other countries wait their turn in a designated spot for their load, i.e., container, to be picked up by a straddle carrier (Figure 7.1b). The straddle carrier holds the load under the operator and moves it (Figure 7.1c and d) to a temporary hold area (see Figure 7.1e) from where it is loaded onto ships. Containers are usually held for 2 days in this area. When they are ready to be loaded onto ships, two types of mobile overhead gantry cranes that move on tracks and have special container-holding attachments lift the containers from above and take them to another location where automated guided vehicles (AGVs) are waiting to receive the load (Figure 7.1e). The two types of mobile overhead gantry cranes are shown in Figure 7.1f and g. A fleet of AGVs then transport the containers to tower cranes (Figure 7.1f). The tower cranes can move along tracks and be positioned very close to the loading area of the ships. Moreover, one of their arms can be tilted upward at a 90° angle when not in use (Figure 7.1f and g). Using overhead cranes, the containers are picked up from the AGV and transported one by one to the ship deck (Figure 7.1h). While the figures illustrate how ships are loaded, unloading is done in a similar manner—only the steps are reversed. Effective use of AGVs, cranes, and trucks allows ECT to load or unload a ship in about 1 day.

7.2 INTRODUCTION

In the preceding chapters, we have focused on how machines, workstations, and auxiliary/support services need to be placed relative to one another in order to maximize material flow–related manufacturing efficiencies. Now, it is time to turn our attention to the systems that actually transfer material between machines, workstations, and support services. These systems are referred to as material-handling systems (MHSs). The facility design problem includes not only the selection and placement of manufacturing equipment but also that of the MHS. In a typical manufacturing company, a number of material-handling devices (MHDs) are used in combination to form a MHS. It is not uncommon to see the same product handled by various types of MHDs between various stages of processing. For example, heavy equipment such as industrial trucks are used to transport the material from central storage to the preliminary processing area because raw material is generally bulky. For example, in the case of a transformer manufacturing company, sheet metal is cut into smaller panels at the sheet metal–processing area. These panels are moved via fork-lift trucks to a transformer housing fabrication area. After the panels are welded into a housing area and the electrical assembly is placed inside the housing, they gradually become bulkier and heavier and, hence, require transportation via hoists.

The major function of MHSs and MHDs is to transport parts and material between various stages of processing. Thus, they do not add "value" to the product and can be considered a "necessary waste," according to the just-in-time (JIT) or lean philosophy. There are instances where MHSs perform additional "production" functions, however. For example, in a coated abrasive plant, paper is sent through a mile-long heating-cum-drying chamber via special MHDs. The purpose of the

MHD is not only to transport the paper through the heating chamber but also to expose its abrasive-coated side to facilitate quick drying. Similarly, in the automated mailroom-handling system shown in Figure 7.2, palletizers and sorters in the automated bundle and pallet-handling system allow many printed advertisement and late breaking news or sports inserts to be included in the newspaper bundle quickly and efficiently. In addition to bringing the newspaper and inserts to the mailroom,

(a)

(b)

(c)

FIGURE 7.1 MHSs in action. (a) Docks constructed on reclaimed land, (b) truck arrival with container, and (c) straddle carrier removing container from truck. (*Continued*)

(d)

(e)

(f)

(g)

(h)

FIGURE 7.1 (*Continued*)   MHSs in action. (d) Straddle carrier moving container to holding area, (e) mobile gantry crane, (f) fleet of AGVS, (g) overhead crane, and (h) containers loaded on ship. (Courtesy of Europe Combined Terminals B.V.)

the MHS shown also sorts and palletizes the bundles. Figure 7.3 illustrates application of conveyors in sorting cartons. In the package-handling facilities shown in Figure 7.3a, divert placements route packages to the appropriate side (takeaway) conveyors. These side conveyors are fitted with extendable dock conveyors (not shown but similar to the one in Figure 7.12) from which it is easy to load the packages directly onto waiting trucks.

Material handling is best summarized by Tompkins et al. (2010) as an activity that uses the "right method to provide the right amount of the right material at the right place, at the right time, in the right sequence, in the right position, and at the right cost." This definition tells us that we

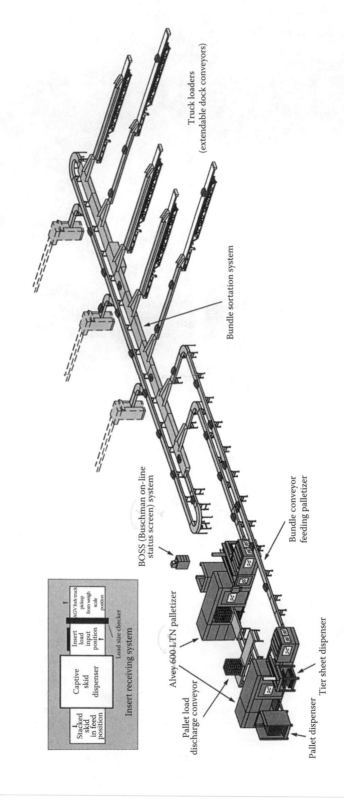

FIGURE 7.2 Automated MHS with palletizers, sorters, and combiners. (Courtesy of FKI Logistex.)

(a)   (b)

FIGURE 7.3 Conveyors used in sortation applications. (a) Sortation conveyors with diverters and (b) belt conveyors in sortation application. (Courtesy of Vanderlande Industries and Bastian Solutions.)

should look at the material-handling function from a broader systems perspective instead of as a simple material transfer activity. Thus, it may sometimes be advantageous to route a part in progress via a longer conveyor if the processing rate of a succeeding process is slower than its preceding one. In this case, the MHS is used for part transportation and line balancing. Of course, this should be done only if one or more of the following conditions hold: (1) processing rate of the succeeding station cannot be decreased; (2) it is not desirable to increase the processing rate of the preceding station; and (3) increasing conveyor length is relatively inexpensive and space is available.

While taking a systems approach to material handling, we must also bear in mind that, depending on the product handled, anywhere from 20% to 40% of a product cost can be attributed to material-handling cost. Our objective in designing an MHS must not only be to minimize design and operational costs but also to come up with a system that supports all the activities on the shop floor effectively.

Apple (1977) has suggested the use of the "material-handling equation" in arriving at a material-handling solution. As shown in Figure 7.4, it involves seeking thorough answers to six major questions—why (select material-handling equipment), what (is the material to be moved), where and when (is the move to be made), how (will the move be made), and who (will make the move). It should be emphasized that all the six questions are extremely important and should be answered satisfactorily. Otherwise, we may end up with an inferior material-handling solution. In fact, it has been suggested that analysts come up with poor solutions because they jump from the "what" to the "how" question (Apple, 1977).

The material-handling equation can be specified as *Material + Move = Method* as shown in Figure 7.4. Very often, when the *material* and *move* aspects are analyzed thoroughly, it automatically uncovers the appropriate material-handling *method*. For example, analysis of the type and characteristics of *material* may reveal that the material is a large unit load on wooden pallets. Further analysis of the logistics, characteristics, and type of *move* may indicate that 6 meters of load/unload lift is required, distance traveled is 50 meters, and some maneuvering is required while transporting the unit load. This suggests that a forklift truck would be a suitable MHD. Even further analysis of the method may tell us more about the specific features of the forklift truck. For example, a narrow-aisle forklift truck, with a floor load capacity of 1/2 ton, and so on.

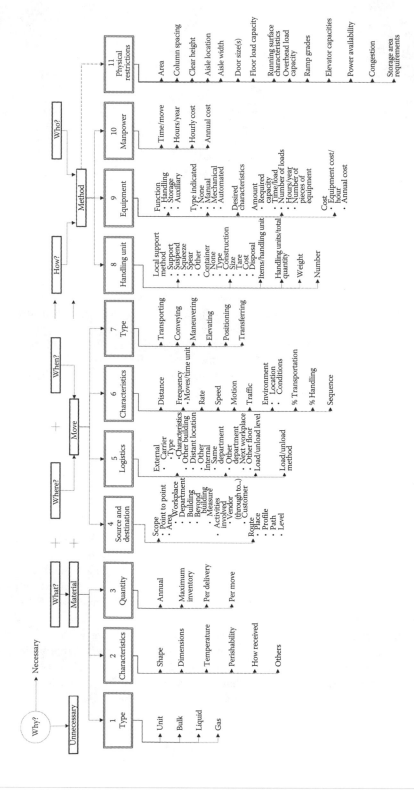

FIGURE 7.4 Material-handling equation. (Apple, J.M.: *Plant Layout and Material Handling*. 1977. Copyright Wiley-VCH Verlag GmbH & Co. KGaA.)

7.3 MULTIMEDIA-BASED EDUCATIONAL SOFTWARE MODULE FOR LEARNING THE 10 PRINCIPLES*,†

Material handling is a vital function in a manufacturing or distribution system. In fact, it is a key link between processes in a system. Efficient handling of material allows a manufacturing or service system to operate at high levels of productivity. U.S. companies invest over scores of billions of dollors annually in material-handling technology. The Department of Commerce has identified material handling as one of the fastest growing segments of the world economy. It is, therefore, extremely important for readers to understand the principles of material handling, as well as the design, implementation, operation, and control of MHSs so that they can assure the cost-effectiveness of this investment. It is for this reason that the two multimedia-based educational software modules were developed with funds from the National Science Foundation (www.nsf.gov) and the support MHI (www.mhi.org), as well as a number of companies in the United States and other parts of the world. The modules are available at http://sundere.okstate.edu/downloadable-software-programs-and-data-files.

The first module is titled "10 Principles of Materials Handling" and is described in this section. The second titled "Design and Analysis of Integrated Materials Handling" is described in Chapter 8. A major objective of the modules is to provide instructors and students a mechanism to integrate descriptive and factual material with analytical models. Both were developed with the participation of experts, learning theorists, project evaluation experts, and multimedia computer programmers (Figure 7.5). The modules introduce students to (1) the 10 principles of material handling, (2) three categories of material-handling equipment, (3) illustrate industrial applications of a major material-handling device in each category, and (4) teach the problem-solving process through an extensive series of models, algorithms, problems, and solutions, which illustrate the creative application of quantitative tools and design problems in solving technology selection and technology analysis

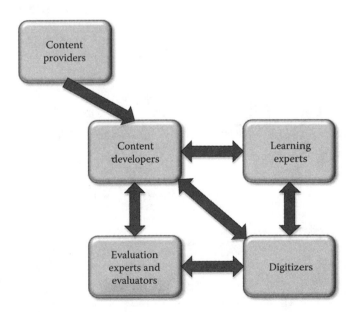

FIGURE 7.5 Multimedia educational software development team.

* This material is based on the work supported by the National Science Foundation under Grant No. EEC 9872506. Any opinions, findings, and conclusions or recommendations expressed in this material are those of the author(s) and do not necessarily reflect the views of the National Science Foundation.
† Parts of this section and figures have been reprinted from Heragu et al. (2003).

problems. Both include extensive simulations of various dynamic processes involved in material handling, including animated simulations of real-world implementations of engineering designs. Each analytical model includes variables, including temporal, spatial, mechanical, and material ones, which the student can manipulate in order to see changes flow through different configurations of material-handling processes and in order to achieve optimal or near-optimal solutions. Learners can work on the modules both synchronously (in the real-time context of the classroom) or asynchronously (at their own pace). We now describe the "10 Principles" module.

Information on the 10 principles is structured in four layers, namely *Discover*, *Explore*, *Contrast*, and *Extend*. In addition, there is another layer called *Integrate* (see Figure 7.6). Each principle is introduced via the first four layers. All the layers include rich media, including, text, voiceovers, animation, video clips, play spaces, and an interactive question and answer format. The modules have been designed so that multiple learning styles can be accommodated. For example, users can move from one layer to another seamlessly. When the "10 Principles" module is launched, the user is taken to the *Introduction* screen from which an introductory video containing a message from the project director can be launched. The bottom of this screen has icons that allow the user to proceed to the

- Goals and objectives screen
- Module navigation screen that includes animated demonstrations illustrating how a user can navigate through the various principles, layers, glossary and access the introduction layer again, if necessary. This screen also includes links to the previous and next screens and links to turn the music on or off and exit from the module.
- Learning screen that uses animated demonstrations to explain in detail the purpose of the various layers and how to use the play space, notebox, multiple perspectives feature, and various media launches.

The *Introduction* layer explains that there are several ways of entering the "10 Principles" module. Depending on interests, background experience, and study requirements, the learner can choose

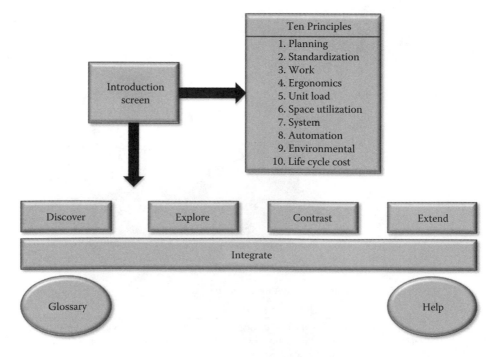

FIGURE 7.6 Information design for "10 Principles of Materials Handling."

a sequence that fits his or her needs. For example, a learner can view the principles in any order, move freely between layers in any order within each principle, and within each layer, move from one principle to another.

The purpose of the *Discover* layer is to let the user discover a principle on his or her own by watching a situation that to an expert clearly illustrates application of that principle. In this layer, no questions are posed to the learner. Instead, a definition is presented along with a rich animation and video clip of a real-world application. The user is only instructed to watch the animation and video clip and make a note of the aspects of the animation and video clip that stand out. A "notebox" is provided throughout for this purpose—to make notes and to answer questions that are posed in subsequent layers.

Next, the *Explore* layer allows the learner to explore the principle in more depth, by examining key aspects of the principle. Each principle has three or more key aspects. In addition to providing text of the key aspects on the screen, the explore layer attempts to provide a deeper understanding of the principle by using two-dimensional or three-dimensional animations, still graphics, video clips, audio files, multiple perspectives, and interactive "play spaces." In all the layers, the learner is posed a question and asked to note answers in a "notebox." To aid the student in answering the questions, multiple perspectives on the question are provided. The "multiple perspectives" feature discusses key aspects or points from different perspectives—those of a manager, field engineer, professor, and student. The perspectives appear in the intended sequence, and the learner sees photographic images of the perspective fade in and out as that perspective is heard. Often, when the learner clicks on a perspective icon, key ideas are reinforced through additional models and theories.

The interactive play space permits the user to interact with a mathematical model by adjusting input parameters (within an allowed range) and study the impact of these parameters on the model output (Figure 7.7). By allowing the user to modify parameters and to study the effect of the input parameters on the model output, the learner is able to question or understand why the input

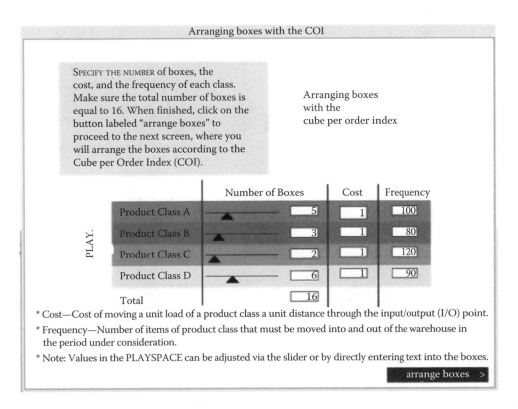

FIGURE 7.7   Interactive "play space" in "10 Principles of Materials Handling."

parameter changes produce the resulting changes in the model output. For example, in the case of the space utilization principle, the user interacts with a cube per order index (COI) model discussed in more detail in Chapter 8. After entering or adjusting the input parameters (including the number of input/output (I/O) transactions generated by an item at the warehouse I/O point, size of the item, and cost of handling), the user is taken to a screen where he or she is asked to rearrange the boxes and arrive at the optimum solution whose value is displayed prominently. A learner without any exposure to the COI policy will at first try to rearrange the boxes at random. Each rearrangement leads to a new objective function value that could be closer to or further away from the optimum value. A little practice teaches the user that moving some boxes closer to the I/O point improves the solution and moving some others makes it worse. At this point, a typical user begins to question why this happens and goes back to the data input screen in an attempt to understand. When the learner recognizes that boxes generating the most I/O transactions and occupying the least space are ideal candidates to place near the I/O point of the warehouse, he or she begins to understand the underlying principles of the COI policy. Most users are then likely to interact with the play space a few more times to confirm their observation. Next, the user is taken through a mathematical treatment of the COI policy model, and multiple perspectives offer more insight into the model and the policy.

Users do not need to know details of the mathematical model, but the play space may motivate them to learn the principles underlying the model. For example, after interacting with a queuing network model and coming up with a near-optimal solution by trial and error, the user may be interested in learning more about analytical (queuing network) models and how they might be advantageous compared to empirical (simulation) models.

(a)

(b)

(c)

(d)

FIGURE 7.8 Sample screen shots from four layers. (a) Discover layer, (b) explore layer, (c) contrast layer, and (d) extend layer.

After gaining a deeper understanding on a principle, the user is reintroduced to the animation and video clip shown in the *Discover* layer. This time, a voice explains details of the animation and video clip. Questions are again posed to help check the user's level of understanding. The learner is then taken to a *Contrast* layer, where the learner is presented two cases (both based on real-world case studies) that illustrate improper and proper application of the principle being studied. Using graphic images, video clips or animations, text screens, and audio, details of each application are presented in such a way that a student with a thorough understanding of the principle will be able to identify the consequences of improper application of the principle and the benefits of properly applying it. Once again, targeted questions and multiple perspectives are used to help the student understand drawbacks of the improper application, how they might be overcome, as well as the good aspects of the proper application. In the case of the planning principle, the two cases illustrate improper and proper application of the principle in installing airport baggage-handling systems— one in Denver and the other in Oslo, Norway. Every layer, including the *Contrast* layer, has questions that the user is asked to answer via the notebox.

The *Extend* layer extends the principle to other domains. All the principles are explained for the most part in a manufacturing, warehouse or service system context. The *Extend* layer discusses the same principle in a different context. Proper utilization of space is extremely important in countries where there is an extremely high premium on space. For example, automated lifts and robots are used to park cars in tight spaces in Japanese parking garages. Sample screen shots of the four layers are provided in Figure 7.8.

The last layer, called the *Integrate* layer, is intended to help the learner integrate the various principles. Design and operational decisions that address a particular principle impact one or more of the other principles. This layer attempts to provide the learner with an understanding of the factors that impact two or more principles. This layer uses a rich case study that shows how multiple principles interact with one another. In addition, the module has a self-contained glossary section including all the terms and definitions used in the 10 principles. This section can be accessed by clicking on the "book" icon contained in the right of every screen or by clicking on hyperlinked texts within the layers.

7.4 MATERIAL-HANDLING PRINCIPLES

In this section, we discuss the 10 principles of material handling. Many of these, for example, the planning, system, life cycle cost, and environmental principles, are general management and operation principles and apply to not only material handling but also facilities design and planning in general. The 10 principles of materials are listed in Table 7.1 along with their definitions. These principles were developed by MHI. We strongly recommend that the reader peruse the module before continuing with the remaining sections. The module introduces key concepts and topics such as JIT or lean manufacturing, reverse logistics, ergonomics, simplifying, combining, or eliminating work elements, information and product flow integration, strategies for successful automation, minimizing pollution, recycling waste, product life cycle cost minimization, preventive maintenance, and equipment replacement analysis. The definition of each principle tells us its meaning. Although all the 10 principles are extremely important, the unit load principle and space utilization principles deserve special discussion. They are discussed in the next two subsections. The key aspects for all other principles are presented next. They were developed by the MHI.

Key aspects of the planning principle

1. The plan should be developed as a consultation between the planner(s) and all who will use and benefit from the equipment to be employed.
2. Success in planning large-scale material-handling projects generally requires a team approach involving management, engineering, computer and information systems, finance, and operations.

TABLE 7.1
The 10 Principles of Materials Handling

Principle	Definition
Planning	A material-handling plan is a prescribed course of action that is defined in advance of implementation, specifying the material, moves, and the method of handling.
Standardization	Standardization is a way of achieving uniformity in the material-handling methods, equipment, controls, and software without sacrificing needed flexibility, modularity, and throughput.
Work	The measure of work is material-handling flow (volume, weight, or count per unit of time) multiplied by the distance moved.
Ergonomic	Ergonomics is the science that seeks to adapt work and working conditions to suit the abilities of the worker.
Unit load	A unit load is one that can be stored or moved as a single entity at one time, regardless of the number of individual items that make up the load.
Space utilization	Effective and efficient use must be made of all available space.
System	A system is a collection of interdependent entities that interact and form a unified whole.
Automation	Automation is a technology for operating and controlling production and service activities through electromechanical devices, electronics, and computer-based systems with the result of linking multiple operations and creating a system that can be controlled by programmed instructions.
Environmental	The environmental principle in material handling refers to conserve natural resources and minimizing the impact of material-handling activities on the environment.
Life cycle cost	Life cycle costs include all cash flows that occur between the time the first dollar is spent on the material-handling equipment or method until its disposal or replacement.

3. The material-handling plan should reflect the strategic objectives of the organization as well as the more immediate needs.
4. The plan should document existing methods and problems, physical and economic constraints, and future requirements and goals.
5. The plan should be flexible and robust so that sudden changes in the process will not make the plan unusable.

Key aspects of the standardization principle

1. The planner should select methods and equipment that can perform a variety of tasks under a variety of operating conditions and in anticipation of changing future requirements.
2. Standardization applies to sizes of containers and other load-forming components as well as operating procedures and equipment.
3. Standardization, flexibility, and modularity must not be incompatible.

Key aspects of the work principle

1. Simplify processes by combining, shortening, or eliminating unnecessary moves to reduce work.
2. Consider each pickup and setdown or placing material in and out of storage, as distinct moves and components of distance moved.
3. Design layouts and develop methods, and sequences, that simplify and reduce work.

Key aspects of the ergonomic principle

1. Equipment should be selected so that repetitive and strenuous manual labor is eliminated and that users can operate effectively.

2. The ergonomics principle embraces both physical and mental tasks.
3. Using ergonomics will improve production and reduce errors. The material-handling workplace and the equipment employed to assist in that work must be designed so that they are safe for people.

Key aspects of the systems principle

1. Systems integration encompasses the entire supply chain including reverse logistics. The chain includes suppliers, manufacturers, distributors, and customers.
2. At all stages of production and distribution, minimize inventory levels as much as possible.
3. Information flow and physical material flow should be integrated and treated as concurrent activities.
4. Materials must be easily identified in order to control their movement throughout the supply chain.
5. Meet customer requirements regarding quantity, quality, and on-time delivery and fill orders accurately.

Key aspects of the automation principle

1. Preexisting processes and methods should be simplified or reengineered before any efforts at installing mechanized or automated systems.
2. Consider computerized MHSs wherever appropriate, for effective integration of material flow and information management.
3. In order to automate handling, items must have features that accommodate mechanization.
4. Treat all interface aspects in the situation as critical to successful automation.

Key aspects of the environmental principle

1. Design containers, pallets, and other products used in material handling so that they are reusable or biodegradable.
2. Systems design must include the by-products of material handling.
3. Hazardous material requires special handling considerations.

Key aspects of the life cycle costs principle

1. Life cycle costs include capital investment; installation, setup, and equipment programming; training, system testing, and acceptance; operating, maintenance, and repair; and recycle, resale, and disposal.
2. Plan for preventive, predictive, and periodic maintenance of equipment. Include the estimated cost of maintenance and spare parts in the economic analysis.
3. Prepare a long-range plan for equipment replacement.
4. In addition to measurable cost, other factors of a strategic or competitive nature should be quantified when possible.

7.4.1 Unit Load Principle

A unit load may be defined as a number of items or bulk materials that are arranged so that they can be picked up and delivered as one load (Apple, 1977). It is implied that a unit load is too large for manual material handling. Generally, it is true that larger the unit load, lower will be the cost per unit handled. However, there may be other factors that may favor a smaller unit load. Examples include cost of unitizing and de-unitizing, space required for material handling, material-handling carrier payload considerations, work-in-process inventory costs, storage, and return of empty pallets or containers used to hold the unit load.

The design of a unit load is not a trivial task. In addition to the size and shape of the unit load, we must determine the devices and types of pallets or containers that will be used for handling the unit load, stacking shape, whether the items will be shrink-wrapped or strapped, whether and what types of automatic palletizers and de-palletizers will be used to create and dismantle unit loads, and so on.

Apple (1977) suggests the following seven-step procedure to design a unit load:

- Determine whether the unit load concept is applicable
- Select the unit load type
- Identify the most remote source of a potential unit load
- Determine the farthest practicable destination for the unit load
- Establish the unit load size
- Determine the unit load configuration
- Determine how to build the unit load

Containers and pallets are typically used to hold the items. These can be made of metal, plastic, or wood, although wooden pallets are commonly seen in manufacturing and distribution companies for handling a unit load (see Figure 7.9). Some types of pallets can be folded flat when not in use to conserve space. In addition to containers and pallets, tots, crates, drums, and bins are also used to handle the unit load. With respect to handling unit loads, once again several options are available. These largely depend on whether the unit loads are on pallets or containers or some other holding device. If the unit load is to be lifted via hooks on the container, cranes or hoists are used; if the unit load is to be lifted from underneath a pallet, forklift trucks are used; if the load is to be squeezed between two lifting surfaces, clamp trucks are used; and so on.

Depending on whether pallet packing is done by manufacturers or distributors, the objectives may be different (Askin and Standridge, 1993). If it is done by the latter, the packages tend to be of different dimensions. This problem, which is similar to the well-known cutting stock or bin-packing problem, has been studied by Hodgson (1982) among others. In the manufacturer's pallet-packing problem, the objective is to select the optimal size and shape of the box and pallet so that the number of units packed (which are usually identical) is maximized. Of course, there are constraints on size, shape, and weight of boxes and pallets.

A unit load is one that can be stored or moved as a single entity at one time, regardless of the number of individual items that make up the load. The key aspects of this principle are

FIGURE 7.9 Wooden pallet used for handling unit loads. (Courtesy of PECO Pallet, Inc.)

1. Less effort and work are required to collect and move many individual items as a single load than to move many items one at a time.
2. Load size and composition may change as material and product move through various stages of manufacturing and the resulting distribution channels.
3. Large unit loads are common both before manufacturing as raw materials and after manufacturing as finished goods.
4. During manufacturing, smaller unit loads, sometimes just one item, yield less in-process inventory and shorter item throughput times.
5. Smaller unit loads are consistent with manufacturing strategies that embrace operational objectives such as flexibility, continuous flow, and lean manufacturing.

The first and fourth key aspects indicate the trade off between large and small unit loads. Whereas large loads minimize the effort and work required, they also increase the work-in-process inventory and (due to Little's law discussed in Chapter 13 and included in the "10 Principles" multimedia software module) product flow times. Notice that the first item in a large unit load consisting of 100 items must wait for the remaining 99 to accumulate before it can be moved to its destination. The second and third key aspects indicate that the size of the unit load handled varies from small as a to medium to large product moves through various stages of processing from raw materials to finished product. In the "10 Principles" module, a car assembly plant is used to illustrate how the load size changes from one process to another. The fifth key aspect indicates that smaller lot sizes—often as small as one unit—are consistent with the JIT or lean.

The *Extend* layer in the Unit Load principle section of the "10 Principles of Material-handling" software module indicates the challenges faced by common carriers such as United Parcel Service in transporting large unit loads—a whale, a panda, and Hummer vehicles.

7.4.2 Space Utilization Principle

The space utilization principle states that effective and efficient use must be made of all available cubic space. The key aspects of this principle are

1. In work areas, cluttered, unorganized spaces, and blocked aisles must be eliminated.
2. In storage areas, the objective of maximizing storage density must be balanced against accessibility and selectivity.
3. When transporting loads within a facility, always consider using overhead space.

When items are block stacked one on top of another on the floor rather than on shelves, warehouse space is lost. This kind of stacking occurs in staging areas where items remain briefly for inspection. Storing or retrieving less than a full block results in unutilized space. This loss of warehouse space is called honeycombing loss and relates to the first key aspect. With respect to the second key aspect, the two objectives of dense storage and accessibility are conflicting. One can be achieved only at the expense of the other. When items are going to be in the warehouse for a long time, dense storage is appropriate, but when items are in the warehouse only briefly, accessibility and selectivity become important considerations for storage. Figure 7.10 shows a cold storage facility that uses mobile racks that can be stacked very close to each other maximizing storage density. One can maximize storage density by using mobile racks, horizontal and vertical carousels, and specially designed modular drawers, but then it takes longer to get to the goods. The third key aspect reminds us that warehouse space is three dimensional, so a designer must not forget to transport material in the overhead space where appropriate.

FIGURE 7.10   Mobile racks used in a beer storage facility. (Courtesy of SpaceSaver Corporation.)

7.5   TYPES OF MHDs

A number of types of MHDs are now available for moving material between stations on a shop floor. Most of the equipment move materials via material-handling paths on the shop floor, but there are some (e.g., cranes, hoists, and overhead conveyors) that utilize the space above the machines. The choice of a specific MHD depends on a number of factors including cost; the shape, weight, size, and volume of the loads; space availability; and types of workstations. We must ascertain how well a specific MHD fits in the overall system.

- Does it permit flexibility?
- Is it inexpensive and easy to maintain?
- Can it be integrated with existing systems?
- Does it significantly increase manufacturing efficiency?

In the remainder of this section, we list each type of MHD and provide illustrations. For further information on these MHDs, the reader may consult a number of excellent sources including the educational resources link in MHI's website (http://mhi.org/cicmhe/resources) and links contained therein, including the link to the material-handling multimedia bank at Laval University, Quebec, Canada as well as technical publications such as *Modern Materials Handling* (http://www.mmh.com), *Material Handling and Logistics* (http://mhlnews.com), *Plant Engineering* (http://www.plantengineering.com), and *IE Solutions* (www.iienet2.org). In addition to product information and advertisements from MHD manufacturers, these magazines often compare features of MHDs (see for example, Buyer's Guide for forklift trucks and AGVs in April 1996 issue of *IE Solutions*).

The seven basic types of MHDs can be classified as follows:

1. Conveyors
2. Palletizers
3. Trucks
4. Jibs, cranes, and hoists
5. Robots
6. AGVs
7. Warehouse material-handling systems (WMHSs)

7.5.1 CONVEYORS

Conveyors are typically fixed path MHDs. This means that we should consider this type of equipment only when the volume of parts/material to be transported is large. It is also advisable to use the conveyors only when material transported is relatively uniform in size and shape. Depending on the surface of the conveyors, we can alphabetically classify this MHD as

* Accumulation conveyor
* Belt conveyor
* Bucket conveyor
* Can conveyor
* Chain conveyor
* Chute conveyor
* Gravity conveyor
* Power- and- free conveyor
* Pneumatic or vacuum conveyor
* Roller conveyor
* Screw conveyor
* Slat conveyor
* Towline conveyor
* Trolley conveyor
* Wheel conveyor

Many conveyors are illustrated in Figures 7.11 through 7.28. Our list is by no means complete, and variations are possible. For example, belt conveyors may be classified as troughed belt conveyors (used for transporting bulky material such as coal) and magnetic belt conveyors (used for moving ferrous material against gravitational force). For the latest product information on conveyors and other types of material-handling equipment, we strongly encourage the reader to refer to recent issues of *Material Handling and Logistics* and *Modern Materials Handling*. These publications and websites have not only articles illustrating the use of material-handling equipment but also numerous product advertisements. In fact, *Modern Materials Handling* also prepares an annual buyer's guide that lists addresses of manufacturers by equipment category. Links to the websites of material-handling equipment system designers, manufacturers, users, and consultants can also be found via www.mhi.org (see Exercise 7).

Figure 7.11 shows an accumulation conveyor that accumulates units in the feeder conveyor lines until they are ready for transportation to the trunk conveyor lines that extend into the trailers. An extendable dock conveyor is shown in Figure 7.12. The retractable portion of the conveyor in Figure 7.12 goes all the way into the truck (parked at the dock) to minimize lifting by the operator and permits fast and easy unloading. Figure 7.13 shows a belt conveyor that allows us to regulate the gaps between cartons (for sorting purposes) by varying belt speeds. Because the downstream belt has a higher speed than the upstream belt, gaps between the cartons are gradually reduced.

FIGURE 7.11   Accumulation conveyor. (Courtesy of Bastian Solutions.)

FIGURE 7.12   Extendable dock conveyor. (Courtesy of DPD, Aschaffenburg, Germany.)

FIGURE 7.13   Belt conveyor. (Courtesy of Bastian Solutions.)

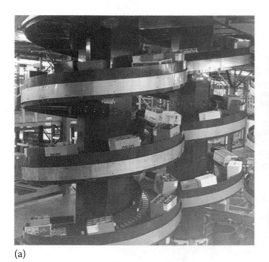

(a)   (b)

FIGURE 7.14   (a) Chute conveyor and (b) tilt-tray conveyors. (Courtesy of Bastian Solutions.)

FIGURE 7.15   Gravity roller conveyor. (Courtesy of Bastian Solutions.)

Chute conveyors are typically used to transport material from a higher level to a lower level—for example, suitcases and bags sent from the aircraft to baggage claim areas located in the basement of an airport. Figure 7.14 is a tilt-tray conveyor that drops its contents by tilting the tray and is used in small package, letter, and parcel-handling applications. Figure 7.15 shows a gravity roller conveyor. In a conventional (unpowered) accumulation conveyor, velocity of the cartons can only be controlled by varying the inclination of the conveyor. Similarly, line pressure buildup can be controlled only by regulating carton entry into the conveyor. For a given inclination and a given carton entry rate, power conveyors allow the operator to control the velocity of the cartons and the line pressure buildup. They use actuators and brake modules for this purpose.

Power- and- chain conveyors are typically overhead conveyors with carriers transported via chains. They have two tracks—powered and unpowered. When it is necessary to disengage a carrier—for example, to store a part or to process it—a mechanism allows the carrier to be disengaged from the power track and mount it on the unpowered track. Material flow can then be diverted to other areas of the shop floor via other MHDs. An inverted power- and- free conveyor is illustrated

FIGURE 7.16   Power and free conveyors. (Courtesy of Jervis B. Webb, Company.)

FIGURE 7.17   Curved roller conveyor. (Courtesy of Bastian Solutions.)

in Figure 7.16. The 730,000 square feet General Motors painting facility in Moraine, Ohio, uses almost 7 miles of inverted power and free conveyors.

Curved roller conveyors are illustrated in Figure 7.17. This conveyor allows boxes coming from a feeder line to join a main conveyor line. A skid conveyor utilized in the assembly and painting of a southeastern truck manufacturer is illustrated in Figure 7.18. Figures 7.19 and 7.20 show two of several variations of the overhead trolley conveyor, in which equally spaced trolleys are supported from an overhead chain and hook system. The hook itself moves on a track and is pulled by the chain. Overhead trolley conveyors are useful in painting applications.

Many applications have a combination of two or more conveyors (see Figures 7.21 through 7.28). Figure 7.21 illustrates how items arriving via several side roller conveyors and feeding into a main

FIGURE 7.18   Skid conveyor. (Courtesy of Jervis B. Webb, Company.)

(a)   (b)

FIGURE 7.19   (a) Overhead conveyor used for car assembly and (b) overhead conveyor with automobile bumpers. (Courtesy of Gould Communications and Dematic Corp.)

belt conveyor can be sorted in a desired fashion. The divert placements in the roller conveyor shown in Figure 7.22 show how cartons from the main line can be diverted into side conveyor lines for sorting. Figures 7.23 and 7.24 illustrate the sortation application of conveyors. In Figure 7.23, loads coming down on belt conveyors are diverted to a wheel conveyor automatically by "pop-up" skate wheels located between belts. Instead of "pop-up" mechanisms, sometimes push diverters are used to divert cartons from one conveyor line to another. In Figure 7.24, cartons coming down on a slat conveyor are diverted to a roller conveyor. Sorting is automatically done, and each customer receives all the required units in a specified shipping lane. Gravity rollers and skate wheel conveyors feed items to an accumulation conveyor in Figure 7.25. Figure 7.26 shows the use of belt conveyors to transport containers at three different levels. Observe that one conveyor is at an upward inclination, another at a downward inclination, and a third (accumulation) conveyor that is horizontal.

FIGURE 7.20   Overhead trolley conveyor. (Courtesy of Gould Communications.)

(a)

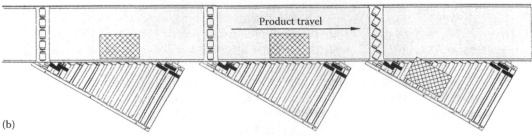

(b)

FIGURE 7.21   Use of "pop-up" minirollers to regulate entry of items from side roller conveyor lines to a central belt conveyor. (a) Consolidation application and (b) top view. (Courtesy of FKI Logistex.)

(a)

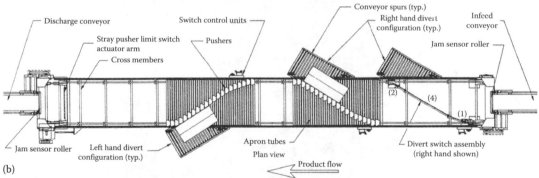

(b)

FIGURE 7.22 Use of divert placements to regulate entry of items from a central conveyor to side conveyor lines. (a) Sortation application and (b) top view. (Courtesy of FKI Logistex.)

FIGURE 7.23 Use of "pop-up" skate wheels to divert cartons to desired lanes. (Courtesy of FKI Logistex.)

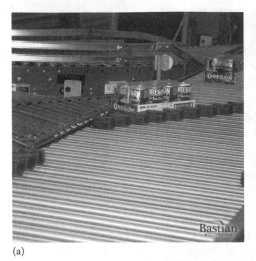
(a)

(b)

FIGURE 7.24 Conveyors used in sortation applications. (a) Sliding shoe conveyor diverting cases and (b) sliding shoe diverting cartons. (Courtesy of Bastian Solutions.)

FIGURE 7.25 Gravity rollers and skate wheel conveyors feeding items to an accumulation conveyor. (Courtesy of Bastian Solutions.)

Like the belt conveyors shown in Figure 7.13, the two-line roller-belt conveyor combination shown in Figure 7.27 allows a manufacturer to regulate the gaps between cartons for sortation purposes. Another multilane conveyor is shown in Figure 7.28.

The device illustrated in Figure 7.29 allows items to be diverted into or away from a main conveyor line. The heavy-duty turntable device shown can rotate items 360° for transportation to auxiliary conveyors. Figure 7.30 shows a load centering device.

The functioning and use of most of the conveyors listed above can be readily understood by examining Figures 7.11 through 7.25. However, there are a few that need further discussion. Towline conveyor is a fixed-path MHD that allows trolleys, trucks, carts, dollies, or even AGVs to be temporarily

FIGURE 7.26   Conveyors feeding containers at three levels. (Courtesy of Vanderlande Industries.)

FIGURE 7.27   Combination of belt and roller conveyor. (Courtesy of FKI Logistex.)

connected to in-ground towlines. Sometimes, the towlines may be above ground as seen in ski mountains, where trolley and towline conveyors are used to transport people to the top of a hill.

The reader may have seen a pneumatic or vacuum conveyor in "drive-through" lanes at banks that are located 10 or more meters away from the building. Such a conveyor is generally used for transporting lightweight materials such as mail and small packets, in offices and laboratories.

FIGURE 7.28   Multilane conveyor.

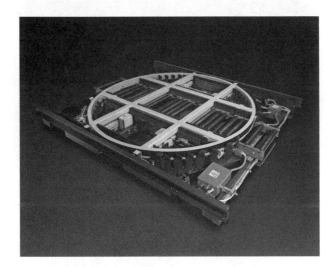

FIGURE 7.29   Auxiliary device for conveyors. (Courtesy of Pentek.)

FIGURE 7.30   Load centering device. (Courtesy of Pentek.)

7.5.2 PALLETIZERS

Palletizers are high-speed automated equipment used to palletize unit loads coming from production or assembly lines. With operator-friendly touch screen controls, they palletize at the rate of 100 cases per minute (see Figure 7.31), palletize two lines of cases simultaneously, or simultaneously handle multiple products (see Figure 7.32).

FIGURE 7.31 High-speed palletizer. (Courtesy of C&D Skilled Robotics.)

FIGURE 7.32 Palletizer capable of handling multiple products with different dimensions. (Courtesy of Europack.)

7.5.3 PALLET-LIFTING DEVICES

There are several types of pallet lifting and tilting devices for loading and unloading pallets from pallet trucks and raising or lowering heavy cases to desired heights (see Figures 7.33a). One type, the scissor lift (shown in Figure 7.33b), is mobile and has room for an operator to ride on it. Other devices not only allow loading and unloading of pallet trucks from three sides but also serve as a ramp. Still others tilt so that operators can access items in a container without straining their backs.

7.5.4 TRUCKS

Industrial trucks are also extensively used in manufacturing facilities. Their main advantage is that the material-handling path need not be fixed. They are particularly useful when the unit load moved varies frequently in size, shape, and weight. When the volume of the parts or material moved is low, the number of trips required for each part is relatively small. Such situations favor the use of trucks as MHDs even more. Industrial trucks vary widely in features, costs, and functionality. Some are easily maneuverable, some can lift loads up to 10 meters, some can move material in narrow aisles, and so on. It is, therefore, not surprising that there are more than 20 types of industrial trucks. Examples of these are listed as follows:

- Hand truck
- Forklift truck
- Pallet truck
- Platform truck
- Counterbalanced truck
- Tractor–trailer truck
- AGVs

(a)   (b)

FIGURE 7.33 Pallet-lifting device. (a) Manual device and (b) scissor-lift device. (Courtesy of Air Technical Industries.)

Some trucks are shown in Figure 7.34. A few of their applications are illustrated in Figure 7.35. The truck in Figure 7.36 has an attachment that allows it to lift and deposit small containers in a shipyard. A taxonomy of industrial trucks is shown in Table 7.2.

Depending on whether the truck is powered or manual, electric powered or gas powered, side loading or front loading, we can develop an even more detailed classification of industrial trucks. This is left as Exercise 10. Although AGVs are listed in this section, they are discussed thoroughly in Section 7.6.

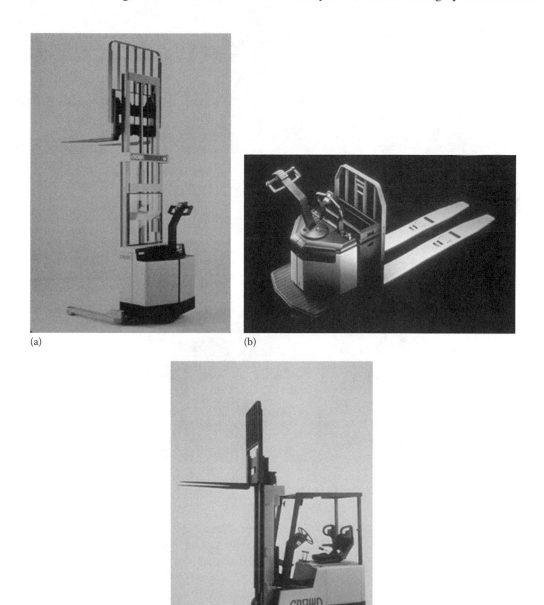

(a)

(b)

(c)

FIGURE 7.34 Examples of industrial trucks. (a) Walkie pallet truck, (b) platform pallet truck, and (c) counter balanced truck. (Courtesy of Crown Corporation.)

(a)

(b)

(c)

FIGURE 7.35   Some applications of industrial trucks. (a) Walkie pallet truck, (b) platform pallet truck, and (c) counter balanced truck. (Courtesy of Crown Corporation.)

FIGURE 7.36   Truck used in shipyard operations. (Courtesy of Europe Combined Terminals B.V.)

Walkie stackers such as the ones shown in Figure 7.34a are capable of handling several hundred pounds of load with lift heights of 10–20 meters. They are available with outriggers, straddles, and other attachments. The operator can steer the self-powered truck that can move forward or backward in a variety of speeds. This type of truck is desirable when material movements are relatively short, requires low lifts, and volume is relatively low. Walkie pallet trucks are used when material is handled over relatively short distances and for unit loads. The rider versions (Figures 7.34b and 7.35b) have room for an operator to stand and are used for pallet handling, low-level order picking, or low-volume transportation. Stockpickers are useful for higher level (up to 15 meters) order picking and have standing room for the operator to place items on a platform or fork. Counterbalanced trucks have sitting room for the operator and can handle unit loads via forklifts up to 15 meters. The rider reach/straddle and turret trucks (Figure 7.34c) are useful when the aisle is narrow and the lift height is above 6 meters. Tow tractors are useful in applications where material is stored on carts and towed. For example, baggage loaded on aircrafts is handled via carts towed by trucks.

TABLE 7.2

Taxonomy of Industrial Trucks Shown in Figures 7.34 and 7.35

Truck Type	Material					Move								Method					
	Type		Weight			L/UL Method		L/UL Level			Utilization			Operation		Physical Restrictions			
	UL	LUL	<5k	5k–8k	>8k	SL	EL	GL	<30 feet	30–45 feet	<20%	20%–50%	>50%	Rider	Walkie	Ramp	Rough	Smooth	Noise
Walkie stackers	X		X			X			X				X		X	X		X	
Tow tractors	X				X	X		X			X	X		X					
Walkie pallet	X		X			X		X				X			X	X		X	
Rider pallet	X				X	X		X				X	X	X		X		X	
Stock pickers	X		X			X				X		X	X	X				X	
Counter balanced	X		X			X			X			X		X		X		X	
Turret	X		X			X				X		X	X	X		X		X	

The trucks that pull the carts are typically bigger than those in Figure 7.35b, but the principle of material transfer is the same.

7.5.5 ROBOTS

Robots are programmable devices that resemble the human arm. They are also capable of moving like the human arm and can perform functions such as pick- and -place, load and unload (see Figure 7.37a). Robots typically have six basic motions or degrees of freedom, consisting of three arm and body and three wrist motions. Of course, by putting the robot on a track, we can provide additional axes of motion. By fitting them with special devices such as a camera, they can be used to inspect defective parts. They can be used for manufacturing activities that are hazardous for the human to do, e.g., welding (Figure 7.37b), painting, or in areas that cause fatigue, e.g., repetitive assembly operations. They are also used extensively for material handling, especially in small manufacturing cells that are dedicated to the production of a family of parts. Typically, machines in a cell are arranged in circular or U shape. This allows the robot to access all the machines in one sweep. Figure 7.38 shows the use of a robot fitted with vacuum suction cups in palletizing boxes. The robot follows an algorithm to arrange boxes on a pallet. Given a set of boxes of different shapes, it attempts to maximize the number of boxes loaded on the pallet. Robots can be classified as follows:

- Point-to-point robots
- Contouring or continuous path robots
- Walkthrough or teach robots
- Leadthrough or teach pendant robots
- Hydraulic robots
- Servo-controlled robots

(a)

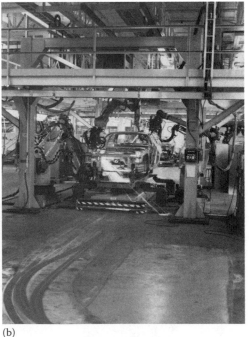

(b)

FIGURE 7.37 Use of (a) pick-and-place robots and (b) welding robots. (Courtesy of Bastian Solutions and Gould Communications.)

FIGURE 7.38 Articulated arm type robot designed to palletize products. (Courtesy of C&D Skilled Robotics.)

A point-to-point robot can move only between two prespecified points. It is useful in load or unload, pick-and-place, and in spot welding operations. On the other hand, a contouring robot can move along any number of points on a path. Thus, by choosing a large number of closely spaced points, we can have the robot move along a smooth curve. Such robots are useful in spray painting, continuous welding, and object grasping operations. The control and memory requirements for contouring robots are obviously much greater than that of point-to-point robots as we are concerned with only two points in the latter whereas multiple points in the former. A walkthrough or teach robot is one that can be programmed by manually moving the robot arm through a sequence of moves. The robot then "memorizes" these moves and repeats them continuously until it is reprogrammed again. Variations of this, for example, the teach pendant robot, are available. The pendant is a handheld device that we use to program the robot's movements. Once we teach a set of moves to the robot using the pendant, it executes these until reprogrammed. In addition, robots that can be programmed off-line are also available. Such systems allow us to increase robot utilization as virtually no time is lost in teaching the robot. Programming for a new move is generally done while the robot is still executing its previous one. Based on the type of drive system employed, robots can be classified as hydraulic robots, servo-controlled robots, or pneumatic robots. Hydraulic robots require more space due to the drive system but can be used for applications that require greater strength and speed (Groover and Zimmers, 1984). Servo-controlled robots are driven by DC stepping motors and are more accurate than hydraulic robots.

7.5.6   AGVs

AGVs have become popular especially in the past 25 years and will continue to so in the years to come. There are fewer AGV installations in the United States compared to Europe. AGVs are beginning to find use in a variety of manufacturing and distribution companies—from newspaper companies to automobile companies (see Figure 7.39a through c for illustrations). The AGV in Figure 7.39a is used for transporting equipment in a hospital, the one in Figure 7.38b handles large steel parts, and the one in Figure 7.39c is used as a tow truck. When AGVs are not in use, they typically wait at a charging station where the on-board batteries are charged.

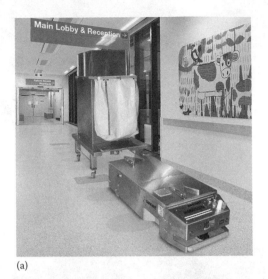

(a)

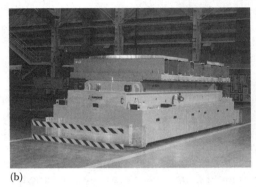

(b)

(c)

FIGURE 7.39 Use of AGVs in service and manufacturing industries. (a) AGV used in a hospital, (b) AGV transporting large block of metal, and (c) AGV used as a tow-truck. (Courtesy of Savant Automation.)

An AGV is a variable path device whose movement can be controlled in a number of ways. Wires or special tapes embedded on the shop floor transmit radio signals, or strategically positioned radars send signals to receivers mounted on AGVs. The tapes behave much like the embedded wires with the added advantage that they can be easily removed and placed elsewhere, thus allowing us to vary the AGV path as frequently as desired. Using wires, tapes, or radars, we can thus specify when, how, and where the AGV should move. When fully automated, these can be used for material-handling activities in an unmanned environment (see Figure 7.40). In some applications, pick-and-place robots mounted on AGVs have been used for automated loading, transportation, and unloading. The material-loading surface can be of the platform type, forklift type, or pallet type. In addition, AGVs can be integrated with other MHDs and also are effective in integrating material-handling activities in different cells. For example, AGVs can be attached to towlines for certain material-handling operations and thus be used in place of industrial trucks (Askin and Standridge, 1993). The deck configurations can be modified to allow the AGV to be used in conjunction with other handling equipment, e.g., a conveyor and a forklift truck.

Figure 7.40 illustrates the use of AGVs in shipyard operations. A fleet of AGVs traverse an area with embedded wires and follow a fixed, unidirectional, circular path adjacent to the ship docking area and are computer controlled from a central tower. The load, or container, is then picked up by a gantry crane with hoists and loaded on the ship deck.

FIGURE 7.40 AGV and gantry crane used for loading containers on ships. (Courtesy of Europe Combined Terminals B.V.)

7.5.7 Hoists, Cranes, and Jibs

These MHDs are preferred when the parts to be moved are bulky and require more space for transportation. Because the space above machines is typically utilized only for carrying power and coolant lines, there is abundant room to transport bulky material. Hoists, cranes, and jibs make use of the space above the machines. Thus, material movement need not interfere with production activities or those of production personnel. However, these MHDs are very expensive and time consuming to install and require elaborate foundations and support. Hoists transport material in a vertical direction using large hooks. This lifting device may be manual, electrical, or pneumatic (see Figure 7.41). On the other hand, cranes transport material in a horizontal direction using hooks mounted on large beams (see Figures 7.42 through 7.44). The beams themselves can move horizontally, thus allowing a crane to access any point on the shop floor. Of course, this access is limited to a certain range, which depends on the length of the horizontal beams and the space they can traverse. Cranes may be supported overhead or on the floor via vertical beams. The latter is called a gantry crane (Figure 7.43) and the former is referred to as a bridge crane (Figure 7.44). Jibs appear much like a gantry crane without one of the vertical support beams

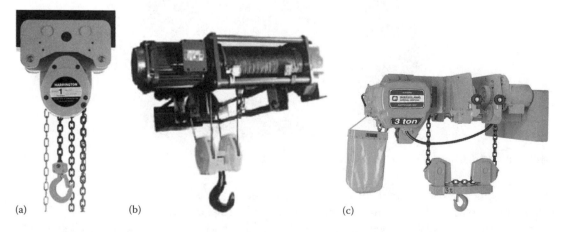

(a)　　　　(b)　　　　(c)

FIGURE 7.41 (a) Manual hoist, (b) electric hoist, and (c) pneumatic hoist. (Courtesy of Harrington and Ingersoll-Rand.)

FIGURE 7.42   Crane attachment. (Courtesy of Bushman Equipment.)

FIGURE 7.43   Gantry crane. (Courtesy of LK Goodwin Co.)

(Figure 7.45). The overhead arm can also be rotated, providing access to all points within a circle, whose diameter is twice the arm's length. Once again, further classification of cranes is possible. For example, tower cranes, such as those in Figure 7.46, used in multistoried building construction sites and shipyards, stacker cranes such as those used in storage and retrieval of items in warehouses, and so on.

(a) (b)

FIGURE 7.44 Bridge crane. (a) Dock loading application and (b) machine-shop application. (Courtesy of LK Goodwin Co.)

(a) (b)

FIGURE 7.45 Jib cranes. (a) Simple jib crane and (b) manufacturing application. (Courtesy of LK Goodwin Co.)

7.5.8 WAREHOUSE MATERIAL-HANDLING DEVICES

Warehouse MHDs are typically referred to as storage and retrieval systems. If they are automated to a high degree, they are called automated storage and retrieval systems (AS/RS). A warehouse (including an AS/RS) was illustrated in Figure 1.14. The primary function of an AS/RS is to store and retrieve materials in addition to transporting them between the pick or deposit (P/D) stations and the storage location of the materials. They can be classified in a number of ways. For example, we can classify AS/RS based on the type of load they can handle. Thus, we have unit load AS/RS for storing or retrieving palletized unit loads, mini-load AS/RS for storing or retrieving small parts, and

(a)   (b)

FIGURE 7.46   Tower cranes. (a) Construction cranes and (b) shipyard overhead crane. (Courtesy of Europe Combined Terminals B.V.)

even micro-load AS/RS for storage, retrieval, and handling of even smaller products such as circuit card components. In addition, we have man-on-board AS/RS in which a person does the storing and retrieving, deep-lane AS/RS that consist of vehicles that can move in narrow aisles and therefore provide more storage space than other comparable systems, and storage carousels in which the storage bins themselves revolve allowing items to be picked at the end of the aisle. Because aisle space is not required in a storage carousel system, more storage space is available as the racks can be placed very close to one another. A more detailed discussion of warehouses is provided in Chapter 8.

7.6   AGV SYSTEMS

Although we have seen a more detailed classification of MHSs, for purposes of modeling and analyzing these, it is convenient to have the classification presented as follows (Johnson and Brandeau, 1992):

- Synchronous systems
- Asynchronous systems

Synchronous systems are typically used in continuous processes or in heavy traffic, discrete parts environments. Asynchronous systems, on the other hand, are used in light traffic, discrete part environments and where material-handling flexibility is desired. Examples of synchronous systems include conveyors and asynchronous systems include AGV system, AS/RS, forklift trucks, monorails, cranes, and hoists. Because AS/RS is typically used to transfer material into and out of storage and because we have a detailed chapter on warehousing, their design and analysis is covered in the next chapter. Beginning with the next section, we provide specific design and analysis models with numerical examples for synchronous and asynchronous (other than AS/RS) systems. In the remainder of this section, we overview some of the design and control problems encountered in AGV systems.

1. Material flow network, i.e., travel paths and travel directions of the material-handling devices
2. Location of P/D points in the material flow network
3. Number and type of AGVs
4. Assignment of AGVs to material transfer requests
5. AGV routing and dispatching
6. Strategies for resolving route conflicts

These problems must be addressed so that the volume of material transferred (or AGV throughput rate) is maximized and these costs are minimized.

- Purchase, maintenance, and operating costs of AGVs, computer control devices, and the material flow network
- Inventory costs and production equipment idle costs incurred due to excessive material transfer and wait times

The design and control problems listed above are rather difficult to solve. In some cases, it is so difficult to even formulate a realistic mathematical model that we often have to resort to simulation to design and control AGV systems. As the reader probably knows (or will find out in a later section in this chapter), simulation is a tool that can be used to *predict* the performance of a *given* system. It cannot design a system by itself. The user has to provide various design scenarios and evaluate each of these with the simulation tool to determine which of the scenarios performs best with respect to some selected criteria. Clearly, the number of possible scenarios is so large that it is impossible to evaluate all of them. Hence, we adopt the following sequential procedure to solve the above design and control problems.

Determine the layout and location of P/D points: The models and algorithms we studied in Chapters 4, 5, and 10 are useful for determining the layout of machines, but what about the P/D points? Depending on the production equipment configuration, this may be already known to us. For example, the load and unload section of the production equipment maybe located and accessible only from one side of the equipment, and thus, this may restrict us to locate the P/D points on this side of the equipment. If there are no system constraints in locating the P/D points, we could use some models in the literature that determine the location of machines and P/D points simultaneously (see, for example, Heragu, 1990).

Determine the material flow network including the travel paths and travel directions of the AGVs: There are some integer programming models to optimally determine the material flow network, given an existing layout and a P/D point location. These can be solved optimally only for small-sized problems. Hence, heuristic algorithms have been developed. The models and heuristic algorithms are reviewed in Rajagopalan and Heragu (1997).

Determine the number and type of AGVs: There are deterministic and probabilistic models for determining the number and type of AGVs (e.g., Maxwell and Muckstadt, 1982; Srinivasan et al., 1994). The problem with deterministic models is that they assume all the required information (e.g., material transfer requests) is known with certainty. However, the MHS environment is highly stochastic. The difficulty with stochastic models is that they often overestimate or underestimate the required number of AGVs, mainly because it is difficult to estimate the vehicle congestion and empty (unloaded) travel times. Because the deterministic and probabilistic models have certain drawbacks, the three-stage approach outlined in Chapter 10, namely, to employ a deterministic model (e.g., the model in Maxwell and Muckstadt, 1982) to come up with an initial number or a lower bound on the required number of vehicles, estimating the throughput rate using a stochastic model (such as the one in Srinivasan et al., 1994), refining this number to achieve a desired level of throughput, and conducting further analysis of this refined solution via simulation, could be used.

Determine the assignment of AGVs to material transfer requests: Even for this problem, deterministic models such as assignment, transportation, and mixed-integer programming models and probabilistic models can be used to determine how the AGVs can be assigned to departmental moves. See Das and Heragu (1988) and Johnson and Brandeau (1992) for a brief overview of deterministic and stochastic models, respectively.

Determine AGV routing and dispatching rules and develop strategies for resolving route conflicts: Simulation tends to be the dominant tool used to determine and study the impact of AGV routing and dispatching as well as conflict resolution rules, although there are some analytical models available (see Malmborg and Shen, 1994, for example). Once again, we refer the reader to the paper by Johnson and Brandeau (1992) for an overview of the various simulation-based studies that have considered various strategies.

It is perhaps apparent to the reader that there are numerous models and algorithms available for each of the problems listed above. Rather than discuss all of these, we refer the interested reader to the following survey papers that describe the various approaches: Johnson and Brandeau (1992), Noble and Tanchoco (1993), Sinriech (1995), and Heragu and Rajagopalan (1996).

Three material flow network designs are shown in Figures 7.47 through 7.49. In the first, we have P/D points in which the pick-up and drop-off points are located at the same place (see points 2, 4, and 5) and those that are separated (see points 1, 3, and 6). The network shown in Figure 7.47 assumes unidirectional travel and has multiple loops, whereas the one in Figure 7.48 has a single loop but allows bidirectional travel. Although bidirectional travel offers more flexibility in material transfer, the operational control of such systems poses more difficulties. Figure 7.49 shows an interesting design that is based on the cellular manufacturing concept. Just as we partition a large system into mutually independent subsystems using the cellular manufacturing concept, we partition a large material flow network into several nonoverlapping, single loop, unidirectional networks that allow for parts to be transferred between any pair of pick-up and drop-off points. This configuration (presented by Bozer and Srinivasan, 1989) was called the tandem configuration and is illustrated in Figure 7.47. While it has some disadvantages, for example, a part may have to be handled over several loops before it reaches its destination, the configuration shown in Figure 7.49 offers several advantages. Without employing bidirectional vehicles, the configuration offers material transfer flexibility. Congestion is eliminated, and the vehicle management and traffic management control system is less complicated than in a conventional system such as the one shown in Figure 7.47.

7.7 MODELS FOR MATERIAL-HANDLING SYSTEM DESIGN

The design of a manufacturing system involves solving a number of problems in a certain sequence. It typically involves selection of the required type and number of production/manufacturing equipment as well as MHDs, grouping of appropriate machines into cells, and determining the layout of the machines within each cell and the layout of the cells. Obviously, this is not the only way in which a system is to be designed. Sometimes, it may be advantageous to put off the MHD selection problem until after the layout of machines, and, hence, the type and number of production or

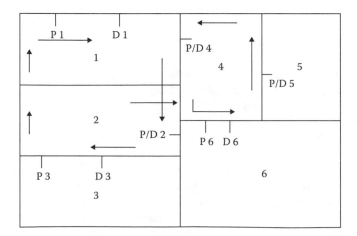

FIGURE 7.47   Conventional and unidirectional material flow network configuration with multiple single loops.

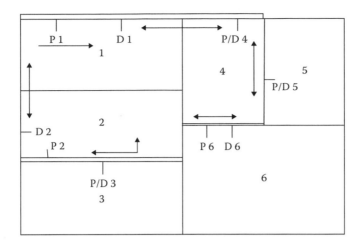

FIGURE 7.48 Conventional network configuration with a single loop and bidirectional travel.

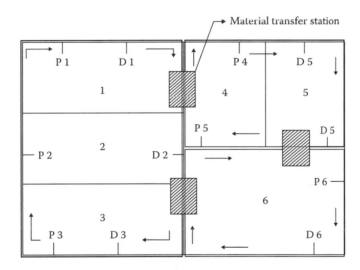

FIGURE 7.49 Network configuration with multiple nonoverlapping single loops and unidirectional travel.

manufacturing equipment has been determined. For example, a U-line or circular layout may dictate use of a robot or conveyor only for material handling; a straight-line layout may call for an AGV and so on. In other situations, it may be advantageous to select the production equipment and MHDs at the same time.

7.7.1 Rule of Thumb Approach

TranSystems has provided an online cost calculator that can be accessed at http://www. werc.org/assets/1/Publications/TranSystems%20Rules%20of%20Thumb%202010%20v2.pdf. This calculator is based on the work originally done by Gross & Associates and can be used to estimate capital equipment costs for material-handling equipment based on the selected layout, application, warehouse technologies, and operating systems (see Figure 7.50). Although this calculator is meant to provide a rough cost range for commonly used material-handling equipment, it is a useful tool for companies exploring alternate material-handling options.

Handling equipment			Handling equipment: continued		
Electric lift trucks (with battery and charger)			Rail guidance cost per foot installed		$12–$17
By type and capacities					
Walkie pallet jack—48 inch forks			Battery changing/charging systems		
	4,000 lb.	$3,500–$5,500	Charging system base price		$25,000–$40,000
	6,000 lb.	$6,500–$8,500	Additional cost for each		$450–$850
Rider pallet jacks—48 inch forks			Battery position		
	6,000 lb.	$9,000–$11,000	Propane counterbalanced life trucks		
	8,000 lb.	$9,500–$13,000			
Rider pallet jacks—96 inch forks				3,000 lb.	$18,000–$21,000
	6,000 lb.	$9,000–$12,000		4,000 lb.	$20,000–$26,000
	8,000 lb.	$9,500–$13,000		5,000 lb.	$21,000–$27,000
Low level order picker truck	$12,000–$20,000			6,000 lb.	$24,000–$29,000
Tugger	$8,000–$12,000				
Walkie straddle stacker			Lift truck options		
	3,000 lb.	$12,000–$16,000	Carton clamp		$8,000–$13,000
Walkie reach truck			Paper roll clamp		$9,000–$18,000
	3,000 lb.	$16,000–$20,000	claw clamp to pick layers of cartons		$18,000–$31,000
Counterbalanced life truck			Ship-sheet-push/pull		$7,000–$12,000
	3,000 lb.	$23,000–$28,000	Side-shifter		$1,500–$ 2,500
	4,000 lb.	$24,000–$29,000	Drum handler		
	5,000 lb.	$26,000–$30,000	(single non-hydraulic)		$ 1,500–$2,000
	6,000 lb.	$29,000–$33,000	Drum handler		
Narrow-aisle reach truck			(double non-hydraulic)		$2,500–$4,000
	3,000 lb.	$25,000–$31,000	Drum handler		
	4,000 lb.	$27,000–$34,000	(single hydraulic)		$6,000–$9,000
	4,500 lb.	$28,000–$37,000	Drum handler		
Narrow-aisle deep reach truck			(double hydraulic)		$7,000–$10,000
	2,500 lb.	$26,000–$34,000	Lift height sensors		$500–$2,000
	3,200 lb.	$27,000–$35,000	Mast camera monitors		$3,000–$4,000
Side loader truck			Truck mounted scales		$500–$6,000
	6,000 lb.	$90,000–$100,000	Automatic guided vehicles (AGV's)		
	10,000 lb.	$95,000–$120,000	Agv unit-load carrier with Roller deck (includes agv controller, wireless or wired path, batteries, charges, etc.)		$200,000–$300,000
Four directional truck					
	4,500 lb.	$45,000–$60,000			
Order picker truck			Agv tow vehicle (includes agv controller, wireless or wired path, batteries, chargers, etc.)		$150,000–$200,000
	3,000 lb.	$23,000–$39,000			
Wire guidance package	$ 6,500–$ 8,500				
Turret truck—operator up			Agv tow trailer without roller bed		$3,000–$5,000
w/Wire guidance	3,000 lb.	$80,000–$110,000			
Swing mast truck			Agv tow trailer with Non-powered roller bed		$12,000–$16,000
w/wire guidance	4,000 lb.	$50,000–$85,000			
Narrow aisle articulating			Agv tow trailer with powered roller bed		$20,000–$25,000
Truck	4,000 lb.	$55,000–$75,000			
Truck guidance			Agv with mast and forks (outrigger type)		$250,000–$350,000
Wire guidance cost per foot Installed	$3–$5		Agv pick-up and delivery (p&d) stations		$ 9,000–$15,000

FIGURE 7.50   Excerpt from warehousing and manufacturing pricing guidelines. (Courtesy of TranSystems.)

7.7.2   PROBABILISTIC APPROACH

Deterministic models assume that all the required data are known with certainty. In reality, we do not always have this luxury. For example, we do not know exactly how many units of each part type are going to be produced, and therefore we do not know exactly what the material-handling requirements are. Similarly, we may assume that the service rate of the material-handling device is deterministic, but, in actuality, the service rate may indeed be probabilistic because of the process variability, device failure, maintenance, operator unavailability, and other factors. For example, examination of past service data may indicate that the service time is exponentially distributed with a mean of 5 minutes. Under such circumstances, queuing-based and simulation models are useful tools in estimating the system performance. Obviously, each has its advantages and disadvantages. Simulation models, for example, are time consuming to develop but can be used to model highly complex scenarios. Queuing-based models, on the other hand, can be developed and solved rather easily. Another major advantage is that the primary type of data required for such models is the arrival rates of parts and service rates of MHDs. With the availability of software for educational use (e.g., RAQS, Kamath et al., 1995, available at http://www.okstate.edu/cocim/raqs/index.html and MPA, Meng et al., 2008, available at http://sundere.okstate.edu/downloadable-software-programs-and-data-files), use of these models for obtaining estimates of key performance measures for manufacturing and MHSs is rather straightforward. However, modeling situations in which there are nonexponential service times, limited waiting buffers, and complex dispatching rules is not trivial with these models, and we often have to resort to simulation models.

7.7.2.1 Queuing Model for MHS Design

As discussed in Chapter 13, queuing and queuing network models give us estimates of (1) the time a part (or customer) spends in queue, (2) the time a part spends in system, (3) the average number of parts in queue, (4) the average number of parts in system, and (5) the utilization of server (MHS or machine). Before discussing application of queuing models for MHS design, we caution the reader that queuing-based models are descriptive, rather than prescriptive. In other words, they do not prescribe how the system is to be designed, but they describe how the system is going to behave for a given system design. For example, they do not tell us the optimal number of operators and trucks to be assigned to the material-handling problem, but for a given configuration, they help determine the expected idle time for operators and forklift trucks. If we consider a set of feasible solutions (system designs), we can use queuing-based models to determine which of these performs best with respect to some specified criteria as illustrated in Example 7.1.

Example 7.1

A bottled water producer in New England has a large warehouse adjoining the bottling facility. Pallets of bottled water cases have to be delivered from the palletizer to the warehouse. Due to the lift requirements, surface of the corridor connecting the bottling facility to the warehouse, ramps, costs, and other considerations, the company has decided to use forklift trucks for delivery of the pallets to the warehouse. These can be leased from a manufacturer. It has been determined that it takes an average of 15 minutes for a forklift truck to travel from the palletizer to the warehouse, unload the pallet, and return to the palletizer in the bottling facility. An operator is required to assist in the loading operation, and this takes 12 minutes per pallet on average. Two or three such operators are available. Given that the inter-arrival and operator service times follow an exponential distribution, operator and forklift leasing hourly costs are $20 and $50, respectively, and that the company wants to lease five trucks, determine whether two or three operators should be assigned in order to minimize the operator *and* forklift truck idle time.

Solution

The aforementioned problem can be modeled as a finite source queuing model in which there are a prespecified number of servers and finite number of customers arriving for service according to known arrival and service time distributions. Using the notation in Chapter 13, we can ascertain that for the first configuration, $c = 2$, $N = 5$, $\lambda = 4$, and $\mu = 5$. Using the following formula (reproduced from Chapter 13 for convenience), we find that $c_1 = 4$, $c_2 = 6.4$, $c_3 = 7.68$, $c_4 = 6.14$, and $c_5 = 2.46$.

Multiple server model:

$$c_n = \begin{cases} N!\lambda^n/((N-n)!n!\mu^n) & \text{for } n = 1,2,\ldots,S \\ N!\lambda^n/((N-n)!S!S^{n-S}\mu^n) & \text{for } n = S+1, S+2,\ldots,C \\ 0 & \text{for } n > C \end{cases}$$

Using the general formula $P_n = c_n P_0$, i.e., $P_1 = 4P_0$, $P_2 = 6.4P_0$, $P_3 = 7.68P_0$, $P_4 = 6.14P_0$, $P_5 = 2.46P_0$, and $P_0 + P_1 + P_2 + P_3 + P_4 + P_5 = 1$, we find that

$$P_0 = 0.036, \; P_1 = 0.145, \; P_2 = 0.231, \; P_3 = 0.278, \; P_4 = 0.222, \; P_5 = 0.088$$

The aforementioned values of P_i ($i = 0, 1, \ldots, 5$) and the following formulae (from Table 13.3) allow us to calculate the average time spent in queue by a customer, i.e., forklift truck:

$$L_q = \sum_{n=S}^{n} (n - S) P_n; \quad L = L_q + \sum_{n=0}^{S-1} n P_n + S\left(1 - \sum_{n=0}^{S-1} P_n\right); \quad W_q = \frac{L_q}{(\lambda(N - L))}$$

$$L_q = 0 * P_2 + 1 * P_3 + 2 * P_4 + 3 * P_5$$

$$= 0.278 + 2 * 0.222 + 3 * 0.088 = 0.986 \text{ trucks}$$

$$L = 0.986 + 0 * P_0 + 1 * P_1 + 2(1 - P_0 - P_1)$$

$$= 0.986 + 0.145 + 2(1 - 0.036 - 0.145) = 2.769 \text{ trucks}$$

$$W_q = \frac{0.986}{4 * (5 - 2.769)} = 0.110 \text{ hours}$$

Also, because the fraction of time an operator is idle is given by $P_0 + 0.5P_1$ (why?—see Exercise 16a), we can determine the operator idle time to be 0.109 hours. The idle cost when two operators are assigned is therefore

(Number of operators) $*$ (Idle time per operator) $*$ (Hourly cost per operator)
+ (Number of Trucks) $*$ (Average waiting time in queue for truck)
$*$ (Hourly leasing cost for truck) $= 2 * 0.109 * 20 + 4 * 0.11 * 50 = \$26.36$

Similarly, it can be verified (see Exercise 20) that when three operators are assigned to assist in loading the forklift trucks, the average time spent in queue by a forklift truck and the operator idle time is 0.0197 and 0.2904 hours, respectively. Notice that the fraction of time an operator is idle for the three operator case is given by $P_0 + 0.67P_1 + 0.33P_2$ (why?—see Exercise 16b). The idle cost when three operators are assigned is

(Number of operators) $*$ (Idle time per operator) $*$ (Hourly cost for operator)
+ (Number of Trucks) $*$ (Average waiting time in queue for truck)
$*$ (Hourly leasing cost for truck) $= 3 * 0.2904 * 20 + 4 * 0.0197 * 50 = \$21.36$

Thus, it is worthwhile increasing truck utilization by incurring additional labor costs. As mentioned in Chapter 13, as long as the inter-arrival, i.e., travel and loading times are independent with mean $1/\lambda$, the above results will hold no matter what the distribution of the interarrival time is. However, when the service time is nonexponential, exact analytical results are not available, and we have to resort to analysis via approximate queuing models or exact analysis via simulation.

7.7.3 STATIC AND PROBABILISTIC APPROACH

In this section, we discuss a spreadsheet-based software (AGV-DST) developed for MHI that uses two models—a static, transportation model to determine the minimum number of AGVs vehicles and a probabilistic model to estimate the average number of vehicles—required to move material between workstations in a given layout (Meller et al., 2005). The Microsoft Excel–based software is available at http://sundere.okstate.edu/Book and requires the Solver Add-in feature to be installed. The user enters basic data such as number of workstations, loaded and unloaded flow (measured in trips per unit time) between them, AGV speed, maximum desired utilization, load and unload times, vehicle availability per hour, congestion, and delay factors.

Example 7.2

For the input data provided in Figure 7.52a and b, use the AGV-DST spreadsheet software to estimate the minimum and average number of AGVs required for a six workstation job shop.

Solution

After entering the required data in AGV-DST, when the user clicks on the "Calculate Results" icon, the program produces a intermediate and summary results including minimum and average number of AGVs required (Figure 7.53a and b, respectively).

7.8 OPERATIONAL ASPECTS OF MATERIAL-HANDLING SYSTEM

Once the required MHDs have been purchased, at an operational level, it is important to ensure that there are no performance or operational problems. Managers and supervisors want to be sure that MHDs are not underutilized, parts are not spending an inordinate amount of time waiting to be transferred, and so on. In addition, schedulers would like to know how to assign departmental moves to the various MHDs, what dispatching rules to use for each MHD, and the sequence in which jobs or orders are to be transported or picked. Just as in the design domain, there are several approaches one could use to address planning, operational, and system performance issues. In the remainder of this chapter, we discuss deterministic and probabilistic approaches to MHS operational problems.

7.8.1 Deterministic Approach

There are very few deterministic models that address operational problems in MHSs. Hassan et al. (1985) presented an integer programming model for selecting MHD and assigning them to departmental moves. Assuming that the layout of the machines (and hence, the rectilinear distance between their centers) is available, they estimate the travel time and costs. Empty trips are not taken into consideration. Given a set of candidate-handling equipment, they develop a model that minimizes the sum of operating and capital costs. The model allows only one equipment to perform each move. They also developed a heuristic construction algorithm to solve the integer programming model. The algorithm implicitly considers maximizing equipment utilization and minimizing variations in equipment types.

7.8.2 Probabilistic Approach

As in the case of MHS design, a majority of the models used to study operational performance measures of general manufacturing systems or MHSs are queuing-based or simulation models. The latter is discussed in more detail in Chapter 13. We focus on queuing and queuing network models in this section.

7.8.2.1 Queuing and Queuing Network Models for MHS Performance Analysis

As discussed in Chapter 13, queuing network models represent networks of queues. These are generally more difficult to solve than queuing models. Hence, we study them last. First, we discuss application of some basic queuing models.

Example 7.3

An AGV that has a custom-made loading deck is dedicated to transport high-precision discrete parts in unit loads between two workstations. After delivering its load to the second workstation, the AGV immediately returns empty to the first one to pick up the next one waiting in the output queue of the first workstation. Analysis of past data indicates that the time between two successive load pickups follows an exponential distribution with a mean of 6 minutes. Because the transport path is fixed and "noise" in the transit time due to external factors is negligible, the sum of the AGV loading, transportation, and unloading time can be treated as being deterministic and equal to 5 minutes. Determine the utilization of the AGV and the average number of parts waiting for transportation in the output queue of the first workstation.

Solution

The above problem may be modeled as an M/G/1 queuing model. Also, $\lambda = 10$, $\mu = 12$. We know that the utilization ρ for the AGV is given by $\lambda/\mu = 10/12 = 0.83$, and the average number of unit loads waiting in the output queue for transportation is given by $L_q = (\lambda^2\sigma^2 + \rho^2)/(2(1 - \rho))$ (see Chapter 13). Because the service time is deterministic, $\sigma^2 = 0$ and therefore $L_q = 0.83 * 0.83/2(1 - 0.83) = 2.08$ unit loads. The above system appears to be operating reasonably well because the utilization of the AGV is not too high and there are only two loads waiting at the output queue.

Example 7.4

Two AGVs are used for material transportation in a flexible manufacturing cell consisting of four workstations—A, B, C, and D. Two part types are processed in this cell. Part type 1 is processed on A, B, C, and A in that sequence. Part type 2 is processed on D, C, and B. All the material transfers between machine pairs are done by the AGVs. Any of the AGVs can be used to transfer any of the part types. The part types are transported on pallets whose numbers are limited based on their expected annual production volume. Four pallets are available for part type 1 and three for part type 2. After a part is processed, the same pallet is used to load the next part. Thus, the four pallets for part type A and three for part type B keep circulating within the system. The setup is illustrated in Figure 7.51. The processing times for each part type on each machine are provided in Table 7.3. Determine the utilization of AGV and workstation A and the average number of units of part type A on or waiting for the AGV.

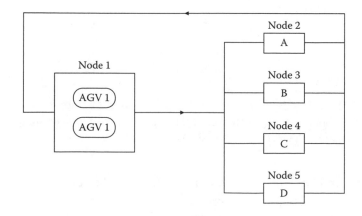

FIGURE 7.51 Queuing network representation of the problem described in Example 7.4.

Solution

This problem may be modeled as a closed queuing network and solved heuristically using the mean value analysis (MVA) algorithm discussed in Chapter 13. Recall that the MVA procedure in Section 13.4.4 involves calculating the following three items—throughput time, production rate, and queue length. Using the notation in Chapter 13 for N^r, i.e., the number of pallets for part type r, $N^1 = 4$ and $N^2 = 3$ for our problem. Each node, except the AGV node, has a single server. Label the node corresponding to the AGV as 1, node corresponding to workstation A as 2, and so on (see Figure 7.51). Based on the service times given in Table 7.3, we can write the service rates as follows:

$$\mu_1^1 = 20; \ \mu_2^1 = 5; \ \mu_3^1 = 10; \ \mu_4^1 = 7.5; \ \mu_4^1 = 0; \ \mu_1^2 = 12; \ \mu_2^2 = 0; \ \mu_3^2 = 15; \ \mu_4^2 = 30; \ \mu_4^2 = 6.67$$

Once again, using the notation in Chapter 13 for W_{ir}, L_{ir}, v_{ir}, and X_{ir}, i.e., the average waiting plus service time per visit of part type r to machine i, the average number of parts of type r at machine i, the production (or arrival) rate for part type r, and the expected number of visits of part type r to machine i, respectively, and the following formulae, we are now ready to apply the iterative MVA algorithm presented in Chapter 13.

$$W_{ir} = \frac{1}{\mu_i^r} + \left[\frac{N^r - 1}{N^r}\right]\left[\frac{L_{ir}}{\mu_i^r}\right] + \sum_{i \neq r} \frac{L_{ir}}{\mu_i^r}$$

$$X_r = \frac{N^r}{\sum_{i=1}^{m} v_{ir} W_{ir}}$$

$$L_{ir} = X_r(v_{ir} W_{ir})$$

The steps of the MVA algorithm have been carried out in a spreadsheet and are shown in Table 7.4. After seven iterations, we find that all the L_{ir} values in the last iteration converge to within 4% of the values found in the previous one. Hence, we terminate the algorithm.

To determine the average number of units of part type A on or waiting at the AGV, we look up the value for L_{11} in the last iteration, which is equal to 0.88. Because there are four pallets for part type A, this means that roughly 22% (0.88/4) of the time, the pallets are not in production. To calculate the AGV utilization, we must find out the utilization due to both part types. We know that, in general, the formula $\rho = \lambda/s\mu$ yields the utilization. However, because a part type can make more than one visit to a machine or AGV, the arrival rate must be adjusted accordingly. Thus, the AGV utilization is the sum of the utilizations due to part types A and B.

TABLE 7.3

Processing Times for Two Part Types on Four Workstations and the AGV

| Part Type | AGV | Workstation | | | |
		A	B	C	D
1	3	12	6	8	—
2	5	—	4	2	9

TABLE 7.4

Iterations of the MVA Algorithm

Iteration Number	Step Number	Part Type	AGV	Workstation A	B	C	D	No. of Pallets
				Transportation and processing times				
		1	3	12	6	8	—	4
		2	5	—	4	2	9	3
				No. of visits				
		v_{i1}	3	2	1	1	0	
		v_{i2}	2	0	1	1	1	
				Service rates				
		μ_i^1	20	5	10	7.5	0	
		μ_i^2	12	0	15	30	6.67	
Iteration 1	Step 1	L_{i1}	1	1	1	1	0	
		L_{i2}	0.75	0	0.75	0.75	0.75	
	Step 2	W_{i1}	0.10	0.35	0.23	0.26	0.00	
		W_{i2}	0.13	0.00	0.20	0.18	0.23	
	Step 3	X_1	2.70					
		X_2	3.46					
	Step 4	L_{i1}	0.81	1.89	0.61	0.70	0.00	
		L_{i2}	0.89	0.00	0.69	0.63	0.78	
Iteration 2	Step 2	W_{i1}	0.10	0.48	0.19	0.22	0.00	
		W_{i2}	0.13	0.00	0.16	0.14	0.23	
	Step 3	X_1	2.37					
		X_2	3.83					
	Step 4	L_{i1}	0.73	2.29	0.45	0.53	0.00	
		L_{i2}	0.98	0.00	0.61	0.54	0.87	
Iteration 3	Step 2	W_{i1}	0.10	0.54	0.17	0.20	0.00	
		W_{i2}	0.13	0.00	0.14	0.12	0.24	
	Step 3	X_1	2.25					
		X_2	4.00					
	Step 4	L_{i1}	0.71	2.44	0.39	0.46	0.00	
		L_{i2}	1.03	0.00	0.56	0.46	0.95	
Iteration 4	Step 2	W_{i1}	0.11	0.57	0.17	0.19	0.00	
		W_{i2}	0.13	0.00	0.13	0.10	0.24	
	Step 3	X_1	2.21					
		X_2	4.06					
	Step 4	L_{i1}	0.70	2.50	0.37	0.43	0.00	
		L_{i2}	1.05	0.00	0.53	0.43	0.99	
Iteration 5	Step 2	W_{i1}	0.11	0.57	0.16	0.19	0.00	
		W_{i2}	0.13	0.00	0.13	0.10	0.25	
	Step 3	X_1	2.19					
		X_2	4.07					

(*Continued*)

TABLE 7.4 (*Continued*)
Iterations of the MVA Algorithm

Iteration Number	Step Number	Part Type	AGV	Workstation A	B	C	D	No. of Pallets
	Step 4	L_{i1}	0.70	2.52	0.36	0.42	0.00	
		L_{i2}	1.06	0.00	0.52	0.41	1.02	
Iteration 6	Step 2	W_{i1}	0.11	0.58	0.16	0.19	0.00	
		W_{i2}	0.13	0.00	0.13	0.10	0.25	
	Step 3	X_1	2.19					
		X_2	4.08					
	Step 4	L_{i1}	0.70	2.53	0.35	0.41	0.00	
		L_{i2}	1.06	0.00	0.51	0.40	1.03	

$$\frac{X_1 v_{11}}{2\mu_{11}} + \frac{X_2 v_{12}}{2\mu_{12}} = \frac{1.56 * 3}{2 * 20} + \frac{3.05 * 2}{2 * 12} = 0.371.$$

In a similar manner, utilization of the workstations can also be determined (Exercise 21). From the aforementioned solution, it appears that we could do with just one AGV because the utilizations are fairly low. To get familiarized with the MVA method, we suggest that the reader resolve the problem on a spreadsheet assuming there is only one AGV (see Exercise 22).

7.9   SUMMARY

In this chapter, we provided an introduction to the various types of material-handling devices available today. We discussed deterministic, stochastic, and knowledge-based models for dealing with the design and operational problems encountered in MHSs. We provide a brief overview of simulation models and a more detailed discussion of queuing and queuing network models in Chapter 13. Due to the advances in computer technology, simulation will become a dominant analysis tool for manufacturing and MHSs. When used in conjunction with deterministic as well as queuing and queuing network–based models, we can use simulation models even more efficiently.

7.10   REVIEW QUESTIONS AND EXERCISES

1. Define material handling. Explain the importance of material handling in manufacturing and distribution activities.
2. List and discuss the 10 principles of material handling.
3. *Project*: Learn the 10 principles of material handling available at http://sundere.okstate.edu/downloadable-software-programs-and-data-files. Explore the case study presented in the *Integrate* layer and answer the following questions:
   (a) Which principles were applied in the facility illustrated in the *integrate* layer
   (b) Which principles were not applied?
   (c) Were any applied incorrectly.
   Provide sound justification for your answers.
4. What are some of the functions that MHSs provide in addition to material transportation? Explain.

5. Explain in your own words what is meant by the unit load concept. Discuss the seven-step procedure of unit load design discussed in Section 7.4.1.
6. List the advantages and disadvantages of each type of MHD listed in Section 7.5. Mention the kind of material-handling applications for which each is suitable.
7. *Project*: Consult relevant sources (e.g., *Material Handling Industry, Modern Materials Handling, Material Handling and Logistics, Plant Engineering, IE Solutions*) and develop a detailed classification of MHDs. (Note: the list should be more comprehensive than in Section 7.5.) Within each major classification, make at most two subclassifications. Explain in sufficient detail with illustrations of each type of MHD and provide at least one name and full address (telephone, facsimile numbers, e-mail, and Internet addresses) of the equipment manufacturer for each MHD type.
8. *Project*: Repeat Exercise 7 for conveyors.
9. Are palletizers material-handling equipment? Explain your answer.
10. *Project*: Repeat Exercise 7 for industrial trucks.
11. Make a list of six applications of robots. Briefly explain each.
12. Why are AGVs popular in material handling?
13. Repeat Exercise 7 for hoists, cranes, and jibs.
14. List and explain some of the more important design and operational problems encountered in AGV systems.
15. *Project*: For each of the following problems, collect relevant literature, examine the models and algorithms available, and discuss the advantages and disadvantages of each.
    (a) Material flow network
    (b) P/D point location in material flow networks
    (c) AGV selection
    (d) AGV assignment
    (e) AGV routing and dispatching.
16. Consider Example 7.1.
    (a) Explain why the fraction of time an operator is idle is $P_0 + 0.5P_1$ for the two operator case.
    (b) Explain why the fraction of time an operator is idle is $P_0 + 0.67P_1 + 0.33P_2$ for the three operator case.
17. Consider Example 7.1. Show that when three operators are assigned, the average time spent in queue by a forklift truck and the operator idle times are 0.0197 and 0.2904 hours, respectively.
18. A warehouse manager wants to determine the number of operators to assign to an unloading operation. (The loading operation is done automatically.) The material handling is done by six AGVs, and the average time to pick up a load and travel to the operator for unloading is 16 minutes. The average unloading time is 10 minutes. Given that the hourly rental–leasing costs for each AGV is $55 and the operator's hourly labor costs (including fringe benefits) is $35, determine how many operators are to be assigned to minimize operator and AGV idle time. (Assume exponential service time and interarrival distributions.)
19. Using a push diverter, units are loaded from a conveyor on an AGV deck automatically. The AGV picks up the load and delivers to another area where it is manually unloaded. Assuming the load, transit, and unload times can be treated as being deterministic and that it is 6 minutes and the arrival rate of units at the conveyor is 8 per hour, determine utilization of the AGV.
20. If the arrival rate increases to 7.5 per hour in Exercise 19, what problems occur? Justify your answer.
21. Determine the AGV utilization for Example 7.3.
22. Consider Example 7.3. Assuming there is only one AGV, determine

   (a) Utilization of the AGV and the four workstations A, B, C, and D.

   (b) Average waiting time and queue length of the two-part types at the AGV as well as the four workstations.

   (c) Compare the solution you get in parts a and b with the solution in Example 7.3 and Exercise 21. Which solution is better? Explain.

23. *Project*: Visit a nearby facility (manufacturing or service) that has a relatively extensive MHS. After interviewing relevant personnel in the facility and collecting necessary data, apply the "material-handling equation" in arriving at a material-handling solution. Compare your solution with the current MHS installation. Explain to the appropriate facility personnel and your class, details of your solution. If it is different, justify your solution.

8 Storage and Warehousing

8.1 AUTOMATED STORAGE AND RETRIEVAL SYSTEMS IN ACTION

Phoenix Pharmaceuticals, a German pharmaceutical company founded in 1994, has a 20,000 square meters warehouse in Herne, Germany (Figure 8.1). This warehouse, which has an annual turnover of $400 million, receives pharmaceutical supplies from 19 plants all across Germany and distributes them to area drug stores. Phoenix has a 30% market share and is a leader in the pharmaceutical business. Due to competitive and other business reasons, the company must fill each order from drug stores and ship it in less than 30 minutes. There are roughly 87,000 items stored in the warehouse of which 61% is pharmaceutical and the remainder are cosmetic supplies. The number of picks range anywhere from 150 to 10,000 in any given month. If Phoenix did not have warehouses located at strategic locations, it will obviously not be able to fill and ship orders and respond to its customers adequately. It is not only very costly for the company to ship the pharmaceutical supplies from the plant to each drug store directly but also not possible to do so because of the distances.

Order picking in Phoenix is done using the following three levels of automation:

1. Manual order picking using flow racks
2. Semiautomated order picking using an automatic dispensing system
3. Full automation using a robotic order picker

Incoming customer orders are printed on high-speed printers, and the orders are attached manually to containers and sent via conveyors to manual order-picking areas (Figure 8.2). Here, operators pick items specified in an order from flow racks, fill the container, and send it to shipping areas from where it is sent to the customer (i.e., drug stores). Order picking in Phoenix is done manually for bulky items that are not suitable for the AS/RS.

Semiautomated order picking is used for small items (e.g., a box containing a few dozen aspirin tablets, nasal spray medicine, and cold and cough remedies), which are stacked up on the outside of automatic vertical dispensers in their respective columns (Figure 8.3). The dispenser has several columns—one for each brand of medicine picked. The dispensers are inclined over a conveyor forming an A shape, and a computer-controlled mechanism dispenses items specified in an order from their respective columns onto the moving conveyor belt (Figure 8.4). The items then proceed to the end of the conveyor line, where they are dropped into a waiting container. Each container corresponds to a specific order. The totes are at a lower level than the conveyor line. Hence, there is no need for manual handling of the picked items. A light signal (Figure 8.3) tells the operators when items need to be replenished—typically when the item has reached or gone below its safety stock level. The automatic dispensing mechanism is very effective for picking a large variety of items for which the picking frequency is medium. The replenishment is manual, but dispensing is automated. It is relatively inexpensive, and the order picking is done at a much faster rate than the manual order picking. The degree of accuracy is also very high. However, it can usually be used only for handling relatively small items.

The third level of order picking in Phoenix is done via an expensive robotic order picker. Phoenix has two sets of robots—one for storage and another for retrieval. The retrieval robots (Figure 8.5) pick items from narrow aisles whose width is just a little over that of the robot (Figure 8.6). Equipped with computers and optical scanners, the robot retrieves items specified in an order and puts them into one of several compartments (see the circular compartmentalized drum in the middle

FIGURE 8.1   Phoenix Pharmaceuticals warehouse, Germany. (Courtesy of Phoenix Pharmaceuticals.)

FIGURE 8.2   Customer orders placed inside containers and sent to manual picking areas. (Courtesy of Phoenix Pharmaceuticals.)

FIGURE 8.3   Items replenished on an automatic vertical dispenser. (Courtesy of Phoenix Pharmaceuticals.)

FIGURE 8.4 Items are dispensed onto a conveyor from their respective storage columns. (Courtesy of Phoenix Pharmaceuticals.)

FIGURE 8.5 Retrieval order pickers. (Courtesy of Phoenix Pharmaceuticals.)

of Figure 8.7). Each compartment corresponds to a customer order. The required items are picked from their respective locations, loaded onto the compartments, and taken to a conveyor line (see the right side of Figure 8.5), where they are dropped into waiting bins. Each compartment in the circular drum has a metal flap at the bottom that automatically opens and allows all the items in an order to be dropped into its specified bin.

The storage robot has a deck that can hold large bins (Figure 8.8). Items to be stored in racks are put into these bins, which are then loaded on the robot deck one at a time. A robot arm plunges into the bin and picks items using vacuum suction cups (again one at a time—see Figure 8.9). The items are then put into their respective storage bins using robot arms equipped with optical scanners (Figure 8.10). The bins are then transported and stored by the robot.

FIGURE 8.6   Robots used for storing bins in racks automatically. (Courtesy of Phoenix Pharmaceuticals.)

FIGURE 8.7   Computer and compartmentalized storage bins on a robot. (Courtesy of Phoenix Pharmaceuticals.)

FIGURE 8.8   Storage robot. (Courtesy of Phoenix Pharmaceuticals.)

FIGURE 8.9 Using vacuum suction cups, the robot picks items from a bin. (Courtesy of Phoenix Pharmaceuticals.)

FIGURE 8.10 Optical scanners help a robot arm store items in their respective bins. (Courtesy of Phoenix Pharmaceuticals.)

8.2 INTRODUCTION

Many manufacturing and distribution companies maintain large warehouses to store in-process inventories or components received from an external supplier. Businesses that lease storage space to other companies for temporary storage of material also own and maintain a warehouse. In the former case, it has been argued that warehousing is a time-consuming and non-value-adding activity (waste). Because additional paperwork and time are required to store items in storage spaces and retrieve them later when needed, the just-in-time (JIT) or lean manufacturing philosophy suggests that one should do away with any kind of temporary storage and maintain a pull strategy in which items are produced only as and when they are required. They should be produced at a certain stage of manufacturing, only if they are required at the next stage. Moreover, the quantity produced should directly correspond to the amount demanded at the next stage of manufacturing. JIT or lean manufacturing philosophy requires that the same approach be taken toward components received from suppliers. The supplier is considered as another (previous) stage in manufacturing. However, in practice, because of a variety of reasons, including the need to maintain sufficient inventory of items because of the unreliability of suppliers and to improve customer service and respond to their needs quickly, it is not possible, or at least not desirable, to completely do away with temporary storage.

Consider the following situation in Nike—a manufacturer of athletic wear. Nike built a large distribution warehouse in Belgium in the mid-1990s (see Figure 12.1). One of their main business objectives is to serve 75% of their customers within 24 hours. Without appropriate warehousing facilities, it is impossible for Nike to achieve this objective because many of their manufacturing plants and suppliers are overseas—in the Far East!

"Members club" stores such as Sam's Club, Costco, and BJ's Warehouse Club have found a niche in the consumer retailing business in the past decade. These stores provide memberships to businesses and their employees and allow only members to shop in their stores. They generally sell merchandise in bulk and directly out of their warehouse, eliminating the need to build and maintain costly retail stores. While this significantly reduces overhead costs for the warehouse, for the consumer, it typically costs less to shop in such stores than in traditional malls because he or she buys in bulk. The primary function in such warehouse stores is not warehousing but retailing!

The aforementioned two examples amply demonstrate the need for establishing warehouses to satisfactorily service end customers despite the lack of value-added services in many of them. This chapter is devoted to warehouse and storage design and planning.

8.3   WAREHOUSE FUNCTIONS

There are several reasons for building and operating warehouses. In many cases, the need to provide better service to customers and be responsive to their needs seems to be the primary reason. While it may seem that the only function of a warehouse is warehousing, i.e., temporary storage of goods, in reality, many other functions are performed. Some of the more important ones are listed and briefly discussed as follows (Kulwiec, 1980):

Temporary storage of goods: To achieve economies of scale in production, transportation, and handling of goods, it is often necessary to store goods in warehouses and release them to customers as and when the demand occurs.

Put together customer orders: Warehouses, for example, the Nike distribution center (DC) in Figure 12.1, receive shipments in bulk from several sources and, using an automated or manual material-handling system (MHS), put together individual customer orders and ship them directly to the customers.

Serve as a customer service facility: Because warehouses ship goods to customers and therefore are in direct contact with them, a warehouse can serve as a customer service facility, handle replacement of damaged or faulty goods, conduct market surveys, and even provide after-sales service. For example, many Japanese electronic goods manufacturers let warehouses handle repair and after-sales service in North America. United Parcel Service (UPS) handles not only the transportation and distribution of laptops (notebooks) for Toshiba Corporation but also their repair in its Supply Chain Solutions facility near its air-hub in Louisville, KY, using UPS employees (Figure 8.11).

Protect goods: Because warehouses are typically equipped with sophisticated security and safety systems, it is logical to store manufactured goods in warehouses to protect against theft, fire, floods, and weather elements.

FIGURE 8.11   United Parcel Service air-hub in Louisville, KY.

Segregate hazardous or contaminated materials: Safety codes may not allow storage of hazardous materials near the manufacturing plant. Because no manufacturing takes place in a warehouse, this may be an ideal place to segregate and store hazardous and contaminated materials.

Perform value-added services: Many warehouses routinely perform several value-added services such as packaging goods, preparing customer orders according to specific customer requirements, inspecting arriving materials or products, testing products not only to make sure they function properly but also to comply with federal or local laws, and even assembling products. Clearly, inspection and testing do not add value to the product. However, we have included them here because they may be a function necessitated by company policy or federal regulations.

Inventory: Because it is difficult to forecast product demand accurately, in many businesses, it may be extremely important to carry inventory and safety stocks to allow them to meet unexpected surges in demand. In such businesses, not being able to satisfy a demand when it occurs may lead to a loss in revenues or, worse yet, may severely impact customer loyalty toward the company. Also, companies that produce seasonal products, e.g., lawn mowers and snow throwers, may have excess inventory left over at the end of the season and have to store the unsold items in a warehouse.

8.4   MATERIAL-HANDLING AND STORAGE SYSTEMS USED IN WAREHOUSES

A typical warehouse consists of the following two main elements:

1. Storage medium
2. MHS

Of course, there is a building that encloses the storage medium, goods, and the storage/retrieval (S/R) system. Because the main purpose of the building is to protect its contents from theft and weather elements, it is made of a strong, lightweight material. Warehouses come in different shapes, sizes, and heights depending on the number of factors (see Exercise 6), including the kind of goods stored inside, volume, and type of S/R systems used. The Nike warehouse in Laakdal, Belgium, covers a total area of 0.15 million square meters. Its high-bay storage is almost 30 meters in height, occupies roughly half of the total warehouse space, and is served by a total of 26 person-on-board stacker cranes. On the other hand, a "members club" store may have a total warehouse space of 25,000 square meters with a building height of 12 meters.

8.4.1 Storage Medium

The primary advantage of using storage racks is that it allows the maximization of space utilization in a warehouse. Each rack has a number of storage spaces (which typically have a square or rectangular cross section) that are arranged from top to bottom and left to right. These storage spaces are used for storing items in the warehouse. There are several types of storage racks available for storing items. These are illustrated in Figures 8.12 through 8.16 and explained next.

Stacking frames: Pallets are sometimes stacked using vertical steel frames that can be attached to the four corners of a standard wooden or steel pallet (Figure 8.12a). Horizontal beams that can be attached to the four vertical frames allow pallets to be stored on them. When the horizontal beams are attached to vertical steel frames, they look like an inverted stool. This design allows pallets, boxes, or cartons to be loaded directly one above the other. These types of racks are not only collapsible for easy storage when not in use but also allow additional racks to be stored above them, thereby utilizing vertical storage space.

Cantilever racks: Cantilever beams supported from vertical members allow much storage flexibility (Figure 8.12b). Although these are typically used for storage of long loads such as rods, tubes, and pipes, they can be fitted with decks and used as a regular rack to store smaller loads.

Selective rack: For every storage bay, this rack has a pair of vertical uprights, horizontal beams, and cross-braces for stability (Figure 8.12c). The bay length depends on the length of the horizontal beam. This type of rack is called selective because it makes every load accessible (or be selected) as illustrated in Figure 8.13a. In this figure, two single racks are placed back to back. When two

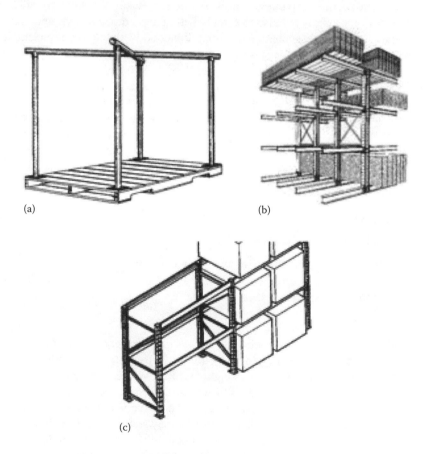

(a)   (b)

(c)

FIGURE 8.12 (a) Pallet frame, (b) cantilever rack, and (c) steel rack. (Reprinted with permission from "Advanced Material Handling" perpared by MHI.)

single-deep selective racks are placed back to back, we get a double-deep selective rack design, which reduces the aisle space requirement significantly. Both designs permit multiple levels of vertical storage and are generally fixed in one place in the warehouse. A variation of this design allows the racks (which are on wheels or floor-mounted rails) to be moved from one location to another. These are called mobile racks, and examples are shown in Figure 8.13b. The example shown in

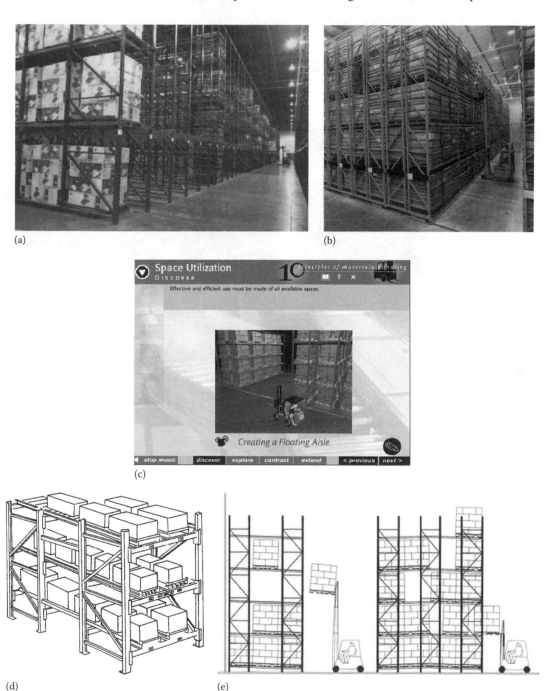

FIGURE 8.13 (a) Single-deep selective racks, (b) mobile racks, (c) screenshot of "10 principles" module, (d) flow rack, (e) push-back rack. *(Continued)*

(f)

(g)

(h)

FIGURE 8.13 (*Continued*)   (f) tunnel and push-back racks, (g) drive through rack, and (h) warehouse application. (Courtesy of Steel King Industries, Inc. Storax, Inc., BARPO Group Plc.)

Figure 8.13b is a narrow-aisle system capable of accommodating turret trucks. It has pickup and drop-off stations at the end of each rack. Mobile rack systems are ideal for use in food freezer facilities or other facilities where the turnover is low. They permit efficient use of warehouse space because entire racks can be moved from one end of the rail to another, thus allowing racks to be stacked up very close to each other with no aisle space. When items need to be stored or retrieved, the racks are moved one at a time until the required rack is accessible. A demonstration of this is available in the "Discover" and "Explore" layers of the space utilization principle in the "10 Principles" module available at http://sundere.okstate.edu/Book. In fact, Figure 8.13c is a snapshot of a screen from that module.

Flow rack: Items are loaded from one end of the flow rack on sloping wheels or rollers and retrieved from the other (Figure 8.13d). Whereas the lift truck moves to the load to pick it up, the loads come to the lift truck, person, or other device for pickup in a flow rack. Perhaps, the reader has

(a)   (b)

FIGURE 8.14 Racks used for storage in automated warehouse. (a) Racks for storing totes and (b) pallet racks. (Courtesy of Jervis B. Webb, Company and Savoye Logistics.)

FIGURE 8.15 Rack-supported building. (Courtesy of Savoye Logistics.)

seen such a design in the dairy department of a grocery store where milk, yogurt, and other dairy products are stored on flow racks. These racks are used for accumulating items at point of use and provide high-density storage.

A variation of the flow rack is called a push-back rack. Whereas items are retrieved on a first-in first-out basis (FCFS) in flow rack, it is done on a last-in first-out (LIFO) basis in the push-back rack (Figure 8.13e). Items are loaded on a rail-guided carrier from the front of the rack. While the vehicle is depositing the load, the previously stored loads are pushed back one position to the rear to make room for the new load. The new load is always at the front. When the front load is removed, the remaining loads advance forward due to a gentle slope in the racks. Figure 8.13e illustrates a combination of a tunnel flow and a push-back rack. A single-deep flow rack at the ground level in Figure 8.13e has a tunnel allowing forklift truck movement so that these forklift trucks can access items on either side of the tunnel. Figure 8.13f and 8.13g shows racks in typical warehouses. In addition to the aforementioned, there are modular drawers such as the ones in Figure 8.13e used for storing small parts such as drills, pins, washers, and screws. They offer extremely high-density

FIGURE 8.16   Storage and retrieval vehicles. (a) S/R device, (b) Aisle-captive AS/RS, and (c) Aisle-tot-aisle AS/RS. (Courtesy of Jervis B. Webb, Company of Vanderlande Industries.)

storage. Bins and containers that can be stacked one above the other serve as transportation and storage devices. When the parts need to be transported to point of use, the entire bin or container is transported via mechanical equipment.

Racks for automated storage and retrieval systems: The rack shown in Figure 8.14a can hold bins for storing small items. It is a narrow aisle with just enough room for a robotic retrieval system to store and retrieve items. Similarly, the one shown in Figure 8.14b has just enough room for a person-aboard S/R system to access cartons that are stored on regular racks. In fact, there is nothing special about the racks in Figure 8.14 other than the fact that they are specifically designed for use with automated systems. In recent years, high-rise warehouses that are rack supported are becoming very common. Rather than having freestanding high-rise racks, these warehouses are supported by racks. Thus, the racks have dual function—storage and building support (Figure 8.15). For tax purposes, the building is considered to be "operating equipment" because it is attached to storage racks. It, therefore, qualifies for depreciation at a faster rate as well as investment tax credit (Kulwiec, 1980). Typically, operating equipment have higher rates of depreciation than property not considered to be operating equipment (e.g., building shells whose only function is to enclose a warehouse). Kulwiec (1980) also points out that another advantage of rack-supported buildings is that they can be constructed quickly as many activities can be done simultaneously.

8.4.2 Storage and Retrieval Systems

Two types of S/R systems are typically seen in practice (Kulwiec, 1980):

1. "Person-to-item" system in which the storage racks are stationary and a person or machine goes to the storage location of items and picks them.
2. "Item-to-person" system in which the items come to the end of an aisle (from their storage location) where an operator or a transportation mechanism (e.g., conveyor) sends the items to their point of use. The flow racks discussed in Section 8.4.1 fall into this category and so do storage carousels (such as those in a food vending machines or at a dry-cleaning establishment) and mini-load systems.

S/R systems can be manual or automated. The degree of automation can vary from simple automation (e.g., flow racks with powered wheels or rollers) to full automation (e.g., the robotic retrieval system discussed in Section 8.1). Because partial or fully automated S/R systems are common and also referred to many times in this chapter, we will use the well-known acronym AS/RS for such systems in the remainder of this chapter.

There are several types of S/R systems in use. Chapter 7 described several types of trucks that can be used in S/R operations. The S/R device in Figure 8.16a is capable of maneuvering in narrow aisles, may be counterbalanced, can reach loads three to four levels high, and may have standing room for an order picker. In an aisle–captive AS/RS, each S/R device is dedicated to a specific aisle, as shown in Figure 8.16b. In an aisle-to-aisle AS/RS as shown in Figure 8.10c however, the S/R devices can go from one aisle to any other. This design permits the use of fewer vehicles that aisles, resulting lower equipment costs, but also provide lower throughput. A fleet of six automated guided vehicles (AGVs), each equipped with a 7-meter mast stack and retrieve loads from three levels in the Boston Globe warehouse, where inserts get automatically placed in the newspaper. The person-on-board S/R vehicle in Figure 8.16c requires very little aisle space and therefore allows high-density storage. It has guide rollers at its base, which engage on special tracks located in the aisle. The mast also has rollers that engage on overhead guide tube permitting stability and alignment while the vehicle is in the aisle. This aisle-to-aisle system is useful in high-speed full pallet or selective order picking. In addition to these types of S/R systems, there are several others that are widely used in practice. They can be classified as

- Storage carousels
- Mini-load AS/RS
- Logistacker AS/RS
- Robotic order picker
- High-rise AS/RS

FIGURE 8.17   Storage carousels. (Courtesy of MHI.)

Storage carousels can rotate around vertical or horizontal axes. In the former case, items are typically stored in wire baskets, tubs, or small bins and are usually not enclosed from the top (see Figure 8.17). In the latter case, not only are the bins enclosed from all sides, but also the carousel system itself is fully enclosed with an opening near the bottom for retrieval of bins. The revolving motion is actuated via foot pedals or push buttons. The advantage of carousels is that there is no need for aisles as the items come to the operator. Thus, warehouse space utilization can be maximized.

Mini-load AS/RS is a fully enclosed system with narrow aisles that have just enough room for an S/R machine that can handle mini- or micro-loads. It is equipped with an optical scanner that allows it to scan codes on the front of storage bins (Figure 8.18a). The S/R machine moves to locations specified in a pick list (or order), retrieves the items in a sequence determined either by an operator entering information into the computer at the end of the aisle or by a sophisticated algorithm (such as the ones in Section 8.7.1 programmed into the computer controlling the AS/RS), and brings the picked items or order to the operator station. The AS/RS throughput rate is fairly high because it usually has two independent motors permitting motion in the two axes at different speeds. Also, while the operator is unloading one bin, the S/R machine picks up another and transports it back to its storage location. Four AS/RS are shown in Figure 8.18a through d.

The *Logistacker AS/RS* is an alternative to the conventional mini-load AS/RS and is useful in order-picking operations requiring high throughput. Whereas the traditional mini-load AS/RS can only pick one item at a time, the Logistacker AS/RS can pick as many as 12 totes at once (see Figure 8.19). The core of the Logistacker AS/RS is a 6-meter long liftable conveyor (see the overhead fixture in Figure 8.19) that is mounted on a gantry crane and can thus access any aisle and any level within an aisle. The liftable conveyor is of the same length as the aisle and can accomplish storage or retrieval of totes in one pass via grippers. In a retrieval transaction, the retrieved totes are deposited at the I/O point located overhead. From here, conveyors take the totes away to their designated order fulfillment or sortation stations. The main advantage of the Logistacker AS/RS is that multiple totes are retrieved at once, and this offers high throughput gains. In a dual command mode, where storage transactions are paired with retrieval transactions, the Logistacker AS/RS is able to store and retrieve up to 1500 totes per hour. It should be noted that the Logistacker AS/RS is

FIGURE 8.18 Automated storage and retrieval system applications. (a) Micro-load storage and retrieval and (b) pallet storage and retrieval. (c) Mini-load storage and retrieval and (d) tire storage and retrieval. (Courtesy of Vanderlande Industries.)

not very efficient if the number of totes to be picked from each layer is 4 or less or when the required throughput is less than 200 totes per hour (see Rouwenhorst et al., 2000).

The *robotic order picker* is similar to the mini-load AS/RS in its functional capabilities, but the storage and retrieval is done by a robot that moves on the floor, as shown in Figure 8.20. Because the robot arm can only extend to a limited height, such systems typically cannot service racks that are more than 3 meters tall.

High-rise AS/RS are automated systems in which the order picking is done manually or by automated and computer-controlled S/R devices (see Figure 8.21a through d). The S/R machine

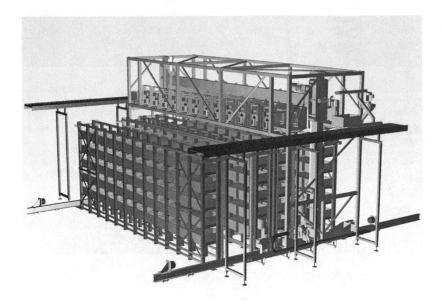

FIGURE 8.19   Logistacker ASRS. (Courtesy of Blue Storage B.V.)

FIGURE 8.20   Robotic order picker. (Courtesy of Phoenix Pharmaceuticals.)

may be aisle captive, i.e., dedicated to a specific aisle. For example, the Nike warehouse shown in Figure 8.21a has 26 stacker cranes that are dedicated one to each aisle. Like the mini-load systems, the high-rise AS/RS could be powered by two independent motors. The AS/RS in Figure 8.21 are all aisle-captive systems because the cranes are restricted to move within an aisle. Whereas, the AS/RS in the S/R devices in Figures 8.21a–d can go up or down using a mast.

Figure 8.22 illustrates the use of a combination of the various types of S/R equipment in warehousing activities. Lift trucks are used to replenish stock in the pallet racks, which are then manually loaded on conveyors that take away the items to accumulation, sortation, and packaging operations via conveyors from the upper and lower levels as shown on the right side of Figure 8.22. In addition, as illustrated in the left side of the figure, a person aboard an order-picking vehicle picks items from the pallet rack and takes them to order pick or sortation areas via conveyors (not shown), where the goods are automatically sorted and manually put into boxes or cartons and shipped.

FIGURE 8.21 High-rise AS/RS. (a) AS/RS used in Nike warehouse and (b) AS/RS for pallet storage. High-rise AS/RS. (c) AS/RS for large pallets and (d) pallet racks for AS/RS. (Courtesy of Vanderlande Industries and Unarco Rack.)

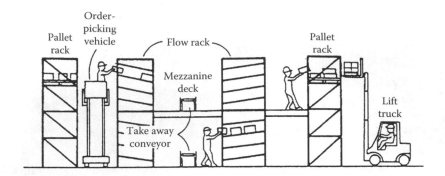

FIGURE 8.22 Interface among different types of S/R equipment. (Courtesy of MHI.)

8.5 AUTONOMOUS VEHICLE STORAGE AND RETRIEVAL SYSTEMS

The typical AS/RS system has a dedicated storage crane for each aisle. The crane may store and retrieve unit loads automatically or a person aboard a crane may do the order picking. This design typically requires a crane to be dedicated to one aisle. Within an aisle, however, the crane can travel the entire depth (from the racks at the front all the way to the back) and the entire height of an aisle (from the lowermost rack to the rack in the topmost tier) and accomplish a storage or retrieval transaction in minutes.

Recently, material-handling equipment manufacturers have begun to design, develop, and implement alternative configurations to the traditional AS/RS design. One such manufacturer is Savoye Logistics in France. This company makes autonomous vehicles and builds warehouse systems that utilize these vehicles to satisfy the dual objectives of achieving dense storage and fast throughput.

Called the autonomous vehicle storage and retrieval system (AVSRS), this warehouse MHS has a number of vehicles that can carry unit loads autonomously on rails (Figure 8.23a and b). The vehicles travel on rails to execute horizontal motion required to reach a storage or retrieval location within a warehouse tier or level (Figure 8.23c). To execute vertical motion, they make use of lifts that transport them to the required tier. Once they arrive at the tier, they again travel to storage or retrieval locations on rails (see Figure 8.23d). The vehicle movement can be

(a)

(b)

FIGURE 8.23 Autonomous vehicle storage and retrieval systems. (a) Autonomous vehicle and (b) rails for autonomous vehicle.   (*Continued*)

FIGURE 8.23 (*Continued*) Autonomous vehicle storage and retrieval systems. (c) Storage locations, (d) warehouse designs showing lifts, (e) unit load on the right about to be retrieved, (f) unit load just retrieved, (g) forks on autonomous vehicle and (h) lift for transporting autonomous vehicle to tires. (*Continued*)

(i)   (j)

FIGURE 8.23 (*Continued*)   Autonomous vehicle storage and retrieval systems. (i) Warehouse management system and (j) exterior view of warehouse. (Courtesy of Savoye Logistics.)

seen from Figure 8.23e and f. Note that the vehicle approaching the unit load on the far right of Figure 8.23e and actually picking up the unit load in Figure 8.23f. The vehicle has forks (Figure 8.23g) to withdraw a pallet into the vehicle or release it for depositing into a storage location. The lift is shown in Figure 8.23h and the warehouse management system (WMS) in Figure 8.23i. The WMS is capable of real-time control either from the warehouse or a remote location and provides real-time information concerning lift, vehicle, storage location status— idle or occupied—utilization, cycle times, throughput rate, and so on. A warehouse installation is shown in Figure 8.23j.

The AVSRS design is highly flexible. We can have any vehicle choose any available lift or designate vehicles to specific lifts. We can restrict vehicles to specific tiers or make them tier captive. Vehicles can be easily added, removed, or taken away for maintenance without stopping order-picking operations. The warehouse can be built in a modular fashion and expanded or contracted as the need arises. The biggest advantage of the AVSRS is the ability to provide a high degree of flexibility in the design configuration without sacrificing throughput or warehouse productivity. It has been installed in many locations in Europe. For example, in Beauvais, northern France, Savoye Logistics has installed the AVSRS for a 50,0000 pallet position Unilever warehouse that is run by DHL. Savoye Logistics provides WMS support.

8.6   WAREHOUSE DESIGN

Just as in any other facility, there are a myriad of problems that must be addressed while designing a warehouse. Of course, many of these problems are different from those seen in manufacturing facilities, but there are some that are common to both. For example, the problem of deciding how many and where to locate warehouses is similar to that encountered in manufacturing facilities design. However, the design of racks to store goods is unique to warehouse design. In this section, we list and discuss the various design problems in warehouses.

8.6.1   WAREHOUSE LOCATION

Location is perhaps the first and foremost problem that must be addressed in warehouse design, given that it has been decided to build warehouse(s) to improve customer service or for other reasons. General location problems are the subject of Chapter 9, so we will not address this problem here but list some of the specific decisions that need to be made with respect to warehouse location, which are as follows:

1. How many warehouses must be built?
2. Where should each be built?
3. How large should each warehouse be?

Of course, answers to these questions will depend on the number of factors, including location of manufacturing plants, customer bases, desired level of customer service, renting, leasing or construction costs, taxes, insurance, and other factors discussed in Chapter 9. Once the aforementioned questions are addressed, the designer then focuses on each warehouse and begins addressing design problems within each.

8.6.2 Overall Layout of Warehouse

The warehouse layout depends on the items stored, space available in the warehouse, height, storage medium used, S/R methods, the layout of road and rail tracks around the warehouse, and other factors. Typical warehouse layouts seen in a "members club" store and a high-rise automated warehouse are illustrated in Figures 8.24 and 8.25. In Figure 8.24, customers enter from the lower left corner of the store, pick up shopping carts or dolly, visit the desired aisles, load the cart or dolly with items, proceed to the cashier (C), and exit the warehouse. Because there is not much shipping done in such stores, but mainly receiving, the receiving area is shown in the upper right corner of Figure 8.24 with some space set aside for staging. As staging is not a value-adding activity and occupies valuable space, it is desirable to take the incoming goods directly to the storage area or order-picking positions. This is true for any warehouse. If the warehouse can work with its suppliers and have them prepare special tags or barcodes that identify the product, their exact location in the warehouse can be determined as and when the goods are received by the receiver using handheld scanners. The items can then be sent directly to their storage location. Warehouses have a staging area mainly to identify the product and determine where it is to be stored.

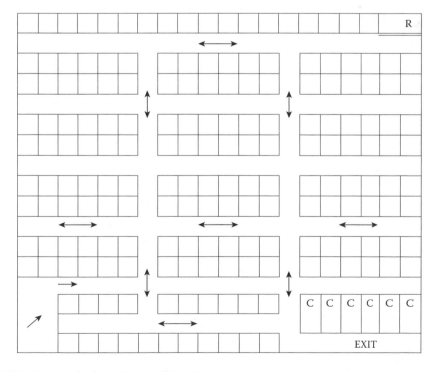

FIGURE 8.24   Layout of a "members club" warehouse store.

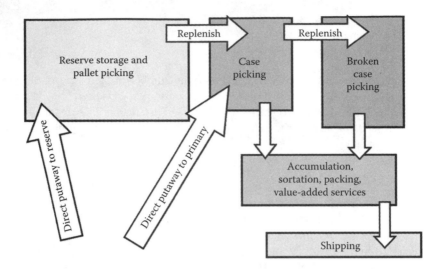

FIGURE 8.25   Layout of a high-rise automated warehouse. (Courtesy of John S. Usher.)

The high-rise warehouse shown in Figure 8.25 has separate areas for order picking and sortation as well as shipping and receiving.

8.6.3   LAYOUT AND LOCATION OF DOCKS

Some typical dock layouts are illustrated in Figure 8.26. The layout and location of docks also depend on number of factors. Some of these are as follows:

- Will pickup by retail customers be allowed?
- Will the warehouse be serviced by trucks of limited type and size, or will there be a variety of trucks (large and small) from different freight carriers?

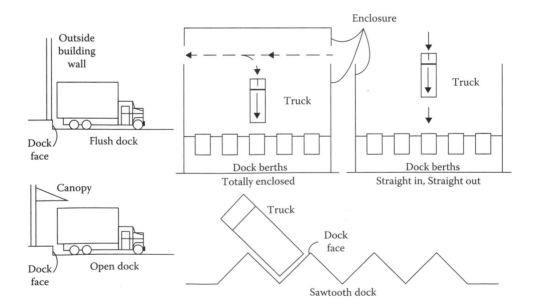

FIGURE 8.26   Typical dock layouts. (Reprinted with permission from "Advanced Materials Handling", prepared by MHI.)

- Should shipping and receiving be combined or should they be separated?
- The layout of the road/rail network.
- Room available for maneuvering trucks and operating warehouse doors/gates.

In addition to these, the number of docks also has a major impact on the layout and location of docks. The number of docks required must be carefully determined by considering several factors. Some of these are listed as follows (Kulwiec, 1980):

- Can shipping and receiving be arranged at different times in the day (for example, shipping in the morning and receiving in the afternoon, or vice versa)? This allows us to have fewer docks and better utilization of dock space as well as dock personnel and facilities.
- Average and peak number of trucks or rail cars handled per day.
- Average and peak items per order.
- Seasonal highs and lows.
- Types of load handled, their sizes, shapes, whether they are cartons, cases, or pallets.
- Is protection from weather elements required while loading/unloading trucks?

8.6.4 Rack Design

Determination of the length and width of the warehouse depends on the number of items to be stored, the number of storage spaces required, the number of rows and columns of racks, and the rack height. In the following discussion, we discuss a mathematical model (presented in Askin and Standridge, 1993) that allows us to determine two key parameters—length and width of the warehouse—given some information concerning racks. Suppose that we wish to find the number of rows and columns of racks required to hold a maximum of n items in (n) equal-area square storage spaces. If we know the number of rows and columns of rack spaces required to hold the n items, we can easily calculate the length and width of the warehouse by allowing the required space for aisles.

Without loss of generality, let us assume that the racks are arranged so that their longer side is parallel to the horizontal axis (as shown in Figure 8.24). If we define x, y to be the required number of columns and rows of rack spaces, respectively, and it is known that the total required aisle length is equal to a multiple of the sum of the horizontal rack spaces, say a, and that the total required aisle width is equal to a multiple b of the sum of the vertical rack spaces, then the length and width of the warehouse are equal to $ax + x$ and $by + y$, respectively. If we wish to minimize the average one-way distance traveled from the customer entrance, the model may be written as follows:

Model 1

$$\text{Minimize } \frac{x(a+1)+y(b+1)}{2}$$

$$\text{Subject to } xyz \geq n$$

$$x, y \leq 0 \text{ and integer}$$

Because the distance traveled (from the lower left corner or any other corner in the warehouse) varies linearly between a maximum of $\{x(a + 1) + y(b + 1)\}$ and a minimum of 0, the average distance traveled is given by $\{x(a + 1) + y(b + 1)\}/2$. The objective function, therefore, minimizes the average distance traveled. The first (nonlinear) constraint ensures that the total number of storage spaces available in z levels (equal to xyz) exceeds the minimum required value n. Because the number of rows and columns of rack spaces have to be integers, this is enforced by the last constraint. For the warehouse shown in Figure 8.24, the multipliers a and b can be approximately determined as 1/9 and 6/11, respectively (see Exercise 10). Actually, the multipliers must be slightly higher to account for storage space lost to entrance, exit, and cashier areas as well as restrooms, employee lounge, docks, and other service areas.

The aforementioned model may be solved approximately by relaxing* integer restrictions on x, y and noting that under such a relaxation, the first constraint will be an equality at optimality, i.e., $xyz = n$. Using this equality to solve for x and substituting the resulting value in the objective function gives rise to an unconstrained minimization problem in one variable—y. It can be shown that the unconstrained minimization objective is convex (see Exercise 11). Therefore, taking the derivative of the new objective with respect to y, setting the resulting equation to 0 and solving for y will yield an optimal value of the variable. Using this value and the equation $xyz = n$, the other variable's value may be readily found as illustrated next:

$$xyz = n \Rightarrow x = \frac{n}{yz}$$

The unconstrained objective is therefore

$$\frac{n(a+1)/yz + y(b+1)}{2}$$

Taking the derivative with respect to y, setting the resulting equation to 0, and simplifying, we get

$$\frac{-n(a+1)}{2y^2z} + \frac{b+1}{2} = 0$$

Further simplification, rearrangement, and use of the equation $xyz = n$ yields the following two formulae for x and y:

$$y = \sqrt{\frac{n(a+1)}{z(b+1)}} \quad \text{and} \quad x = \sqrt{\frac{n(b+1)}{z(a+1)}} \tag{8.1}$$

Formulae (8.1) give insight into the effect of the aisle space multiples a, b on the warehouse shape. It is easy to see that different values of a, b yield different values of x, y, and hence different values for the length and width of the warehouse. For example, if $a = b$ and the input point from which travel originates is at one of the corners, it can be seen that $x = y$ and therefore the warehouse has a square shape (see Exercise 12). Similarly, it can be seen that moving the travel origination point or, in the case of the warehouse in Figure 8.24, moving the customer entrance to different locations along the exterior of the warehouse leads to different dimensions for the warehouse (see Exercises 14 through 19).

Example 8.1

Consider the warehouse layout shown in Figure 8.27. Assuming the point from where travel originates is at the lower left corner, and assuming reasonable values for the aisle space multipliers a, b, determine the length and width of the warehouse required to accommodate 2000 square storage spaces of equal area when there are three levels, four levels, and five levels.

Solution
Assuming Figure 8.27 is drawn to scale, a reasonable value for horizontal aisle space multiplier a can be determined as 0.5; for b, it can be approximated as 0.2. Using Equation 8.1, for the three-level case, x, y can be calculated as

---

* Such a relaxation does not severely affect the optimality of the solution obtained.

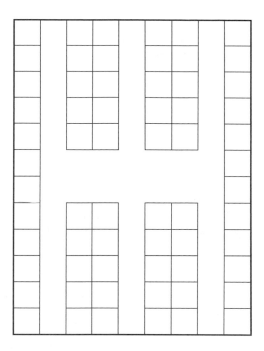

FIGURE 8.27   Layout of a warehouse.

$$y = \sqrt{\frac{2000(0.5+1)}{3(0.2+1)}} \approx 29 \quad \text{and} \quad x = \sqrt{\frac{n(b+1)}{z(a+1)}} \approx 24$$

This solution gives a total storage of 24 × 29 × 3 = 2088. Due to rounding up, we get 88 more storage spaces than required. If this is not enough to cover the area required for lounge, customer entrance and exit, and other areas, the aisle space multipliers a and b must be increased appropriately and the x and y values recalculated. Similarly for the four-level and five-level cases, the building dimensions can be determined as 25 × 20 units and 18 × 23 units, respectively. In addition, it is easy to calculate the average distance traveled for all the three cases (see Exercises 20 through 25).

8.6.5   DESIGN MODEL FOR WAREHOUSE SPACE ALLOCATION*

The two primary functions of warehouse include (1) temporary storage and protection of goods and (2) providing value-added services such as fulfillment of individual customer orders, packaging of goods, after-sales services, repairs, testing, inspection, and assembly. A warehouse is generally divided into the following areas to perform the aforementioned functions: reserve storage area, forward area, and cross-docking area. The reserve area is where goods are held until they are required for shipment to the customer or for performing value-added services or order collation. The latter is typically done in the forward area. The forward area could also be used to store fast movers that do not occupy much space. Cross-docking refers to the process in which items, cartons, or pallet loads are taken directly from the receiving trucks to the shipping trucks. It provides a fast product flow and reduces or eliminates the costs associated with handling and holding inventory. Typically, conveyors or forklift trucks are used to transfer materials from receiving to shipping in the cross-docking area.

The design of warehouse is a complex problem. It includes a large number of interrelated decisions among the warehouse processes, warehouse resources, and warehouse organizations.

---

* This section has been reprinted from Heragu et al. (2005).

Rouwenhorst et al. (2000) classify the warehouse design and planning problems into three levels of decisions—strategic, tactical, and operational. At the strategic level, there are several decisions that must be made. These range from determining the number of warehouses, their size, and locations to designing and building the warehouse to select the material handling equipment necessary for achieving the desired throughput rate. It also includes determining the functional areas in the warehouse and their sizes, designing the process flow determination of the warehouse layout, and selecting the WMS. At the tactical level, the main concerns include determination of the number of personnel to operate the system, allocation of products to the functional areas, developing order picking and replenishment policies, capacity planning, and others. At the operational level, the concerns include the selection of routing policies, the determination of batch size, the dock assignment, short-term (daily or weekly) work force assignment, and the task assignment.

After the warehouse location, number, and size have been determined, the warehouse designer may want to determining what storage areas are to be included and the size of each so that an appropriate MHS can be selected and the warehouse laid out. Although determining the size of each functional area is a strategic level problem, it depends on another tactical level problem, namely, how the products will be distributed among the functional areas. The latter is the product allocation problem. Thus, a joint solution of the functional area size determination and product allocation problems is desirable. However, the general approach undertaken by practitioners is to solve the two problems sequentially by generating multiple alternatives for the functional area size problem and then determine how the products can be allocated to each of these alternatives. In this paper, we develop a higher level model that jointly determines the functional area sizes and the product allocation in a way that minimizes the total material-handling cost. The output of the model serves as base for further detailed warehouse design. It should be noted that although the internal sizing problem in a warehouse is a strategic problem, the relative size of the functional areas may change over the lifetime of the warehouse depending on shifting trends in product demand.

In this section, we consider warehouse configurations that include a subset of the following five functional areas: receiving, shipping, staging for cross-docking operation, reserve, and forward. In the receiving area, pallet loads or individual cartons of products are received. If necessary, they are staged for a short period and then either moved to the shipping area directly (cross-docking operation) or to the storage area. In the shipping area, picked order items are readied (e.g., shrink-wrapped, packed) and staged (if necessary) for shipping to the next destination. In the staging area for cross-docking, products are sorted and accumulated for further outbound operations. Reserve area is a storage area for bulky product items that typically reside in the warehouse for a relatively longer duration. Normally, the reserve area uses high-density storage equipment to achieve the goal of high space utilization. Forward area is a relatively smaller storage area typically used for fast order picking or performing value-added operations or order collation. Thus, the following material flows are possible in a warehouse (see Figure 8.28):

Flow 1: Receiving → cross-docking → shipping
Flow 2: Receiving → reserve area → shipping
Flow 3: Receiving → reserve area → forward area → shipping
Flow 4: Receiving → forward area → shipping

Flow 1 is the cross-docking operation. Upon receipt, product items are either put into a staging area for a short period and then moved to the shipping area or directly moved to the shipping area. The received products are typically presorted at suppliers' facilities. The operation here is simply to pass on the product to a customer or the next facility in the supply chain. A number of companies use this strategy for efficient operation and management of the supply chain (Editor, 1996c).

Flow 2 is a typical warehouse operation. Products are stored in the reserve area and order-picking operation is performed as required. It is assumed that typically only those items that remain in the warehouse for relatively extended periods and shipped as is (or with minimal value-added

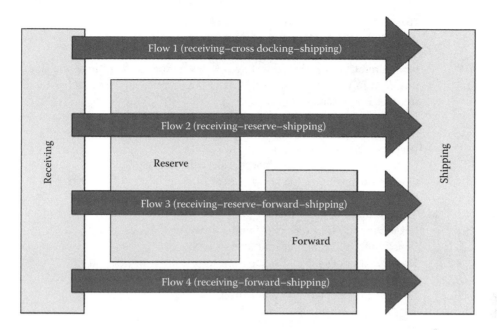

FIGURE 8.28 Typical product flows in a warehouse.

operations) will be allocated to the reserve storage area. Flow 3 is also a typical warehouse operation. Products are first stored in the reserve area typically in pallet loads, broken into smaller loads (cartons or cases), and then moved to the forward area for fast order picking, order consolidation, or performing value-added operations. Flow 4 can be thought of as another form of cross-docking operation. Products are received and then are directly put into forward area to perform the order consolidation. This type of operation is usually seen in the supplier warehouses or when there is a need to consolidate large orders.

We present a mathematical model that determines the flow to which each product must be assigned and, as a result, the size of the functional areas within the warehouse. The models make the following assumptions:

1. The available total storage space is known.
2. The expected time a product spends on the shelves is known. This is referred to as the dwell time.
3. The cost of handling each product in each flow is known.
4. The dwell time and cost have a linear relationship.
5. The annual product demand rates are known.
6. The storage policies and material-handling equipment are known, and these affect the unit handling and storage costs.

In formulating the model, the following notation is used:

Parameters

i Number of products $i = 1, 2, \ldots, n$
j Type of material flow; $j = 1, 2, 3, 4$
λ_i Annual demand rate of product i in unit loads
A_i Order cost for product i
P_i Price per unit load of product i

p_i	Average percentage of time a unit load of product i spends in the reserve area if product is assigned to material flow 3
q_{ij}	1 when product i is assigned to material flow $j = 1$, 2 or 4
$\lceil d_i \rceil + 1$	when product i is assigned to flow $j = 3$, where d_i is the ratio of the size of the unit load in the reserve area to that in the forward area and $\lceil d_i \rceil$ is the largest integer greater than or equal to d_i
a, b, c	Levels of space available in the vertical dimension in each functional area, a, cross-docking; b, reserve; and c, forward
r	Inventory carrying cost rate
H_{ij}	Cost of handling a unit load of product i in material flow j
C_{ij}	Cost of storing a unit load of product i in material flow j per year
S_i	Space required for storing a unit load of product i
TS	Total available storage space
Q_i	Order quantity for product (in unit loads)
T_i	Dwell time (in years) per unit load of product i
LL_{CD}, UL_{CD}	Lower and upper storage space limit for the cross-docking area
LL_F, UL_F	Lower and upper storage space limit for the forward area
LL_R, UL_R	Lower and upper storage space limit for the reserve area

Decision variables

$$X_{ij} = \begin{cases} 1 & \text{if product } i \text{ is assigned to flow type } j \\ 0 & \text{otherwise} \end{cases}$$

α, β, γ Proportion of available space assigned to each functional area

α, cross-docking; β, reserve; γ, forward

Model 1

$$\text{Minimize } 2\sum_{i=1}^{n}\sum_{j=1}^{4} q_{ij}H_{ij}\lambda_i X_{ij} + 0.5\sum_{i=1}^{n}\sum_{j=1}^{4} q_{ij}C_{ij}Q_i X_{ij} \tag{8.2}$$

$$\text{Subject to } \sum_{j=1}^{4} X_{ij} = 1 \quad \forall i \tag{8.3}$$

$$0.5\sum_{i=1}^{n} Q_i S_i X_{i1} \leq a\alpha TS \tag{8.4}$$

$$0.5\sum_{i=1}^{n} Q_i S_i X_{i2} + \sum_{i=1}^{n} p_i Q_i S_i X_{i3} \leq b\beta TS \tag{8.5}$$

$$0.5\sum_{i=1}^{n} (1 - p_i)Q_i S_i X_{i3} + 0.5\sum_{i=1}^{n} Q_i S_i X_{i4} + \leq c\gamma TS \tag{8.6}$$

$$\alpha + \beta + \gamma = 1 \tag{8.7}$$

$$LL_{CD} \leq a\alpha TS \leq UL_{CD} \tag{8.8}$$

$$LL_{CD} \leq a\alpha TS \leq UL_{CD} \tag{8.9}$$

$$LL_F \leq c\gamma TS \leq UL_F \tag{8.10}$$

$$X_{ij} = 0 \text{ or } 1 \quad \forall i,j \tag{8.11}$$

The dwell time is the average duration a product stays in the shelf and is assumed to be known or can be estimated by the warehouse manager. In fact, based on annual product demand order cost A_i, price per unit load of product P_i, and carrying cost rate r, a simple economic order quantity (EOQ) formula, $Q_i = \sqrt{2A_i\lambda_i/rP_i}$, can be used to determine the optimal order quantity Q_i, as well as the average time a unit load of a product spends on the shelves. For example, because the time between two successive replenishments is Q_i/λ_i, the average dwell time per unit load of product i is $T_i = 0.5Q_i/\lambda_i$. Note that $0.5Q_i = \lambda_i T_i$, and this value or another reasonable estimate must be used in the objective function (8.2), which minimizes the total cost of handling the average annual loads of each product assigned to its respective area and the corresponding annual storage costs. The reader should not confuse storage costs with inventory holding costs. While inventory holding costs depend only on the value of the inventory, they are the same whether the inventory is in reserve or forward or cross-docking area. Storing costs, on the other hand, depend on the area in which the product is stored, and these costs tend to carry a premium for the cross-docking and forward areas (because these are considered prime real estate in a warehouse) and are relatively not that expensive for the reserve area. Of course, the handling costs are different (in fact, the opposite) for these areas, and thus our model trades off storage costs against handling costs. Note that X_{ij} tells us whether or not product i is assigned to flow j, and $0.5Q_iX_{ij}$ gives us the average number of the corresponding unit loads in inventory.

The model implicitly assumes that the unit load size for each product is not dependent on the flow to which the product is assigned. In general, the size of a unit load for a product i that remains in one area is equal to that received from the supplier. The exception is for products assigned to flow 3 because these products have different unit load sizes in the two areas pertaining to flow 3. The unit load size of products assigned to flow 3 could be equal to that received from the suppliers in the reserve area but different when handled in the forward area. This occurs because a pallet load is broken down to cases or cartons in the forward area and is relevant only for products assigned to flow 3. We introduce d_i to denote the ratio of the size of the unit load in the reserve area to that in the forward area, and q_{ij} for $j = 3$ accounts for the fact that product i is handled $d_i + 1$ times. H_{ij} and C_{ij} should therefore correspond to aggregate handling and storage costs for $j = 3$. The model also implicitly assumes that for a product assigned to flow 3, the unit load size decreases as it moves from the reserve to the forward area. If necessary, this assumption can be relaxed and a more general model can be developed rather easily.

The model also assumes that each product incurs two material-handling transactions—one for receiving and another for shipping, regardless of the area it is assigned to. If products assigned to a particular flow require more than two (or only one) material-handling transactions, the coefficient of the corresponding terms in the objective function must be appropriately weighted. For example, in some cases, the products assigned to the combined forward/reserve flow may incur three transactions, one for receiving at the reserve area, another for shipping to forward area, and a third for shipping. If this is the case, that term must have a coefficient of 3. Constraint (8.3) ensures that each product is assigned to only one type of material flow. If the same product could be allocated to multiple flows due to different demand patterns, then our model requires that the manager to estimate the percentage of this product that could be assigned to two or more of the four flows. For modeling purposes, additional versions of this product are then created (depending on how many flows this product could be assigned to) with the demand data appropriately reduced. For example, let us assume that 70% of a certain product whose demand is 10,000 units per year on average is likely to be assigned to the reserve storage area, and another 30% to cross-docking. For modeling purposes, an additional version of this product is therefore created with demand equal to 7000 and 3000 for the two products. Note that although the manager may assume that the product is likely to be split on a 7:3 ratio to reserve/cross-docking, the model may provide a different assignment based on the total costs. Constraints (8.4) through (8.6) ensure that the space constraints for the cross-docking, reserve, and forward areas are met. The right-hand side includes three additional variables whose sum is required to be 1 (constraint 8.7). This is to ensure that 100% of the space available is allocated to the three areas. Constraints (8.8) through (8.10) serve to enforce upper and lower limits on the space that can be allocated to the cross-docking, forward, and reserve areas.

We believe much of the input data such as the type and number of products, annual demand for each, order cost, unit price, and carrying cost rate are readily available to the warehouse designer. The storage cost C_{ij} is typically a function of the size of a product's unit load, warehouse leasing, or construction costs per square foot, as well as the type of shelving used in each area encompassing flow j. C_{ij} and H_{ij} for flow 3 must be aggregated to account for the fact that a pallet load could be broken down into cases or cartons. The cost of handling a unit load of product i in each flow j is a function of the product size, its handling characteristics, and the MHS used in the area(s) included in flow j. Chapter 3 provides a simple formula to estimate these costs based on labor, nonlabor, and prorated capital recover costs of the MHS. Adapting this formula to the warehouse application is left as an exercise (see Exercise 29).

The model (8.2) through (8.11) can be solved optimally using available mixed-integer programming software, and even large-scale problems with 75,000 items can be solved.

Example 8.2

A warehouse handles six classes of products. The average annual demand, order cost, price, and space required per unit load for each class of product is shown in Table 8.1. Assuming a carrying cost rate of 10%, total available space of 100,000 square meters, determine how the products should be allocated to each of the four areas. Assume the handling and storage costs given in Table 8.2, an upper and lower bound of 75,000 and 35,000 square meters for reserve and forward areas and a maximum of 15,000 square meters for the cross-docking area. Also, assume $a = b = c = 1$, $p_i = 0.2$ and 1 for products 3 and 6, 0 for all others, and $q_{i3} = 1$ for $i = 1, 2, \ldots, 6$.

TABLE 8.1
Product-Specific Input Data for Numerical Example

Products	1	2	3	4	5	6
Annual demand	10,000	15,000	25,000	2,000	1,500	95,000
Order cost	50	50	50	50	50	150
Price/unit load	500	650	350	250	225	150
Space required	10	15	25	10	12	13
p_i	0	0	0.2	0	0	1

TABLE 8.2
Handling and Storage Costs for Each Area–Product Combination

Area/product	1	2	3	4	5	6
1	0.0707 (20)	0.0203 (15)	0.0267 (4)	0.3354 (5)	0.4083 (15)	0.0726 (20)
2	0.0849 (5)	0.2023 (5)	0.0428 (20)	0.559 (4)	0.6804 (25)	0.0871 (5)
3	0.1061 (10)	0.2023 (10)	0.0054 (1)	1.0062 (5)	1.2248 (45)	0.1088 (10)
4	0.0778 (15)	0.2023 (10)	0.0481 (9)	0.0671 (1)	0.8165 (30)	0.0798 (15)

Solution

Using the price, demand, order cost, and carrying cost rate, it is easy to determine these order quantities—141.42, 151.91, 267.26, 89.44, 81.75, and 1378.41—for the six products, respectively, using the EOQ formula and to calculate the dwell time per unit load of each product using the T_i formula provided earlier (see Exercise 30a). For a given product, we assume the dwell time is the same irrespective of the flow to which it is assigned. After all, the dwell time is a function of the product demand and order quantity. Based on the aforementioned and the costs given in Table 8.2, it is easy to set up the model developed earlier (Exercise 30a). Solution of the model yields an assignment of products 2 and 5 to flow 1 (cross-docking), products 1 and 6 to flow 2 (reserve), and products 3 and 4 to flows 3 and 4, respectively, at an annual handling and storage cost of $26,300. Because of the lower handling and storage costs in the reserve area, 98.5% of the available space is assigned to this area.

8.6.6 Spreadsheet-Based Tool for Designing AS/RS

In this section, we present a spreadsheet-based tool for the design of an AS/RS. Called ExASRS, the software developed by the Material Handling Industry is available via http://sundere.okstate. edu/Book. ExASRS iteratively attempts to develop an efficient AS/RS configuration that meets storage and throughput requirements using a minimum number of aisles. Based on user-provided input values of

- Number of AS/RS cranes
- Number of storage and retrieval transactions per time unit
- Maximum storage volume
- Storage location height and width
- S/R crane velocity and acceleration in the horizontal and vertical directions
- Pickup and deposit times
- Rack type and shape
- Percentage of dual command transactions

ExASRS calculates the minimum number of aisles, length, and height of racks as well as the number of rows and columns per rack that can achieve the operational (throughput) requirements. It also allows the user to perform sensitivity analysis by altering the number of aisles and determining its impact on storage volume, throughput, and total cost.

Example 8.3

A warehouse manager wants to determine the minimum number of aisles to use in her warehouse with 10,000 pallet positions. The hourly storage and retrieval transactions are 100 pallets per hour. It is anticipated that 75% of the AS/RS transactions are dual command transactions. In other words, a storage transaction will be paired with a retrieval transaction 75% of the time. Other information concerning number of AS/RS cranes, storage location dimensions, crane speed, and

Enter the following information

	Parameter	Value
1	Number of storage/retrieval shuttles	3
2	Number of storage requests per hour	100
3	Number of retrieval requests per hour	100
4	Maximum storage volume (units)	10,000
5	Storage cell height with clearances (ft)	9
6	Storage cell width with clearances (ft)	6
7	S/R machine horizontal velocity (ft/min)	600
8	S/R machine vertical velocity (ft/min)	1,000
9	S/R machine horizontal accel/decel (ft/s^2)	10
10	S/R machine vertical accel/decel (ft/s^2)	10
11	Pick-up time (s)	5
12	Deposit time (s)	5
13	% Dual command ops	75.00
14	Maximum allowable utilization (%)	90.00
15	Rack shape	Square-in-time(1) ▼
16	Rack type	Single deep(1) ▼

FIGURE 8.29   Input file for ExASRS.

rack shape are given in Figure 8.29. Determine the minimum number of aisles that will allow the manager to minimize cost and achieve the desired throughput.

Solution
When the data in Figure 8.29 is input to ExASRS and the program is executed, the output shown in Figure 8.30a is generated. Note that two aisles with 53 rows and 48 columns per rack allow us to meet the throughput requirements. Performing sensitivity analysis (Figure 8.30b), we see that adding one more aisle increases average throughput from 217 to 366 per hour.

8.6.7   STORAGE POLICIES

There are five main policies that are used to store incoming items in the warehouse. The simpler policy, called *random storage policy*, is to store an incoming item in any available location. If more than one location is available for storage, theoretically, the incoming item has an equal probability of being assigned to any of the available locations. In practice, however, it gets assigned to the closest available space. As pointed out in Francis et al. (1992) and Tompkins et al. (2010), storage and retrieval under the random policy are not so "purely" random in practice. Operators tend to store or retrieve items from the closest location.

The *dedicated policy*, on the other hand, requires assignment of items to prespecified locations that depend on the item type being stored. Each of the two policies has its own advantages and disadvantages. For example, for the same volume and frequency of S/R operations, the random policy requires less storage space than the dedicated policy. This is primarily because under the latter policy, the space dedicated to each item must equal the maximum level of inventory of that item. Of course, the maximum level is observed as soon as the item is replenished. With the random policy, however, the maximum aggregate inventory level is lower because the products are typically replenished at different times (see Exercise 33). In other words, because the products are replenished at different times, the maximum aggregate inventory level tends to be lower than the sum of

Results

Number of aisles	2
Length of rack (ft)	288.00
Height of rack (ft)	477.00
Number of columns per rack	48
Number of rows per rack	53

Display result details ☑

Storage openings per aisle	5088
Storage openings required per aisle	5000.00
Avg time per request (s)	29.87
Time available per request (s)	32.40
Avg throughput (requests/h)	216.93
Required throughput (requests/h)	200.00
Avg time per request (s) FCFS	29.87
Avg time per request (s) mod. FCFS	26.23
Avg time per request (s) nearest neighbor	23.56
Avg time per request (s) mod. nearest neighbor	22.92
Th (s)	28.80
Tv (s)	28.62
T (s)	28.80
b	0.99
T(SC) (s)	50.00
T(DC) (s)	74.57
T(QC) (s)	123.72
T(STC) (s)	172.88

(a)

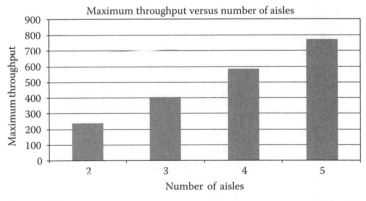

Number of aisles	Max throughput	Avg throughput (number required per hour)
2	241	217
3	407	366
4	587	528
5	773	696

(b)

FIGURE 8.30   Results produced by ExASRS. (a) Average time results and (b) throughput and aisle results.

the individual item's maximum inventory levels. Although the random policy requires less storage space, one of the problems is that if there are numerous items to be stored, the retrieval takes more time as some time is lost in searching the item's location (more so for items that are frequently delivered). Thus, throughput rate (i.e., the number of S/R transactions per unit time) of the system goes down, the S/R equipment will not be effectively utilized, the warehouse tends be disorderly with items scattered all over (sometimes even put aside in aisles), and so on. The reader is encouraged to identify other advantages and disadvantages of the two policies (Exercise 32).

The third policy—*cube-per-order index (COI) policy*—is operationally very simple and a widely used policy for allocating storage space to items in a warehouse. It was first proposed by Heskett (1963). COI for an item is defined as the ratio of the item's storage space requirement (its cube) to the number of S/R transactions for that item. The COI policy can be summarized as follows. List the items in a nondecreasing order of their COI. Allocate the first item in this list to the required number of storage spaces that are nearest to the input/output (I/O) point. Allocate the second item in the list to the required number of storage spaces that are next closest to the I/O point and so on until all the items are allocated. Thus, the COI policy puts items having a large number of S/R requests and requiring less storage space near the I/O point. The COI policy has been investigated by several researchers. For references and results, see Malmborg and Bhaskaran (1990).

The fourth policy, called the *class-based storage policy*, is based on Pareto's observation that a small percentage of a country's (or world's) population has the most wealth. Conversely, a large percentage of the population has the least wealth. This phenomenon, called the Pareto effect (named after the nineteenth century economist Vilfredo Pareto who observed it in wealth distribution), is seen in many spheres of life. For example, a company making multiple products may generate most of its revenues (say, 80%) from 20% of its products. Twenty percent of the students in an industrial engineering class may take up 80% of the instructor's office hours! In a warehouse, 80% of the S/R activity is due to 20% of the items, 15% due to 30% of the items, and the remaining 5% of the S/R activity is due to 50% of the items. Thus, we classify incoming items into one of three classes—A, B, and C—depending on the level of S/R activity it generates. If it is between 0% and 5% of the total S/R activity, the item is a class C item; if it is between 5% and 20%, the item is a class B item. Otherwise, it is a class A item.

To minimize the time spent in storage and retrieval, class A items must be stored closest to the input/output point, class B to the next closest location, and so on. Although each class of items has a dedicated storage space, an item can be stored randomly in any available storage space in the location assigned to its class, just as any item in a dedicated policy can be randomly stored in any available space dedicated to that item. It is perhaps apparent to the reader that the random and dedicated policies are two extremes, and the other two discussed so far fall somewhere in between. For example, if all the items were grouped into one class in the class-based storage policy, what we have is randomized storage. On the other hand, if we have as many classes as items in a class-based storage policy, we have dedicated storage policy.

The *shared storage policy* also falls between the two extremes of random storage and dedicated storage. As in the random storage operating policy, the same storage space may hold different items over time. However, the allocation of items to storage spaces is not random but carefully controlled. Fast-moving items are stored in spaces near the I/O point. Slow-moving items are stored in spaces farther away from the I/O point. Because the replenishment of items may not be instantaneous but occur at a constant rate, the time spent in inventory may vary a lot even for the same product. Also, because different items may reach their maximum inventory levels at different times, by properly allocating items to storage locations using the shared storage policy, we can increase system throughput and improve space utilization. This policy is being used in practice, and warehouse managers rely on their experience, intuition, and some rules of thumb in determining allocation of items to storage locations. Very few theoretical results are available for the shared storage policy (see, for example, Goetschalckx and Ratliff, 1990), and hence, we do not discuss this further. In the following discussion, we discuss models useful in determining values of design variables under the remaining four operating policies.

8.6.7.1 Design Model for Dedicated Storage Policy

Consider the following warehouse storage problem. A warehouse has p I/O points through which m items enter and leave the warehouse. The items are stored in one of n storage spaces or locations. Each location requires the same storage space. It is known that item i requires S_i storage spaces. Ideally, we would like $\sum_{i=1}^{m} S_i = n$. However, if the left-hand side is less than that of the right-hand side, we can always add a dummy product $(m + 1)$ to take up the remaining $n - \sum_{i=1}^{m} S_i$ spaces.

We will assume throughout our discussion that the aforementioned equality holds. (Of course, if the right-hand side is less than the left-hand side, there is no feasible solution to the problem as there are not enough storage spaces to accommodate all the items.)

There are f_{ik} trips of item i through I/O point k. The cost of moving a unit load of item i to and from I/O point k is c_{ik} and the distance of storage space j from I/O point k is d_{kj}. Given the aforementioned and a binary decision variable x_{ij} that specifies whether or not item i is assigned to storage space j, we are ready to formulate a model to assign the items to storage spaces in such a way that the cost of moving the items in and out of the I/O points is minimized. This is shown next in model 2. Note that the objective spreads the cost of moving item i in and out of storage evenly to all the S_i spaces occupied by the item. The cost for each item–space pair is calculated as a product of three factors: (1) frequency of trips made to each I/O point, (2) travel distance to each I/O point, and (3) cost per unit distance. It assumes that the cost per unit distance may vary for each I/O point–storage space combination.

Model 2

$$\text{Minimize} \sum_{i=1}^{m} \sum_{j=1}^{n} \frac{\sum_{k=1}^{p} c_{ik} f_{ik} d_{kj}}{S_i} x_{ij} \tag{8.12}$$

$$\text{Subject to} \sum_{j=1}^{n} x_{ij} = S_i \quad i = 1, 2, \ldots, m \tag{8.13}$$

$$\sum_{i=1}^{m} x_{ij} = 1 \quad j = 1, 2, \ldots, n \tag{8.14}$$

$$x_{ij} = 0 \text{ or } 1 \quad i = 1, 2, \ldots, m; j = 1, 2, \ldots, n \tag{8.15}$$

Substituting

$$w_{ij} = \frac{\sum_{k=1}^{p} c_{ik} f_{ik} d_{kj}}{S_i}$$

the objective may be written as

$$\text{Minimize} \sum_{i=1}^{m} \sum_{j=1}^{n} w_{ij} x_{ij} \tag{8.16}$$

The preceding model appears similar to the quadratic assignment problem (QAP) in Chapter 10, except that the objective function is linear and the first "assignment" constraint is more general than that seen in the QAP—the right-hand side is not 1 but S_i. If the right-hand side were 1 for constraint (8.13), then the preceding model would be the well-known assignment problem. It is called the generalized assignment problem for obvious reasons. Like the transportation and assignment problems, the optimal solution will always be such that x_{ij}'s are integers. (The reasons have been discussed elsewhere in the book, but also see Exercise 35.) Hence, integer constraint (8.15) can be replaced by a nonnegativity constraint. Because the generalized assignment problem is a special case of the transportation problem (notice that the transportation problem has B_j instead of 1 in the right-hand side of constraint (8.14)), it can be solved using the transportation algorithm. The latter is computationally efficient. It can be verified in Chapter 11 that the transportation algorithm examines the rows and columns of the cost (w_{ij}) matrix and only does addition or subtraction in each iteration. Even if we have several thousand storage spaces, the problem can be solved quickly on a computer.

Example 8.4

Consider the warehouse layout shown in Figure 8.31. There are three I/O points through which four items enter and leave the warehouse. All three I/O points are located on the periphery of the building, with the first equidistant from storage spaces 14 and 15, second from 5 and 9, and the third from storage spaces 2 and 3. The frequency of trips for each item—I/O point combination and the number of storage spaces required for each of the four items—is shown in Figure 8.32. Also shown in parentheses in the table is the cost of moving a unit load of item i to/from I/O point k. The distance of storage space j from I/O point k is given in Figure 8.33. Determine how the four items are to be allocated to the 16 storage spaces if the travel cost between the I/O points and the storage spaces is to be minimized.

Solution
First, we calculate the $w_{ij} = s$ using the formula. For example, w_{39} is calculated as

$$\frac{[c_{31} * f_{31} * d_{19} + c_{32} * f_{32} * d_{29} + c_{33} * f_{33} * d_{39}]}{2} = \frac{[96(4)(3) + 15(7)(1) + 85(9)(4)]}{2} = 2158.5$$

Values calculated for the other w_{ik}'s are shown in Figure 8.34. Using these values, a generalized assignment model may be set up and solved (see Exercise 39) to get the item to storage space assignment illustrated in Figure 8.35, the cost of this solution is $24,038.3. Figure 8.35 shows the assignment of items to the spaces depicted in Figure 8.31. Notice that items 3 and 4 are placed together, whereas 1 and 2 are split in two locations.

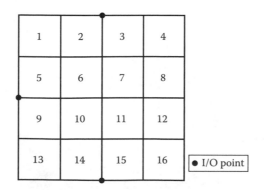

FIGURE 8.31   Layout of the warehouse in Example 8.4.

	1	2	3	S_i
1	150(5)	25(5)	88(5)	3
$[f_{ik}(c_{ik})] = 2$	60(7)	200(3)	150(6)	5
3	96(4)	15(7)	85(9)	2
4	175(15)	135(8)	90(12)	6

FIGURE 8.32 f_{ik} and c_{ik} values for four items and three I/O points and S_i values for the four items.

	1	2	3	4	5	6	7	8	9	10	11	12	13	14	15	16
1	5	4	4	5	4	3	3	4	3	2	2	3	2	1	1	2
$[d_{kj}] = 2$	2	3	4	5	1	2	3	4	1	2	3	4	2	3	4	5
3	2	1	1	2	3	2	2	3	4	3	3	4	5	4	4	5

FIGURE 8.33 d_{kj} values for 3 I/O points and 16 storage spaces.

	1	2	3	4	5	6	7	8	9	10	11	12	13	14	15	16
1	1626.7	1271.7	1313.3	1751.7	1481.7	1126.7	1168.3	1606.7	1378.3	1023.3	1065.0	1503.3	1316.7	961.7	1003.3	1441.7
2	1020.0	876.0	996.0	1380.0	996.0	852.0	972.0	1356.0	1092.0	948.0	1068.0	1452.0	1308.0	1164.0	1284.0	1668.0
3	1830.0	1308.0	1360.5	1987.5	1968.0	1446.0	1498.5	2125.5	2158.5	1636.5	1689.0	2316.0	2401.5	1879.5	1932.0	2559.0
4	2907.5	2470.0	2650.0	3447.5	2470.0	2032.5	2212.5	308.0	2212.5	1775.0	1955.0	2752.5	2135.0	1697.5	1877.5	2675.0

FIGURE 8.34 w_{ij} values for 4 items and 16 storage spaces.

FIGURE 8.35 Values for 4 items and 16 storage spaces.

8.6.7.2 Design Model for COI Policy under Certain Conditions

Consider a special case of the design model for dedicated storage policy in which all the items use the I/O points in the same proportion and the cost of moving a unit load of item i is independent of the I/O point. Let us define P_k as the percentage of trips through I/O point k, $k = 1, 2, ..., p$ (for any item, because all the items use the I/O points in the same proportion). Due to the additional constraints included in this model, there is no need for the first subscript in f_{ik} and c_{ik}. Therefore, f_{ik}, c_{ik} can be replaced by f_i, c_i, respectively, and the model may be formulated as follows.

Model 3

$$\text{Minimize} \sum_{i=1}^{m} \sum_{j=1}^{n} \frac{\sum_{k=1}^{p} c_i f_i P_k d_{kj}}{S_i} x_{ij} \tag{8.17}$$

Subject to constraints (8.13) through (8.15)

Substituting $w_j = \sum_{k=1}^{p} P_k d_j$, the objective may be written as

$$\text{Minimize} \sum_{i=1}^{m} \sum_{j=1}^{n} \frac{c_i f_i}{S_i} w_j x_{ij} \tag{8.18}$$

The model given by expressions (8.18) and constraints (8.13) through (8.15) is even easier to solve than model 2 and does not require the use of the transportation algorithm. It involves rearranging the "cost" term $c_i f_i / S_i$ for each item i, $i = 1, 2, ..., m$ and "distance" term w_j for each storage space j, $j = 1, 2, ..., n$ in nonincreasing and nondecreasing order, respectively, and matching the item, say i^*, corresponding to the first element in the ordered "cost" list with the storage spaces corresponding to the first S_{i^*} elements in the ordered "distance" list, the second item, say l^*, with the storage spaces corresponding to the next S_{l^*} elements in the ordered "distance" list, and so on until all the items are assigned to all the storage spaces.* Notice that this is exactly what the COI policy does, except it calculates the inverse of the "cost" term, COI, that is, storage space requirement divided by the cost incurred as a result of handling item i (or frequency of item i), and orders the elements in nondecreasing order of their COI values, thereby producing the same result as the aforementioned algorithm.

The previous model is captured as an interactive "playspace" in the "10 Principles of Materials Handling" module as part of the space utilization principle. The user can set the input data, namely, c_i, f_i, and S_i for four classes of products (see Figure 8.36a) and in the next screen is presented with an initial starting solution as well as the optimal solution for the problem created by the user (Figure 8.36b). The objective of this exercise is to move the products to minimize the total cost of storing and retrieving products from the warehouse through the only I/O point. The reader is encouraged to create problems and find an optimal solution for each (see Exercises 40 and 41).

The COI index policy is used even in allotting parking spaces in a shopping mall. For example, the manager of a large shopping mall applied the COI model to designate parking spaces to employees, customers, and service vehicles and came up with a qualitative chart and parking space designation shown in Table 8.3.

---

* Notice from Chapter 11 (Section 11.2.1) that arranging the cost and distance vectors in nonincreasing and nondecreasing order, respectively, and taking a product of the two vectors will provide a lower bound on the material-handling cost function. This algorithm can be shown to be optimal (Francis, 1967).

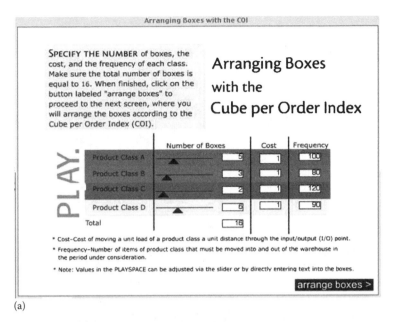

(a)

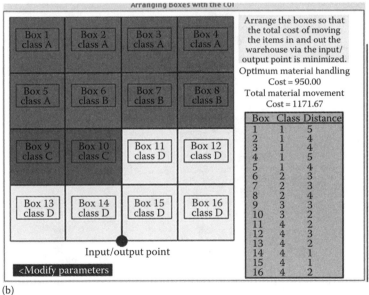

(b)

FIGURE 8.36 Interactive "playspace" for the COI storage policy model in the "space utilization" principle of the "10 Principles" educational module. (a) COI problem setup in "10 principles" module and (b) initial solution generated in "10 principles" module.

Example 8.5

Consider Example 8.3. Ignoring the c_{ik} and f_{ik} data and assuming that

1. The four items use the three I/O points in equal proportion
2. The total number of pallets of each item moved per time period are 100, 80, 120, and 90, respectively
   and
3. The cost of moving a load of each item through a unit distance is $1.00

TABLE 8.3

Qualitative Assessment of Parking Spaces in a Shopping Mall

Vehicle Category	Size Classification	Number of Transactions per Day	Parking Space Assignment
Customer vehicles	Small–medium	Medium	Close to mall entrances
Employee vehicles	Small–medium	Few	Far away from mall entrances
Service and delivery vehicles	Large	Many	Very close to mall entrances

determine the optimal assignment of the items to the storage spaces using the algorithm developed in this section.

Solution

The unsorted w_j values for the data are shown in Figure 8.37 and sorted values in Figure 8.38. Sorting the $c_i f_i/S_i$ values in nonincreasing order, we get 60, 33.33, 16, 15 corresponding to items 3, 1, 2, and 4, respectively. Thus, the optimal storage space assignment is as follows (also see Figure 8.39):

Item 1 to storage spaces 2, 5, and 7
Item 2 to storage spaces 1, 3, 9, 11, and 14
Item 3 to storage spaces 6 and 10
Item 4 to storage spaces 4, 8, 12, 13, 15, and 16

j	1	2	3	4	5	6	7	8	9	10	11	12	13	14	15	16
w_j	3.0	2.6	3.0	4.0	2.6	2.3	2.6	3.6	2.6	2.3	2.6	3.6	3.0	2.6	3.0	4.0

FIGURE 8.37   Unsorted w_j values for the data in Example 8.5.

j	6	10	2	5	7	9	11	14	1	3	13	15	8	12	4	16
w_j	2.3	2.3	2.6	2.6	2.6	2.6	2.6	2.6	3.0	3.0	3.0	3.0	3.6	3.6	4.0	4.0

FIGURE 8.38   Sorted w_j values for the data in Example 8.5.

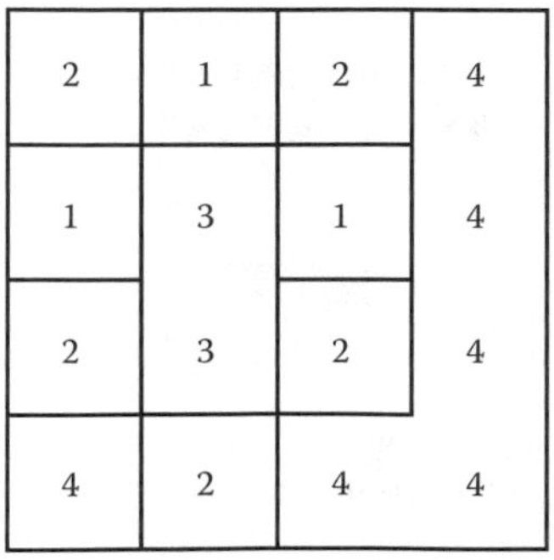

FIGURE 8.39   Optimal assignment of 4 items to 16 storage locations.

8.6.7.3 Design Model for Random Storage Policy

Consider the following random storage policy. Items coming into a warehouse are stored randomly in one of several storage spaces that are empty and available. Each empty storage space has an equal probability of being selected for storage. As pointed out previously, the storage or retrieval may not be purely random. However, we will assume a "pure" random policy to make the modeling easier. Because each storage location has an equal likelihood of being accessed for storage or retrieval, the problem of interest here can be defined as follows. Given that n storage spaces are required, determine the storage space layout so that the total expected travel distance between each storage space and p I/O points is minimized. Following the algorithm in the previous section, we can determine the sum of the distances of each storage space with each I/O point as $\sum_{k=1}^{p} d_{kj}$, arrange the spaces in nondecreasing order of the sum of these distances, and pick the n closest storage spaces. Note that n depends on the inventory levels of all the items, and as demonstrated in Exercise 33, the total number of spaces n would be less than that required under the dedicated policy.

Example 8.6

It is desired to determine the storage space layout for 56 storage spaces in a rectangular warehouse with dimensions 140 meters × 70 meters (see Figure 8.40). A random storage policy will be used, and the designer wants to minimize the total distance traveled. Each storage space is a 10 square meter × 10 square meter, and the I/O point is located in the middle of the south wall.

Solution

Calculating the distance of all the potential storage spaces to the single I/O point and arranging them in nondecreasing order, we obtain the layout shown in Figure 8.41 and find that the largest

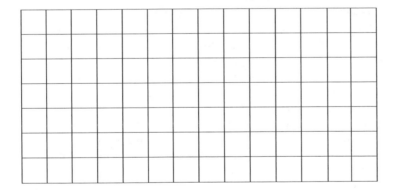

FIGURE 8.40 Warehouse layout with potential storage spaces.

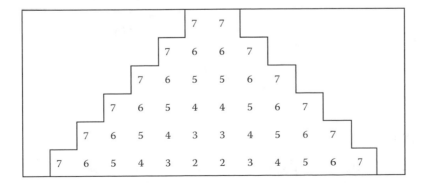

FIGURE 8.41 Storage space layout for Example 8.6.

distance traveled is 70 meters. The average distance traveled can be computed by summing the total distance traveled (2800) by the number of storage spaces (56) as 50 meters.

8.6.8   Travel Time Models

In Example 6, we calculated the average distance traveled as the sum of the distances of each individual storage space (to the I/O points) divided by the number of storage spaces:

$$\frac{\sum_{k=1}^{p}\sum_{j=1}^{n}d_{kj}}{n}$$

When the number of storage spaces is large, calculating the average distance can be tedious. If the storage spaces are small relative to the total area, we can approximate the average distance traveled by assuming that the storage spaces are continuous points on a plane and using the following integral. Notice in the following integral that we are assuming the warehouse (with dimensions X and Y) is in the first quadrant, that there is only one I/O point at the origin and lower left corner of the warehouse, and that the distance metric of interest is rectilinear:

$$\int_{0}^{X}\int_{0}^{Y}\frac{1}{A}(x+y)\,dxdy$$

If there are two or more I/O points, the distance metric is something other than rectilinear, and there are no restrictions on where the warehouse is to be located in a plane, this expression may be easily modified (see Exercises 44 through 46). While the average distance traveled for a given warehouse shape may be determined using expressions such as the one shown previously, one of the questions of interest to the warehouse designer may be the following: What warehouse shape minimizes the expected one-way travel time? Francis and White (1974) show how the optimal warehouse shape can be determined for various configurations (i.e., number of I/O points, distance metric, etc.) and that these shapes range from a diamond to circle to a trapezium!

Models that minimize warehouse construction costs in addition to average travel distance are more realistic than the ones seen so far. Further, if we constrain warehouse shape to be a rectangle (because most warehouses are rectangular) and obtain its dimensions so that construction and travel distance costs are minimized, then such a solution will be more meaningful to the designer. Such a model is discussed in detail in Francis and White (1974), and we provide a brief discussion next.

It is required to find the optimum dimensions of a rectangular warehouse of area A so that the expected one-way travel costs and construction costs are minimized. Assume that the construction cost is a function of the perimeter of the rectangle, say $r[2(a + b)]$, where r is the unit (perimeter) distance construction cost, a and b are the warehouse dimensions. There is one I/O point located at the origin, and the lower left corner has coordinates (p, q). Assuming c is the cost incurred for each unit distance traveled, the cost expression to be minimized is

$$2r(a+b)+c\int_{p}^{p+a}\int_{q}^{q+b}\frac{1}{A}\left(|x|+|y|\right)dxdy$$

Francis and White (1974) show that the optimal value of a and b given that

1. The I/O point must lie on or outside the warehouse's exterior walls, i.e., $p \geq 0$ and
2. The warehouse area must be A square units (i.e., $ab = A$)

is given by the following two expressions, respectively:

$$a = \sqrt{A\left(\frac{c+8r}{2c+8r}\right)} \quad \text{and} \quad b = \sqrt{A\left(\frac{2c+8r}{c+8r}\right)}$$

All the models discussed so far implicitly assume that each transaction involves one storage or retrieval, and thus the transaction cycle is a *single command* cycle. In other words, the person or the machine that does the storage or retrieval executes only one storage or retrieval transaction and returns to a reference point (say the I/O point) before executing the next transaction. However, if multiple transactions are made in each trip, for example, the person or the machine stores an item in one location, moves to another to retrieve an item, and returns back to the I/O point, the system throughput rate can be significantly increased. Depending on how many transactions are performed in each trip, we have *dual command* or *multiple command* cycles. Incorporating *dual command* cycles in the travel time models is more involved but can be done for some configurations (see Exercise 46).

We have only listed some of the basic problems in warehouse design, and there are several others that can be considered. We refer the interested reader to the papers by Ashayeri and Gelders (1985), Bozer and White (1990), Hodgson and Lowe (1982), Matson and White (1982), Kulwiec (1982), and Gray et al. (1992), among others for a treatment of these additional design considerations.

8.7   WAREHOUSE OPERATIONS

The manager of a warehouse faces numerous operational problems. A partial list is as follows:

- What is the sequence in which orders are to be picked?
- What is the routing policy to be used for order picking?
- How frequently are orders to be picked from the high-rise storage area? Do we batch pick operations or do we pick whenever an order comes in?
- Is there a limit on the number of items picked? If so, what is the limit?
- How are operators to be assigned to stacker cranes?
- How do we balance the picking operator's workload?
- Do we release items from the stacker crane into the sorting stations in batches or as soon as the item is picked?

Among the operational problems in the aforementioned list, problems related to order picking appear a number of times. The reader may have correctly guessed that one of the major problems in a warehouse operation is that of order picking. This is indeed true. Surveys have shown that order picking consumes over 50% of the activities in a warehouse (Tompkins et al., 2010). It is therefore not surprising that order picking is the single largest expense in warehouse operations (Gray et al., 1992; Koenig, 1980; Ogburn, 1984). Hence, this entire section is devoted to order picking.

In a typical high-volume order-picking operation, several thousands of orders are picked per day. As indicated previously, the devices used for order picking range from unit-load/mini-load/person-on-board AS/R systems to stacker cranes. Sometimes, these devices operate almost continuously, and the transaction (order picking and storage) rates are usually high (Bozer et al., 1990). Because the construction and operation of such systems are expensive, managers are interested in maximizing throughput capacity among other parameters. Throughput can be maximized by making changes in the design and operation of the warehouse. For example, the number of S/R devices may be increased. However, not only is this solution costly, but also it may not be practical to add more devices due to space or other constraints. Hence, if throughput can be maximized by making operational changes, such solutions are usually less expensive to implement and are therefore preferred. For example, we may increase the speed of the devices or the minimize order-picking time. One of the ways of doing the latter is to determine the optimal sequence in which items in an order are to be picked. In the following section, we discuss three heuristic algorithms for the order-picking sequence problem.

8.7.1 Order-Picking Sequence Problem

There are two basic picking methods—*order picking* and *zone picking*. In order picking, an operator is responsible for picking all the items in an order, whereas in zone picking, an operator assigned to a zone is responsible only for picking items in an order that are within his or her zone. The order picker may have to travel throughout the warehouse in order picking, but travel is restricted to the assigned zone in zone picking. This section deals with order picking and not zone picking.

The order-picking problem that arises in automated and semiautomated warehouses and DCs can be stated as follows. An AS/RS or a person on board an AS/RS retrieves items as listed in an order from their respective locations. The items in a list are called picks. Beginning from a pickup/delivery (P/D) or I/O location, the AS/RS retrieves the picks listed in the order in a particular sequence and then places the picked orders back at the I/O point. The operations are reversed when items are to be stored in the rack spaces. In order to increase throughput of the system, it is desirable to choose a sequence so that the time required to pick items in a list or order is minimized. Although most systems have a single I/O point, it is conceivable that there may be one point for pickup and another for delivery. If the reader is familiar with the traveling salesman problem (TSP), he or she may be thinking that the order-picking problem may be modeled as a TSP. If so, the reader is right. However, the distance metric is not the Euclidean metric for reasons mentioned in the following paragraph.

The AS/R machine typically has two independent motors that allow movement in the horizontal and vertical directions simultaneously. Therefore, the time required to travel from a point with coordinates (x_i, y_i) to another with coordinates (x_j, y_j) depends on the horizontal and vertical distances between the two points as well as the horizontal and vertical speeds of the two motors—h, v—respectively. Mathematically, it can be stated as

$$\max\left[\frac{|x_i - x_j|}{h}, \frac{|y_i - y_j|}{v}\right]$$

Once again, it may be obvious to the reader that, from the distance metrics discussed in Chapter 3, the order-picking problem can be modeled as a TSP with the Tchebychev metric. We will refer to this problem as the TTSP. A mathematical formulation for the TTSP is as follows:

$$\text{Minimize} \sum_{i=1}^{m} \sum_{j=1/j\neq i}^{n} d_{ij} w_{ij}$$

$$\text{Subject to} \sum_{i=1, i\neq j}^{n} w_{ij} = 1 \quad j = 1, 2, \ldots, n$$

$$\sum_{j=1, j\neq i}^{n} w_{ij} = 1 \quad i = 1, 2, \ldots, n$$

$$u_i - u_j + n w_{ij} \leq n - 1 \quad \text{for } 2 \leq i \neq j \leq n$$

$$d_{ij} = \max\left[\frac{|x_i - x_j|}{h}, \frac{|y_i - y_j|}{v}\right] \quad \text{for each } i, j; i \neq j$$

$$w_{ij} = 0 \text{ or } 1 \quad \text{for each } i, j; i \neq j$$

$$u_i = \text{are arbitrary real numbers} \quad \text{for each } i$$

It is assumed in this model that there are n points including the I/O point. The decision variable w_{ij} takes on binary values (0 or 1) depending on whether a location i is visited immediately after a location j in the order-picking sequence. The objective minimizes the total length of the order-picking "tour." The first two constraints ensure that each arc in the optimal tour has exactly two end points—onc on either side of it. The third set of constraints ensures that the optimal solution has no subtours. The fourth constraint specifies that the distance metric is Tchebychev.

Although optimal algorithms exist for solving the TSP or the TTSP, they are generally not preferred as they both are known to be NP-complete. Moreover, because order picking is a planning problem that is to be addressed on a frequent basis (e.g., hourly) and because a quick solution is required, heuristics instead of optimal algorithms arc used to solve the order-picking problem (Goetschalckx and Ratliff, 1988). We therefore discuss two heuristic algorithms for solving it.

Like the layout algorithms, algorithms for the TSP or TTSP can be classified as (Golden et al., 1980)

- Construction algorithms
- Improvement algorithms
- Hybrid algorithms

Construction algorithms are those in which the tour is constructed, usually, one at a time, until a complete tour is developed. In improvement algorithms, a given initial solution is improved, if possible, by exchanging the positions of two or more points in the initial tour. As discussed in Chapter 5, two improvement strategies are possible. We may either consider all possible exchanges and choose the one that results in the largest savings and continue this process until no further improvement is possible; or as soon as an exchange results in savings, we may make the exchange and examine other possible exchanges (of course, continuing until we cannot improve the solutions any further). The former is called steepest descent 2-opt or 3-opt depending on whether 2 or 3 points are considered for exchange and the latter is called greedy 2-opt or 3-opt. In general, it is not advantageous to exchange more than three points at a time (Lin and Kernighan, 1973). Hybrid algorithms employ a construction algorithm to obtain an initial solution and improve it using an improvement algorithm. Literature surveys on the TSP and TTSP problems are available from a number of other sources. For example, an excellent discussion of models and algorithms for the TSP problem is presented in Lawler et al. (1985). For an extensive computational analysis and performance comparison of several TSP heuristic algorithms, the reader may refer to Golden et al. (1980) and Adrabinski and Syslo (1983). For the TTSP problem, a survey and computational comparison of several efficient algorithms has been done in Goetschalkx and Ratliff (1988), Bozer et al. (1990), and in Heragu et al. (1994). In the next two sections, we discuss a heuristic and a software for the TSP.

8.7.1.1 Convex Hull Heuristic Algorithm

In this section, the convex hull heuristic for the TSP is discussed (Golden et al., 1980). This heuristic has been shown to be optimal under certain conditions and is very efficient. This algorithm was first introduced by Or (1976) and independently by Stewart (1977). It was later modified by Allison and

Noga (1984) and Goetschalckx (1983). This algorithm consists of three phases. In the first phase, the convex hull of the points is determined using the algorithm in Akl and Toussaint (1978). Assuming the P/D point to be the origin, $(x_o = 0, y_o = 0)$ the algorithm performs the following:

1. It determines the two points with maximum x and y coordinates. These points are referred to as x_{max} and y_{max}. It then deletes the points that are inside the polygon formed by the extremal points x_{max}, y_{max} and the origin O. The boundary of the rack and the boundary of the polygon define three regions (see Figure 8.42). (In case $x_{max} = y_{max}$, only two regions are formed, and the following explanation can easily be extended to that case.)

2. For each region, the algorithm constructs the convex path between the two extremal points as explained next:

    (a) It first sorts the points in regions 1 and 2 in ascending order of their x-coordinates and then the points in region 3 in descending order of x-coordinates.

    (b) Starting with the first extremal point, for every three consecutive points i, $i + 1$, and $i + 2$, it computes the following quantity V until the other extremal point is reached.

$$V = (y_{i+1} - y_i)(x_{i+2} - x_{i+1}) + (x_i - x_{i+1})(y_{i+2} - y_{i+1})$$

If $V \leq 0$, a convex hull with points i, $i + 1$, and $i + 2$ cannot be generated (Figure 8.43); otherwise, it is possible to generate a convex hull (Figure 8.44). Using some or all of the sorted points in regions

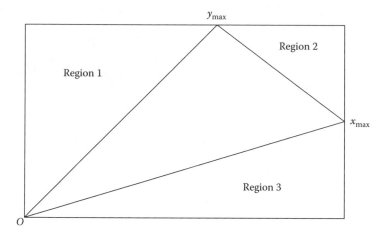

FIGURE 8.42   Formation of three regions in phase 1 of the convex hull heuristic.

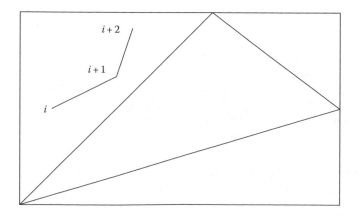

FIGURE 8.43   Three points not on a convex hull.

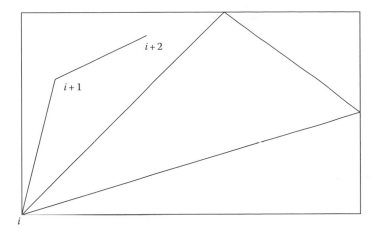

FIGURE 8.44   Three points on a convex hull.

1, 2, and 3, three at a time, the algorithm generates a convex hull. This hull is referred to as a subtour. Points that are not in the subtour are considered for inclusion in phases 2 and 3.

In the second phase, points that can be included in the subtour without increasing the cost are inserted. Such free insertion points lie on a parallelogram that has two adjacent points in the subtour as its corner (Goetschalckx and Ratliff, 1988). In the third phase, points that were not included in the subtour in phases 1 and 2 are included using a minimal insertion cost criteria. It has been shown that if there are no points left for insertion in phase 2 or 3, the subtour is optimal. However, when there are some points available for insertion in phase 3, they can be inserted one at a time in two ways; this leads to two variants of the convex hull algorithm. They are both heuristic algorithms—greedy hull (GH) and steepest descent hull (SDH)—and are explained as follows:

1. *GH*: The cost of inserting a randomly selected point i between each pair of points in the current subtour is determined. The point is inserted between those two points, that result in the least increase in the total cost. The aforementioned procedure is repeated until all the points are included in the subtour.
2. *SDH*: The minimum insertion cost is found for all the points that are not in the current subtour. Then, the point that has the least insertion cost is inserted between the corresponding pair of points in the subtour. The preceding procedure is repeated until there are no more points left for insertion.

We may also use 2-opt, simulated annealing, genetic algorithm, and the other general purpose techniques studied in Chapter 11 for the order-picking problem. The application is straightforward. For example, given an initial tour, the 2-opt algorithm exchanges the position of two points and finds the cost associated with the new tour. This is done for all possible pairs, i.e., $n(n-1)/2$ pairs, and the one with the least cost replaces the initial tour. Starting with this as the initial tour, the preceding procedure of examining the $n(n-1)/2$ two-way exchanges and replacing the initial tour with the least cost tour is continued until it is no longer possible to improve the current initial tour. Instead of checking all the possible exchanges, the greedy 2-opt heuristic algorithm replaces the initial tour with the first better tour encountered. The stopping criterion is the same as that for the 2-opt heuristic algorithm, that is, the preceding procedure is continued until no 2-way exchange leads to an improvement in the initial solution. Similarly, the simulated annealing algorithm discussed in Chapter 11 may be modified as follows for order picking.

Simulated annealing for order picking:

    Step 0: Set S = initial solution; z = corresponding objective function value (OFV); r = cooling factor; T_{in} = initial temperature; $T = T_{in}$; $T_{fin} = 0.1T_{in}$.

    Step 1: Randomly select points i and j in S and exchange their positions. If the resulting solution S' has an OFV $z' \leq z$, set $S = S'$ and $z = z'$; otherwise, set $S = S'$ with probability $e^{-\delta/T}$.

    Step 2: Repeat step 1 until the number of new solutions examined is equal to 16 times the number of neighbors.

    Step 3: Set $T = rT$. If $T > T_{fin}$, go to step 1; otherwise, return S and STOP.

8.7.1.2 Software for Solving the TSP*

We now discuss a software called tsp_distribute, which includes many of the algorithms included in Chapters 5 and 11. These include 2-opt, greedy 2-opt, 3-opt, greedy 3-opt, simulated annealing, and genetic algorithms. The neural network algorithm included in tsp_distribute is not discussed in this book but is based on the self-organizing feature map algorithm. Information about the latter can be found in Ritter et al. (1992).

The software allows the user to read data previously stored in a data file, or it can randomly generate one for any number of cities provided by the user. The educational version only allows the user to generate data up to 100 cities, however. The data input is very simple. The user provides the (x, y) coordinates for each city in the problem. Once the data are read from a data file or randomly generated by the program, the user must first solve it using a greedy algorithm that greedily constructs a starting solution by linking each city (beginning with the first) to the closest city and repeating the procedure until all the cities are linked in one giant tour that does not include any subtours. An output screen of the software is shown in Figure 8.45.

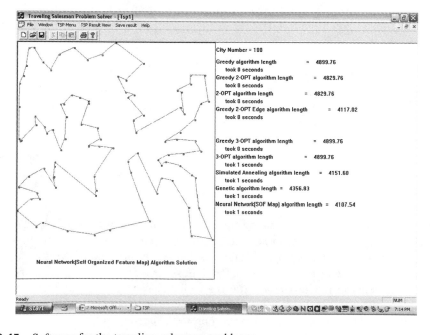

FIGURE 8.45   Software for the traveling salesman problem.

---

* The author is grateful to Dr. Byung-In Kim of POSTECH, Korea, for allowing use of his software developed via the http://sundere.okstate.edu/Book website.

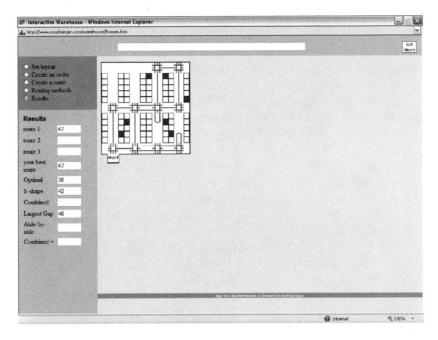

FIGURE 8.46 Snapshot of the interactive software in www.roodbergen.com.

The algorithm displays how the cities are linked in the giant tour using the most recently invoked algorithm, but it also shows the costs of the solutions obtained via the other algorithms on the right side of the screen. It is rather easy for the user to have the city links displayed for any of the solutions generated via the multiple algorithms.

8.7.2 ROUTING PROBLEM

Once the items to be picked in a list have been identified, the warehouse manager dispatches the order pickers to travel to the items, pick them, and return to the I/O point. Obviously, the sequence in which the items are picked and the routing strategy used to pick the items in the list determine how fast the order picker is able to return to I/O point. The faster he or she returns, the more orders can be picked. There are several routing strategies available in the literature. An interactive website located at www.roodbergen.com allows the user to create routing problems and test the alternate routing strategies for these. For example, the user can set a layout, including number of aisles, number of storage locations per aisle, and the I/O point location, randomly create an order of a specified size, manually create a route for the order picker, and compare this with optimal, S-shaped, largest gap, and other strategies. Information on these strategies and the interactive software can be accessed at www.roodbergen.com. The reader is encouraged to visit the website and try several routing strategies and compare them with the optimal and S-shaped strategies. A snapshot of the website is presented in Figure 8.46.

8.8 AUTOMATIC IDENTIFICATION

Automatic identification (Auto ID) is a general term given to a variety of technologies that allow machines, e.g., bar code readers, to identify objects and facilitate data capture seamlessly, accurately, and automatically. Smart cards, voice-recognition systems, optical character recognition, magnetic, chemical, biometric technologies, and radio-frequency identification (RFID) are all examples of Auto ID technologies. RFID has received much attention in the popular media. Companies such as Wal-Mart and federal agencies such as the Department of Defense (DoD) have required their suppliers to comply with specific product tagging and tracing requirements in the supply chain through RFID or other means.

RFID and Global Positioning Systems (GPS) are two key technologies becoming increasingly implemented in many logistical operations. Although RFID is based on principles that have been applied for a few decades now (e.g., use of radio frequencies in airline control and communication systems), it has generated substantial interest in the logistics industry primarily due to the Wal-Mart and DoD mandates. GPS has also been heavily deployed in many sectors of the logistics industry.

8.8.1   RFID

The RFID system consists of tags that contain a micro-chip and contain a unique identification code (Figure 8.47). The tag has an antenna to send and receive signals from a reader (Figure 8.48). The reader, which also has an antenna, sends out electromagnetic waves through the antenna. The tag receives the signal and is charged with sufficient energy to send back a response. The reader decodes the response waves into a unique digital ID, which is then entered into a remote server. Whereas bar code readers can only identify a brand of products, for example, all *Special K* cereal boxes of a certain size are identified via the same bar code, the RFID tag allows us to track each *Special K* box individually. Companies have identified numerous benefits with the ability to track units individually. One of the biggest advantages is that, unlike bar-coding systems, RFID does not require a line of sight for identification. Furthermore, it helps in tracking and tracing specific shipments throughout the supply chain, providing distribution visibility to the customer, recalling products manufactured on a certain day, plant, or lot for safety and other reasons, minimizing theft, loss, shrinkage, and maximizing availability of products on the shelf.

Although RFID systems have many advantages, they also have some disadvantages. In many applications, the cost is a barrier to their implementation. Products containing liquid or metals are difficult to handle. For example, because metal reflects RFID signals, special tags and devices are required for such applications. Tags can be passive, semi-active, or active. Passive tags have no power source of their own and are relatively inexpensive. Semi-active and active tags have their own power source, but the power source is used to broadcast a signal in the latter. It is also more expensive than semi-active tags.

RFID systems have been employed in numerous applications. Tags are used to record marathon runners' times. They have been used as highway toll tags, to track livestock, and bags by airline companies. Companies such as Metro in Germany are experimenting with stores that have

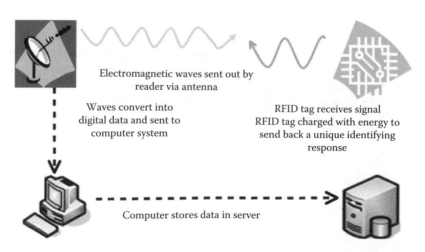

FIGURE 8.47   Radio-frequency identification technology. (Courtesy of RFID Journal LLC.)

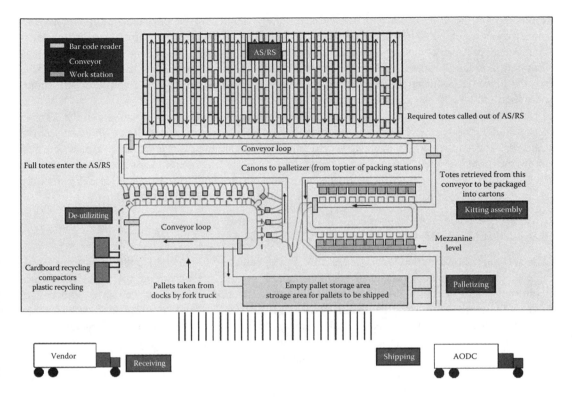

FIGURE 8.48 Layout of the virtual distribution center.

electronic product codes (EPCs) on all their products, allowing the store manager to get greater insight into the impact of product display, identifying purchasing patterns, for example, to see which items are purchased together, and offer more convenience to the customer. With EPCs on all products, it is possible to do away with checkout stations in a store. The customer simply has to pick up the items, put them in a bag or cart, and walk in between readers so the items can be charged to the customer's credit card automatically without the customer having to wait! Readers interested in the technology, its limitations, and applications are encouraged to visit www.rfidjournal.com.

Companies in the logistics business have adopted the technologies (especially RFID) more because of mandate than choice. As a result, many organizations use the RFID for simple functions such as product tagging and tracing and allow customers to track a shipment as it moves through the supply chain. They have not quite determined how to benefit from them. We believe, however, that the real-time information provided by these two systems can be effectively used to design and operate logistical systems on a real-time basis in direct response to new, constantly changing information that is being supplied by the two technologies.

In many systems, including logistical systems, the information based on which a planning decision is made (for example, scheduling trucks for customer deliveries) is significantly different from the information that exists at the time the decision is executed, thereby severely limiting the value of the original decision. Companies can use the RFID and GPS technologies and supporting models and decision support systems to develop effective and efficient supply chains that transfer products seamlessly and smoothly over the entire distribution network, even in the face of fast-changing information, and disturbances caused by forces internal or external to the system (e.g., hurricane-related disruptions, instability in fuel supply, and disruptions due to hostile activity).

In addition to the popular silicon chip–based RFID system, there are competing "chipless" RFID systems. These are yet to be proven for widespread applications, but one particular technology works as follows. Special metallic fibers called "taggents" embedded on paper, plastic, fabric, or just about any nonmetallic surface can be used to uniquely identify the paper or plastic surface on which they arc impregnated. When a radio frequency is sent from a reader with minimal power source, the "taggents," because of their unique orientation, lengths, and density, send back interference patterns that can be decoded into a unique ID (see www.inkode.com). Thus, every time the paper label is read, it gives back the unique ID allowing the tote, box, passport, event ticket, check, or other object on which the label is placed to be uniquely identified.

Chipless RFID technology can be applied to passports, classified documents, checks, currency, and software, thereby preventing their fraudulent use or duplication. It can enhance security, tracking, and control of physical assets and classified documents. Because the "taggents" are relatively inexpensive (a few pennies per ID) and do not require line of sight for identification, they can be embedded on a shipping invoice inside cardboard cartons containing high-value items to ensure there is no theft, tampering, pilferage, or other unauthorized opening of the cartons.

8.9 MULTIMEDIA CD FOR DESIGNING A DC*

The software module titled "Analysis and Design of Integrated Material Handling Systems" available at http://sundere.okstate.edu/Book is designed to integrate knowledge of material handling technology with modeling skills. It makes use of self-paced, multimedia supported exercises to increase knowledge of AS/RS and conveyors and illustrates material handling applications of analytical modeling through hands-on design and problem solving. The module is targeted to three groups of individuals: (a) intermediate level learners including juniors and seniors in undergraduate engineering programs, (b) learners with strong modeling skills but limited knowledge of technology, and (c) working professionals without adequate modeling skills but strong knowledge of current technology and practice.

The multimedia module is modeled after an actual DC and is called Teen Fashion Center, Inc. Each working day, the DC receives 650 pallets of garments from manufacturers in Asia, Latin America, Europe, and the United States and serves 2250 retail outlets in the eastern United States. Twenty-eight thousand totes containing individual Stock Keeping Units (SKUs) must be stored in the AS/RS system. The primary receiving functions in this DC include unpacking, inspecting, and storing SKUs. Shipping functions include picking, consolidating, palletizing, and dispatching store orders. There are five "areas" in this DC where different material handling functions are carried out (Figure 8.49). For example, in the deunitizing area, the material flow functions are

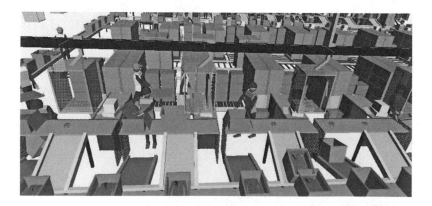

FIGURE 8.49 Snapshot of an animation in the deunitizing section of the multimedia module.

* This material is based on the work supported by the National Science Foundation under Grant No. EEC 9872506. Any opinions, findings, and conclusions or recommendations expressed in this material are those of the author(s) and do not necessarily reflect the views of the National Science Foundation.

receive, unpack, inspect, sort tote, and store in the ASRS. Similarly, in the kit assembly area, kits must be put into cartons and orders assembled; in the pallet storage area, outbound pallets must be staged and shipped.

Materials arriving from suppliers are unpacked, inspected, and put into plastic totes, and the totes are sent to the reserve area. When store orders arrive each week, the required items in each store order are put together or "kitted" by requesting the appropriate totes from the reserve area and placing the required number of each item requested in each store order into cartons. Each day, 780 retail store orders (on average) are put together and palletized at the DC. The pallets are consolidated based on their destinations and dispatched via 78 truck loads.

The "Analysis and Design of Integrated Material Handling Systems" module allows the user to understand the material handling functions in the five areas of the DC, select appropriate material handling technology for each of the five areas, configure them, and analyze the performance of their design with respect to important operational performance criteria under alternate demand scenarios. The five areas are

1. Deunitizing area
2. AS/RS area
3. Kitting area
4. Palletizer area
5. Pallet storage area

The three exercises for the user in each area are

1. *Select* technology
2. *Configure* subsystem
3. *Analyze* interface

In the deunitizing area, operators unpack cartons, put materials into totes, and put away the empty cartons on carriers circulating on overhead conveyors. After gaining an understanding of the material-handling functions in this area, the user is asked to select specific technology for the overhead conveyor system by choosing among available speed and capacity alternatives. If the parameters selected cause failed load attempts, for example, the operator is ready to put an empty carton on the overhead conveyor system but no carrier is available, the user can specify how the failed loads are to be handled. The module has rich animation (see Figure 8.49) and audio that help users understand the pros and cons of their design choices.

In the configure subsystem exercises, the user is asked to configure the system so that the design performs well with respect to operational performance measures. For example, in the deunitizing area, the user can specify how many workstations to keep open. Each workstation that is open requires an operator. If too many workstations are open, the labor cost increases and operators are idle. But the throughput rate (number of cartons opened and totes filled per hour) increases. On the other hand, if too few workstations are open, cartons recirculate and cause congestion in the conveyor system, because the operators are busy and therefore unable to open the cartons, unpack, and put materials into totes the first time cartons arrive at their station. Thus, the user sees the trade-offs of the alternate technology selections and system configurations.

In the analyze interface, the design choices made by the user are tested against multiple demand scenarios. Just prior to the back-to-school season and holiday season, the demand is very high but relatively low during the summer season. Exercises in the analyze interface section allow the user to understand when a system designed for peak periods is preferred over or systems designed for average demand periods.

Each of the remaining four areas has exercises categorized into the three sections: select, configure, and analyze. More details on the module are available in Hales et al. (2007), and the module can be accessed via http://sundere.okstate.edu/Book.

8.10   SUMMARY

This chapter introduced the reader to warehouse systems and discussed the various activities, that take place in such systems. Several types of storage media and S/R systems were discussed with illustrations. Some of the more important warehouse design problems were discussed. Where appropriate, mathematical models were used to aid in the design process. Although it is not out intention to give an exhaustive treatment of the models used in warehouse design, we have provided some of the basic models that will help the reader begin to understand some of the more sophisticated ones found in the literature. Several storage policies and models for use with these policies were discussed. The dominant problem in warehouse operations—order picking—was studied. Three algorithms to solve the order-picking sequence problem were discussed. Because the warehouse design and planning models do not consider all the problems simultaneously, it should be emphasized that the solutions obtained via these models be used only as a benchmark with which to compare other solutions.

8.11   REVIEW QUESTIONS AND EXERCISES

1. List and explain why warehouses are necessary in manufacturing and service organizations.
2. List and explain ten functions of a warehouse.
3. *Project*: Consult relevant sources in your library (e.g., *Modern Materials Handling, Material Handling and Logistics*, and *IE Solutions*) and develop a detailed classification of storage media used in warehouses. (*Note*: The list should be more comprehensive than in Section 8.4.) Explain in sufficient detail (with the help of illustrations) each type of MHS, its advantages and disadvantages, and provide at least one name and full address (telephone, facsimile numbers, e-mail, and Internet addresses) of the manufacturer for each type of storage medium.
4. *Project*: Repeat Exercise 3 for the three types of S/R equipment—manual, semiautomated, and automated.
5. *Project*: Visit a nearby warehouse. Interview the manager and appropriate warehouse personnel to ascertain five reasons why the warehouse was built and five most important functions of the warehouse. Then, draw a layout showing the storage media, S/R system, entrances, exits, and other support services of the warehouse. Analyze the warehouse operation and identify strengths and weaknesses of the warehouse design and operation. Prepare a detailed report discussing the previous issues and present your findings to your class.
6. List and explain the factors that determine the size and shape of a warehouse.
7. List and explain the various design problems encountered in a warehouse.
8. Discuss the various factors that affect the warehouse location decision.
9. Discuss the various factors that affect the layout and location of docks in a warehouse.
10. Show that the values of the multipliers a, b for the warehouse in Figure 8.27 are 1/9, 6/11, respectively.
11. Show that the unconstrained minimization function in Section 8.6.4 is convex. (*Hint*: You may use differential calculus to establish this fact or verify the convex shape of the function by plotting the function for various values of y on a spreadsheet and then show that the function only increases on either side of the minimum.)
12. Suppose $a = b$ in model 1. Show that the warehouse is a square.
13. Suppose $a = 2b$ in model 1. Determine the dimensions of the warehouse.
14. Assuming the point where travel originates is in the middle of one of the exterior horizontal sides of the warehouse in Figure 8.27, determine the dimensions of the warehouse in terms of n, z, a, b.
15. Assuming the point where travel originates is in the middle of one of the exterior vertical sides of the warehouse in Figure 8.27, determine the dimensions of the warehouse in terms of n, z, a, b.
16. Suppose $a = b$ in Exercise 14. Determine the dimensions of the warehouse.
17. Suppose $a = 2b$ in Exercise 14. Determine the dimensions of the warehouse.

18. Suppose $a = b$ in Exercise 15. Determine the dimensions of the warehouse.
19. Suppose $a = 2b$ in Exercise 15. Determine the dimensions of the warehouse.
20. Determine the average distance traveled for the three cases described in Example 8.1. Verify that the building dimensions for the 4 level and 5 level case are 25×20 units and 18×23 units, respectively.
21. Suppose $a = b$ in Example 8.1. Determine the dimensions of the warehouse and the average distance traveled.
22. Suppose $a = 2b$ in Example 8.1. Determine the dimensions of the warehouse and the average distance traveled.
23. Assuming the point where travel originates is in the middle of one of the exterior vertical sides of the warehouse in Figure 8.27, determine the dimensions of the warehouse and the average distance traveled.
24. Suppose $a = b$ in Exercise 23. Determine the dimensions of the warehouse and the average distance traveled.
25. Suppose $a = 2b$ in Exercise 23. Determine the dimensions of the warehouse and the average distance traveled.
26. The dimensions of a rectangular warehouse are to be determined for a company. It is known that the warehouse must accommodate a minimum of 1500 storage spaces with one cubic meters volume. The storage spaces will be used to hold pallets that are of dimensions 0.9 meters × 0.9 meters × 0.9 meters. Assuming that the horizontal and vertical aisle space multipliers a, b are 3 and 2, respectively, and that the I/O point for the warehouse is to be located in the middle of the east wall, determine dimensions of the warehouse with
    (a) 3 levels of stacking
    (b) 4 levels of stacking
    (c) 3 levels of stacking
27. Repeat Exercise 26 assuming $a = 2$, $b = 1$.
28. Repeat Exercise 26 assuming $a = 3$, $b = 2$.
29. A formula for prorating the capital and operating costs of a MHS was provided in Chapter 3 (see Equations 3.2 through 3.4). Develop a similar formula to determine the prorated warehouse material handling costs H_{ij} in the objective function (8.2) in Section 8.6.5.
30. (a) Find Q_i and T_i for Example 8.2. Verify that you get the same results shown in the solution to Example 8.2.
    (b) Set up an integer programming model to simultaneously determine the product allocation and area determination for the three areas shown in Figure 8.28 using the data in Example 8.2 and the results from Exercise 30a. Solve the model using any available integer programming software. Verify that you get the same results shown in the solution to Example 8.2.
31. A warehouse handles six classes of products. The average annual demand, order cost, price, and space required per unit load for each class of product as well as the handling and storage costs are given in Tables 8.4 and 8.5. Assuming a carrying cost rate of 5% and total available

TABLE 8.4
Product-Specific Input Data for Exercise 31

Products	1	2	3	4	5	6
Annual demand	1,000	5,000	5,000	9,000	25,000	15,000
Order cost	75	65	25	150	60	50
Price/unit load	250	450	550	50	225	500
Space required	20	25	15	20	2	23
p_i	0.2	0.2	0	0.2	0	0

TABLE 8.5

Handling and Storage Costs for Each Area–Product Combination for Exercise 31

Area/Product	1	2	3	4	5	6
1	0.0607 (10)	0.203 (15)	0.267 (12)	0.354 (5)	0.4083 (12)	0.726 (10)
2	0.0649 (5)	0.4023 (5)	0.0228 (6)	0.259 (4)	0.7804 (20)	0.871 (15)
3	0.161 (6)	0.1223 (103)	0.054 (15)	0.1062 (5)	1.0248 (25)	0.188 (10)
4	0.778 (12)	0.023 (14)	0.481 (10)	0.0671 (10)	0.0165 (15)	0.798 (10)

TABLE 8.6

Scheduled Receipts of 5 Products for 12 Periods

Period	Product				
	A	B	C	D	E
1	20	12	66	22	97
2	15	8	15	22	12
3	30	4	16	25	88
4	12	6	17	21	66
5	14	7	18	18	79
6	60	1	19	14	55
7	17	12	15	23	9
8	20	40	16	36	25
9	21	13	17	30	96
10	22	12	18	22	90
11	23	12	19	89	90
12	23	12	15	22	88

space of 120,000 square meters, determine how the products should be allocated to each of the four areas. Assume an upper and lower bound of 55,000 and 45,000 square meters for reserve and forward areas and a maximum of 10,000 square meters for the cross-docking area. Also, assume $a = b = c = 1$, $p_i = 0.2$ and 1 for products 1, 2, and 4, 0 for all others and $q_{i3} = 1$ for $i = 1$, 2, …, 6. Determine how the six products are to be allocated to the three areas (four flows) and the size of each area in the warehouse. How much of the total space is utilized? Set up an integer programming model and solve it using any available software.

32. Discuss five advantages and disadvantages of the random and dedicated storage policies. Explain under what circumstances we might prefer one over the other.

33. A warehouse specializing in five high-turnover consumer items is scheduled to receive the quantities shown in Table 8.6 in the next 12 periods. The warehouse manager likes to maintain a quantity of safety stock for each product in each period equal to 10% of the scheduled receipt of that product for the next period. Determine the total storage spaces required under the dedicated and random storage policies. Which policy requires fewer storage spaces? Why?

34. What are the similarities between the dedicated and class-based storage policies? What are the differences? Explain.

35. Explain why even when the {0, 1} integer restrictions on the x_{ij} variables are relaxed, the solution to model 2 will always be an integer solution.

36. Consider the warehouse layout shown in Figure 8.50. There are four I/O points through which six items enter and leave the warehouse. The four I/O points are located in the four corners of the building. The frequency of trips for each item—I/O point combination as well as the number

1	2	3	4	5	6
7	8	9	10	11	12
13	14	15	16	17	18
19	20	21	22	23	24
25	26	27	28	29	30

FIGURE 8.50 Layout of a warehouse with 30 storage spaces.

		1	2	3	4	S_i
	1	200(6)	125(8)	300(9)	88(5)	5
	2	100(6)	200(3)	150(6)	100(10)	5
$[f_{ik}(c_{ik})]$	3	150(9)	150(5)	125(12)	250(7)	4
	4	150(12)	230(5)	60(7)	30(4)	4
	5	96(4)	15(7)	40(12)	85(9)	4
	6	175(15)	135(8)	85(10)	90(12)	8

FIGURE 8.51 f_{ik} and c_{ik} values for six items and four I/O points and S_i values for the six items.

of storage spaces required for each of the six items—is shown in Figure 8.51. Also shown in parentheses in the table are the cost of moving a unit load of item *i* to/from I/O point *k*. The distance of storage space *j* from I/O point *k* is given in Figure 8.52. Determine how the 6 items are to be allocated to the 30 storage spaces if the travel cost between the I/O points and the storage spaces is to be minimized.

37. Consider Exercise 36. Ignoring the c_{ik} and f_{ik} data and assuming that
    (a) The six items use the four I/O points in equal proportion.
    (b) The total number of pallets of each item moved per time period is 150, 80, 120, 100, 200, and 90, respectively.
    (c) The cost of moving a load of each item through a unit distance is $1.00. Determine the optimal assignment of items to storage locations.
38. Solve Exercise 37 using the transportation algorithm. Verify that the answer is the same as that obtained in Exercise 37.

FIGURE 8.52 d_{kj} values for 4 I/O points and 30 storage spaces.

39. Set up a generalized assignment model for Example 8.3 and solve it using LINDO or the transportation algorithm. Verify that the answer you get is the same as that obtained in Example 8.3.
40. Create a dataset for the COI policy "playspace" in the space utilization principle of the "10 Principles of Materials Handling" software. Find the optimal solution by trial and error.
41. Create a dataset using the data in Exercise 37 for the COI policy "playspace" in the space utilization principle of the "10 Principles of Materials Handling" module. Find the optimal solution by trial and error. Verify that the solution you obtained in Exercise 37 is the same as that you obtained using the software.
42. Determine the storage space layout for 156 storage spaces in a rectangular warehouse with dimensions 200 meters × 100 meters. A random storage policy will be used, and the designer wants to minimize the total distance traveled. Each storage space is a 10 meters × 10 meters square, and the I/O point is located in the middle of the west wall.
43. Repeat Exercise 42 assuming the warehouse dimensions are 300 meters × 300 meters. Is the solution you get different from that obtained in Exercise 42? Why or why not? Explain.
44. Assuming there are two I/O points in the lower left and upper right corners, respectively, of a warehouse, which has dimensions 100 meters × 300 meters and the distance metric is Euclidean, set up and solve the expression for the average distance traveled.
45. Assuming that there is one I/O point in the lower left and corner of a warehouse, which has dimensions X meters by Y meters, items are located randomly, that the storage and retrieval is done by a machine that can travel in the x and y directions independently with speeds of h and v, respectively, and that $X \geq Yh/v$, set up and solve the expression for the average distance traveled.
46. Repeat Exercise 45 assuming dual command cycle.
47. List and explain the various operational problems encountered in a warehouse.
48. Differentiate between order picking and zone picking. What are the advantages and disadvantages of each? Explain under what circumstances each might be used.
49. A particular warehouse order consists of six items to be picked. Coordinates of the I/O point and the six storage locations are (10, 10), (15, 15), (19, 20), (12, 20), (10, 17), (17, 14), and (19, 13), respectively. Assuming the S/R machine has horizontal and vertical speeds of 20, 10 units per second, respectively, set up a mathematical model to determine the optimal picking sequence. Solve the model using LINDO or any other available integer programming software.
50. Solve the order-picking sequence problem described in Exercise 49 using both variations of the convex hull algorithm. Is the solution obtained different from that obtained in Exercise 49? Which solution is optimal? Explain.
51. Solve the order-picking sequence problem described in Exercise 49 using both variations of the 2-opt algorithm—greedy 2-opt and steepest descent 2-opt. Is the solution obtained different from that obtained in Exercise 49? Which solution is optimal? Explain.

52. Solve the order-picking sequence problem described in Exercise 49 using the simulated annealing algorithm. Is the solution obtained different from that obtained in Exercise 49? Which solution is optimal? Explain.

53. What are the advantages and disadvantages of
    (a) Solving the TTSP model using a suitable integer programming algorithm
    (b) Solving it using the convex hull 2-opt or simulated annealing algorithms

54. *Project*: Consult relevant sources in your library and review 15 algorithms used for the TSP and TTSP. At least two of these must be optimal algorithms. Discuss the advantages and disadvantages of each. Develop computer codes for 10 heuristic algorithms for the TSP/TTSP, test and compare their performance using test problems available in the literature (e.g. Bozer et al., 1990; Heragu et al., 1994), and discuss the merits of each.

55. Randomly generate a TSP using the tsp-distribute software available at http://sundere.okstate. edu/Book for these problem sizes
    (a) 10 cities
    (b) 25 cities
    (c) 50 cities
    (d) 100 cities
    Solve each of these problems using all the algorithms provided in the software. Explain why the 2-opt and 3-opt algorithms provide good solutions for the smaller sized problems but not the others.

56. Solve Exercise 49 using the tsp_distribute software. Is the best solution you obtained (from among the solutions generated using the multiple algorithms in tsp_distribute) different from that in Exercise 49? Why or why not?

9 Logistics and Location Models

9.1 MOTIVATING CASE STUDY

Logistics and distribution is a significant part of the ever-increasing service sector of the U.S. economy. It is estimated to represent 12% of the U.S. gross domestic product (GDP) and is four times that worldwide. Many sectors of the logistics and distribution (LoDI) economy are exploding. For example, U.S. companies in the third-party logistics (3PL) business had revenues exceeding $158 billion in 2014. This industry alone has seen 14% annual growth (compounded) since 1996. According to IBM's vice-president of On-Demand Innovation Services, "Marketplace trends are driving fundamental changes in the current business environment. Traditional company value chains and business models must be dramatically transformed, decisions must be made dynamically, and business processes must be able to quickly adapt to execute those decisions. Organizations must be designed to respond immediately to unpredictable economic fluctuations."

To support its worldwide missions during times of war and peace, the U.S. Department of Defense (DoD) operates the largest and most complex logistics and distribution system in the world. In fiscal year 2014, the Defense Logistics Agency (DLA) moved goods and provided services worth almost $38 billion throughout the world. These goods are required for peacetime and wartime military operations as well as humanitarian missions and emergency preparedness and include food, clothing, medical supplies, weapon systems, repair parts, and other items required for humanitarian missions. In order to ensure that public funds are utilized and managed most efficiently and services are provided most effectively, the DLA continuously seeks ways and means to reduce costs and improve delivery times in its operations.

In many ways, the city of Louisville owes its existence and growth to the falls of the Ohio River. Because riverboats that transported goods up and down the Ohio River in both directions could not navigate the 26 feet drop in elevation occurring over 1½ miles (called the water falls, but in reality these are cascading rapids), the goods had to be offloaded from downstream boats and loaded on to upstream boats and vice versa. This led to the construction of wharfs and warehouses to store the goods exchanged between the upstream and downstream boats, which eventually led to the expansion of the city. When rail networks were built in the United States in the late nineteenth and early twentieth centuries and interstate highway networks in the 1950s, because of its central location, Louisville not only kept its position as a major logistics and distribution hub, but also began expanding as a major manufacturing and industrial center. Attracted by the resources at Louisville and its geographical location, United Parcel Service (UPS) built its airline operations headquarters in Louisville and underwent a billion dollar expansion in 2000 that resulted in the Worldport (worldwide air hub shown in Figure 9.1). UPS expanded again in 2010 with a second, $1 billion investment that has increased its size to a 3,000,000 square feet facility.

The greater Kentuckiana region surrounding Louisville is also referred to as the "auto alley" due to the fact that 10% of the cars and trucks sold in the United States are produced in and around Louisville. Many auto suppliers for the Toyota, Corvette, Ford, and other area assembly plants are also located in the vicinity. Louisville itself is a major metropolitan city within 600 miles of two-thirds of the U.S. population and is considered an important logistics hub.

FIGURE 9.1   Aerial view of UPS Worldport.

There are two trends in logistics and distribution that affect the greater Louisville area significantly:

1. Increased transportation expected along the Mississippi and Ohio Rivers as well as the interstate highway networks attributable to the logistics and distribution of alternate fuels (ethanol, biodiesel, etc.), pharmaceuticals, autoparts, and materials for the aluminum industry.
2. Goods are produced in the midwest, but an overwhelming majority is shipped to the U.S. coasts. One in eight consumers in the United States is in California.

Louisville is ideally suited as a manufacturing, consolidation, and distribution hub for the goods. There is increased congestion in transportation networks in general, but particularly in the Kentuckiana region. In the short term, this presents a challenge. But in the long term , it presents opportunities for increasing transport efficiency and transportation throughput at current or slightly expanded capacity levels.

9.2   INTRODUCTION

We discussed several design problems in Chapter 1. One of the problems that must be addressed in the early stages of facility design is facility location, which is the focus of this and the next chapter, along with the broader topics of supply chain management (SCM) and logistics management. The facility location problem itself is part of a broader management problem called supply chain management (SCM). Managers and analysts have shifted their focus from just the manufacturing plant to the entities that the plants interact with, for example, suppliers, warehouses, and customers. SCM, which covers the entire supply chain from the raw material supplier to the end customer, has therefore received much attention in the past decade.

Logistics management, an important component of SCM, can be defined as the management of transportation and distribution of goods (Russell and Taylor, 1995). The term "goods" include raw materials or subassemblies obtained from suppliers as well as finished goods shipped from plants to warehouses or customers. Logistics management includes all distribution and transportation activities from suppliers through to customers. The cost of logistics as a percentage of gross domestic product (GDP) in many European Union countries ranges from 9% to 20% in many industries.

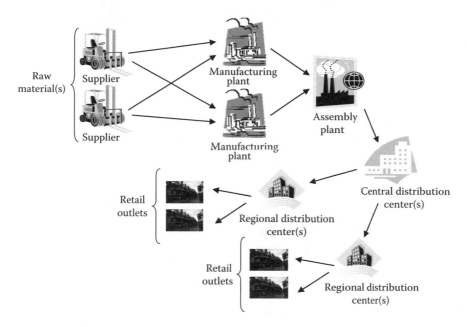

FIGURE 9.2   Entities in a supply chain.

Similar to the definition of material handling in Chapter 7, logistics management is the management of a series of macro-level transportation and distribution activities, with the main objective of delivering the right amount of material at the right place at the right time at the right cost using the right methods. The decisions typically encountered in logistics management concern facility location, transportation, and goods handling and storage.

A supply chain includes all the entities involved in making and distributing goods or services to the end consumer (Figure 9.2). When product service and recycling are considered, the supply chain includes the flow of goods or materials from the end consumer back to the manufacturing plants via 3PL providers. The typical entities in a supply chain include

- Raw material supplier
- Subassembly or component suppliers
- Manufacturing plants
- Warehouses
- Distribution centers
- Logistics service providers
- Retail outlets
- End consumers

Logistics service providers are vital entities in the supply chain. Many companies outsource this function to 3PL companies so that the distribution of raw materials, assemblies, components, and finished goods is managed by companies whose core competency is in providing logistics services. For example, Toshiba outsourced the distribution and also repair of its "Notebook" laptops to the supply chain division of UPS in Louisville, KY. Many 3PL companies also work with manufacturing plants to manage the inventories at various stages of the supply chain.

There are several important considerations in the design and operation of a supply chain. Some of these are (Ghiani et al., 2004)

- Warehousing
- Distribution channels
- Freight transportation
- Freight consolidation
- Transportation modes

Warehousing: Managers must determine whether to operate centralized or decentralized warehouses, whether to engage the services of a third-party provider for warehousing function, and whether to eliminate the need for a warehouse and ship directly to customer. Each strategy has its own advantages and disadvantages. For example, centralized warehouses allow better control of the storage and shipment of goods. Decentralized warehouses can quickly react to changes in demand patterns. Allowing a 3PL provider to manage the inventory and distribution functions allows a company to focus more of its resources in the areas in which it has core competencies. Shipping directly to the customer eliminates the costs associated with storage, but it may not be feasible for all industries.

Distribution channels: As part of its overall business strategy, a company must determine how to get the products to the end consumer. Should it utilize a wholesaler? A retailer? Or ship goods directly to consumers? There are examples of companies in the same industry that utilize different channels in meeting customer demands. Whereas Rubbermaid sells its products through retail outlets, Tupperware bypasses that channel and sells directly to the consumer.

Freight transportation and transportation modes: Multiple modes of transportation are available for shipping goods from manufacturer to consumer. Surface transport includes good transportation via ship, rail, and road. Containers are typically designed so that they can utilize multiple modes. For example, containers transported via ships are unloaded at a port and loaded on a rail for long-haul shipment across the country and then on a truck for shipment within a locality or region. See the standardization principle in the "10 Principles of Materials Handling" module at http://sundere.okstate.edu/Book for more information about this.

Freight consolidation: Due to the differences in shipment volume and frequency between the customers served by a 3PL provider, there may be cost advantages in consolidating shipments from multiple sources so that multiple less-than-truckload (LTL) shipments to the same destination can be combined into a full truckload (TL) shipment.

9.3 LOGISTICS, LOCATION, AND SUPPLY CHAIN

In general, there are three levels of decisions that managers must make in designing a supply chain. At the long-term or strategic level, the decisions include facility location, selecting distribution channels, deciding whether to subcontract logistics services to TPL providers, and so on. These decisions typically cover many years, have long-term effects, and once made, cannot be changed easily. At the medium-term or tactical level, the decisions impact the operations for a shorter period of time—typically, months or quarters. These decisions include determining staffing levels for the next quarter, determining whether to have LTL or TL shipments, whether to enter into short-term leases for warehouse space, and others. At the shorter-term or operational level, the decisions include when to schedule pickups and deliveries, staff assignment to specific tasks, and other day-to-day operational decisions. In the remainder of this chapter and next, we will focus on one of the more important strategic problems—the facility location problem.

In the manufacturing context, a facility is where raw materials, processing equipment, and people come together to make a finished product. Proper location of a facility will go a long way in improving its overall effectiveness. A good facility location must satisfy these objectives:

1. It must be as close as possible to raw material sources and customers.
2. Skilled labor must be readily available in the vicinity of a facility's location.
3. Taxes, property insurance, construction, and land prices must not be too high.
4. Utilities must be readily available at a reasonable price.
5. Local, state, and other government regulations must be conducive to business.
6. Business climate must be favorable and the community must have adequate support services and facilities such as schools, hospitals, and libraries, which are important to employees and their families.

It is often extremely difficult to find a single location with all the aforementioned characteristics at the desired level. For example, a location in Silicon Valley may offer a highly skilled labor pool for a software design company, but construction and land costs may be too high. Similarly, another location may offer low tax rates and minimal government regulations, but may be too far from the raw materials source or customer base. Thus, the facility location problem becomes one of selecting a site (among several available alternatives) that optimizes a weighted set of objectives. In the next section, we discuss what these objectives typically are. We conclude this section by providing several classifications of the location and logistics management problems.

Logistics management problems can be classified as

1. Location problems
2. Allocation problems
3. Location–allocation problems

Location problems involve the determination of the location of one or more new facilities in one or more of several potential sites. Obviously, the number of sites must be at least equal to the number of new facilities being located. The cost of locating each new facility to each of the potential sites is assumed to be known. It is made up of the fixed cost of locating a new facility at a particular site plus the operating and transportation cost of serving customers from this facility–site combination.

Allocation problems assume that the number and location of facilities is known *a priori* and attempt to determine how each customer is to be served. In other words, given the demand for goods at each customer center, the production or supply capacities at each facility, and the cost of serving each customer from each facility, the allocation problem determines how much each facility is to supply each customer center.

Location–allocation problems involve determining not only how much each customer is to receive from each facility, but also the number of facilities, their locations, and capacities.

Facility location problems can be classified as single-facility and multifacility location problems. As the name implies, single-facility location problems deal with the optimal determination of the location of a single-facility and multifacility location problems deal with the simultaneous location determination of more than one facility. Generally, single-facility location problems are location problems, but multifacility problems can be location as well as location–allocation problems (Hax and Candea, 1984).

Another classification of location problems depends upon whether the set of possible locations for a facility is finite or infinite. If a facility can be located anywhere within the confines of a geographic area, then the number of possible locations is infinite. Such a problem is called a continuous space location problem. In contrast, discrete space location problems have a finite feasible set of sites in which to locate a facility. The continuous space problem assumes that the transportation costs are proportional to the distance. Because facilities can be located anywhere in a 2D space, sometimes the optimal location provided by the continuous space model may be infeasible. For example, a continuous space model may locate a manufacturing facility in the middle of a lake!

In this chapter, we examine important factors involved in facility location, and we discuss models for discrete and continuous location problems. Most of these are single-facility location models. We also present a facility location case study.

9.4   IMPORTANT FACTORS IN LOCATION DECISIONS

In practice, there are a multitude of factors that have an impact on location decisions. The degree of impact of these factors depends upon whether the scope of our location problem is international, national, statewide, and community-wide. For example, if we are trying to determine the location of a manufacturing facility in a foreign country, factors such as political stability, foreign exchange rates, business climate, duties, and taxes play a greater role. If the scope of the location problem is restricted to a few communities, then factors such as community services, property tax incentives, local business climate, and local government regulations are more important than others. Some of the most important factors in location decisions are listed in Table 9.1.

If we examine the inputs required to manufacture a product or provide a service, two things stand out—people and raw materials. For a location to be effective, it must therefore be in close proximity to relatively less expensive, skilled labor pools and raw materials sources. One of the reasons for electronics and software companies locating in Silicon Valley is the availability of highly skilled computer professionals. Similarly, many U.S. companies have opened manufacturing facilities in Mexico and the Far East to take advantage of lower labor wage rates. Many companies look for labor pools with higher productivity, a strong work ethic, and absence of unionization.

With respect to raw materials, some industries find it more important to be close to raw materials sources than others. These tend to be industries for which raw materials are bulky or otherwise

---

TABLE 9.1

List of Factors Affecting Location Decisions

Factor

Proximity to source of raw materials

Cost and availability of energy and utilities

Cost, availability, skill, and productivity of labor

Government regulations at the federal, state, county, and local levels

Taxes at the federal, state, county, and local levels

Insurance

Construction costs and land price

Government and political stability

Exchange rate fluctuation

Export, import regulations, duties, and tariffs

Transportation system

Technical expertise

Environmental regulations at the federal, state, county, and local levels

Support services

Community services—schools, hospitals, recreation, and so on

Weather

Proximity to customers

Business climate

Competition-related factors

expensive to transport. Companies that have implemented just-in-time (JIT) or lean manufacturing strategies also like to be located near supplier factories so that frequent deliveries can be made. This allows them to maintain low levels of inventories and thereby reduce costs. Other inputs that have an impact on location decisions include cost and availability of energy and utilities, land prices, and construction costs.

In addition to the aforementioned input-related factors, there is an output-related factor that plays an important role in evaluation of location—proximity to customers. This factor is important because the product shelf life may be short, the finished product may be too bulky or require special care during transportation, duties and tariff may be too high necessitating the facility location to be close to the market area, and so on.

9.5 TECHNIQUES FOR DISCRETE SPACE LOCATION PROBLEMS

Our focus in this chapter is primarily on the single-facility location problem. We provide both discrete space and continuous space models. The single-facility for which we seek a location may be the only one that will serve all the customers, or it may be an addition to a network of existing facilities that are already serving customers.

9.5.1 QUALITATIVE ANALYSIS

The location scoring method is a very popular subjective decision-making tool that is relatively easy to use. It consists of the following steps:

Step 1: List all the factors that have an impact on the location decision.
Step 2: Assign appropriate weights (typically between 0 and 1) to each factor based on the relative importance of each.
Step 3: Assign a score (typically between 0 and 100) to each location with respect to each factor identified in step 1.
Step 4: Compute the weighted score for each factor for each location by multiplying its weight by the corresponding score.
Step 5: Compute the sum of the weighted scores for each location and choose a location based on these scores.

Although step 5 calls for a location decision made solely on the basis of the weighted scores, these scores were arrived at in a subjective manner, and hence, a final location decision must also take objective measures such as transportation costs, loads, and operating costs into consideration. Quantitative methods that take into consideration some of these measures are discussed in Section 9.5.2.

Example 9.1

A payroll processing company has recently won several major contracts in the midwest region of the United States and central Canada and wants to open a new, large facility to serve these areas. Because customer service is so important, the company wants to be as near its customers as possible. Preliminary investigation has shown that Minneapolis, Winnipeg, and Springfield, IL, would be the three most desirable locations and the payroll company has to select one of these three. A subsequent thorough investigation of each location with respect to eight important factors has generated the raw scores and weights listed in Table 9.2. Using the location scoring method, determine the best location for the new payroll processing facility.

TABLE 9.2

Factors and Weights for Three Locations

		Location		
Weights	Factors	Minneapolis	Winnipeg	Springfield
0.25	Proximity to customers	95	90	65
0.15	Land and construction prices	60	60	90
0.15	Wage rates	70	45	60
0.10	Property taxes	70	90	70
0.10	Business taxes	80	90	85
0.10	Commercial travel	80	65	75
0.08	Insurance costs	70	95	60
0.07	Office services	90	90	80

TABLE 9.3

Weighted Scores for the Three Locations in Table 9.2

	Location		
Weighted Score	Minneapolis	Winnipeg	Springfield
Proximity to customers	23.75	22.50	16.25
Land and construction prices	9.00	9.00	13.50
Wage rates	9.50	6.75	9.00
Property taxes	7.00	9.00	7.00
Business taxes	8.00	9.00	8.50
Commercial travel	8.00	6.50	7.50
Insurance costs	5.60	7.60	4.80
Office services	6.30	6.30	5.60
Sum of weighted scores	**78.15**	**76.65**	**72.15**

Solution

Steps 1, 2, and 3 have already been completed for us. We now need to compute the weighted score for each location–factor pair (step 4), add these weighted scores, and determine the location based on these scores (step 5).

From the analysis in Table 9.3, it is clear that Minneapolis is the best location based on the subjective information in Table 9.2. Of course, as mentioned before, objective measures must be considered, especially because the weighted scores for Minneapolis and Winnipeg are close.

9.5.2 Quantitative Analysis

There are several quantitative techniques available for the discrete space, single-facility location problem. Each is appropriate for a specific set of objectives and constraints. For example, the so-called minimax location model is appropriate for determining location of an emergency service facility, where the objective is to minimize the maximum distance traveled between the facility and any customer. Similarly, if the objective is to minimize the total distance traveled, the transportation model is appropriate. In this section we discuss the transportation model in detail and briefly describe the minimax location model. The transportation model is typically used to determine the optimal distribution of goods between a given set of plants and a known set of warehouses so

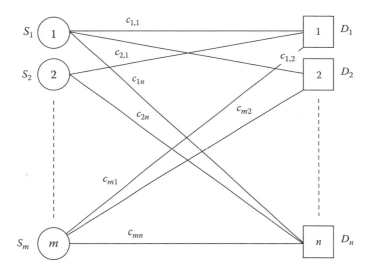

FIGURE 9.3 Graphical representation of a transportation problem.

as to minimize the overall cost of transporting goods between the plants and warehouses, while satisfying customer demand and plant supply constraints (Figure 9.3). The reader may question: the existence of a location problem the set of plants including their location is given. Before we answer this question, we present the transportation model and an algorithm for solving it. We use the following notation:

Parameters

m Number of plants
n Number of warehouses
c_{ij} Cost of transporting a unit from plant i to warehouse j
S_i Supply capacity of plant i (in units)
D_j Demand at warehouse j (in units)

Decision variable

x_{ij} Number of units transported from plant i to warehouse j

The transportation model is

$$\text{Minimize} \sum_{i=1}^{m} \sum_{j=1}^{n} c_{ij} x_{ij} \tag{9.1}$$

$$\text{Subject to} \sum_{j=1}^{n} x_{ij} \leq S_i \quad i = 1, 2, \ldots, m \tag{9.2}$$

$$\sum_{i=1}^{m} x_{ij} = D_j \quad j = 1, 2, \ldots, n \tag{9.3}$$

$$x_{ij} \geq 0 \quad i = 1, 2, \ldots, m, \ j = 1, 2, \ldots, n \tag{9.4}$$

The objective function (9.1) minimizes the total cost of transporting the goods from the plants to the warehouses. Constraint (9.2) ensures that the total number of units leaving each plant does not exceed its capacity. Constraint (9.3) ensures that the total number of units received at each warehouse is at least equal to the demand at that location. If the total demand equals the total supply, we have a balanced transportation problem. If the total demand exceeds the total supply (or the total supply exceeds the total demand), we can add a dummy plant (warehouse) that produces (absorbs) the excess demand (capacity) to convert the unbalanced problem to a balanced transportation problem. In the balanced transportation problem, all the inequality signs ($\leq$ in constraint (9.2) and $\geq$ in constraint (9.4)) can be replaced by equality signs. Constraint (9.4) ensures that the number of units transported from each plant to each warehouse be nonnegative. Although integer restrictions are not imposed on the x_{ij} variables, it should be noted that these are automatically satisfied in the optimal solution provided the S_i and D_j values are all integers. This can be proved by verifying the following three factors:

1. The transportation simplex algorithm (TSA) provided next is an optimal algorithm that converges in a finite number of steps.
2. We can always begin the TSA with a feasible initial solution such that all x_{ij} variables have integer values, provided the S_i and D_j values are all integers and a feasible solution exists for the given transportation problem.
3. The TSA involves only additions and subtractions, so we only have a feasible integer solution at each iteration.

Transportation simplex algorithm (TSA)

Step 1: Check whether the transportation problem is balanced or unbalanced. If balanced, go to step 2. Otherwise, transform the unbalanced transportation problem into a balanced one by adding a dummy plant (if the total demand exceeds the total supply) or a dummy warehouse (if the total supply exceeds the total demand) with a capacity or demand equal to the excess demand or excess supply, respectively. Thus, transform all the $\geq$ and $\leq$ constraints to equalities.

Step 2: Set up a transportation tableau by creating a row corresponding to each plant including the dummy plant and a column corresponding to each warehouse including the dummy warehouse. Enter the cost of transporting a unit from each plant to each warehouse (c_{ij}) in the corresponding cell (i, j). Enter 0 cost for all the cells in the dummy row or column. Enter the supply capacity of each plant at the end of the corresponding row and the demand at each warehouse at the bottom of the corresponding column. Set m and n equal to the number of rows and columns, respectively, and all $x_{ij} = 0$, $i = 1, 2, \ldots, m$; and $j = 1, 2, \ldots, n$.

Step 3: Construct a basic feasible solution using the Northwest corner method.

Step 4: Set $u_1 = 0$ and find $v_j, j = 1, 2, \ldots, n$ and $u_i, i = 1, 2, \ldots, n$ using the formula $u_i + v_j = c_{ij}$ for all basic variables.

Step 5: If $u_i + v_j - c_{ij} \leq 0$ for all nonbasic variables, then the current basic feasible solution is optimal; stop. Otherwise, go to step 6.

Step 6: Select the variable $x_{i^*j^*}$ with the most positive value $u_{i^*} + v_{j^*} - c_{i^*j^*}$. Construct a closed loop consisting of horizontal and vertical segments connecting the corresponding cell in row i^* and column j^* to other basic variables. Adjust the values of the basic variables in this closed loop so that the supply and demand constraints of each row and column are satisfied and the maximum possible value is added to the cell in row i^* and column j^*. The variable $x_{i^*j^*}$ is now a basic variable and the basic variable in the closed loop that now takes on a value of 0 is a nonbasic variable. Go to step 4.

Several aspects of the TSA are worth a discussion. However, we must first define three terms used in the algorithm—basic variable, nonbasic variable, and basic feasible solution. Consider a balanced transportation model with $m+n$ constraints and mn variables. It is easy to show that we have a redundant constraint because we know that the total supply equals total demand. So, we really have $m+n-1$ constraints. Set any $mn-m-n-1$ variables equal to 0 and solve for the remaining $m+n-1$ variables. The solution obtained is considered a basic feasible solution if all the $m+n-1$ variables whose values we solved for are nonnegative. The variables whose values were set to 0 are called nonbasic variables and the remaining $m+n-1$ variables are basic.

Step 2 involves setting up a transportation tableau. The values at the end of each row and at the bottom of each column indicate the supply capacities at the plants and demand at the warehouses, respectively. The Northwest corner method mentioned in step 3 is a simple method to identify a basic feasible solution. It begins by assigning the maximum possible value to the cell in the Northwest corner. This value is equal to the minimum of the supply capacity or the demand in the corresponding row or column, respectively—i.e., $x_{11}=\min\{S_1, D_1\}$. If $x_{11}=S_1$, cross out row 1, and set $x_{21}=\min\{S_2, D_2-x_{11}\}$. Otherwise, cross out column 1, and set $x_{12}=\min\{S_1-x_{11}, D_2\}$. Repeat this procedure of crossing out a row or column and setting the subsequent x_{ij} value equal to the remaining supply in row i or the remaining demand in column j, until all the rows and columns are crossed out. For a balanced transportation problem, this will occur in at most $m+n-1$ steps. Sometimes the remaining supply in row i is also equal to the remaining demand in column j. When this occurs we have a degenerate basic feasible solution, and we set x_{ij} equal to this value and (arbitrarily) set $x_{i,j+1}$ to 0 and repeat the procedure treating $x_{i,j+1}$ as a basic variable.

The closed loop in step 6 must be constructed so that

1. There are at least four line segments.
2. The loop begins and ends in the cell corresponding to the nonbasic variable x_{ij}.
3. Each horizontal and vertical line segment except the ones beginning and ending at the cell corresponding to the nonbasic variable must pivot on the first basic variable encountered.

We now illustrate solution of a transportation model using the TSA with this numerical example.

Example 9.2

Seers Inc. has two manufacturing plants at Albany and Little Rock supplying Canmore brand refrigerators to four distribution centers in Boston, Philadelphia, Galveston, and Raleigh. The company has seen a surge in demand of this brand of refrigerators and this surge is expected to last for several years into the future. Seers Inc. has therefore decided to build another plant in Atlanta. The expected demand at the four distribution centers and the maximum capacity at the Albany, Atlanta, and Little Rock plants are given in Table 9.4. Determine the total cost of transporting goods from the three plants to the four distribution centers.

TABLE 9.4
Costs, Demand, and Supply Information for Example 9.2

	Boston	Philadelphia	Galveston	Raleigh	Supply Capacity
Albany	10	15	22	20	250
Little Rock	19	15	10	9	300
Pittsburgh	17	8	18	12	400
Atlanta	21	11	13	6	400
Demand	200	100	300	280	

Solution

Step 1: Because the total supply capacity (950 units) exceeds the total demand (880 units), transform the unbalanced transportation problem into a balanced one by adding a dummy warehouse with demand equal 70 units.

Step 2: Set up a transportation tableau as shown in Table 9.5. Set $m = 3$, $n = 5$.

Step 3: A basic feasible solution constructed using the Northwest corner method is shown in Table 9.6.

Step 4: Set $u_1 = 0$ and find v_j, $j = 1, 2, ..., n$ and u_i, $i = 1, 2, ..., n$ using the formula $u_i + v_j = c_{ij}$ as shown in Table 9.7. The c_{ij}'s are shown only for the basic variables in the table.

Step 5: Because $u_i + v_j - c_{ij} > 0$, for the nonbasic variable in cell (3, 2), go to step 6.

Step 6: Because the nonbasic variable in cell (3, 2) is the only one for which $u_i + v_j - c_{ij} = 7 > 0$, select the variable x_{32}. Construct a *closed loop* consisting of horizontal and vertical segments connecting the corresponding cell in row 3 and column 2 to other basic variables. Adjust the values of the basic variables in this closed loop so that the supply and demand constraints of each row and column are satisfied and the maximum possible value is added to the cell in row 3 and column 2, as shown in Table 9.8. The variable x_{32} is now a basic variable and the basic variable in the closed loop in cell (2, 2) that now takes on a value of 0 is a nonbasic variable. Notice that we have a

TABLE 9.5

Transportation Tableau for Seers, Inc. Assuming the Plant Is Located in Atlanta

	Boston	Philadelphia	Galveston	Raleigh	Dummy	Supply Capacity
Albany	10	15	22	20	0	250
Little Rock	19	15	10	9	0	300
Atlanta	21	11	13	6	0	400
Demand	200	100	300	280	70	

TABLE 9.6

Basic Feasible Solution Constructed Using the Northwest Corner Method for Transportation Data in Table 9.5

	Boston	Philadelphia	Galveston	Raleigh	Dummy	Supply Capacity
Albany	200	50				250
Little Rock		50	250			300
Atlanta			50	280	70	400
Demand	200	100	300	280	70	

TABLE 9.7

Finding u_i and v_j, Values

	Boston	Philadelphia	Galveston	Raleigh	Dummy	u_i
Albany	10	15				0
Little Rock		15	10			0
Atlanta			13	6	0	3
v_j	10	15	10	3	−3	

TABLE 9.8
Improving the Basic Feasible Solution in Example 9.2

	Boston	Philadelphia	Galveston	Raleigh	Dummy	Supply Capacity
Albany	200	50				250
Little Rock		0	300			300
Atlanta		50		280	70	400
Demand	200	100	300	280	70	

TABLE 9.9
Optimal Solution for Example 9.2

	Boston	Philadelphia	Galveston	Raleigh	Dummy	Supply Capacity
Albany	200				50	250
Little Rock			300			300
Atlanta		100		280	20	400
Demand	200	100	300	280	70	

degenerate basic feasible solution, so we assume the cell variable x_{22} is basic and that variable x_{33} is nonbasic. The cost of the current solution is $7980.

Repeating steps 4, 5, and 6, it is easy to obtain the optimal solution shown in Table 9.9 with a cost of $7780 (Exercise 12).

To answer the question why we study the transportation problem in a chapter devoted to facility location, consider the following problem. We have m plants in a distribution network that serve n customers. Due to an increase in demand at one or more of these n customers, it has become necessary to open an additional plant. There are p possible sites where the new plant could be located. To evaluate which of the p sites minimizes the distribution (transportation) costs to the greatest extent, we can set up p transportation models, each with n customers and $m+1$ plants, where the $m+1$th plant corresponds to the new location being evaluated. Solution of the model will tell us not only the distribution of goods from the $m+1$ plants (including the new one from the location being evaluated) but also the cost of such a distribution. The location that yields the least overall distribution cost would be the one where the new facility is to be located. This is illustrated in Example 9.3.

Example 9.3

Consider Example 9.2. In addition to Atlanta, Seers Inc. is considering another location—Pittsburgh. The expected demand at the four distribution centers and the maximum capacity at the Albany and Little Rock plants are given in Table 9.10. Notice that a plant of any size can be built at both locations, so that the theoretical capacity of plants at the new locations is infinite. Determine which of the two locations, Atlanta or Pittsburgh, is suitable for the new plant. Seers Inc. wishes to utilize all of the capacity available at its Albany and Little Rock locations.

Solution

First, we must set up two transportation models, first assuming that the new plant will be located at Atlanta and the second at Pittsburgh. These two problems are displayed in Tables 9.11 and 9.12.

TABLE 9.10

Costs, Demand, and Supply Information for Example 9.3

	Boston	Philadelphia	Galveston	Raleigh	Supply Capacity
Albany	10	15	22	20	250
Little Rock	19	15	10	9	300
Atlanta	21	11	13	6	No limit
Pittsburgh	17	8	18	12	No limit
Demand	200	100	300	280	

TABLE 9.11

Transportation Model with Plant in Atlanta

	Boston	Philadelphia	Galveston	Raleigh	Supply Capacity
Albany	10	15	22	20	250
Little Rock	19	15	10	9	300
Atlanta	21	11	13	6	330
Demand	200	100	300	280	880

TABLE 9.12

Transportation Model with Plant in Pittsburgh

	Boston	Philadelphia	Galveston	Raleigh	Supply Capacity
Albany	10	15	22	20	250
Little Rock	19	15	10	9	300
Pittsburgh	17	8	18	12	330
Demand	200	100	300	280	880

Notice that the maximum capacity of the new plant required at either location is 330 because the capacity at Albany and Little Rock is to be fully utilized.

Solving the Pittsburgh model using the TSA (see Exercise 13) yields a total cost of \$9510. Formulating the Atlanta problem as a transportation model and solving it using the simplex algorithm, we get a total cost of \$7980. We should have used the TSA (designed for transportation models, which are special cases of linear programming models) instead of the simplex algorithm (designed for general linear programming models), but to illustrate how to formulate transportation models, we use the latter approach. The LINGO computer input and output are shown.

Pittsburgh Model

```
Data:
    M = 3;
    N = 4;
Enddata
Sets:
```

```
     Plants/1..M/: Supply;
     Customers/1..N/: Demand;
     Links(Plants,Customers): Cost,Volume;
Endsets

Data:
     Supply= 250 300 330;
     Demand= 200 100 300 280;
     Cost= 10 15 22 20
           19 15 10 9
           17 8 18 12;
Enddata
     ! Objective function;
     Min= @SUM(Links(i,j): Cost(i,j)*Volume(i,j));
     ! Constraints;
     @FOR (Plants(i): @SUM(Customers(j): Volume(i,j))<=Supply(i));
     @FOR (Customers(j): @SUM(Plants(i): Volume(i,j))>=Demand(j));
   End


   Global optimal solution found.
   Objective value:                          9510.000
   Total solver iterations:                        13

                         Variable        Value
                         VOLUME( 1,  1)   200.0000
                         VOLUME( 1,  2)   50.00000
                         VOLUME( 2,  3)   300.0000
                         VOLUME( 3,  2)   50.00000
                         VOLUME( 3,  4)   280.0000
```

Atlanta Model

```
Data:
     M= 3;
     N= 4;
Enddata
Sets:
     Plants/1..M/: Supply;
     Customers/1..N/: Demand;
     Links(Plants,Customers): Cost,Volume;
Endsets

Data:
     Supply= 250 300 330;
     Demand= 200 100 300 280;
     Cost= 10 15 22 20
           19 15 10 9
           21 11 13 6;
Enddata
     ! Objective function;
     Min= @SUM(Links(i,j): Cost(i,j)*Volume(i,j));
     ! Constraints;
     @FOR (Plants(i): @SUM(Customers(j): Volume(i,j))<=Supply(i));
     @FOR (Customers(j): @SUM(Plants(i): Volume(i,j))>=Demand(j));
   End
```

```
Global optimal solution found.
Objective value:                              7980.000
Total solver iterations:                            14

                    Variable         Value
            VOLUME( 1,  1)        200.0000
            VOLUME( 1,  2)         50.00000
            VOLUME( 2,  3)        300.0000
            VOLUME( 3,  2)         50.00000
            VOLUME( 3,  4)        280.0000
```

In both cases, the distribution pattern remains the same with Boston receiving all of its shipments from Albany; Galveston from Little Rock; Raleigh from the new plant and Philadelphia receiving 50 units from Albany and the new plant. However, because the Atlanta location minimizes the cost, the decision is to construct the new plant at Atlanta.

The location of an emergency facility such as a fire station, police station, or hospital can be treated as a minimax location problem if the objective is to minimize the maximum distance traveled or *cost* incurred by any customer. We now show how such a problem can be solved trivially by inspecting a cost or distance matrix. Consider the problem of locating a fire station in a county. Assume we have m possible sites and that the county can be divided into n locations based on population. The distance between the centroid of each of the n locations and each potential site for the fire station is available as shown in the distance matrix illustrated in Figure 9.4. To find the optimal location, we simply scan each row for the maximum distance and choose the site corresponding to the row that has the smallest maximum distance. This location ensures that the maximum distance that needs to be traveled by any customer is minimized (see Exercise 19).

9.6 HYBRID ANALYSIS

A disadvantage of the qualitative method discussed in Section 9.5.1 is that a location decision is made based entirely on subjective evaluation. While this disadvantage was overcome in the quantitative method discussed in the previous section, a problem is that it does not allow us to incorporate unquantifiable factors that have a major impact on the location decision. For example, the quantitative techniques can easily consider transportation and operational costs. But intangible factors such as attitude of a community toward businesses, potential labor unrest, reliability of auxiliary service providers, and public dissatisfaction with job outsourcing, although important in choosing

Location

$$
\text{Site} \begin{bmatrix}
d_{11} & d_{21} & \cdots & d_{1n} \\
d_{21} & d_{22} & \cdots & d_{2n} \\
\cdot & \cdot & \cdots & \cdot \\
\cdot & \cdot & \cdots & \cdot \\
\cdot & \cdot & \cdots & \cdot \\
d_{m1} & d_{m2} & \cdots & d_{mn}
\end{bmatrix}
$$

FIGURE 9.4   Distance between potential sites and multiple locations.

a location, are difficult to capture. Thus, what we need is a method that incorporates subjective as well as quantifiable cost or other factors.

In this section, we discuss a multiattribute, single-facility location model based on the ones presented by Brown and Gibson (1972) and Buffa and Sarin (1987). This model classifies the objective and subjective factors important to the specific location problem being addressed as

- Critical
- Objective
- Subjective

While the meaning of the latter two factors is obvious, the meaning of critical factors needs further discussion. In every location decision, there is usually at least one factor that determines whether or not a location would be considered further for evaluation. For example, if water is extensively used in a manufacturing process, e.g., brewery, then a site that does not have adequate water supply now or in the future is automatically removed from further consideration. It should be noted that some factors can be objective and critical or subjective and critical. As an example, adequacy of skilled labor may be a critical factor as well as a subjective factor.

After the factors are classified, they are assigned numerical values as follows:

$$CF_{ij} = \begin{cases} 1 \text{ if location } i \text{ satisfies critical factor } j \\ 0 \text{ otherwise} \end{cases}$$

OF_{ij} = cost of objective factor j at location i
SF_{ij} = numerical value assigned (on a scale of 0–1) to subjective factor j for location i
w_j = weight assigned to subjective factor j ($0 \leq w_j \leq 1$)

Let us assume that we have m candidate locations, p critical, q objective, and r subjective factors. We can then determine the overall critical factor measure (CFM_i), objective factor measure (OFM_i), and subjective factor measure (SFM_i) for each location i as follows:

$$CFM_i = CF_{i1}CF_{i2}\cdots CF_{ip} = \prod_{i=1}^{p} CF_{ij} \quad \text{for } i = 1, 2, ..., m \tag{9.5}$$

$$OFM_i = \frac{\max_i\left[\sum_{j=1}^{q} OF_{ij}\right] - \sum_{j=1}^{q} OF_{ij}}{\max_i\left[\sum_{j=1}^{q} OF_{ij}\right] - \min_i\left[\sum_{j=1}^{q} OF_{ij}\right]} \quad \text{for } i = 1, 2, ..., m \tag{9.6}$$

$$SFM_i = \sum_{j=1}^{r} w_j SF_{ij} \quad \text{for } i = 1, 2, ..., m \tag{9.7}$$

The location measure LM_i for each location is then calculated as

$$LM_i = CFM_i\left[\alpha OFM_i + (1-\alpha)SFM_i\right] \quad \text{for } i = 1, 2, ..., m \tag{9.8}$$

where α is the weight assigned to the objective factor measure. Notice that even if one critical factor is not satisfied by a location i, then CFM_i and hence LM_i are equal to 0. The OFM_i values are calculated so that the location with the maximum $\sum_j OF_{ij}$ would get an OFM_i value of 0 and the one with the least $\sum_j OF_{ij}$ value would get an OFM_i value of 1. Equation 9.8 assumes that the objective factors are cost based. If any of these factors are profit based, then a negative sign has to be placed in front of each objective factor and Equation 9.4 can still be used. This works because maximizing a linear profit function z is the same as minimizing $-z$. (Example 9.4 illustrates how profit-based as well as cost-based objective factors can be incorporated in the same model.)

After LM_i is determined for each candidate location, the next step is to select the one with the greatest LM_i value. Because the α weight is subjectively assigned by the user, it may be a good idea for the user to evaluate the LM_i values for various appropriate α weights, analyze the trade-off between objective and subjective measures, and choose a location based on this analysis.

Example 9.4

Mole-Sun Brewing company is evaluating six candidate locations—Montreal, Plattsburgh, Ottawa, Albany, Rochester, and Kingston—for constructing a new brewery. There are two critical, three objective, and four subjective factors that the management wishes to incorporate in its decision making. These factors are summarized in Table 9.13. The weights of the subjective factors are also provided in the table. Determine the best location if the subjective factors are to be weighted 50% more than the objective factors.

Solution

Clearly, $\alpha = 0.4$ so that the weight of the subjective factors $(1 - \alpha = 0.6)$ is 50% more than that of the objective factors. To determine OFM_i, we first determine the $\sum_j OF_{ij}$ values, as shown in Table 9.13, and then ascertain the maximum and minimum of these to be -35 and -95, respectively. From these two values, it is easy to calculate OFM_i values using Equation 9.6. For example, for the Montreal location, $OFM = [-35 - (-67)]/[-35 - (-95)] = 0.53$. Using the CFM_i, OFM_i, and SFM_i values, we get the location measure LM_i using Equation 9.8. The values are shown in Table 9.14. Based on an α value of 0.4, the Plattsburgh location seems favorable. However, as the weight of objective factors, α, increases to more than 0.6, the Montreal location becomes attractive.

9.7 TECHNIQUES FOR CONTINUOUS SPACE LOCATION PROBLEMS

Continuous space location models determine the optimal location of one or more facilities on a 2D plane. The obvious disadvantage is that the optimal location suggested by the model may not be a feasible one for a variety of reasons. For example, it may be in the middle of a water body—river, lake, or sea. Or the optimal location may be in a community that prohibits such a facility. Despite this drawback, these models are useful because they lend themselves to easy solution. Furthermore, if the optimal location is infeasible, techniques that find the nearest feasible and optimum locations are available.

In Chapter 3, we discussed the most important and widely used distance metrics: Euclidean, squared Euclidean, and rectilinear. We introduce three single-facility location models, each incorporating a different distance metric along with solution methods or algorithms for these models. Because the optimal solution for a continuous space model may be infeasible, where available, we also discuss techniques that enable us to find feasible and optimal locations.

9.7.1 MEDIAN METHOD

As the name implies, the median method finds the median location (defined later) and assigns the new facility to it. It is used for single-facility location problems with rectilinear distance. Consider m

TABLE 9.13

Critical, Subjective, and Objective Factor Ratings for Six Locations for Mole-Sun Brewing Company, Inc.

	Critical		Objective			Factors			
							Subjective		
Location	Water Supply	Tax Incentives	Revenue	Labor Cost	Energy Cost	Community Attitude 0.3	Ease of Transportation 0.4	Labor Unionization 0.25	Support Services 0.05
Albany	0	1	185	80	10	0.5	0.9	0.6	0.7
Kingston	1	1	150	100	15	0.6	0.7	0.7	0.75
Montreal	1	1	170	90	13	0.4	0.8	0.2	0.8
Ottawa	1	0	200	100	15	0.5	0.4	0.4	0.8
Plattsburgh	1	1	140	75	8	0.9	0.9	0.9	0.55
Rochester	1	1	150	75	11	0.7	0.65	0.4	0.8

TABLE 9.14

Location Analysis for Mole-Sun Brewing Company, Inc., Using the Hybrid Method

	Critical		Objective				Subjective				
Location	Water Supply	Tax Incentives	Revenue	Labor Cost	Energy Cost	Sum of Objective Factors	Community Attitude 0.3	Ease of Transportation 0.4	Labor Unionization 0.25	Support Services 0.05	SFM_i	LM_i
Albany	0	1	185	80	10	−95	0.5	0.9	0.6	0.7	0.7	0
Kingston	1	1	150	100	15	−35	0.6	0.7	0.7	0.8	0.67	0.4
Montreal	1	1	170	90	13	−67	0.4	0.8	0.2	0.8	0.53	0.53
Ottawa	1	0	200	100	15	−85	0.5	0.4	0.4	0.8	0.45	0
Plattsburgh	1	1	140	75	8	−57	0.9	0.9	0.9	0.6	0.88	0.68
Rochester	1	1	150	75	11	−64	0.7	0.65	0.4	0.8	0.61	0.56

facilities in a distribution network. Due to marketplace reasons, e.g., increase in customer demand, it is desired to add another facility to this network. The interactions between the new facility and existing ones are known. The objective of this problem is to locate the new facility to minimize the total interaction cost between each existing facility and the new one.

At the macro level, this problem arises, for example, when deciding where to locate a warehouse that is to receive goods from several plants with known locations. At the micro level, this problem arises when we must add a new machine to an existing network of machines on the factory floor. Because the routing and volume of parts processed on the shop floor is known, interaction (in number of trips) between the new machine and existing ones can be easily calculated. Other non-manufacturing applications of this model are given in Francis et al. (1992). Consider the following notation:

f_i = traffic flow between new facility and existing facility i
c_i = cost of transportation between new facility and existing facility i per unit
x_i, y_i = coordinate points of existing facility i

The median location model is then to

$$\text{Minimize } TC = \sum_{i=1}^{m} c_i f_i \left[|x_i - \bar{x}| + |y_i - \bar{y}| \right] \tag{9.9}$$

where
TC is the total cost of distribution
$\bar{x}, \bar{y}$ are the optimal coordinates of the new facility

Because the $c_i f_i$ product is known for each facility, it can be thought of as a weight w_i corresponding to facility i. In the rest of this chapter, we therefore use the notation w_i instead of $c_i f_i$. We can now rewrite Equation 9.9 as

$$\text{Minimize } TC = \sum_{i=1}^{m} w_i |x_i - \bar{x}| + \sum_{i=1}^{m} w_i |y_i - \bar{y}| \tag{9.10}$$

Because the x and y terms can be separated, we can solve for the optimal $\bar{x}$ and $\bar{y}$ coordinates independently. Here is the median method:

Median method

Step 1: List the existing facilities in nondecreasing order of the x coordinates.
Step 2: Find the jth x coordinate in the list (created in step 1) at which the cumulative weight equals or exceeds half the total weight for the first time, i.e.,

$$\sum_{i=1}^{j-1} w_i < \sum_{i=1}^{m} \frac{w_i}{2} \quad \text{and} \quad \sum_{i=1}^{j} w_i \geq \sum_{i=1}^{m} \frac{w_i}{2} \tag{9.11}$$

Step 3: List the existing facilities in nondecreasing order of the y coordinates.
Step 4: Find the kth y coordinate in the list (created in step 3) at which the cumulative weight equals or exceeds half the total weight for the first time, i.e.,

$$\sum_{i=1}^{k-1} w_i < \sum_{i=1}^{m} \frac{w_i}{2} \quad \text{and} \quad \sum_{i=1}^{k} w_i \geq \sum_{i=1}^{m} \frac{w_i}{2} \tag{9.12}$$

The optimal location of the new facility is given by the jth x coordinate and the kth y coordinate identified in steps 2 and 4, respectively.

Four points about the model and algorithm are worth mentioning. First, the total movement cost—that is, the OFV of Equation 9.10—is the sum of movement cost in the x and y directions. These two cost functions are independent in the sense that the solution of one does not have an impact on the solution of the other. Moreover, both cost functions have the same form. This means that we can solve the two functions separately using the same basic procedure, as done in the median method.

Second, in step 2, the median method determines a point on the 2D plane such that no more than one-half of the total traffic flow cost is to the left or right of the point. In step 4, the same is done so that no more than one half of the total traffic flow cost is above or below the point. Thus, the optimal location of the new facility is a median point.

Third, it can be shown that any other x or y coordinate will not be that of the optimal location's coordinates. In other words, the median method is optimal. We offer an intuitive explanation. Because the problem can be decomposed into the x-axis and y-axis problems and solved separately, let us examine the x-axis problem, i.e., the following x-axis movement cost function:

$$\sum_{i=1}^{m} w_i |x_i - \bar{x}|$$

Suppose the facilities are arranged in nondecreasing order of their x-coordinates as shown in Figure 9.5. Let us assume that the x coordinate at which the cumulative weight exceeds* half the total weight is the point shown as x_j in the figure. (The cumulative weights are shown below the respective coordinates in Figure 9.5. For coordinate x_j, we indicate the fact that the cumulative weight exceeds half the total weight). Let us also assume that the optimal x coordinate of the new facility falls at the coordinate indicated as x^* in Figure 9.5. For every unit distance we move to the left of x^*, the x-axis movement cost decreases by more than half the total weight and increases by less than half the total weight, because the facilities to the left of x^* have a combined weight exceeding half the total weight and, therefore, those to the right of x^* must have a combined weight of less than half the total weight. Because every unit distance movement to the left improves our cost function, it is beneficial to keep moving to the left until we have reached the x_j coordinate. Any further movement to the left only increases the total cost. Thus, x_j must be the optimal coordinate for the new facility. In a similar manner, we can establish the result for the optimal y coordinate (see Exercise 26).

Fourth, these coordinates could coincide with the x and y coordinates of two different existing facilities or possibly one existing facility. In the latter case, the new facility has to be moved

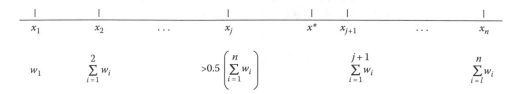

FIGURE 9.5   Illustration of optimality of the median method.

* We can prove the case where the cumulative weight equals half the total weight in a similar manner (see Exercise 25).

to another location because it cannot be located on top of an existing one! To determine alternate feasible and optimal locations, the contour line method can be used (see Francis and White, 1974). We now introduce a numerical example to illustrate the median method.

Example 9.5

Two high-speed printers are to be located in the fifth floor of an office complex that houses four departments of the Social Security Administration. Coordinates of the centroid of each department as well as the average number of trips made per day between each department and the printers' yet-to-be-determined location are known and given in Table 9.15. Assume that travel originates and ends at the centroid of each department. Determine the optimal location—the x and y coordinates—for the printers.

Solution

We use the median method to get this solution:

Step 1:

Department Number	x Coordinates in Nondecreasing Order	Weights	Cumulative Weights
3	8	8	8
1	10	6	14
2	10	10	24
4	12	4	28

Step 2: Because the second x coordinate—namely, 10—in the list is where the cumulative weight equals half the total weight of $28/2 = 14$, the optimal x coordinate is 10.

Step 3:

Department Number	y Coordinates in Nondecreasing Order	Weights	Cumulative Weights
1	2	6	6
4	5	4	10
3	6	8	18
2	10	10	28

Step 4: Because the third y coordinate in the list is where the cumulative weight exceeds half the total weight of $28/2 = 14$, the optimal y coordinate is 6. Thus, the optimal coordinates of the new printers are (10, 6).

TABLE 9.15

Centroid Coordinates and Average Number of Trips to Printers in Example 9.5

Department Number	x Coordinate	y Coordinate	Average Number of Daily Trips to Printers
1	10	2	6
2	10	10	10
3	8	6	8
4	12	5	4

Although the median method is the most efficient algorithm for the rectilinear distance, single-facility location problem, we present another method for solving it, because this method is used in Chapter 12 for location of multiple facilities. It involves transforming the nonlinear unconstrained model given by Equation 9.10 into an equivalent linear, constrained model using techniques from Chapter 10. Consider the following notation:

$$x_j^+ = \begin{cases} (x_i - \bar{x}) & \text{if } (x_i - \bar{x}) > 0 \\ 0 & \text{otherwise} \end{cases} \tag{9.13}$$

$$x_j^i = \begin{cases} (\bar{x} - x_i) & \text{if } (x_i - \bar{x}) \leq 0 \\ 0 & \text{otherwise} \end{cases} \tag{9.14}$$

We can observe that

$$|x_i - \bar{x}| = x_i^+ + x_i^- \tag{9.15}$$

$$x_i - \bar{x} = x_i^+ - x_i^- \tag{9.16}$$

A similar definition of y_i^+, y_i^- yields

$$|y_i - \bar{y}| = y_i^+ + y_i^- \tag{9.17}$$

$$y_i - \bar{y} = y_i^+ - y_i^- \tag{9.18}$$

Thus, the transformed linear model is

$$\text{Minimize} \sum_{i=1}^{n} w_i(x_i^+ + x_i^- + y_i^+ + y_i^-) \tag{9.19}$$

$$\text{Subject to } x_i - \bar{x} = x_i^+ - x_i^- \quad i = 1, 2, \ldots, n \tag{9.20}$$

$$y_i - \bar{y} = y_i^+ - y_i^- \quad i = 1, 2, \ldots, n \tag{9.21}$$

$$x_i^+, x_i^+, y_i^+, y_i^- \geq 0 \quad i = 1, 2, \ldots, n \tag{9.22}$$

$$\bar{x}, \bar{y} \text{ unrestricted in sign} \tag{9.23}$$

For the objective function (9.19) to be equivalent to Equation 9.10, one of the requirements is that the solution be such that either x_i^+ or x_i^-, but not both, be greater than 0. (If both are, then the values of x_i^+ and x_i^- do not satisfy their definition in Equations 9.11 and 9.12.) Similarly, only one of y_i^+, y_i^- must be greater than 0. Recall that this is true for the LMIP models in Chapter 10, where we enforce this by introducing a binary variable z_{ij}. Fortunately, these conditions are automatically satisfied in the preceding linear model. This can be easily verified by contradiction. Assume that in the solution to the preceding transformed model, x_i^+ and x_i^- take on values p and q, where p, q, >0. We can immediately observe that such a solution cannot be optimal, because one can choose another set of values for x_i^+, x_i^- as follows:

$$x_i^+ = p - \min\{p,\, q\} \quad \text{and} \quad x_i^- = q - \min\{p,\, q\} \tag{9.24}$$

and obtain a feasible solution to the model that yields a lower objective value than before, because the new x_i^+, x_i^- values are smaller than their previously assumed values. Moreover, at least one of the new values of x_i^+ or x_i^- is zero according to the preceding expression. This means that the original set of values for x_i^+, x_i^- could not have been optimal. Using a similar argument, we can show that either y_i^+ or y_i^- will take on a value of 0 in the optimal solution (see Exercise 33).

The model described by expressions (9.19) through (9.23) can be simplified by noting that x_i can be substituted as $x_i - x_i^+ + x_i^-$ from equality (9.14) and the fact that x_i is unrestricted in sign. y_i may also be substituted similarly resulting in a model with $2n$ fewer constraints and variables. We now set up a constrained LP model similar to the one for Example 9.5 and solve it using LINGO. The solution obtained has a total cost of 92 and is the same as that produced by the median method. Notice that XBAR, XPi and XNi in the model below stand for $\bar{x}$, x_i^+ and x_i^- respectively. Also, only one of XPi, XNi and YPi, YNi take on positive values. If XPi is positive in the optimal solution, it means that the new facility is to the left of existing facility i according to Equations 9.11 and 9.12. Similarly, if YPi is positive, the new facility is below existing facility i. Obviously, XBAR and YBAR give us the coordinates of the new facility's optimal location.

```
Data:
    N=4;
Enddata
Sets:
    Facility/1..N/: W, X, Y, XP, XN, YP, YN;
Endsets

Data:
    W=6 10 8 4;
    X=10 10 8 12;
    Y=2 10 6 5;
Enddata
! Objective function;
Min= @SUM(Facility(i): W(i)*(XP(i)+XN(i)+YP(i)+YN(i)));
! Constraints;
        @FOR (Facility(i): X(i)-Xbar=XP(i)-XN(i);
                           Y(i)-Ybar=YP(i)-YN(i));
End


    Global optimal solution found.
    Objective value:                        92.00000
    Total solver iterations:                       5

                    Variable        Value
                        XBAR       9.00000
                        YBAR      6.000000
                     XP( 4)      2.000000
                     XN( 3)      2.000000
                     YP( 2)      4.000000
                     YN( 1)      4.000000
                     YN( 4)      1.000000
```

9.7.2 GRAVITY METHOD

In some location problems, the distance function may not be linear, but nonlinear. If it is quadratic, as is the case in some location problems, then the problem of determining the optimal location of the new facility is rather simple. To understand the method of solving such problems, consider the following objective function for single-facility location problems with a squared Euclidean distance metric:

$$\text{Minimize } TC = \sum_{i=1}^{m} c_i f_i \left[\left(x_i - \bar{x} \right)^2 + \left(y_i - \bar{y} \right)^2 \right] \tag{9.25}$$

As before, we substitute $w_i = f_i c_i$, $i = 1, 2, \ldots, m$ and rewrite the objective function as

$$\text{Minimize } TC = \sum_{i=1}^{m} c_i f_i \left(x_i - \bar{x} \right)^2 + \sum_{i=1}^{m} c_i f_i \left(y_i - \bar{y} \right)^2 \tag{9.26}$$

Because this objective function can be shown to be convex, partially differentiating TC with respect to $\bar{x}$ and $\bar{y}$, setting the two resulting equations to 0 and solving for $\bar{x}$ and $\bar{y}$ provides the optimal location of the new facility:

$$\frac{\partial TC}{\partial \bar{x}} = 2 \sum_{i=1}^{m} w_i \bar{x} - 2 \sum_{i=1}^{m} w_i x_i = 0 \tag{9.27}$$

$$\therefore \bar{x} = \frac{\sum_{i=1}^{m} w_i x_i}{\sum_{i=1}^{m} w_i} \tag{9.28}$$

$$\frac{\partial TC}{\partial \bar{y}} = 2 \sum_{i=1}^{m} w_i \bar{y} - 2 \sum_{i=1}^{m} w_i y_i = 0 \tag{9.29}$$

$$\therefore \bar{y} = \frac{\sum_{i=1}^{m} w_i x_i}{\sum_{i=1}^{m} w_i} \tag{9.30}$$

It is easy to see that the optimal locations $\bar{x}$ and $\bar{y}$ are simply the weighted averages of the x and y coordinates of the existing facilities. Hence, this method of determining the optimal location is popularly known as the center-of-gravity or gravity or centroid method.

If the optimal location determined using the gravity method is infeasible, we may again use the contour line method discussed in Francis and White (1974) to draw contour lines from neighboring points to determine a feasible, near-optimal location. However, the contour lines will not be lines but a circle through the point under consideration, that has the optimal location as its center! For a proof of this, see Francis and White (1974). Thus, if the gravity method yields an optimal location $(\bar{x}, \bar{y})$ that is infeasible in terms of locating the new facility, all we need to do is find a feasible point (x, y) having the shortest Euclidean distance to $(\bar{x}, \bar{y})$ and place the new facility at (x, y).

Example 9.6

Consider Example 9.5. Suppose the distance metric to be used is squared Euclidean. Determine the optimal location of the new facility using the gravity method.

TABLE 9.16

Calculation of the Optimal Coordinates of the New Facility in Example 9.5

Department i	x_i	y_i	w_i	$w_i x_i$	$w_i y_i$
1	10	2	6	60	12
2	10	10	10	100	100
3	8	6	8	64	48
4	12	5	4	48	20
Total			**28**	**272**	**180**

Solution

Table 9.16 shows the $\sum_i w_i x_i$, $\sum_i w_i y_i$, and $\sum_i w_i$ values. Using these and Equations 9.28 and 9.30, it is easy to conclude that the optimal coordinates $(\bar{x}, \bar{y})$ are $272/28 = 9.71$, and $180/28 = 6.43$, respectively. If this location is not feasible, we find another point that has the nearest Euclidean distance to (9.7, 6.4) feasible and locate the copiers there.

9.7.3 WEISZFELD METHOD

In the RENT-ME single-facility location problem presented in Chapter 10, we discuss a mechanical analog and its computer simulation and see how it can be used to solve the single-facility location problem with Euclidean distance (see Figure 10.1). We now discuss how the model may be solved analytically. The objective function for the single-facility location problem with Euclidean distance can be written as

$$\text{Minimize } TC = \sum_{i=1}^{m} c_i f_i \sqrt{(x_i - \bar{x})^2 + (y_i - \bar{y})^2} \tag{9.31}$$

As before, substituting $w_i = c_i f_i$ and taking the derivative of TC with respect to $\bar{x}$ and $\bar{y}$, setting the derivatives to zero, and solving for $\bar{x}$ and $\bar{y}$ yield

$$\frac{\partial TC}{\partial \bar{x}} = \frac{1}{2} \sum_{i=1}^{m} \frac{w_i \left[2(x_i - \bar{x}) \right]}{\sqrt{(x_i - \bar{x})^2 + (y_i - \bar{y})^2}}$$

$$= \sum_{i=1}^{m} \frac{w_i x_i}{\sqrt{(x_i - \bar{x})^2 + (y_i - \bar{y})^2}} - \sum_{i=1}^{m} \frac{w_i \bar{x}}{\sqrt{(x_i - \bar{x})^2 + (y_i - \bar{y})^2}}$$

$$= 0 \tag{9.32}$$

$$\therefore \bar{x} = \frac{\displaystyle\sum_{i=1}^{m} \frac{w_i x_i}{\sqrt{(x_i - \bar{x})^2 + (y_i - \bar{y})^2}}}{\displaystyle\sum_{i=1}^{m} \frac{w_i}{\sqrt{(x_i - \bar{x})^2 + (y_i - \bar{y})^2}}} \tag{9.33}$$

$$\frac{\partial TC}{\partial \overline{y}} = \frac{1}{2} \sum_{i=1}^{m} \frac{w_i \left[2 \left(y_i - \overline{y} \right) \right]}{\sqrt{\left(x_i - \overline{x} \right)^2 + \left(y_i - \overline{y} \right)^2}}$$

$$= \sum_{i=1}^{m} \frac{w_i y_i}{\sqrt{\left(x_i - \overline{x} \right)^2 + \left(y_i - \overline{y} \right)^2}} - \sum_{i=1}^{m} \frac{w_i \overline{y}}{\sqrt{\left(x_i - \overline{x} \right)^2 + \left(y_i - \overline{y} \right)^2}}$$

$$= 0 \tag{9.34}$$

$$\therefore \overline{y} = \frac{\displaystyle\sum_{i=1}^{m} \frac{w_i y_i}{\sqrt{\left(x_i - \overline{x} \right)^2 + \left(y_i - \overline{y} \right)^2}}}{\displaystyle\sum_{i=1}^{m} \frac{w_i}{\sqrt{\left(x_i - \overline{x} \right)^2 + \left(y_i - \overline{y} \right)^2}}} \tag{9.35}$$

Because $\sqrt{\left(x_i - \overline{x} \right)^2 + \left(y_i - \overline{y} \right)^2}$ appears twice in the denominators in Equations 9.33 and 9.35, the solution of $\overline{x}, \overline{y}$ is not defined when $\overline{x} = x_i$ and $\overline{y} = y_i$ for some i. This means that if the new facility's optimal coordinates coincide with that of an existing facility, Equations 9.33 and 9.35 are not defined because the denominators take a value of 0. We therefore cannot use them in computing the optimal coordinates $\overline{x}, \overline{y}$. The possibility of the optimal location of the new facility coinciding with that of an existing facility is rare in practice, but it cannot be ruled out. Hence, we apply another method for solving the single-facility Euclidean distance problem. Although (theoretically) optimal algorithms do not exist for this problem, there is a method due to Weiszfeld (1936) that is guaranteed to converge to the optimal location. This iterative algorithm is relatively straightforward and described in the following.

Weiszfeld Method

Step 0: Set iteration counter $k = 1$:

$$\overline{x}^k = \frac{\displaystyle\sum_{i=1}^{m} w_i x_i}{\displaystyle\sum_{i=1}^{m} w_i} \qquad \overline{y}^k = \frac{\displaystyle\sum_{i=1}^{m} w_i x_i}{\displaystyle\sum_{i=1}^{m} w_i}$$

Step 1:

$$\overline{x}^{k+1} = \frac{\displaystyle\sum_{i=1}^{m} \frac{w_i x_i}{\sqrt{\left(x_i - \overline{x} \right)^2 + \left(y_i - \overline{y} \right)^2}}}{\displaystyle\sum_{i=1}^{m} \frac{w_i}{\sqrt{\left(x_i - \overline{x} \right)^2 + \left(y_i - \overline{y} \right)^2}}}$$

$$\overline{y}^{k+1} = \frac{\displaystyle\sum_{i=1}^{m} \frac{w_i y_i}{\sqrt{\left(x_i - \overline{x} \right)^2 + \left(y_i - \overline{y} \right)^2}}}{\displaystyle\sum_{i=1}^{m} \frac{w_i}{\sqrt{\left(x_i - \overline{x} \right)^2 + \left(y_i - \overline{y} \right)^2}}}$$

Step 2: If $\overline{x}^{k+1} = \overline{x}^k$ and $\overline{y}^{k+1} \approx \overline{y}^k$, stop. Otherwise, set $k = k + 1$ and go to step 1.

TABLE 9.17
Coordinates and Weights for Four Departments in Example 9.7

Department Number	x_i	y_i	w_i
1	10	2	6
2	10	10	20
3	8	6	8
4	12	5	4

Notice that the initial seed for $\bar{x}$ and $\bar{y}$ was obtained via Equations 9.28 and 9.30, which were used in the gravity method. Although the aforementioned method is theoretically suboptimal, it provides $\bar{x}$ and $\bar{y}$ values that are very close to the optimal. For practical purposes, the algorithm works very well and can be readily implemented on a spreadsheet.

If the optimal location provided by the Weiszfeld method is not feasible, we may once again use the contour line method to draw contour lines and then choose a suitable, feasible, near-optimal location for the new facility. However, the methods for drawing contour lines for the Euclidean distance metric, single-facility location problem are not exact methods. These approximate methods basically compute TC for a given point (x, y), choose a neighboring x (or y) coordinate, and search for the y (or x) coordinate that yields the same TC value previously computed. This procedure is repeated until we come back to the starting point.

Example 9.7

Consider Example 9.5. Assuming the weight of department 2 is not 10, but 20 and that the distance metric to be used is Euclidean, determine the optimal location of the new facility using the Weiszfeld method. Data for this problem are shown in Table 9.17.

Solution

Using the gravity method, the initial seed can be shown to be (9.8, 7.4). With this as the starting solution, we can apply step 1 of the Weiszfeld method repeatedly until we find that two consecutive $\bar{x}, \bar{y}$ values are equal.

As shown in Table 9.18, this occurs in the 25th iteration. For convenience, the total costs at the first 10th, 20th, and 25th iterations are also shown in Table 9.18. The reader should note that the optimal location for this problem, (10, 10), is the same as that of an existing facility, i.e., department 2. This is no accident and occurs because department 2's weight is more than one-half of the cumulative weights. In fact, when a facility i's weight is greater than or equal to one-half of the sum of the weights for all the remaining facilities, the new facility's optimal location will be the same as that of facility i. This is true under the rectilinear as well as Euclidean distance metrics as demonstrated by Examples 9.6 and 9.7.

9.8   FACILITY LOCATION CASE STUDY*

We now present a relocation project undertaken by a small facility. A small manufacturing company that is currently located in a university technology park has witnessed major growth since introducing an innovative technology into the marketplace. Its owner now wants to relocate to a new location where a bigger facility can be built. In January, she hired senior Industrial and Systems Engineering (ISE) students at the university to investigate several potential locations and select one of them that suits her needs best.

---

* Based on an actual project.

TABLE 9.18
Several Iterations in the Weiszfeld Method for Example 9.7

Iteration Number	x	y	TC
1	9.7	7.8	113.4
2	9.7	8.2	19.9
3	9.8	8.4	19.8
4	9.8	8.7	109.9
5	9.8	8.9	109.1
6	9.9	9.0	108.5
7	9.9	9.2	108.0
8	9.9	9.3	107.6
9	9.9	9.4	107.2
10	9.9	9.5	106.9
11	9.9	9.6	106.7
12	10	9.6	106.5
—	—	—	—
20	10	9.9	105.6
—	—	—	—
25	10	10	105.5

The student group adopted the following five-step approach, which is based on the hybrid analysis discussed in Section 9.5.1.

Step 1—Determination of requirements: Conduct an interview with the owner and facility manager to determine these company-specific requirements for the new facility:

1. The company will relocate to New York or Vermont.
2. At least 15,000 square feet of space is required.
3. A power source of 440 volts, three-phase, 200 ampere electrical service is mandatory in order to power the atomizers used in the manufacturing process.
4. The current rent is $7.50 per square foot per year; the company wants to pay between $3.50 and $4.50 per square foot.
5. The company wants to move within the next 8 months.
6. All suppliers and vendors should be within 100 miles of the facility.
7. A lease of 1–2 years is preferred.
8. There should be adequate room for expansion.
9. An industrial park or shared facility is preferred.
10. The facility should be located close to major highways and airports.
11. A loading bay is required; easy access to the bay is desired.
12. The new facility should be built to suit.
13. The facility maintenance costs should be low or the owner of the building must maintain it.
14. The general condition of the building should be good.
15. The building should not be considered high risk by insurance companies.
16. It is desirable to have secretarial services available nearby.
17. The local and state taxes must be reasonable.

Step 2—Classification of location factors
   Based on the interview with the owner, the ISE students classified the preceding requirements under these three categories:

1. Critical factors
   a. Minimum space requirement
   b. Three-phase, 440 volts, 200 ampere electrical service
   c. Support service providers and vendors within 100 miles
2. Objective factors
   a. Rent
   b. Space rented or leased
   c. Maintenance and insurance costs and taxes
3. Subjective factors
   a. Shared facility
   b. Build to suit
   c. Condition of loading bay
   d. Proximity to airport and major highways
   e. Lease length
   f. Secretarial support
   g. Condition of building

Step 3—Data collection: This step requires the most time, but it is very important and should be done carefully. Information on potential sites and locations was obtained from sources such as the following:

- Chamber of commerce
- Economic development council
- Real estate brokers
- Facility owners

Step 4—Elimination of sites not meeting critical objectives and development of a rating chart: From the 10 sites for which data were collected, it was clear that four did not satisfy one or more of the critical requirements. For the remaining six sites, the ISE student group then devised a chart showing the weights of the objective and subjective factor (see Table 9.19).

Step 5—Site visits and site evaluation: A site visit was conducted for each of the six sites. Data collected for the sites are summarized in Table 9.20. Each factor was then rated as shown in Tables 9.21 and 9.22. After careful evaluation, the six sites were rated as shown in Table 9.23. Based on this evaluation, the Cohoes, NY, site appears to be the best location with Bennington, VT, as a (close) second best location.

TABLE 9.19

Objective and Subjective Factor Weights Data Collected by IME Students for Case Study

Factor	Weight
Rent cost per square foot	0.25
Square meters rental requirement	0.20
Airport and highway proximity	0.10
Building condition	0.06
Build to suit	0.08
Loading bay condition	0.07
Length of lease	0.05
Secretarial services	0.03
Shared facility	0.06
Maintenance, insurance, and taxes as percentage of rent	0.10

TABLE 9.20
Summary of Data Collected for Six Sites in Case Study

Factor/Location	Cohoes, NY	Bennington, VT	Glens Falls, NY	Springfield, VT	Burlington, VT	Albany, NY
Rent cost per square foot	$2.50	$3.75	$4.00	$6.00	$2.50	$6.00
Square meters rental requirement	15,000	15,000	15,000	20,000	22,000	18,000
Airport and highway proximity	Excellent	Good	Good	Average	Good	Excellent
Building condition	Fair	Average	Fair	Good	Excellent	Good
Build to suit (if no, extent of renovation required)	No (high)	No (med)	Yes	No (low)	No	Yes
Loading bay condition	Average	Average	Average	Good	Excellent	Good
Length of lease	Excellent	Excellent	Good	Good	Fair	Fair
Secretarial services	Poor	Good	Poor	Poor	Average	Poor
Shared facility (number of neighbors)	30	14	20	5	1	2
Maintenance, insurance, and taxes as percentage of rent	15	10	25	12	22	30

TABLE 9.21
Development of Rating Scores for Objective Factors

Rent Cost (per square foot)	Rating Score	Square Meters Rental Requirements	Rating Score	Maintenance, Insurance, and Taxes as Percentage of Rent	Rating Score
0–2.49	10	15,000–16,000	3	0–3.99	10
2.5–2.99	9	16,001–19,000	2	4–7.99	9
3.0–3.49	8	19,001–22,000	1	8–9.99	8
3.5–3.99	7			12–15.99	7
4.0–4.49	6			16–19.99	6
4.5–4.99	5			20–23.99	5
5.0–5.49	4			24–27.99	4
5.5–5.99	3			28–31.99	3
6.0–6.49	2			32–35.99	2
6.5 and higher	1			36 and above	1

9.9 SUMMARY

In this chapter, we examined various factors that affect location decisions. We presented several models for the continuous and discrete facility location models. Most of these were applicable to the single-facility location problem. We discussed algorithms for problems with rectilinear, Euclidean, and squared Euclidean location problems. We also presented a facility location case study.

TABLE 9.22
Development of Rating Scores for Subjective Factors

Shared Facility

Number of Neighbors	Rating Score	Loading Bay Condition and Access	Rating Score
0–1	1	Excellent condition, not shared, easy access	9
2–3	2	Good condition, shared bay	7
4–5	4	Average condition, shared bay, difficult access	5
6–7	6	Poor condition, inadequate access	3
8–10	8	No loading bay	1
10 and higher	10		

"Build to Suit," Expenses	Rating Score	Airport/Highway Proximity (Miles)	Rating Score
Borne by landlord	10	Within 10	9
Borne by company, low	3	10–20	7
Borne by company, moderate	2	20–30	5
Borne by company, high	1	30–40	3
		Over 40	1

Length of Lease	Rating Score	Secretarial Services	Rating Score
Very flexible	9	Personnel support, office equipment available nearby	9
1 year mandatory	7	No personnel support, but office equipment available nearby	7
2–3 years mandatory	5	Neighbor will permit use of office equipment	5
3–4 years mandatory	3	Neighbor charges for use of their office equipment, limited availability	3
Over 4 years mandatory	1	No services at all	1

Building Condition	Rating Score	Factors to Consider
Excellent	9	Age of building, last renovation
Good	7	Wood, brick, and/or concrete walls, floor
Average	5	Location in industrial area
Fair	3	Water source in facility (sink, floor drains)
Poor	1	Condition of lighting

9.10 REVIEW QUESTIONS AND EXERCISES

1. Discuss at what stage of facility design the location problem must be addressed.
2. What is the difference between supply chain management, logistics management, and facility location? Explain your answer.
3. Discuss some of the objectives in locating facilities.
4. Explain in one paragraph what is meant by the following:
   (a) Location problems
   (b) Allocation problems
   (c) Location–allocation problems
5. Compare and contrast discrete and continuous location problems.
6. List and explain the important factors that affect location decisions.
7. *Project*: Identify a business that employs 100 or more people and has moved into your area in the past 3 years. List and explain the factors that were considered by the business in

TABLE 9.23

Final Analysis of Six Sites in the Case Study

Factor/Location	Weight	Cohoes, NY	Bennington, VT	Glen Falls, NY	Springfield, VT	Burlington, VT	Albany, NY
Rent cost per square foot	0.25	9	7	6	2	9	2
Square meters rental requirement	0.2	3	3	3	1	1	2
Airport and highway proximity	0.1	9	7	7	5	7	9
Building condition	0.06	3	5	3	7	9	7
Build to suit (if no, extent of renovation required)	0.08	3	2	10	1	1	10
Loading bay condition	0.07	5	5	5	7	9	7
Length of lease	0.05	9	9	7	7	3	3
Secretarial services	0.03	1	7	1	1	5	5
Shared facility (number of neighbors)	0.06	10	10	10	4	1	2
Maintenance, insurance, and taxes as percentage of rent	0.1	7	8	4	7	5	3
Score	1	6.3	5.92	5.51	3.51	5.26	4.23

making the location decision, in order of their importance to the company. Present your findings to your class.

8. A national discount department chain has penetrated the midwest market and opened several retail stores in the area. It wants to set up a warehouse to serve these stores and has evaluated six potential sites. The location factors important to the company, their weights, and the score each site receives for each factor are provided in Table 9.24. Determine which site is to be selected using the scoring method of Section 9.5.1.

9. A manufacturing company wants to open a warehouse to serve 10 retail stores in the southwest. It has evaluated five potential sites. The location factors important to the company, their weights, and the score each site receives for each factor are provided in Table 9.25. Determine which site is to be selected using the scoring method of Section 9.5.1.

10. Use the factor weights in Exercise 9 to answer Exercise 8. Is the solution you obtained different from that in Exercise 8? Why or why not?

TABLE 9.24

Weights and Scores for Six Sites in Exercise 8

Site Number	Location Factor Number (Weight)			
	1 (0.15)	2 (0.2)	3 (0.4)	4 (0.25)
1	90	40	80	70
2	100	30	80	50
3	50	80	85	60
4	60	90	60	75
5	90	30	65	80
6	90	30	85	85

TABLE 9.25
Weights and Scores for Five Sites in Exercise 9

Site Number	Location Factor Number (Weight)			
	1 (0.4)	2 (0.2)	3 (0.3)	4 (0.1)
1	55	20	10	50
2	20	75	80	50
3	90	15	85	100
4	80	95	20	25
5	30	90	25	25

11. Use the factor weights in Exercise 8 to answer Exercise 9. Is the solution you obtained different from that in Exercise 9? Why or why not?
12. Consider Example 9.2. Apply the TSA to find the optimal solution shown in Table 9.9 with a cost of $7,780.
13. Consider Example 9.3. Solve the Pittsburgh transportation model using the TSA. Verify that the optimal solution cost is $9,510. Show the distribution pattern.
14. Consider Example 9.3. Setup a transportation model for the Atlanta transportation problem. Verify that the optimal solution cost is $7,980. Show the distribution pattern.
15. A company wants to locate a facility to supply a key component required in three plants located in Buffalo, Syracuse, and Rochester—all in New York State. Two facilities are already supplying the component to the three plants. It is considering one of two possible sites, site 1 and site 2. The relevant unit transportation cost, demand, and supply information is given in Table 9.26. Determine which of the two sites is to be chosen if the main objective is to minimize transportation costs of the key component.
16. Consider Exercise 15. Solve the problem using the TSA. Is the solution obtained same as that in Exercise 15. Why or why not?
17. The United Shipping Company has three hubs located in Detroit (D), Houston (H), and Boston (B) that handle shipments in pallets that are of uniform size, shape, and weight. These serve five major customers in Atlanta (A), Rochester (R), Pittsburgh (P), Utah (U), and San Francisco (SF). To ease pressure on its existing hubs and also to have a presence in the southwest, the company is considering the addition of one more hub with a capacity of 20,000 pallets in one of the following three cities—Phoenix (Ph), Los Angeles (LA), and Las Vegas (LV). Table 9.27 shows the transportation cost per pallet from each hub (existing and proposed) to each customer area as well as the variable operational cost (per pallet) at each existing and proposed hub. Determine where the new hub is to be built so that operational costs at the hub and distribution costs between hubs and customer areas are minimized.

TABLE 9.26
Cost, Supply and Demand Information for Exercise 15

	Buffalo	Syracuse	Rochester	Supply
Existing facility 1	10	12	8	1000
Existing facility 2	8	12	10	2000
New site 1	6	6	10	5000
New site 2	4	10	12	5000
Demand	3000	3000	2000	

TABLE 9.27
Cost, Capacity, and Demand Figures for Exercise 17

Location	Transportation Costs to Customer Areas per Pallet					Capacity	Handling Cost
	A	R	P	U	SF		
D	17	11	11	30	55	50	2.7
H	9	12	10	15	40	20	2.7
B	17	6	9	23	42	25	3.0
Ph	28	35	30	13	13	40	1.5
LA	55	52	44	9	4	40	2.9
LV	45	50	44	8	5	40	2.0
Demand	35	20	20	18	40		

18. Consider Exercise 17. Solve the problem using the TSA. Is the solution obtained same as that in Exercise 17. Why or why not?
19. A county in the rural part of Texas wants to determine the location of a library. The county can be divided into six population zones. Four potential locations are being considered. The distance from each potential location to each population zone is given in the matrix in Figure 9.6. Determine which location is to be selected if the objective is to minimize the maximum distance to any population zone.
20. Explain the advantage of the hybrid analysis over the qualitative analysis.
21. Consider Exercise 8. Suppose there are two critical factors such that any site not satisfying them will not be considered. In addition, there are two objective factors—the first is revenue based and the second expense based—that the discount department chain would like to consider. The relevant data are provided in Table 9.28. Divide the subjective factor values in Exercise 8 by 100. Assuming the objective factors are worth twice the subjective factors, determine the best location. Is it different from what you got in Exercise 8? Explain.
22. Consider Exercise 21. Assuming the subjective factors are worth twice the objective factors, determine the best location. Is it different from what you got in Exercises 8 and 21? Explain.
23. A company wants to determine which of the six sites listed in Table 9.29 is to be selected for locating a new facility. The objective and subjective factors as well as the performance of the six sites with respect to these factors are also provided in the table. You may assume that all the sites satisfy the critical factors test. Rate the six sites on a scale of 0 to 1 with respect to their relative desirability on each of the four subjective factors. Providing weights of 0.2, 0.3, 0.3, and 0.2, respectively, for the four subjective factors and assuming the objective factor decision weight is 0.8, determine the best location using the hybrid method.

	1	2	3	4	5	6
1	10	15	12	14	8	4
2	4	3	8	16	12	20
3	6	8	12	14	14	12
4	20	12	18	18	12	6

FIGURE 9.6 Distance between four population zones and six potential locations in Exercise 19.

TABLE 9.28

Critical and Objective Factor Ratings for the Six Sites in Exercise 21

Site Number	Critical Factor Number		Objective Factor Number	
	1	2	1	2
1	1	0	100	85
2	1	1	175	185
3	1	1	110	20
4	1	1	200	60
5	1	1	350	80
6	0	1	80	100

TABLE 9.29

Objective and Subjective Factor Ratings for the Six Sites in Exercise 23

Site Number	Objective Factor (Cost in 100,000's)	Labor Availability	Business Climate	Transportation Facilities	Educational Facilities
A	30	Good	Excellent	Fair	Very good
B	45	Fair	Very good	Fair	Excellent
C	25	Very good	Good	Good	Fair
D	28	Fair	Very good	Good	Excellent
E	50	Fair	Excellent	Good	Good
F	35	Excellent	Good	Excellent	Good

24. Explain why continuous space location models are sometimes preferred to the discrete models. Discuss some of the trade-offs of using the continuous or the discrete approach.

25. Using a figure similar to Figure 9.5, intuitively explain how the median method provides the optimal coordinates of the new facility even when

$$\sum_{i=1}^{j} w_j = \sum_{i=1}^{m} \frac{w_j}{2}$$

26. Using a figure similar to Figure 9.5, justify that the y coordinate determined in step 4 of the median method is optimal.

27. A new machine is to be placed in a flexible manufacturing cell consisting of four CNC machines that are located at the following coordinates: (10, 10), (2, 10), (10, 2), and (5, 1). The interactions, i.e., number of trips, between the new machine and the existing ones are 6, 10, 10, and 6 units per time period. Determine the optimal coordinates of the new machine, assuming travel is along vertical and horizontal aisles.

28. Suppose the interaction between the new machine and each existing one in Exercise 27 is 10 units. Where would you expect the new machine to be placed? Does it correspond to the solution obtained in Exercise 27? Explain.

29. A regional telephone is attempting to determine the optimal location for a training facility for its operators. The operators are located in four operations centers and the training facility is to serve these centers. The coordinates of the four centers and the number of operators working in

TABLE 9.30

Coordinates and Number of Operators in Four Centers Described in Exercise 29

| Town | Coordinates | | Number of Operators ('000s) |
	x	y	
Center #1	32	58	6.8
Center #2	49	42	4.3
Center #3	11	69	6.0
Center #4	42	31	3.4

TABLE 9.31

Customer and Trips Data for Exercise 30

Customer	Trips
1	500
2	350
3	550
4	750

each are given in Table 9.30. Assuming the rectilinear metric is the most appropriate distance measure, determine optimal coordinates of the training facility.

30. ABC Printer company leases and services printers to businesses in the tricity area of Albany, Schenectady, and Troy. It has decided to relocate its operations to serve its customers better. The company has four major *customers*—one in each city and another in the state government campus. It is expected that the number of visits the service and marketing departments will have to make to the four customers during each period are shown in Table 9.31. Assume that the travel cost is proportional to the street distance traveled. A rectilinear distance model is used to represent travel because of the perpendicular nature of the streets. The customers are located at (8, 12), (2, 15), (9, 18), and (8, 10), respectively.

(a) Determine the coordinates of the optimal location.

(b) Set up a linear programming model for the problem and solve it using any available software. Verify that the answers from the software are the same as that obtained in (a).

31. Set up a linear programming model for the situation described in Exercise 27. Solve it using LINGO.

32. Repeat Exercise 31 for the situation described in Exercise 28.

33. Consider the LP model described by expressions (9.19) through (9.23). Show that at least one of y_i^+, y_i^- will take on a value of 0 in the optimal solution.

34. Consider Example 9.5. Suppose that the weight of facility 2 is 20 units. Verify that the optimal x, y coordinates are (10, 10) using the median method.

35. Consider Exercise 27. Suppose the interaction of the new machine with the first CNC machine is 30 units rather than 6. Determine the optimal coordinates of the new machine. If it coincides with that of an existing CNC machine, determine a feasible location using the contour line method described in Francis and White (1974).

36. Consider Exercise 27. Suppose the squared Euclidean distance metric is to be used. Determine the optimal location of the new machine. How is it different from that obtained in Exercise 27. Explain.

37. Repeat Exercise 36 for the situations described in Exercises 27, 34, and 35.
38. Consider Exercise 27. Suppose material is moved using overhead material handling equipment and that the Euclidean distance metric is more appropriate. Determine the optimal location of the new machine. Is it different from that obtained in Exercise 27? Explain.
39. Repeat Exercise 39 for the situations described in Exercises 26, 34, and 35.
40. Consider Exercise 30. Assuming the distance metric is not rectilinear but Euclidean, determine the coordinates of the optimal location of ABC's facility. Is it different from what you obtained in Exercise 30? Why or why not? Explain.
41. *Project*: Identify a company in your vicinity that is contemplating a move to a new location. Following the approach in Section 9.4, determine a suitable location for the company that satisfies its needs.

10 Modeling of Design Problems in Facility Logistics

10.1 MODELS

Models may be classified as *physical*, *analog*, and *mathematical models*. The reader has probably seen *physical* models of buildings, dams, stadiums, theaters, or wind tunnels. These models have the same shape and appearance of the real system being studied but are generally much smaller. *Analog* models do not necessarily resemble the real system being studied, yet they provide important information about the system. For example, a graph showing the population increase over time is an analog model. Another example, involves the location of a new facility in an existing network of facilities.* Consider the following two examples.

The RENT-ME car rental company in Chicago has cars available at fine city and two airport locations. It wants to construct a new facility to perform oil changes, regular maintenance and checks, and minor repair work on these cars. The level of interaction of each existing car rental location with the new service facility corresponds to the number of rental cars at that facility. RENT-ME wants to determine the optimal location of the service facility to minimize the overall distance traveled by the cars for servicing.

In its recently passed budget, a local county has been authorized to build a new fire station in or near five towns in the county. The demand in each town for the fire station's services is assumed to be directly proportional to the towns' population. The county must find an optimal location for the fire station.

An analog model for the two aforementioned problems may be set up using tables with holes corresponding to the location of existing facilities (Francis et al., 1992). A string with a weight attached to one of its ends runs through each hole. The weight reflects the interaction of the corresponding facility with the new facility. The other end of each string is tied together into one knot above the table. The knot corresponds to the location of new facility, as shown in Figure 10.1. If we place a pulley at each hole, slide the string over the pulley to eliminate friction at the holes, raise the knot and release it, the strings, attached weights, and the knot will eventually find an equilibrium position and settle there. This position determines the location of the new facility; the length of each string from the corresponding pulley to the knot corresponds to the distance between the corresponding facility and the new one. Thus, the strings are analogous to distances, and the position of the knot is analogous to the location of the new facility. The reader can simulate the weight and string experiment using the computer program LOCDEMO.EXE, which is available at http://sundere.okstate.edu/downloadable-software-programs-and-data-files.*

Mathematical models use mathematical expressions (objective function, constraints, equations, and inequalities) to determine values of the decision variables for the input data provided. A mathematical model is often thought of as an idealized representation due to the limiting assumptions made to represent the problem mathematically. Consider the following simple example. The distance d, traveled during a time period t, is given by the product of velocity v and t. For any set of values of the input parameters, the corresponding value of the decision variable can be readily determined. In this chapter, mathematical models are extensively used to represent the layout, location, and group technology problems.

---

* We thank Ronald Mantel and his students at the University of Twente, Enschede, the Netherlands, for providing us this program.

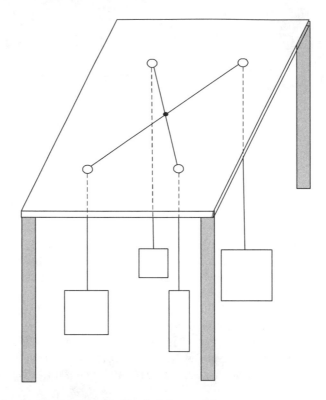

FIGURE 10.1   Analog model of a single-facility location problem.

Mathematical models can be classified as *descriptive* and *prescriptive* (*normative*). Given certain values for a set of decision variables, *descriptive* models predict the performance of the system. *Normative* models, on the other hand, provide us values for a set of decision variables. Thus, a logical approach to system design may involve developing a prescriptive model to determine the values for key decision variables and using these values to run a descriptive model that can evaluate the operational performance of the system. The latter will enable us to fine tune the initial design so that the resulting design achieves better operational performance.

In practice, the modeling step is often ignored. Companies do not have the expertise in modeling facility design problems, and top management is unaware of the significant cost savings that can be achieved by developing *near-optimal* facility designs. The management at many companies is often so busy "fighting fires" that they think they do not have the time to model facility design or planning problems mathematically. In this book we provide real-world examples that highlight the importance of thorough analysis and the proper modeling and solution of complex facility design problems. In almost all the chapters, we also show how mathematical models can help in analyzing and solving facility design and planning problems. After studying this book, the reader should have an appreciation for not only the importance of facility design and planning problems but also the benefits of using mathematical models to analyze these problems.

10.2   ALGORITHMS

After a model is generated, a solution technique must be developed to solve the problem formulated. Solution techniques or procedures are commonly called *algorithms*. We presented several algorithms in Chapter 5. If a given algorithm can produce the optimal solution or at least one optimal

TABLE 10.1

Computation Time Requirement Comparison of Polynomial and Nonpolynomial Algorithms

	Problem Size				
TCF	10	20	40	60	P- or NP-Complete
N	0.001 second	0.002 second	0.004 second	0.006 second	P-complete
N^3	0.001 second	0.008 second	0.064 second	0.216 second	P-complete
2^n	0.001 second	1.0 second	12.7 days	366 centuries	NP-complete

Source: Based on data in Garey, M.R. and Johnson, D.S., *Computers and Intractability: A Guide to the Theory of NP-Completeness*, W.H. Freeman and Company, New York, 1979.

solution (when multiple optimal solutions exist), then it is called an optimal algorithm; otherwise, it is called a heuristic algorithm or simply a heuristic. Note that a good heuristic may find an optimal solution quickly for a problem, but then it cannot confirm that the solution it found is optimal. Some branch-and-bound-based optimal algorithms typically find an optimal solution very quickly but require a significant amount of computer time to verify optimality.

Most facility design and planning problems are computationally difficult. For example, the quadratic assignment problem (QAP) that is used to formulate the layout problem later in this chapter is known to be nondeterministic polynomial (NP)-complete. To understand the term *NP-complete*, we need to understand the time complexity functions (TCFs) of algorithms. The TCF of an algorithm is the greatest amount of time required by the algorithm to solve a given problem. For example, if the TCF of an optimal layout algorithm is n, then it requires 10 times as much computer time to solve a problem with 100 departments as it does to solve a problem with 10 departments. On the other hand, if the TCF of another algorithm is 2^n and it takes 1 second to solve a 20-department problem, it will take 366 centuries to find the solution to a 60-department problem! Some other examples are given in Table 10.1. The first algorithm in this hypothetical situation requires polynomial time, whereas the second requires NP (exponential) time. If it is known that a problem is NP-complete, then we know that no known algorithm can solve this problem optimally in polynomial time. Thus, no known algorithm can solve the QAP, and hence the layout problem, optimally in polynomial time. We discuss this aspect again in Chapter 11. For a detailed discussion of the theory of NP-completeness, see Garey and Johnson (1979).

10.3   GENERIC MODELING TOOLS

This section introduces several generic tools that are available to model design and planning problems. No one modeling tool is superior to the others in every circumstance. Yet three modeling tools—mathematical programming, queuing, and simulation modeling—when used in the right sequence allow the decision maker to incorporate various levels of information and arrive at reasonably good solutions to the design and planning problems.

Mathematical programming is a static tool in the sense that it generally requires deterministic input data. It cannot capture dynamic aspects of the system design and planning problem, such as operator–machine interference, which can be considered in queuing models. However, unlike queuing models, linear programming (LP) is a performance *optimization* tool. Given a set of constraints, for example, capacity constraints and production requirements, LP can tell us the optimal number of machines that should be purchased from a candidate list. Queuing models and simulation, on the other hand, are performance *evaluation* tools.

The *optimal* values of the decision variables, which are provided by the mathematical programming model, should be treated as a preliminary solution to be refined at later stages by queuing and simulation models. For example, consider the decision problem of determining the quantity and type of required manufacturing equipment. We can use mathematical models to do a *rough-cut* determination of the quantity of the required manufacturing equipment. Then, using queuing models we can improve this rough estimate by generating information about key performance measures for the initial design, examining these measures, and suggesting changes to the initial design to improve the performance measures. For example, a suitable queuing model may be used to examine the initial solution (system design) generated by the mathematical model and analyze its performance with respect to a number of criteria including work-in-process inventory buildup, machine utilization, production throughput rate, and job flowtime. Based on the performance analysis, the system designer may suggest changes to the design such as addition of some machines, decommissioning some others, and so on. Using the modified design, the system designer can develop a detailed simulation model and analyze the system's design and operational characteristics.

10.3.1 Mathematical Programming Models

The mathematical programming approach attempts to optimize a quantifiable objective function so that some specified set of constraints is not violated. Typically, the objective function minimizes cost or maximizes profits and the constraints specify a bounded region of feasible solutions to the problem. The problem then becomes one of choosing the feasible solution within the bounded region that optimizes the objective function. A mathematical programming model consists of decision variables and parameters. We seek to determine values of the decision variables that optimize the objective function for the given (deterministic) parameter values. If one or more constraints or the objective function is nonlinear in terms of the decision variables (e.g., a constraint consists of a product of two decision variables, square of one variable, or the inverse of another), the model is a nonlinear model. Conversely, if all the constraints and the objective function are linear functions of the variables, the model is a linear model. If a model requires that any of the variables be integers, it is called linear or nonlinear mixed or pure integer model, depending upon whether or not one or more constraints or the objective function is nonlinear and whether or not some or all of the variables are required to be integers. Generally, it is easier to find an optimal solution to linear models than to mixed-integer or nonlinear models. If a model requires that any of the integer variables be binary (0 or 1) variables, such a model, in many cases, is more difficult to solve than general integer programming models.

10.3.2 Queuing and Queuing Network Models

Queues, or waiting lines, are common in a variety of settings including manufacturing and service systems. Queuing theory is a mathematical analysis technique that can provide estimates of these variables in a waiting line system:

- Number of parts (customers) waiting in a queue
- Time a part (customer) spends waiting in a queue
- Probability that a machine (server) is idle
- Probability of a queue being filled to capacity

Reducing the length of a queue or time spent by a part in a queue results in inventory cost savings in a manufacturing system. Similarly, reducing the length of a queue or time spent by a customer in a queue means providing faster service in that service system. We can achieve lower levels of inventory and higher throughput in one of several ways. For example, we may purchase more equipment, add more servers, redesign the production or service process, or redesign the facility. Each of these options costs time, money, energy, or resources. We need to know which of these alternatives

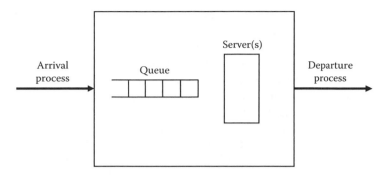

FIGURE 10.2 A queuing system.

produces the greatest improvements at the least cost. Using estimates from a queuing model, we can then decide which (if any) alternative to select to improve system performance.

In a queuing system, customers arrive by some arrival process and wait in a queue for the next available server. In the manufacturing context, customers could be parts and servers could be machines, material handling carriers, or pick-and-place robots. In the service context, customers could be cars and servers could be automatic washing bays. When a server becomes available, a customer is selected for service by some queue discipline. Service is completed by an output process and the customer leaves the system. The arrival and service processes are generally assumed to be random. Figure 10.2 illustrates a simple queuing system.

Both manufacturing and service systems typically contain more than one queuing system. In fact, depending on the number of phases of service, several queuing systems can be networked in a specific manner. Analyzing these queuing networks provides useful information. Queuing systems and queuing networks are discussed in detail in Chapter 13.

10.3.3 SIMULATION APPROACH

Simulation is a popular performance evaluation and modeling tool that is widely used in both manufacturing and nonmanufacturing systems. This powerful approach can be used to analyze complex stochastic systems and to model systems that cannot be handled using either the mathematical programming approach or the queuing network approach. The simulation approach is attractive because it can capture an unlimited number of complexities in one system. In fact, a simulation model can very closely reflect reality, its accuracy limited only by the amount of programming and computer—hardware and software—costs and time. As Ho (1987) put it, the simulation approach is a brute-force, trial-and-error computer experimentation of the real system being studied. A simulation model mimics the operation of a system and enables us to compress or expand time. For example, we could simulate the arrival and service patterns of customers at a bank during a 1-year period, by running a corresponding simulation model for a few minutes on a computer. We could then examine the average customer waiting time or some other performance measure and get a good idea of how the system will perform in the real world.

A simulation model does not provide optimal values for the performance measures. Rather, it generates representative samples of the performance measures and, using these sample points, estimates mean values of the performance measures. The simulation approach can be tailored to any system, can generate estimates of a wide variety of performance measures, and, more important, can evaluate time-variant behavior (Askin and Standridge, 1993).

The simulation approach does have drawbacks, however. Statistical theory tells us that better estimates are obtained from larger sample sizes. The larger the sample size, however, the longer the model must be run on a computer. In addition, developing a simulation model is a time-consuming task. Nevertheless, simulation is a popular modeling tool, because it allows us to visualize operation of a system and do *what-if* analyses and eliminates the need to experiment with the real system. With the availability of object-oriented simulation languages and the increased computational speed and memory-handling capabilities, simulation is extensively used in facilities design, material handling system design, layout, and location.

A simulation model basically consists of three entity classes—resources, transactions, and queues. In the manufacturing context, resources could be machines or material handling systems whose service is required by parts for processing or transportation. If a part has to wait for processing or transportation, it must wait in a queue. The attributes of a resource, transaction, or queue define the state of the system. This state changes whenever an event occurs. Because these events occur at discrete points in time, simulation models are often called discrete simulation models. Several simulation languages are available for developing simulation models. Because simulation modeling is discussed in Chapter 13, we do not provide any further details here about simulation or simulation languages. Overall, simulation is a versatile and powerful tool whose only disadvantage is the enormous amount of effort required to develop, validate, test, and run the model. Therefore, simulation should be used to perform detailed analysis only after a rough or preliminary analysis has been done using LP and queuing approaches and we have reduced the candidate solutions to a small number. This will drastically cut down the modeling effort. Further details on simulation modeling and languages are available in many sources including Law and Kelton (1991), Banks et al. (2009), and Nelson (1993).

The focus of this chapter is on the development of mathematical models for the layout problem. As discussed in Chapter 4, the modeling step is very important in the development of solutions to a problem. In order to develop models that can be solved within reasonable time by generic or specialized algorithms, some simplifying assumptions need to be made. In many applications, such assumptions may be acceptable in the sense that the final solution obtained may be modified rather easily to reflect the actual conditions and such modification does not cause drastic changes to the criteria used to measure the efficiency and effectiveness of the layout. For example, we make the assumption that all departments in a layout problem can only be square or rectangular. Suppose we set up a suitable mathematical model under this assumption, obtain a layout by solving the model and modify the shapes of departments in the final layout to reflect their actual shapes without changing their relative positions. Such a modification will cause the distance between departments to change, and hence, the actual objective function value (OFV)* will be different from what the model solution provides. Typically, this change is not significant and hence our shape assumption may be justified. But, in some cases, modification of the solution to reflect actual conditions could lead to drastic changes in the OFV that the initial assumption in the model may be flawed and unacceptable. Hence, before applying any of the models discussed in this chapter, the reader must exercise caution to see whether the assumptions in the model can be justified in the problem context.

10.4 MODELS FOR THE SINGLE-ROW LAYOUT PROBLEM

We discussed five types of layout in Chapter 3. Although there are a number of objectives in developing a layout, the primary objective is to minimize the cost or time involved in

- Movement of personnel and material between offices, in the case of an office layout
- Transporting work in process, finished parts, materials, and tools between machines or workstations, in the case of machine layout
- Picking items from storage spaces, in the case of a warehouse layout

When the distance between a machine pair is large, material transporters tend to wait until a large unit load is accumulated before moving the load from the first machine to the second. This increases

---

* Assume OFV is the sum of the product of flow and distance calculated for every department pair.

not only the average number of items waiting for the load to be formed and thus the work-in-process inventory, but also the average waiting time per part because each item in the unit load must now wait longer for the last item to arrive. Thus, the machine distances have a direct impact on queuing-based performance measures. As discussed in Chapter 4, in order to develop a solution (layout) that minimizes material handling or transportation costs, we need to model the layout problem. Increasingly, mathematical models are used for this purpose. The model is then solved using a suitable algorithm. The focus of this chapter is on mathematical models used for the facility layout problem. To show the usefulness of these models, we solve some layout problems using commercially available software and show how a first-cut layout is developed.

For purposes of modeling, it is convenient to classify the layout problem into two types:

1. Single-row layout problem
2. Multirow layout problem

The single-row and multirow layout problems are also called as the 1D and 2D space allocation problems, respectively (Simmons, 1969). As the name indicates, the departments are arranged linearly in one row in the single-row layout problem (Figure 10.3). In multirow problems, the departments are arranged linearly in two or more rows. Our definition of multirow layout does not require departments to be arranged along perfect rows. For example, notice that the centroids of rectangular departments in the top row in Figure 10.4 (multirow layout) have identical horizontal coordinates,

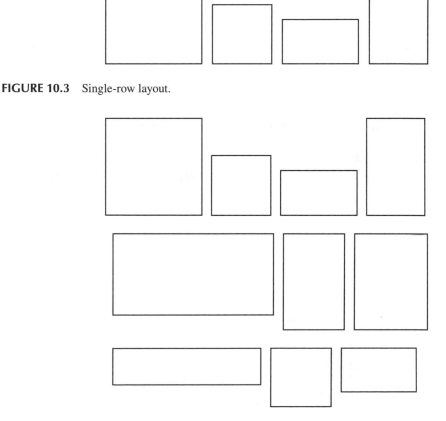

FIGURE 10.3   Single-row layout.

FIGURE 10.4   Multirow layout.

whereas those in the other rows do not. The term "department" in the manufacturing context includes machines, workstations, inspection stations, washing stations, locker rooms, and rest areas. In a service system, it includes offices, lounges, rest rooms, and cafeterias.

One simple example of the single-row layout problem is the arrangement of books on a shelf so that those used more frequently are within easy reach. If we think of the books as *departments*, then we have a single-row layout problem! Of course, if books are to be arranged in more than one shelf, then it is a multirow layout problem. Other single-row layout problems are the layout of machines on one side of the automated guided vehicle (AGV) path on a factory floor and the assignment of incoming aircraft to airport gates (Suryanarayan et al., 1991).

To understand how the airport gate assignment can be treated as a single-row layout problem, consider this example. The Baltimore Washington International airport serves as the northeast hub for Southeast Airlines. Southeast has access to gates at terminal D of this airport. Within a 2-hour time period, several of the company's flights arrive from and depart to various cities. Most passengers on these Southeast flights come in from a city and travel to their final destination via Baltimore. For the others, Baltimore is the final destination. A problem for airline management is to minimize inconvenience to connecting passengers, which is measured by the distance traveled by passengers to board their connecting flights. Southeast must assign the arriving and departing aircraft to specific gates within the 2-hour time period in order to minimize the overall inconvenience to passengers. In other words, it must minimize the overall distance traveled by passengers within the airport by assigning flights with significant interaction to adjacent gates. Typically, the distance between each pair of adjacent gates is the same (see Figure 10.5). Moreover, while determining the distance between gates (d_{ij}), only the distance traveled by passengers *within* the terminal needs to be considered; we may disregard the length of the jetway connecting the aircraft to the terminal, as it is typically the same for each gate, as shown in Figure 10.5. This problem can be modeled as a single-row layout problem in which the "departments" are all of equal length.

The single-row and multirow layout problems are rather easy to understand but very difficult to solve optimally. Models for both are known to be NP-complete (also see Beghin-Picavet and Hansen, 1982; Sahni and Gonzalez, 1976). In this section, a model and algorithm for the single-row problem are discussed. The model is called ABSMODEL 1 and consists of absolute terms in the objective function and constraints. Although nonlinear, this model may be linearized by introducing additional integer and nonnegative variables. We discuss this transformation later. First, we provide a nonlinear model for the single-row layout problem. It assumes the following. (1) The departments are square or rectangular and their shapes are known *a priori*. (2) The departments are arranged along a straight line as shown in Figure 10.3. (3) Their orientation is

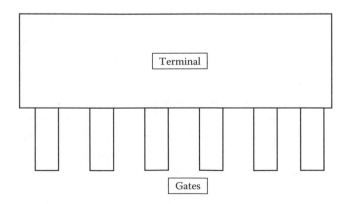

FIGURE 10.5 Assignment of aircraft to gates at an airport.

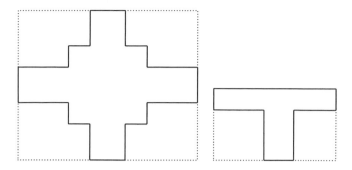

FIGURE 10.6　Approximation of department shapes that are not exact squares or rectangles.

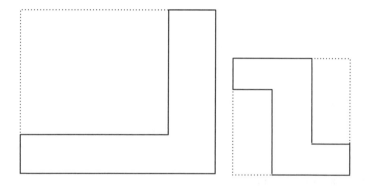

FIGURE 10.7　Treating L- or Z-shaped departments as rectangular or square departments.

known *a priori*. (4) There is no restriction on the shape of the building in which the departments are to be housed.

It may be argued that the first assumption is not realistic, because not all departments or workstations are square or rectangular. However, a close examination of many layout problems will reveal that those departments that are neither exact squares nor rectangles may be treated as being approximately square or rectangular depending on their exact shape. Figure 10.6 shows how to approximate department shapes that are not exact squares or rectangles. Of course, if the shapes of departments that are far from square or rectangular are assumed to be so, as done in Figure 10.7, then the quality of the solution obtained by solving the model will not be good. Fortunately, departments rarely have the shapes shown in Figure 10.7, and not much is lost in terms of optimality by treating departments that are neither rectangular nor square (such as the ones in Figure 10.6) as exact rectangles or squares. Such approximations will greatly simplify the modeling process and, therefore, the solution process as well.

The second assumption listed for ABSMODEL 1 is obvious and needs no explanation. The third and fourth do require some discussion. The orientation of rectangular departments indicates whether the longer side or the shorter side of the department is to be placed horizontally in the building (see Figure 10.8). The distance between rectangular departments depends on their orientation. In many manufacturing systems, machines are to be oriented such that the load and unload points face the aisle to facilitate the loading and unloading of parts. Because these points are fixed for a machine, its orientation is indeed known *a priori*. For square departments, however, orientation does not pose a problem because the distance between departments is the same for any orientation. We discuss the fourth assumption later.

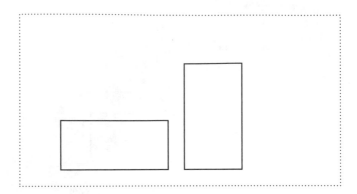

FIGURE 10.8   Two possible orientations of a rectangular department.

10.4.1   ABSMODEL 1

The following is the notation used in ABSMODEL 1:

Parameters

n Number of departments in the problem
c_{ij} Cost of moving a unit load through a unit distance between departments i and j
f_{ij} Number of unit loads between departments i and j
l_i Length of the horizontal side of department i
d_{ij} Minimum distance by which departments i and j are to be separated horizontally
H Horizontal dimension of the floor plan

Decision variable

x_i Distance between center of department i and vertical reference line (VRL)

The parameters and decision variables pertaining to the single-row layout problem are illustrated in Figure 10.9. Because we are considering the single-row layout problem, neither the vertical dimensions of the floor plan nor the vertical side of department i are important, so they are not considered:

$$\text{Minimize} \sum_{i=1}^{n-1} \sum_{j=i+1}^{n} c_{ij} f_{ij} \left| x_i - x_j \right| \tag{10.1}$$

$$\text{Subject to } \left| x_i - x_j \right| \geq 0.5(l_i + l_j) + d_{ij} \quad i = 1, 2, \ldots, n-1 \tag{10.2}$$

ABSMODEL 1 can be used to formulate layout problems with departments of equal or unequal lengths. The objective function of the model (see Expression 10.1) minimizes the total cost involved in making the required trips between departments. Constraint (10.2) ensures that no two departments in the layout overlap. Figure 10.10 illustrates how expression (10.2) satisfies the overlapping constraint. Although nonnegativity constraints are not necessary for this model, they can be included if the user wants a solution in which all the x_i's are positive. In this case, the VRL passes through (or is to the right of) the origin. Whether or not nonnegativity constraints are included, solution to ABSMODEL 1 (i.e., the arrangement of departments) and the OFV will not change, unless the problem has alternative optimal solutions. If the problem has a unique optimal solution, the two solutions obtained with and without the nonnegativity constraints in the model may have different values for the decision variables, but the arrangement of departments and the OFVs will be the same.

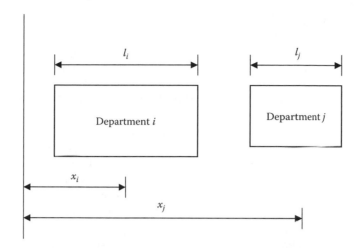

FIGURE 10.9 One possible single-row layout for a problem with unequal department lengths.

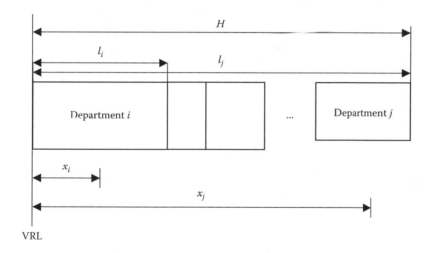

FIGURE 10.10 Illustration of parameters and decision variables for the single-row layout problem.

By not considering overlapping constraints in the vertical dimension, the vertical coordinates of the centers of all departments are implicitly assumed to be the same. Thus, the optimal solution will yield a (horizontal) single-row layout. Moreover, restrictions on the shape and size of the building (the fourth assumption in ABSMODEL 1) are not considered. Thus, we implicitly assume that the building has unlimited area within which we can locate the departments. Of course, because we are minimizing the product of cost, flow, and distance, departments will be

- Arranged horizontally along a single row
- As close to one another as possible in the optimal solution

Although the building size is considered to be infinite in the model, the departments will not be scattered far apart in the optimal solution. If the horizontal building dimension is known, however, and the user wants to include the constraint that the departments are located within the horizontal dimension, then this constraint may be added as shown in the following constraint:

$$H - 0.5(l_i) \geq x_i \geq 0.5(l_i) \quad i = 1, 2, \ldots, n \tag{10.3}$$

Because x_i refers to the distance between the *centroid* of department i and the VRL, constraint (10.3) will be satisfied even when this department is in the left and right extreme positions. This can be verified by examining Figure 10.10. Notice that constraint (10.3) ensures that the VRL coincides with the left of department i when this department is in the left extreme position. In Figure 10.9, H, the horizontal dimension of the floor plan is assumed to be the sum of the lengths of all departments, which is the minimum value required to develop a feasible layout. Of course, we are assuming that the horizontal clearance between each pair of adjacent departments is zero. If this is not the case, H should include that clearance. Note that addition of constraint (10.3) is really not necessary because the objective of the model is to minimize the distance (and hence the cost of transporting material) between departments. A compact layout will be obtained even without including it. If H is greater than or at least equal to the sum of lengths of all departments plus the required clearances, the departments will automatically fall within the boundaries of the building. Constraint (10.3) has been shown here to familiarize the reader with *department-within-building* constraints, which are necessary in multirow layout models.

In general, optimal solutions to nonlinear programming (NLP) problems are hard to find. To understand this, the reader must be familiar with the basic theory of NLP. We provide some basic information that may help the student who is unfamiliar with it. Unlike LP problems, the feasible region for an NLP problem need not be a convex set.* Even if it is convex, the optimal solution to an NLP problem need not be an extreme point of the feasible region. Thus, finding an optimal solution to a general NLP problem may be impossible. There are some special cases of NLP problems for which it is relatively easy to find an optimal solution, however. Examples are an unconstrained NLP problem in which a concave objective function is to be maximized or a convex objective function

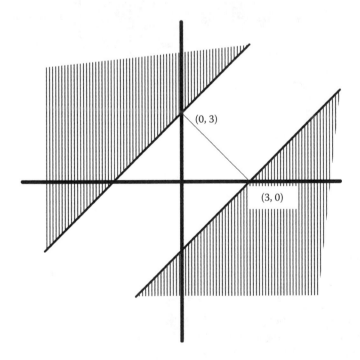

FIGURE 10.11   Set of points that satisfy the constraint in a simple layout problem.

---

* A convex set is a set of points in which the line segment joining any pair of points in the set is wholly contained in that set. For more details, refer to Winston (1994).

is to be minimized, and an NLP problem in which a concave objective function is to be maximized (or a convex objective function minimized) and the feasible region defined by the constraints is a convex set. The latter is a convex set if each constraint is a linear or convex function. Unfortunately, the convex set condition is not satisfied in ABSMODEL 1 or the other nonlinear models presented in this chapter. For example, consider a simple two-machine, single-row layout problem, in which the two machines have a length of 2 and 4 units, respectively. Constraint (10.2) in ABSMODEL 1 for this problem can be written as $|x_1 - x_2| \geq 3$. The set of all points satisfying this constraint is the shaded area in Figure 10.11. Clearly, the line that joins points $(0, 3)$ and $(3, 0)$, both of which are inside the feasible region, is not wholly contained within this region, so it is not a convex set according to the definition given in the footnote. Because this layout problem does not have the convex set property or any other property that makes it easy to find optimal solutions, we have to resort to heuristic local optimum search techniques found in commercially available software packages such as LINGO. This is illustrated in the next example.

Example 10.1

Over the years, business at the TVCR repair shop, which used to repair television (TV) sets, has grown significantly. Now, TVCR repairs other consumer electronic items, such as microwave ovens, computers, and audio systems, in addition to TV sets. TVCR also sells accessories and spare parts for the electronic items it services. During the past year, the owner has received numerous complaints about traffic congestion, misplacement of parts and repair orders, and poor-quality workmanship. The owner has invited a consulting company to study the situation and suggest a remedy. The consulting company has suggested that TVCR expand by building another floor, but the owner of TVCR is not convinced that this high cost recommendation is going to work. She strongly believes that by rearranging the departments within the existing space, the traffic congestion and quality problems can be corrected. She has recruited four industrial and systems engineering students from a local engineering college to study the problems at TVCR and recommend a layout to correct these problems.

After visiting TVCR a few times, the student team collected some data and made these observations:

- The company has one repair technician for TV sets, another for microwave ovens, a third for audio systems, and a fourth for computers.
- Although the repair technicians share some equipment, many pieces of equipment are heavily used by only one repair technician.
- The dimensions of the building are 75 × 15 meters.

The existing layout is shown in Figure 10.12. The student team has recommended that TVCR set up five rooms—one for each repair technician and another for the display of parts, accessories, and the cash register. It has also recommended that a customer service window be created in front of each repair room, so that customers with a specific repair problem can proceed to the appropriate repair area and talk to the concerned repair technician directly. Furthermore, equipment is to be kept in the room where it is used most. For example, if a testing equipment is used most by a TV repair technician, this equipment must be placed in that technician's room. The student team has observed the interaction between each pair of rooms on a typical working day. The interaction depends on how often one technician (e.g., the audio technician) uses a piece of equipment primarily used by another (e.g., the TV technician). The interaction is shown in the trips matrix $[f_{ij}]$ in Figure 10.13. The student team has also determined the room dimensions based on the equipment used in each room and also the work area required for each technician. These dimensions along with the names and numbers of each room are also listed in Figure 10.13. Now, the team must develop a single-row layout model for TVCR.

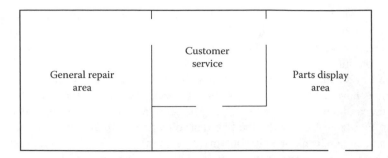

FIGURE 10.12 Existing layout of TVCR.

	Room						Room Number	Room Name	Dimensions (feet)
	1	2	3	4	5				
1	—	12	8	20	0		1	TV	20 × 10
1	12	—	4	6	2		2	Audio	10 × 10
1	8	4	—	10	0		3	Microwave	10 × 10
4	20	6	10	—	3		4	Computer	20 × 10
5	0	2	0	3	—		5	Parts	15 × 10

$[f_{ij}] = $ Room

FIGURE 10.13 Trips matrix and dimensions for five rooms.

Solution

The departments are oriented so that their longer sides are parallel to the length of the building. Because the flow matrix is symmetric, no changes need be made to it.* The cost involved in the movement of personnel is directly proportional to the distance they travel. There is no need to allow for any clearance between rooms.

A computer printout of the LINGO input file corresponding to the ABSMODEL 1 formulation of the aforementioned problem is provided next.

```
Data:
      N = 5;
Enddata

Sets:
      Dept/1..N/: Length, Center;
      ObjFn(Dept,Dept)| &2 #GT# &1: Flow;
Endsets

Data:
      Length = 20 10 10 20 15;
```

---

* A symmetric matrix $[a_{ij}]$ is one in which $a_{ij} = a_{ji}$. If the flow matrix is not symmetric, it can be transformed into one rather easily. This is demonstrated in Example 10.3.

```
Flow = 12 8 20 0
        4 6  2
       10 0
        3;

Enddata
     ! Objective function;
     Min= @SUM(ObjFn(i,j)|j #GT# i: Flow(i,j)*@ABS(Center(i)-Center(j)));
     ! Constraints;
     @FOR (Dept(i)|i #LT# N:
          @FOR (Dept(j)|j #GT# i #AND# j #LE# N:
               @ABS (Center(i)-Center(j)) >= 0.5*(Length(i)+Length(j))));

          Linearization components added:
               Constraints:        80
               Variables:          80
               Integers:           20
```

When the aforementioned model is solved using the LINGO software, the result is a computer output, a part of which is shown in the following text. Note that LINGO automatically linearizes the model and solves it optimally as an integer linear program with an optimal OFV of 1587.50. Note that the value of each variable indicates the horizontal coordinate of the department's center with respect to the VRL. The corresponding single-row layout is shown in Figure 10.14.

```
Global optimal solution found.
Objective value:                              1587.500
Extended solver steps:                             225
Total solver iterations:                          4874

               Variable             Value
               CENTER( 1)          35.00000
               CENTER( 2)          50.00000
               CENTER( 3)           0.000000
               CENTER( 4)          15.00000
               CENTER( 5)          62.50000
```

Including the location of emergency exits, escape routes, and parking access for employees and customers in the mathematical model makes it complex and rather difficult to solve, so these factors were not considered in the model. However, they do need to be incorporated in the final design. The aforementioned solution obtained via LINGO must be modified to comply with local fire codes and incorporate factors not considered by the model. For example, the long corridor in Figure 10.14 may be used as fire escape, and a door for emergency use may be placed at one end of the corridor, with a door for normal use at the other end. Parking space should be provided near that door

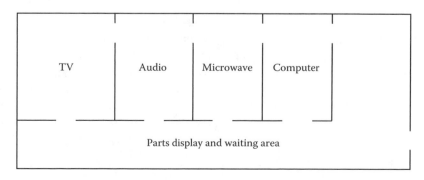

FIGURE 10.14   Single-row layout for TVCR.

if possible. In the next section, a way to linearize the nonlinear ABSMODEL 1 is presented. An advantage of linearizing an NLP model is that the resulting equivalent linear, mixed-integer model may be solved optimally via specialized algorithms.

10.4.2   LMIP 1

To develop an equivalent linear mixed-integer programming (LMIP) model for the single-row layout problem, we introduce additional decision variables:

$$x_{ij}^+ = \begin{cases} x_i - x_j & \text{if } (x_i - x_j) > 0 \\ 0 & \text{otherwise} \end{cases} \tag{10.4}$$

$$x_{ij}^- = \begin{cases} x_j - x_i & \text{if } (x_i - x_j) \le 0 \\ 0 & \text{otherwise} \end{cases} \tag{10.5}$$

$$z_{ij} = \begin{cases} 1 & \text{if } x_i < x_j \\ 0 & \text{otherwise} \end{cases} \tag{10.6}$$

To transform ABSMODEL 1 into a linear model, $|x_i-x_j|$ may be substituted with $x_{ij}^+ + x_{ij}^-$ as given in the following equation:

$$| x_i - x_j | = x_{ij}^+ + x_{ij}^-, \quad i = 1, 2, \dots, n-1; \; j = i+1, \dots, n \tag{10.7}$$

The transformed model is shown next:

$$\text{Minimize} \sum_{i=1}^{n-1} \sum_{j=i+1}^{n} c_{ij} f_{ij} \left(x_{ij}^+ + x_{ij}^- \right) \tag{10.8}$$

$$\text{Subject to} \left(x_{ij}^+ + x_{ij}^- \right) \ge 0.5(l_i + l_j) + d_{ij} \quad i = 1, 2, \dots, n-1; \; j = i+1, \dots, n \tag{10.9}$$

$$x_i - x_j = x_{ij}^+ - x_{ij}^- \quad i = 1, 2, \dots, n-1; \; j = i+1, \dots, n \tag{10.10}$$

For this model to be equivalent to ABSMODEL 1, one of the values of x_{ij}^+ or x_{ij}^- must equal zero. Depending on whether x_{ij}^+ or x_{ij}^- is strictly positive, either x_i is greater than x_j, or x_j is greater than x_i. This constraint is the same as requiring department i to be on either the left of or the right of department j. Unfortunately, this requirement is not automatically satisfied in the aforementioned model (see Exercise 33). To ensure that either x_{ij}^+ or x_{ij}^- is equal to zero, or alternatively, to ensure that department i is either to the left or to the right of department j, additional variables and constraints need to be included as shown in the following LMIP model (LMIP 1). These additions make the model more difficult to solve, but this is the price we have to pay for an optimal solution:

$$\text{Minimize} \sum_{i=1}^{n-1} \sum_{j=i+1}^{n} c_{ij} f_{ij} \left(x_{ij}^+ + x_{ij}^- \right) \tag{10.8}$$

$$\text{Subject to } x_i - x_j + Mz_{ij} \geq 0.5(l_i + l_j) + d_{ij}, \quad i = 1, 2, \ldots, n-1; \; j = i+1, \ldots, n \tag{10.11}$$

$$x_j - x_i + M(1 - z_{ij}) \geq 0.5(l_i + l_j) + d_{ij} \quad i = 1, 2, \ldots, n-1; \; j = i+1, \ldots, n \tag{10.12}$$

$$x_i - x_j = x_{ij}^+ - x_{ij}^- \quad i = 1, 2, \ldots, n-1; \; j = i+1, \ldots, n \tag{10.10}$$

$$x_{ij}^+, x_{ij}^- \geq 0 \quad i = 1, 2, \ldots, n-1; \; j = i+1, \ldots, n \tag{10.13}$$

$$z_{ij} = 0 \text{ or } 1 \quad i = 1, 2, \ldots, n-1; \; j = i+1, \ldots, n \tag{10.14}$$

$$x_i > 0 \quad i = 1, 2, \ldots, n \tag{10.15}$$

Because z_{ij} is a 0 or 1 variable, only one of the constraints (10.11) and (10.12) holds. This means that department i will either be to the left or right of department j in the optimal solution. Moreover, the departments must be positioned so that there is no overlap and the required clearance between them is maintained. Constraint (10.13) enforces nonnegativity for the x_{ij}^+ and x_{ij}^- variables. Constraint (10.14) restricts the z_{ij} variables to take on 0 or 1 values only. Equivalently, one can add the nonlinear constraint given next and eliminate the integer variable z_{ij} to get a formulation with only continuous variables. However, doing so results in an NLP, which is difficult to solve optimally:

$$x_{ij}^+ x_{ij}^- = 0 \quad i = 1, 2, \ldots, n-1, \; j = i+1, \ldots, n$$

Also, as mentioned in the discussion of ABSMODEL 1, adding the constraint that all the x_i variables be greater than zero (constraint (10.15)) does not alter the problem. We include this constraint to exploit the complementary slackness property when solving LMIP 1 using Benders' decomposition algorithm. This is explained in more detail in Chapter 11. In the models presented in the remainder of this chapter, M denotes an arbitrarily large positive number. The next example illustrates how LMIP 1 may be used to formulate a single-row layout problem.

Example 10.2

A furniture manufacturer has expanded his operations and purchased five new machines to be placed in a newly renovated building adjoining an existing facility. Because of the shape of the building and process requirements, the manufacturer has decided to arrange the machines in a single row. The cost to move an average load through unit distance is $1.00. The machine dimensions are listed in Figure 10.15, which also gives the horizontal clearance required between each pair of machines (in meters) as well as the frequency of trips estimated to be made per year for the foreseeable future. The machines are to be arranged so that their longer sides are parallel to the aisle. Formulate a linear programming model similar to LMIP 1 and solve it using an LMIP software.

For this problem, LMIP 1 was set up and solved using LINGO. We present computer printout of the model input and partial solution output next. Variables whose values are not shown in the solution output have a value of 0.

```
Data:
        N = 5;
        M = 999;
Enddata
```

			1	2	3	4	5			1	2	3	4	5
1	25 × 20	1	—	3.5	5.0	5.0	5.0		1	—	25	35	50	0
2	35 × 20	2	3.5	—	5.0	3.0	5.0		2	25	—	10	15	20
3	30 × 30	3	5.0	5.0	—	5.0	5.0		3	35	10	—	50	10
4	40 × 20	4	5.0	3.0	5.0	—	5.0		4	50	15	50	—	15
5	35 × 35	5	5.0	5.0	5.0	5.0	5.0		5	0	20	10	15	—

FIGURE 10.15 Data for Example 10.2.

```
Sets:
      Dept/1..N/: Length, Center;
      ObjFn(Dept,Dept)| &2 #GT# &1: Flow, Clear, XP, XN, Z;
Endsets

Data:
      Length = 25 35 30 40 35;

      Clear = 3.5 5 5 5
               5 3 5
               5 5
               5;

      Flow = 25 35 50 0
              10 15 20
              50 10
              15;

Enddata
      ! Objective function;
      Min= @SUM(ObjFn(i,j)|j #GT# i: Flow(i,j)*(XP(i,j)+XN(i,j)));
      ! Constraints;
      @FOR (Dept(i)|i #LT# N:
         @FOR (Dept(j)|j #GT# i #AND# j #LE# N:
            Center(i)-Center(j)+M*Z(i,j) >= 0.5*(Length(i)+Length(j))+Clear
             (i,j);
            Center(j)-Center(i)+M*(1-Z(i,j))>= 0.5*(Length(i)+Length(j))+Clear
             (i,j);
            XP(i,j) - XN(i,j) = Center(i)-Center(j);
            @BIN(Z(i,j))
      ));

Global optimal solution found.
Objective value:                            13575.00
Extended solver steps:                           199
Total solver iterations:                        3415

                    Variable          Value
                    CENTER( 1)        77.50000
                    CENTER( 2)        110.0000
                    CENTER( 3)        0.000000
                    CENTER( 4)        40.00000
                    CENTER( 5)        151.0000
                     XP( 1, 3)        77.50000
                     XP( 1, 4)        37.50000
                     XP( 2, 3)        111.0000
```

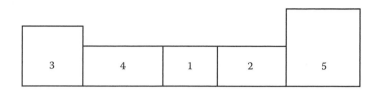

FIGURE 10.16 Optimal machine layout in the new facility of the manufacturing company.

```
XP( 2,  4)        71.00000
XN( 1,  2)        33.50000
XN( 1,  5)        73.50000
XN( 2,  5)        40.00000
XN( 3,  4)        40.00000
XN( 3,  5)        151.0000
XN( 4,  5)        111.0000
 Z( 1,  2)        1.000000
 Z( 1,  5)        1.000000
 Z( 2,  5)        1.000000
 Z( 3,  4)        1.000000
 Z( 3,  5)        1.000000
 Z( 4,  5)        1.000000
```

Note that a value of 999 was used for the M parameter. Using the values of Center(i), XP(i, j), XN(i, j), and Z(i, j) provided by LINGO, we can easily generate the optimal layout shown in Figure 10.16. The clearance between machines is included in the model but not shown in Figure 10.16.

10.5 MODELS FOR THE MULTIROW LAYOUT PROBLEM WITH DEPARTMENTS OF EQUAL AREA

Some applications of the multirow layout problem are the integrated circuit chip layout, the machine layout problem in an automated manufacturing system, typewriter keyboard design, and office layout (Burkard, 1984). Many of these were discussed in Chapter 1. Models have been developed to formulate this problem, such as the QAP (Koopmans and Beckmann, 1957), the LMIP model (Love and Wong, 1976), the quadratic set covering model (Bazaraa, 1975), and the nonlinear model with absolute terms in the objective function and constraints (Heragu, 1988). In this chapter, we present the QAP and the nonlinear model with absolute terms.

10.5.1 QAP

The problem of locating departments with material flow between them was first modeled as a QAP by Koopmans and Beckmann (1957). In a facility layout problem in which there are n departments to be assigned to n given locations, the term *assignment* means matching each department with a specific location *and* vice versa. The QAP formulation requires an equal number of departments and locations. If there are fewer than n—say m—departments to be assigned to n locations, then to use the QAP formulation, we can create $n - m$ dummy departments and assign a zero flow between each of these and all others (including the other dummy departments). If there are fewer locations than departments, the problem is infeasible.

As the name quadratic assignment problem indicates, QAP is an assignment problem in which the objective function is a second-degree function (quadratic) of the variables; that is, it involves the product of two variables. The constraints are linear functions of the variables. An illustration of four locations, four departments, and the assignment of each department to a specific location is shown in Figure 10.17. The following notation is used.

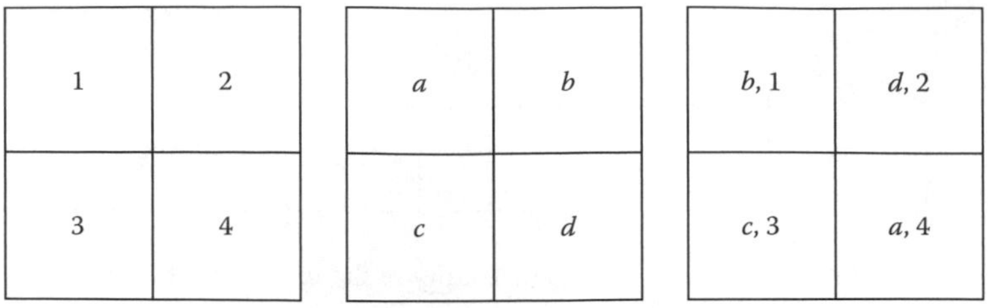

FIGURE 10.17   Four locations, four departments, and the assignment of departments to locations.

Parameters

- n Total number of departments and locations
- a_{ij} Net revenue from operating department i at location j
- f_{ik} Flow of material from department i to k
- c_{jl} Cost of transporting unit load of material from location j to l

Decision variable

$$x_{ij} = \begin{cases} 1 & \text{if department } i \text{ is assigned to location } j \\ 0 & \text{otherwise} \end{cases}$$

The following four assumptions are made in the QAP formulation:

1. a_{ij} includes gross revenues minus the cost of primary input but does not include the transportation cost of material between departments.
2. f_{ik} is independent of the locations of the departments.
3. c_{jl} is independent of the departments.
4. It is cheaper to transport material directly from department i to k than through a third location.

The QAP is then to

$$\text{Maximize} \sum_{i=1}^{n} \sum_{j=1}^{n} a_{ij} x_{ij} - \sum_{i=1}^{n} \sum_{j=1}^{n} \sum_{\substack{k=1 \\ i \neq k}}^{n} \sum_{\substack{l=1 \\ j \neq l}}^{n} f_{ik} c_{jl} x_{ij} x_{kl} \tag{10.16}$$

$$\text{Subject to} \sum_{j=1}^{n} x_{ij} = 1 \quad i = 1, 2, \ldots, n \tag{10.17}$$

$$\sum_{i=1}^{n} x_{ij} = 1 \quad j = 1, 2, \ldots, n \tag{10.18}$$

$$x_{ij} = 0 \text{ or } 1 \quad i, j = 1, 2, \ldots, n \tag{10.19}$$

The objective function maximizes the gross revenues minus the cost of primary inputs and the cost of transferring material between departments. The latter cost is a function of the magnitude of flow between each pair of departments and the distance by which they are separated. Constraints (10.17) and (10.19) ensure that each department i is assigned to exactly one location. Similarly, constraints (10.18) and (10.19) together guarantee that each location j has exactly one department assigned to it. By redefining a_{ij} as the fixed cost of locating department i at location j, we can restate the objective QAP as follows:

$$\text{Minimize} \sum_{i=1}^{n} \sum_{j=1}^{n} a_{ij} x_{ij} + \sum_{i=1}^{n} \sum_{j=1}^{n} \sum_{\substack{k=1 \\ i \neq k}}^{n} \sum_{\substack{l=1 \\ j \neq l}}^{n} f_{ik} c_{jl} x_{ij} x_{kl} \tag{10.20}$$

$$\text{Subject to} \sum_{j=1}^{n} x_{ij} = 1 \quad i = 1, 2, \ldots, n \tag{10.17}$$

$$\sum_{i=1}^{n} x_{ij} = 1 \quad j = 1, 2, \ldots, n \tag{10.18}$$

$$x_{ij} = 0 \text{ or } 1 \quad i, j = 1, 2, \ldots, n \tag{10.19}$$

In particular, if the a_{ij}'s are equal to zero or are identical, then the objective function (10.20) reduces to

$$\text{Minimize} \sum_{i=1}^{n} \sum_{j=1}^{n} \sum_{\substack{k=1 \\ i \neq k}}^{n} \sum_{\substack{l=1 \\ j \neq l}}^{n} f_{ik} c_{jl} x_{ij} x_{kl} \tag{10.21}$$

The QAP consisting of expressions (10.17) through (10.19) and (10.21) is the one most commonly used to model the facility layout problem. There are two special cases of the QAP—the linear assignment problem (LAP) and the traveling salesman problem (TSP). The LAP is concerned with the assignment of each element of a set of objects to a unique element in another set of the same size, so that the total assignment cost is minimized. Thus, in the case of machine–operator assignment, given the cost of assigning each operator to every other machine, we are concerned with assigning each operator to one machine and vice versa to minimize the overall cost of assignment.

Here is the TSP. Given a set of n cities and the distance between each pair of these cities, find the shortest path such that we begin and end in the same city and visit every other city only once—a typical problem that a traveling salesman faces! Thus, even in the TSP, we are concerned with the assignment of each of the n cities to one of n positions in the path so that the total travel distance is minimized. Like the QAP, the TSP and the LAP require the number of objects in the two sets to be the same; if that is not the case, we must create dummy objects in the appropriate set.

Although the QAP has been used frequently to model the facility layout problem, it cannot be used to formulate all types of layout problems. Consider the machine layout problem. Obviously, not all the machines have the same shape or area, so the distance between two locations cannot be determined *a priori* because the locations themselves are not known! The locations are not known because we do not know how the machines are going to be laid out. In fact, this is the very problem we are trying to solve! Such problems can therefore not be formulated as a QAP. Consider the

FIGURE 10.18 Distance between locations depends upon the assignment of departments to the locations (a) Sequential assignment and (b) alternate assignment.

distance between the locations in Figure 10.18a and 10.18b. The distance between any two locations depends upon the sequence of arrangement of all the other machines. Thus, the distance between the same two locations may be different for different layouts. Notice that the distance between locations 1 and 3 in Figure 10.18a is greater than that in Figure 10.18b. This situation does not arise in layout problems in which the departments are all of equal size, however, because the locations all have the same area and hence the distance between any two locations is independent of the departments assigned to them. Because the distance between locations does not change from one layout arrangement to another, using the QAP for layout problems with departments of equal sizes does not cause any problems. The reader can verify this in Exercise 14 at the end of this chapter.

Example 10.3

Due to an aggressive marketing strategy, LonBank Inc., a regional bank, has recently seen an increase in the number of loan applications it processes. Its regional manager has therefore decided to build a new loan center adjacent to one of its branches. Four square offices of equal area are to be located in the new square building. The locations in the building to which departments may be assigned are shown in Figure 10.19. Formulate the problem as a QAP using the data in Figure 10.20, which indicates the trips made by bank personnel and customers between each pair of offices.

Solution

The "trips from–to" matrix in Figure 10.20 is not symmetric; however, we can convert it to trips between matrix and make it symmetric. The matrix $[f_{ij}]$ in Figure 10.20 can be converted into the matrix $[f'_{ij}]$ in which $f'_{ij} = f_{ij} + f_{ji}$ as shown in Figure 10.21.

The rectilinear distances between the locations in Figure 10.19 are given in Figure 10.22. Because all locations have the same area, we assume that each location is a 10 square meter. The cost of making a trip between each pair of locations is assumed to be directly proportional to the distance between the locations. The QAP formulation for this problem can be solved using LINGO. The model input exactly as required by LINGO and a partial solution output are provided next.

```
        ┌──────┬──────┐
        │      │      │
        │  1   │  2   │
        │      │      │
        ├──────┼──────┤
        │      │      │
        │  3   │  4   │
        │      │      │
        └──────┴──────┘
```

FIGURE 10.19 Four locations.

$[f_{ij}]$ =

Office	1	2	3	4
1	—	5	4	8
2	12	—	6	2
3	8	6	—	2
4	3	2	2	—

FIGURE 10.20 Trips from–to matrix.

$[f_{ij}']$ =

Office	1	2	3	4
1	—	17	12	11
2	17	—	12	4
3	12	12	—	4
4	11	4	4	—

FIGURE 10.21 Trips between matrix.

$[d_{ij}]$ =

Site	1	2	3	4
1	—	1	1	2
2	1	—	2	1
3	1	2	—	1
4	2	1		—

FIGURE 10.22 Distance between locations matrix.

```
Sets:
      Dept/1..4/;
      Site/1..4/;
      Flow(Dept,Dept): F;
      Dist(Site,Site): D;
      Assign(Dept,Site): X;
      ObjFn(Dept, Site, Dept, Site);
Endsets

Data:
      D = 0 1 1 2
          1 0 2 1
          1 2 0 1
          2 1 1 0;

      F =  0 17 12 11
          17  0 12  4
          12 12  0  4
          11  4  4  0;

Enddata
      ! Objective function;
      Min= @SUM(ObjFn(i,j,k,l): F(i,k)*D(j,l)*X(i,j)*X(k,l));
      ! Constraints;
      @FOR(Dept(i):
        @SUM(Dept(j): (X(i,j))) = 1);
            @FOR(Dept(j):
              @SUM(Dept(i): (X(i,j))) = 1);
          @FOR(Dept(i):
            @FOR (Dept(j): @BIN(X(i,j)))));

      Global optimal solution found.
      Objective value:                       152.0000
      Extended solver steps:                       47
      Total solver iterations:                    339

                        Variable                Value
                        X( 1, 4)             1.000000
                        X( 2, 3)             1.000000
                        X( 3, 1)             1.000000
                        X( 4, 2)             1.000000
```

Based on the LINGO solution, we can develop the layout shown in Figure 10.23.

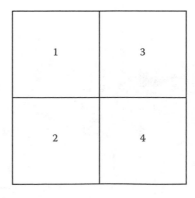

FIGURE 10.23   Layout of four offices in a bank.

10.6 MODEL FOR THE MULTIROW LAYOUT PROBLEM WITH DEPARTMENTS OF UNEQUAL AREA

In this section, we develop a nonlinear model ABSMODEL 2, which, like ABSMODEL 1, consists of absolute terms in the objective function and constraints. However, ABSMODEL 2 is for multirow layout problems in which the departments are square or rectangular and their physical orientation is known *a priori*. As discussed earlier, the shape assumption does not cause serious problems because most department shapes can be approximated as a square or rectangle. The orientation assumption may be difficult to justify, however, because it depends upon the underlying aisle structure. If the aisle structure is known, usually, a machine can be oriented in only one direction—the direction that allows the machine loading and unloading point to be facing and closest to the aisle. Because the aisle structure is not known until after the layout has been determined and we have not yet determined the layout, we cannot fix the orientation of a department or machine. However, if we were to relax this assumption, we would have to introduce many more variables and constraints and make the model even more complex. A trade-off decision has to be made between wanting too much accuracy in the model and being able to solve it. In this chapter, we suggest that the orientation assumption be maintained in the model. We can always examine the resulting solution, modify the initial orientation, if necessary and appropriate, and solve the model again. This procedure can be repeated until we get a layout in which all the machines have "acceptable" orientations. For some problems, this could be a tedious process, but relaxing the orientation constraint in the model would have made it so complex that solving problems with more than six or seven machines would be impossible because of so many integer variables and constraints.

10.6.1 ABSMODEL 2

In addition to the notation for c_{ij} and f_{ij} used in ABSMODEL 1, we define these parameters:

x_i Horizontal distance between center of department i and VRL
y_i Vertical distance between center of department i and horizontal reference line (HRL)
l_i Length of the horizontal side of department i
b_i Length of the vertical side of department i
dh_{ij} Horizontal clearance between departments i and j
dv_{ij} Vertical clearance between departments i and j

Note that l_i and b_i refer not to the length and width of department i, but to the horizontal and vertical sides of the department. Because l_i and b_i are parameters, their values must be known, but then, their values are known only if the department orientations are known *a priori*. This is why we make this assumption in ABSMODEL 2. The parameters, decision variables, and reference lines VRL and HRL relevant to this model are illustrated in Figure 10.24. To keep the presentation simple, it is assumed that the horizontal and vertical clearance between departments are the same. If necessary, this assumption can be relaxed.

Like the other models presented in this chapter, the objective function of ABSMODEL 2 minimizes the total cost of making the required trips between the departments.

$$\text{Minimize} \sum_{i=1}^{n-1} \sum_{j=i+1}^{n} c_{ij} f_{ij} \left(|x_i - x_j| + |y_i - y_j| \right) \tag{10.22}$$

$$\text{Subject to } |x_i - x_j| + Mz_{ij} \geq \frac{1}{2}\left(l_i + l_j\right) + dh_{ij} \quad i = 1, 2, \ldots, n-1; \ j = i+1, \ldots, n \tag{10.23}$$

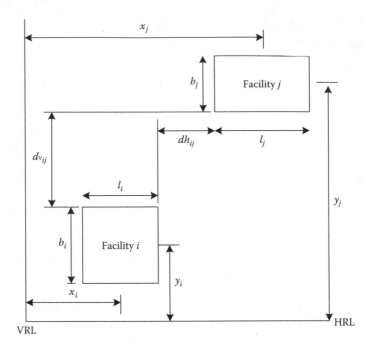

FIGURE 10.24 Illustration of decision variables and parameters for the multirow layout problem with departments of unequal area.

$$| y_i - y_j | + M(1 - z_{ij}) \geq \frac{1}{2}(b_i + b_j) + dv_{ij} \quad i = 1, 2, \ldots, n-1; \ j = i+1, \ldots, n \quad (10.24)$$

$$z_{ij}(1 - z_{ij}) = 0 \quad i = 1, 2, \ldots, n-1; \ j = i+1, \ldots, n \quad (10.25)$$

Constraints (10.23) through (10.25) ensure that departments do not overlap in the horizontal or vertical direction. Constraint (10.25) ensures that z_{ij} takes on values of 0 or 1 only, which in turn ensures that only one of the two constraints (10.23) and (10.24) will hold. As discussed earlier, however, such constraints destroy the convex set property of the feasible solution space region. If the dimensions of the floor plan are known, one may add suitable constraints to ensure that the departments are arranged in such a way that they stay within the boundaries of the floor plan (see Exercise 27).

Example 10.4

An insurance company wants to determine the location of five office cubicles of varying dimensions in the top floor of its headquarters. These offices do not have any interaction with offices in other floors. Work sampling was recently conducted and the number of trips made by personnel between these five offices in a 12-month period was estimated. The estimated trips and office dimensions are provided in Figure 10.25. Use ABSMODEL 2 to formulate the problem.

We make these assumptions.

1. The cost of moving unit distance between any pair of offices is the same.
2. The clearance between the offices in both directions is zero.
3. The offices are oriented so that their longer sides are horizontal.

		Office					Office	Dimensions (feet)
		1	2	3	4	5		
$[f_{ij}] =$ Office	1	—	10	15	20	0	1	25×20
	2	10	—	30	35	10	2	25×20
	3	15	30	—	10	20	3	35×30
	4	20	35	10	—	15	4	30×20
	5	0	10	20	15	—	5	35×20

FIGURE 10.25 Trips matrix and dimensions for five offices.

Here is a partial LINGO input of ABSMODEL 2 for the problem.

```
Data:
      N = 5;
      M = 999;
Enddata

Sets:
      Dept/1..N/: Length, Width, X, Y;
      ObjFn(Dept,Dept) | &2 #GT# &1: Cost, Flow, VClear, HClear, Z;
Endsets

Data:
      Length = 25 25 35 30 35;
      Width  = 20 20 30 20 20;
      VClear = 0 0 0 0
               0 0 0
               0 0
               0;
      HClear = 0 0 0 0
               0 0 0
               0 0
               0;
      Flow   = 10 15 20 0
               30 35 10
               10 20
               15;
      Cost   = 1 1 1 1
               1 1 1
               1 1
               1;
Enddata

      ! Objective function;
      Min= @SUM(ObjFn(i,j)|j #GT# i: Cost(i,j)*Flow(i,j)*(@ABS(X(i)X(j))+@
        ABS(Y(i)-Y(j)))));
            ! Constraints
```

```
@FOR (Dept(i)|i #LT# N:
    @FOR (Dept(j)|j #GT# i #AND# j #LE# N:
        @ABS(X(i)-X(j))+M*Z(i,j)
>= 0.5*(Length(i)+Length(j))+HClear(i,j);
            @ABS(Y(i)-Y(j))+M*(1-Z(i,j))
>= 0.5*(Width(i)+Width(j))+VClear(i,j);
        @BIN(Z(i,j))
));
```

Solving this model using general-purpose NLP algorithms available in most software packages (in our case, LINGO) yields a local optimal solution shown later. In Chapter 11, we discuss another algorithm and a computer program (MULROW.EXE) to solve the model.

```
Local optimal solution found.
Objective value:                        5825.039
Extended solver steps:                       280
Total solver iterations:                   36261

                    Variable        Value
                      X( 1)       812.7531
                      X( 2)       812.7609
                      X( 3)       812.7609
                      X( 4)       785.2609
                      X( 5)      7710.7609
                      Y( 1)       264.2749
                      Y( 2)       244.2749
                      Y( 3)      2110.2749
                      Y( 4)       244.2749
                      Y( 5)       224.2749
                    Z( 1, 2)      1.000000
                    Z( 1, 3)      1.000000
                    Z( 1, 4)      1.000000
                    Z( 1, 5)      1.000000
                    Z( 2, 3)      1.000000
                    Z( 2, 4)      0.000000
                    Z( 2, 5)      1.000000
                    Z( 3, 4)      1.000000
                    Z( 3, 5)      0.000000
                    Z( 4, 5)      1.000000
```

10.6.2 LOOP LAYOUT PROBLEM

In this section, we modify LMIP 1 to formulate a layout problem that arises commonly in automated manufacturing environments—the loop layout problem (LLP). The LLP determines the optimal arrangement of machines or workstations around a loop. Material may be transported along one loop or two loops as shown in Figure 10.26. A typical loop layout has a pick-up and drop-off point through which all the parts enter and exit the system. From the processing sequence of the parts and their production volume, the LLP attempts to determine a layout of the workstations that minimizes the overall material handling costs. Typically, only the internal loop is used for material handling, and it corresponds to a material handling system—for example, loop conveyor, tow line, overhead monorail system, or wire paths of AGVs (Kouvelis and Kim, 1992). Depending on the type of material handling device used, the movement of material may be unidirectional or bidirectional. A study of more than 50 flexible manufacturing systems in Japan found that unidirectional loop layouts are preferred (Jaikumar, 1984). Although bidirectional systems offer more flexibility in material transfer, it is widely recognized that they require more expensive, sophisticated controls and larger investments.

Other models are available for the LLP. However, they are discrete in nature. They assume that *candidate* locations for workstations are available and attempt to find a permutation of these

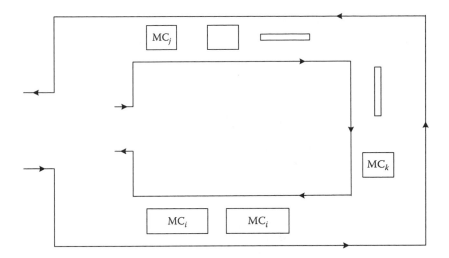

FIGURE 10.26　Loop layout in manufacturing systems.

workstations to minimize an objective function under a given operational strategy. For example, see the QAP-based approaches in Kouvelis and Kim (1992) and Houshyar and McGinnis (1990). These approaches do not take into consideration workstation lengths. Some approaches do take workstation size and shape into consideration *after* their optimal sequence has been determined (for example, see Leung, 1992). Because the candidate locations of workstations are assumed to be known, these models implicitly assume that all workstations have equal area and similar shapes. Otherwise, the candidate locations would be different for each layout and hence would be unknown! In fact, this fundamental flaw has been made by many researchers in applying the QAP to the plant layout problem.

LMIP 2 is fundamentally different on two counts. First, it does not assume candidate workstation locations are known *a priori*. The key variable (x_i) in LMIP 2 is continuous and more accurately represents the distance between workstation centers. Thus, the objective function in LMIP 2 is by far more accurate than what other models use. Second, unlike other models, LMIP 2 is suitable for a variety of operational strategies, such as unidirectional flow, bidirectional flow, backtracking, and no backtracking.

For modeling purposes, we assume that the LLP is equivalent to the single-row layout problem. Thus, we remove the *bends* in Figure 10.26 and turn the loop layout into a straight-line layout. If the material movement is unidirectional, then the LLP is equivalent to a single-row layout problem with no backtracking.

Because we are converting a loop layout into a straight-line layout, the model presented in this section can also be applied for U-line and L-line layouts. We use the following additional notation in model LMIP 2 and assume that material movements occur in a prespecified direction.

L is the length of loop

d_{ij}^+ is the minimum horizontal distance by which departments i and j must be separated

d_{ij}^+ is the maximum horizontal distance by which departments i and j can be separated

h_{ij} is the distance traveled by material handling carrier to visit department j from department i

LMIP 2

$$\text{Minimize } \sum_{i=1}^{n}\sum_{j=1}^{n} c_{ij} f_{ij} h_{ij} \qquad (10.26)$$

$$\text{Subject to } h_{ij} + M(1 - y_{ij}) \geq x_j - x_i \quad \forall i, j \leq n; i \neq j \tag{10.27}$$

$$h_{ij} + M y_{ij} \geq L - \left(x_i - x_j \right) \quad \forall i, j \leq n; i \neq j \tag{10.28}$$

$$x_i - x_j + M y_{ij} \geq d_{ij}^- \quad i, j \leq n; i \neq j \tag{10.29}$$

$$-x_i + x_j - M y_{ij} \geq d_{ij}^- - M \quad i, j \leq n; i \neq j \tag{10.30}$$

$$x_i - x_j - M y_{ij} \leq d_{ij}^+ \quad i, j \leq n; i \neq j \tag{10.31}$$

$$-x_i + x_j + M y_{ij} \leq d_{ij}^+ + M \quad i, j \leq n; i \neq j \tag{10.32}$$

$$y_{ij} = 0 \text{ or } 1; \quad h_{ij} \geq 0 \quad i, j \leq n; i \neq j \tag{10.33}$$

$$x_i > 0 \quad i \leq n \tag{10.34}$$

The objective function minimizes the total material handling cost. Because only one of Equations 10.27 and 10.28 must hold for each department pair, we enforce this using the binary variable y_{ij} and an arbitrarily large positive number M. The binary y_{ij} is defined as

$$y_{ij} = \begin{cases} 1 & \text{if } x_i \leq x_j \\ 0 & \text{otherwise} \end{cases} \quad i, j \leq n; i \neq j \tag{10.35}$$

Also, $L - (x_i - x_j)$ is the maximum distance that the material handling carrier will travel to reach department i from j, and this will occur only when department i is positioned after department j and travel direction is clockwise. This can be verified by the reader by assuming travel in the outer loop is not permitted, removing the bends in Figure 10.26, and examining the distance traveled by the material handling carrier between departments i and j. Moreover, instead of assuming M to be a large positive number, we can set it equal to

$$\sum_i l_i + \varepsilon$$

It is easy to see that this value of M will also ensure that either Equation 10.27 or 10.28 will hold for every department pair i, j. Constraints (10.29) through (10.32) provide upper and lower bounds on the distance between the centers of department pairs. With regard to constraint (10.34), although nonnegativity restrictions are not necessary for the x variables, from the solution standpoint, it is convenient to include the strict inequality because we can exploit the complementary slackness property in solving its dual using the Benders' decomposition approach, explained in detail in Chapter 11.

It is also obvious that the lower bounds for d_{ij}^+, d_{ij}^- are given by the following expressions. These lower bounds are calculated by assuming there is no clearance between departments, that is, each department is *touching* its adjacent ones:

$$d_{ij}^- = \sum_{i=1}^{n} l_i - 0.5(l_i + l_j) \quad i, j \le n, i \ne j \tag{10.36}$$

$$d_{ij}^+ = \sum_{i=1}^{n} l_i - 0.5(l_i + l_j) \quad i, j \le n, i \ne j \tag{10.37}$$

This model can be used to formulate unidirectional U-shaped, L-shaped, or loop layout problems. When the model is solved, it provides the x coordinates of machine centers and hence their sequence. Machines are then positioned along the U-line, L-line, or loop based on this sequence. Of course, when material movement is allowed to occur in both directions as in Figure 10.26, LMIP 1 can be used directly.

10.6.3 LINEAR PROGRAMMING MODEL TO DEVELOP A LAYOUT GIVEN A BLOCK PLAN

All the models studied thus far are useful in determining the block plan of the layout. As mentioned, none of the models can be solved optimally except for relatively small instances of the layout problem, but the heuristics in Chapter 11 can be used to identify near-optimal block plans. In this section, we will study another model that can be used to develop a more detailed layout given a block plan that identifies relative adjacencies between machines. The model is linear and its solution specifies input and output points for each department as well as aisle structure and layout. Assuming the department adjacencies are known greatly simplifies the problem and this is why we are able to solve the model optimally. This model developed by Montreuil (1988) uses this additional notation.

Parameters

L_i^u, L_i^l Upper and lower bounds on the length of department i
W_i^u, W_i^l Upper and lower bounds on the width of department i
P_i^u, P_i^l Upper and lower bounds on the perimeter of department i
HA, VA Set of department pairs adjacent in the horizontal and vertical dimensions, respectively

Decision variables

x_i^u, y_i^u x, y coordinates of upper right corner of department i
x_i^l, y_i^l x, y coordinates of lower left corner of department i

BlockPlan LP

$$\text{Minimize} \sum_{i=1}^{n-1} \sum_{j=i+1}^{n} c_{ij} f_{ij} \left(x_{ij}^+ + x_{ij}^- + y_{ij}^+ + y_{ij}^- \right) \tag{10.38}$$

$$\text{Subject to } x_i - x_j = x_{ij}^+ - x_{ij}^- \quad i = 1, 2, \ldots, n-1; \, j = i+1, \ldots, n \tag{10.39}$$

$$y_i - y_j = y_{ij}^+ - y_{ij}^- \quad i = 1, 2, \ldots, n-1; \, j = i+1, \ldots, n \tag{10.40}$$

$$L_i^l \le \left(x_i^u - x_i^l \right) \le L_i^u \quad i = 1, 2, \ldots, n \tag{10.41}$$

$$W_i^l \le \left(y_i^u - y_i^l \right) \le W_i^u \quad i = 1, 2, \ldots, n \tag{10.42}$$

$$P_i^l \le 2 \left(x_i^u - x_i^l + y_i^u - y_i^l \right) \le P_i^u \quad i = 1, 2, \ldots, n \tag{10.43}$$

$$x_i^l \le x_i \le x_i^u \quad i = 1, 2, \ldots, n \tag{10.44}$$

$$y_i^l \le y_i \le y_i^u \quad i = 1, 2, \ldots, n \tag{10.45}$$

$$x_i - x_j \ge x_i^u - x_i^l - \left(x_j^u - x_j^l \right), \quad i, j \in HA; i \text{ to the right of } j \tag{10.46}$$

$$y_i - y_j \ge y_i^u - y_i^l - \left(y_j^u - y_j^l \right), \quad i, j \in VA; i \text{ above } j \tag{10.47}$$

$$x_i^l \ge x_j^u \quad i, j \in VA \tag{10.48}$$

$$y_i^l \ge y_j^u \quad i, j \in HA \tag{10.49}$$

$$x_i, y_i, x_i^l, y_i^l, x_i^u, y_i^u \ge 0 \quad i = 1, 2, \ldots, n \tag{10.50}$$

This BlockPlan LP model minimizes the total cost of moving material between each pair of departments. It makes numerous assumptions in order to maintain linearity of the model so that it may be solved optimally using the generalized simplex algorithm. Many of these assumptions can be relaxed by adding more constraints or variables, while still maintaining linearity. One example is the implicit requirement that no flow backtracking occurs. Another is the rectangular or square shape assumption for the departments. Similarly, the model can be expanded to include aisles with limits on their width and position. However, one assumption that cannot be relaxed without making the model nonlinear or introducing complicating integer restrictions is the one that requires *a priori* knowledge of the relative positions of the departments and their adjacencies in constraints (10.44) through (10.49). The reader is encouraged to try Exercise 53 to understand usefulness of the model.

Example 10.5

Consider Example 10.4. Assuming the layout has a central corridor of width 5 meters with offices on either side of it, and the restrictions on department sizes as shown in Table 10.2, develop a block plan using the LP model presented in Section 10.6.3. Assume that departments 2, 3, and 4 are in the top row in that order from left to right and departments 1 and 5 are in the bottom row.

Solution

Here is the LINGO model input.

TABLE 10.2

Restrictions on Office Sizes for Example 10.5

Office	L_i^l, L_i^u	W_i^u, W_i^l
1	22, 27	18, 22
2	22, 27	18, 22
3	32, 37	28, 32
4	28, 34	18, 23
5	32, 37	18, 23

```
Sets:
     Dept/1..5/: LU, LL, WU, WL, PU, PL, XU, YU, XL, YL, X, Y;
     ObjFn(Dept,Dept): XP, XN, YP, YN, Cost, Flow;
Endsets

Data:
     LU = 27 27 37 34 37;
     LL = 22 22 32 28 32;
     WU = 22 22 32 23 23;
     WL = 18 18 28 18 18;
     PU = 98 98 138 104 120;
     PL = 80 80 120 92 100;
     Flow =   0 10 15 20 0
             10 0 30 35 10
             15 30 0 10 20
             20 35 10 0 15
              0 10 20 15 0;
     Cost =   0 1 1 1 1
              1 0 1 1 1
              1 1 0 1 1
              1 1 1 0 1
              1 1 1 1 0;
Enddata
     ! Objective function;
     Min= @SUM(ObjFn(i,j):Cost(i,j)*Flow(i,j)*(XP(i,j)+XN(i,j)+YP(i,j)+YN(
       i,j)));
          ! Constraints;
     @FOR (Dept(i):
        @FOR (Dept(j):
              XP(i,j)-XN(i,j) - X(i) + X(j) = 0;
              YP(i,j)-YN(i,j) - Y(i) + Y(j) = 0;
));
              @FOR (Dept(i):
                   XU(i)-XL(i) >= LL(i);
                   XU(i)-XL(i) <= LU(i);
                   YU(i)-YL(i) >= WL(i);
                   YU(i)-YL(i) <= WU(i);
                   2*(XU(i)-XL(i)+YU(i)-YL(i)) >= PL(i);
                   2*(XU(i)-XL(i)+YU(i)-YL(i)) <= PU(i);
                   X(i)>=XL(i);
                   X(i)<=XU(i);
              Y(i)>=YL(i);
              Y(i)<=YU(i);
);
              X(5)-X(1)>=XU(5)-XL(5)-XU(1)+XL(1);
              X(3)-X(2)>=XU(3)-XL(3)-XU(2)+XL(2);
              X(4)-X(3)>=XU(4)-XL(4)-XU(3)+XL(3);
              X(4)-X(2)>=XU(4)-XL(4)-XU(2)+XL(2);
              Y(2)-Y(1)>=YU(2)-YL(2)-YU(1)+YL(1);
```

```
Y(3)-Y(1)>=YU(3)-YL(3)-YU(1)+YL(1);
Y(4)-Y(1)>=YU(4)-YL(4)-YU(1)+YL(1);
Y(2)-Y(5)>=YU(2)-YL(2)-YU(5)+YL(5);
Y(3)-Y(5)>=YU(3)-YL(3)-YU(5)+YL(5);
Y(4)-Y(5)>=YU(4)-YL(4)-YU(5)+YL(5);
XU(2)  <= XL(3);
XU(2)  <= XL(4);
XU(3)  <= XL(4);
XU(1)  <= XL(5);
YU(1)+5 <= YL(2);
YU(1)+5 <= YL(3);
YU(1)+5 <= YL(4);
YU(5)+5 <= YL(2);
YU(5)+5 <= YL(3);
YU(5)+5 <= YL(4);
```

Notice that the last 10 constraints specify that a horizontal aisle of 5 units width be included. The last 20 constraints (including the 6 mentioned in the previous sentence) together enforce the adjacency restrictions—departments 2, 3, and 4 in the top row and departments 1 and 5 in the bottom row in that order. They correspond to constraints (10.46) through (10.49) in the BlockPlan LP model. The other constraints correspond to constraints (10.39) through (10.45). Solving this model using LINGO yields an optimal solution shown as follows. The XP, XN, YP, and YN variables not shown in the partial solution output had a zero value. It is easy to construct a layout using the LINGO solution. The layout is shown in Figure 10.27 along with the location of the coordinates (x, y) for each department.

```
Global optimal solution found.
Objective value:                        6310.000
Total solver iterations:                      20

            Variable           Value
              XU( 1)         310.00000
              XU( 2)         210.00000
              XU( 3)         510.00000
              XU( 4)         810.00000
              XU( 5)         610.00000
              YU( 1)          23.00000
```

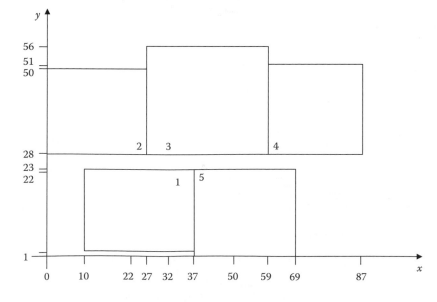

FIGURE 10.27   Layout developed using LINGO Solution in Example 10.5.

YU(2)	50.00000
YU(3)	56.00000
YU(4)	51.00000
YU(5)	23.00000
XL(1)	10.00000
XL(2)	0.000000
XL(3)	210.00000
XL(4)	510.00000
XL(5)	310.00000
YL(1)	1.000000
YL(2)	28.00000
YL(3)	28.00000
YL(4)	28.00000
YL(5)	0.000000
X(1)	32.00000
X(2)	210.00000
X(3)	32.00000
X(4)	510.00000
X(5)	310.00000
Y(1)	22.00000
Y(2)	28.00000
Y(3)	28.00000
Y(4)	28.00000
Y(5)	23.00000
XP(1, 2)	5.000000
XP(3, 2)	5.000000
XP(4, 1)	210.00000
XP(4, 2)	32.00000
XP(4, 3)	210.00000
XP(4, 5)	22.00000
XP(5, 1)	5.000000
XP(5, 2)	10.00000
XP(5, 3)	5.000000
XN(1, 4)	210.00000
XN(1, 5)	5.000000
XN(2, 1)	5.000000
XN(2, 3)	5.000000
XN(2, 4)	32.00000
XN(2, 5)	10.00000
XN(3, 4)	210.00000
XN(3, 5)	5.000000
XN(5, 4)	22.00000
YP(2, 1)	6.000000
YP(2, 5)	5.000000
YP(3, 1)	6.000000
YP(3, 5)	5.000000
YP(4, 1)	6.000000
YP(4, 5)	5.000000
YP(5, 1)	1.000000
YN(1, 2)	6.000000
YN(1, 3)	6.000000
YN(1, 4)	6.000000
YN(1, 5)	1.000000
YN(5, 2)	5.000000
YN(5, 3)	5.000000
YN(5, 4)	5.000000

10.7   DISCUSSION OF MODELS

A fundamental difference between the QAP and the ABSMODELs is that the former requires the locations be known *a priori*, whereas the ABSMODELs do not. In the ABSMODELs, the department centers are specified by their *x* and *y* coordinates; hence, knowing their locations *a priori* is not necessary. Thus, the ABSMODELs are more general. The ABSMODEL for the multirow layout problem with departments of equal area can be shown to be equivalent to the QAP (see Exercise 19). Traditionally, the QAP has been used to formulate the layout problem, but its requirement that the location of departments be known makes it unattractive in certain circumstances. In addition, it requires the departments have equal areas. When the departments are not of equal area, the model has to be modified suitably, thereby rendering it even more complex. Computational results for the QAP indicate that to find an optimal solution even for small-scale problems—for example, a 15-department problem—more than an hour of processing time may be required on a powerful mainframe computer (Burkard, 1984). Even the fast computer processors that are available today have not been able to solve problems with 20 or more departments optimally.

The ABSMODELs can also not be solved optimally for problems of this size, but numerous simple heuristic techniques exist. For example, Heragu and Kusiak (1991), Tam and Li (1991), and van Camp et al. (1992) have solved the ABSMODEL and its variant model using the Lagrangian or penalty methods. Also, Heragu and Alfa (1992) develop a solution approach that involves the use of penalty and simulated annealing algorithms. While these solution approaches, which are heuristics, cannot guarantee optimality, the solution techniques, especially the simulated annealing-based algorithms, have produced the best known solutions for large test problems (e.g., a 90-department problem). The penalty algorithms are presented in Chapter 11 and simulated annealing.

Tam and Li (1991) and van Camp et al. (1992) have developed models that are extensions of the basic ABSMODEL and incorporate additional constraints in developing a facility layout. For example, the model in Tam and Li (1991) assumes the departments have a known length-to-width ratio and their orientation is not fixed. van Camp et al. (1992) take into consideration the dimensions of the building within which departments are placed.

Modeling helps to clarify the problem and takes into consideration those factors that are critical in developing a layout. A model by itself does not provide a solution to a problem. However, algorithms or solution techniques must be developed to solve a model. Some algorithms were discussed in Chapter 5. Advanced algorithms are discussed in Chapter 11.

All the nonlinear models for the single-row and multirow layout problems may be solved using NLP software packages. Although commercial codes are available, many of these cannot generate optimal solutions to any of the models (Brooke et al., 1988). In fact, the final solutions given by the NLP software may be far from optimal. ABSMODELs 1 and 2 can be solved using specialized algorithms, however. One such technique is presented in Chapter 11.

We also discussed techniques to convert ABSMODELs into linear mixed-integer and linear models. Although the LMIP models may be solved optimally using commercial integer programming software packages, due to the presence of a large number of integer variables, this approach is limited to only small-sized problems. Our definition of small-sized problems covers problems with fewer than 12 departments. Hence, specialized optimal and heuristic algorithms need to be developed. This is discussed in Chapter 11.

10.8   REVIEW QUESTIONS AND EXERCISES*

1. Develop a detailed classification of models. If necessary, consult additional sources in the library and develop a detailed classification of models. Define each class of models in one or two sentences.
2. State the advantages and disadvantages of the following three modeling approaches:
   (a) Mathematical programming
   (b) Queuing
   (c) Simulation

---

* Unless otherwise indicated, use the rectilinear distance in computing the cost of a layout.

3. Explain the difference between a space allocation problem and departments-to-locations assignment problem.

4. Figure 10.3 illustrates an application of the single-row layout problem. Assuming appropriate parameters, e.g., clearance between machines, define the variables and formulate a model for the application.

5. Formulate a model for the *aircraft-to-gates* assignment problem illustrated in Figure 10.5. Assume that each pair of adjacent gates is equidistant and that the airline only wants to minimize the inconvenience caused to passengers. Inconvenience is measured as the distance traveled by passengers to board their connecting flights. Thus, passengers terminating their journey at this airport need not be considered.

6. Discuss what additional considerations will render the model you developed for Exercise 5 inappropriate.

7. Is it reasonable to approximate the shapes of departments as squares or rectangles as shown in Section 10.4. Why or why not?

8. Consider Exercise 6 in Chapter 5. Develop a flow matrix that shows the number of "transactions" between each pair of stages. Then, develop a model similar to ABSMODEL 1 assuming each stage of preparation requires the same amount of space. Solve the model using an NLP code (e.g., LINGO or any other available NLP software).

9. In Exercise 8, assume that the local building codes require each set of adjacent stages to be separated by a distance equal to half of each stage's length. Reformulate a model under the new restrictions.

10. Consider Exercise 8 in Chapter 5. Develop a flow matrix that shows the number of trips between each pair of workstations. Then, develop a model similar to ABSMODEL 1 assuming each workstation requires the same amount of space. Solve the model using an NLP code (e.g., LINGO or any other available NLP software).

11. Consider Exercise 9 in Chapter 5. Formulate ABSMODEL 1 and solve it using the LINGO or other NLP software.

12. List real-world layout problems in which the departments are all squares of equal area.

13. Explain the role of models and algorithms in problem solving.

14. Consider Exercise 10 in Chapter 5. Formulate a QAP for the problem of finding optimal assignment of offices. Solve the model using any available NLP software.

15. Consider Exercise 11 in Chapter 5. Formulate the layout problem as a QAP. Solve the model using any available NLP software.

16. If a corridor is required between the two rows of departments in Exercise 14, can the problem still be formulated as a QAP. Why or why not?

17. If a corridor is required between the two rows of departments in Exercise 15, can the problem still be formulated as a QAP. Why or why not?

18. Repeat Exercise 15 using the traffic data in Figure 5.55.

19. Consider the QAP you formulated for the data in Exercise 14. Linearize the model (i.e., develop a model similar to the model in Lawler (1963)). Solve the linear integer model using any IP software package.

20. Repeat Exercise 19 for the data in Exercise 18. Compare the solution you obtained with the one you got in Exercise 18.

21. ABSMODEL 2 was used to model a layout problem with departments of unequal area. Using ideas from ABSMODEL 2, develop another ABSMODEL for layout problems with department of equal area. Include appropriate department-within-building constraints. Compare it with the QAP.

22. Show that the QAP and the ABSMODEL you modeled in Exercise 21 are equivalent.

23. Formulate Exercise 14 using the ABSMODEL you developed in Exercise 21. Solve the model using LINGO or any available NLP software.

24. Formulate Exercise 15 using the ABSMODEL you developed in Exercise 21. Solve the model using LINGO or any available NLP software.

25. Formulate Exercise 18 using the ABSMODEL you developed in Exercise 21. Solve the model using LINGO or any available NLP software.

26. Can Exercise 16 be formulated using the ABSMODEL you developed in Exercise 21? If not, can you formulate an alternative model?

27. Can Exercise 17 be formulated using the ABSMODEL you developed in Exercise 21? If not, can you formulate an alternative model?

28. Assume that the distance measure in Exercises 14 and 15 is Euclidean instead of rectilinear. Formulate new models for the problem and solve using any available NLP software. Is the solution you obtained the same as the ones in Exercises 14 and 15? Explain your answer.

29. Consider Exercise 27 in Chapter 5. Formulate ABSMODEL 2 to minimize the transportation cost. Solve the model using any available NLP software.

30. Assume that the dimensions of the building in which the machines in Exercise 29 are to be housed are 50 × 6 meters. Extend the model you formulated for Exercise 29 by including the department-within-building constraints.

31. How does LMIP 1 differ from ABSMODEL 1? What is the main advantage of the LMIP 1 model? What is its main disadvantage?

32. Using ideas from LMIP 2, transform ABSMODEL 2 into a linear, mixed-integer programming model. How does this model differ from ABSMODEL 2? What is the main advantage of the model you developed? What is its main disadvantage?

33. Consider the flow and clearance matrices in Figure 5.55 that show the number of trips and clearance between each pair of five machines, respectively. Formulate a model using the objective function (10.8) and constraints (10.9) and (10.10). Solve the resulting model using any available LP software package. Can you develop a single-row layout using the x_i values provided by the LP software package. Explain your answer. (Hint: Assume the machine lengths are equal).

34. Using the data in Exercise 33, formulate a model similar to LMIP 1. Solve the model using LINGO, STORM, or any other IP software package. Is the solution you obtained different from that in Exercise 33? Why or why not?

35. Assuming the machine lengths given in the following vector and using the flow and clearance data of Exercise 33, formulate a model using objective function (10.8) and constraints (10.9) and (10.10).

$$\left[l_i \right] = \begin{bmatrix} 6 & 4 & 6 & 3 & 5 \end{bmatrix}$$

Solve the model using LINGO or any other LP software package. Is the solution usable? Why or why not?

36. For the data in Exercise 35, formulate a linear, mixed-integer model (similar to LMIP 1). Solve the model using an IP software package. Is the solution different from the one you obtained in Exercise 35. Why or why not?

37. Develop a loop layout for the data provided in Exercise 33. To do so, formulate a model similar to LMIP 2. Assume $L = 15$. Compare the solution to the one you obtained in Exercise 33.

38. Develop a loop layout for the data provided in Exercise 34. To do so, formulate a model similar to LMIP 2. Assume $L = 35$. Compare the solution to the one you obtained in Exercise 34.

39. Extend the model you constructed in Exercise 33 to ensure that one of x_{ij}^+ or x_{ij}^- is always zero using a nonlinear equation. Solve the resulting model using any available NLP software package. Compare the solution to the one you obtained in Exercise 33.

40. Repeat Exercise 39 using the data in Exercise 35. Compare the solution to the one you obtained in Exercise 35.

41. Consider the flow matrix shown in Table 5.17 that show the number of trips and clearance between each pair of five machines. Using the machine lengths in Table 5.15, formulate a model using the objective function (10.8) and constraints (10.9) and (10.10). Solve the resulting model using any available LP software package. Can you develop a single-row layout using the x_i values provided by the LP software package. Explain your answer. (Hint: Assume the machine lengths are equal).

42. Using the data in Exercise 41, formulate a model similar to LMIP 1. Solve the model using LINGO or any other IP software package. Is the solution you obtained different from that in Exercise 41? Why or why not?

43. Assuming the machine lengths are as given in the following and using the flow and clearance data of Exercise 41, formulate a model using objective function (10.8) and constraints (10.9) and (10.10).

$$\left[l_i \right] = \begin{bmatrix} 3 & 4 & 5 & 6 & 2 \end{bmatrix}$$

Solve the model using LINGO or any other LP software package. Is the solution usable? Why or why not?

44. Using the data in Exercise 43, formulate a linear, mixed-integer model (similar to LMIP 1). Solve the model using an IP software package. Is the solution different from the one you obtained in Exercise 43. Why or why not?

45. Develop a loop layout for the data provided in Exercise 41. To do so, formulate a model similar to LMIP 2. Compare the solution to the one you obtained in Exercise 41.

46. Develop a loop layout for the data provided in Exercise 43. To do so, formulate a model similar to LMIP 2. Compare the solution to the one you obtained in Exercise 43.

47. Extend the model you constructed in Exercise 41 to ensure that one of x_{ij}^+ or x_{ij}^- is always zero using a nonlinear equation. Solve the resulting model using any available NLP software package. Compare the solution to the one you obtained in Exercise 41.

48. Repeat the aforementioned question using the data in Exercise 43. Compare the solution to the one you obtained in Exercise 43.

49. Consider the data in Exercise 14. Formulate a linear mixed-integer model for this problem.

50. Solve the model you formulated in Exercise 49 using any IP software package. Compare it to the solution you obtained in Exercise 14. (Hint: For this exercise and next, introduce additional constraints to ensure that the departments fall within the boundary of the building.)

51. Repeat Exercise 49 for the data in Exercise 10.

52. Repeat Exercise 49 for the data in Exercise 15.

53. Consider Exercise 27 in Chapter 5. Assuming the lower and upper bound on the length and width of each machine is ±1 unit, formulate a BlockPlan model for the layout you generated in Exercise 29. Solve the model using LINGO or any other available LP software and develop a block plan using that solution.

11 Advanced Algorithms for the Layout Problem

11.1 INTRODUCTION

In Chapter 5, we discussed some heuristic algorithms for the layout problem. Now we present two optimal algorithms and several heuristic algorithms for the layout problem. Heuristic algorithms—especially the simulated annealing algorithm, genetic algorithm (GA), and tabu search (TS)—are powerful tools that have been applied to other problems successfully (e.g., traveling salesman problem [TSP], graph partitioning problem, scheduling problems, and others.) We also present an optimal algorithm for the cellular manufacturing system design problem and discuss a method for estimating operational performance measures of a layout.

11.2 OPTIMAL ALGORITHMS*

As discussed in Chapter 5, optimal algorithms are defined to be those that are always guaranteed to produce the best solution (or at least one best solution when there are multiple ones) for a given problem. Heuristic algorithms, or simply heuristics, provide a solution but do not guarantee it to be the best. A good heuristic usually produces the best solution for most small problems.

The definition of the two classes of algorithms raises a natural question. If optimal algorithms can produce the best solution and heuristics are not, why study the latter? Recall the discussion of NP-completeness of layout problems in Chapter 10. Because the layout models are NP-complete, optimal algorithms can produce solutions for only small-sized problems—problems with 20 or fewer departments. For a larger problem, the computer run time is so great that the algorithm is not a practical tool for solving real-world problems. Some optimal algorithms require years of computer run time! This raises a similar question. If optimal algorithms cannot be used in practice, why should we discuss them? The answer lies in the insight that optimal algorithms provide. For example, an optimal algorithm may indicate that the optimal solution has a special property or structure, which can be exploited to develop a good heuristic.

After the layout problem was formulated as a quadratic assignment problem (QAP), a substantial amount of research was done to develop optimal algorithms to solve the model. These optimal algorithms may be divided into three classes:

1. Branch and bound algorithms
2. Decomposition algorithms
3. Cutting plane algorithms

A branch-and-bound algorithm and a decomposition algorithm are described in this chapter. Note that when we discuss optimal algorithms, we do not provide sufficient details to convince the reader that the algorithm will always produce the best solution. We list the steps involved in the algorithm, and the interested reader should consult the relevant papers for proof that the algorithms are optimal.

In Chapter 10, we presented and solved a mixed-integer linear programming model using a general-purpose branch-and-bound algorithm available in many commercial codes such as LINGO.

---

* This chapter (especially the section on optimal algorithms) is intended mainly for graduate classes. Undergraduate classes may choose to omit some or this entire chapter.

(Refer to an introductory operations research textbook such as Winston (1994) to see exactly how a general-purpose branch-and-bound algorithm works. Only, an outline is presented here.) In the general-purpose branch-and-bound algorithm, the integer constraint on the integer variables is relaxed and then the relaxed model—say, LP—is solved using the well-known simplex algorithm. The resulting objective function value (OFV) is the lower bound LB^* (for a minimization problem) and is usually represented as a root node. If the resulting solution has an integer solution for the variables that are restricted to be integers, we are done. If not, integer variables x_i that are not integer in the solution are examined one at a time.

Assume that x_i has a noninteger value equal to V. Two new models (nodes) are formed: LP-1, in which the constraint that x_i is less than or equal to the largest integer less than V is added, and LP-2 in which the constraint that x_i is greater than the smallest integer greater than V is added. The solution of models LP-1 and LP-2 provides new lower bounds on the optimal OFV. (The dual-simplex method may be used to solve LP-1 and LP-2.) Again, if an integer solution is obtained, the algorithm can be terminated if the OFV of the integer solution (which provides an upper bound) is equal to the lower bound. If not, this process of branching from nodes, adding one more constraint to get another node, solving the linear programming (LP) model corresponding to the resulting nodes, checking whether the resulting solution is an integer solution, is continued until an integer solution is obtained and no better integer solution exists. The LP problem corresponding to each node in this general-purpose branch-and-bound algorithm is solved using a simplex or dual-simplex algorithm. This increases computational requirements. Moreover, the lower bound tends to be very low in the early stages of the algorithm. As a result, we may have to examine a large number of nodes before we can identify the optimal solution in this general-purpose branch-and-bound algorithm. This increases memory requirements because we have to keep track of the solutions at numerous nodes. It has been observed in many optimal algorithms, however, that the best solution is found early in the branching process, but its verification is not complete until a large number of solutions have been examined. This means that the computation time to *find* the optimal solution may not be very great, but the time for the algorithm to *terminate* is extremely long. Among the remedies suggested to overcome this problem is to prematurely terminate the branch-and-bound algorithm without verifying that the best solution obtained is indeed an optimal one (Burkard, 1984). Two strategies are presented next:

1. Terminate the enumeration of nodes after a predetermined computation time limit is exceeded.
2. Terminate the search for optimal solution gradually after a certain number of nodes have been examined.

For example, if a better solution is not found after 4 minutes of computation time, the upper bound is decreased by a certain percentage. By doing so, we reach a stage at which the lower bound is equal to the (artificial) upper bound. Because the upper bound is reduced (arbitrarily) at a faster rate, the branch-and-bound algorithm terminates faster. A creative reader may well think of other criteria to terminate the optimal algorithm.

In the next section, we present a special-purpose branch-and-bound algorithm. This algorithm is much more efficient because it examines a smaller number of solutions. Moreover, we solve a linear assignment problem (LAP) at each node using the Hungarian method, which is very fast and efficient.

11.2.1 BRANCH-AND-BOUND ALGORITHM

The branch-and-bound algorithm discussed here was developed by Gilmore (1962) and Lawler (1963) independently. It solves the QAP described in Chapter 10. To describe the algorithm, we must explain the meaning of three terms: *assignment*, *partial assignment*, and *complete assignment*. In a facility layout problem in which n departments are to be assigned to n given locations, the term *assignment* means matching a department with *a* specific location *and* vice versa. A *partial assignment* is one in which a subset of n departments is matched with an equal-sized subset of locations and vice versa. In a *complete assignment*, all the n departments are matched with n locations and vice versa. A complete assignment obtained from a partial assignment must not disturb the partial assignment but only grow from it. Thus, if departments 1, 3, and 4 in a partial assignment are in locations 2, 4, and 5, the assignment of departments 1, 2, 3, 4, and 5 to locations 2, 3, 4, 5, and 1, respectively, is a complete assignment obtained from the partial assignment, but the assignment of departments 1, 2, 3, 4, and 5 to locations 2, 3, 1, 5, and 4 is not.

We begin the optimal algorithm by computing a lower bound LB^* for the optimal OFV. Given the flow and distance matrices $[f_{ij}]$ and $[d_{ij}]$, respectively, consider the new matrix $[w_{ij}]$ in which each entry w_{ij} is equal to the dot product of two vectors $[f_i]$ and $[d_j]$: $[f_i]$ is obtained by removing the element f_{ii} in row i of matrix $[f_{ij}]$ and rearranging the remaining entries in that row in a nonincreasing order. Similarly, $[d_j]$ is obtained by removing the element d_{jj} in row j of matrix $[d_{ij}]$ and rearranging the remaining entries in that row in a nondecreasing order. The lower bound for the optimal OFV—LB^*—is given by solving a LAP with matrix $[w_{ij}]$ as the cost matrix. It is not difficult to prove that this is indeed a lower bound (i.e., a feasible solution with an OFV smaller than LB^* does not exist), but it cannot be done without introducing more mathematical jargon. To keep the number of mathematical expressions in this chapter to a minimum, we do not prove that the solution of the LAP provides a lower bound. The interested reader is referred to a number of references including Gilmore (1962) and Lawler (1963).

We now provide an intuitive justification as to why we get a lower bound on the OFV of the QAP from the calculation of element w_{ij} by sorting $[f_i]$ and $[d_j]$ in a nonincreasing and nondecreasing order, respectively, finding the dot product, and solving the LAP with $[w_{ij}]$ as the cost matrix. Note that in calculating each element w_{ij} of the $[w_{ij}]$ matrix, we allowed each department to take the best possible position. In other words, we may have allowed department i to occupy location j in determining w_{ij}, but when calculating w_{kj} we may have allowed another department k to take the same location j. Thus, the solution obtained by solving a LAP with $[w_{ij}]$ as the cost matrix may have a lower value than the actual cost of even the best feasible layout. At best, the OFV will equal the cost of the corresponding layout.[*] To understand this better, assume that the flow and distance vectors that correspond to the first row of a flow matrix $[f_{ij}]$ and distance matrix $[d_{ij}]$ are [0, 4, 3, 1] and [0, 1, 2, 3], respectively. The dot product of the two vectors (obtained by removing f_{11} and d_{11} and arranging the first vector in a nonincreasing order and second in a nondecreasing order) is equal to 13. Notice that elements in the two vectors did not have to be rearranged. If each element w_{ij} in the $[w_{ij}]$ matrix can be obtained without rearranging the elements in the corresponding vectors (i.e., if the original row vectors $[f_{ij}]$ and $[d_{ij}]$ are prearranged), the OFV of the LAP will equal the material handling cost of a feasible solution. In general, however, the flow and distance data are not always prearranged, so the lower bound obtained usually does not correspond to the cost of a feasible solution.

If the flow and distance matrices are assumed to be symmetric (i.e., the flow from department i to j is equal to the flow in the opposite direction, and the distance from location i to j is the same as that from location j to i), then only the upper triangular matrix may be considered and the OFV of the QAP can be set equal to

$$\sum_{i=1}^{n-1}\sum_{j=i+1}^{n}\sum_{k=1}^{n-1}\sum_{l=k+1}^{n} f_{ij} d_{kl} x_{ik} x_{jl} \tag{11.1}$$

Recall that x_{ik} and x_{jl} are equal to 1 if department i is in location k and department j is in location l, respectively; otherwise these values are 0. A comparison of the QAP objective function (10.21) in

---

[*] This is precisely why we say that solution of the LAP with $[w_{ij}]$ as the cost matrix yields a lower bound on the original problem.

Chapter 10 with expression (11.1) indicates that the OFV given by expression (11.1) is equal to half that of the QAP objective function. This is true only for problems with symmetric flow and distance matrices. It is thus obvious that the lower bound for a QAP in which the OFV is given by expression (11.1) is equal to $1/2LB^*$, where LB^* is calculated as mentioned earlier.

A tight lower or upper bound is one that is as close as possible to the optimal OFV. If a feasible solution is known, the upper bound UB^* can be set equal to the OFV of that solution. Thus, the better the initial solution, the tighter the upper bound. At this stage, we can employ one of the heuristics discussed in Chapter 5 and also later in this chapter and determine a tight upper bound. (Of course, if the user is unwilling to find a solution, the upper bound can be set to infinity.) If LB^* is equal to UB^*, then the initial solution is optimal and the algorithm terminates. Because this usually does not happen, however, especially as the problem size increases, let us examine the case when LB^* is not equal to UB^*.

To keep track of the various computations, it is helpful to construct a graph in which the nodes (or vertices) correspond to a set or subset of solutions and the arcs (or edges) indicate the sequence in which the algorithm moves from one set or subset to another. A graph of nodes and arcs is shown in Figure 11.1. The initial set of all solutions for which the lower bound is known (LB^*) is indicated by the topmost or root node in the figure. The calculation of LB^* may be thought of as step 1 in a three-step algorithm. (The numerical values in Figure 11.1 correspond to those we get when we solve Example 11.1.)

Next, the algorithm computes the lower bound LB_{11}^* for a partial assignment in which department 1 (without loss of generality) is assigned to a location—say location 1. Step 2 of the branch-and-bound algorithm is based on the following simple observation. Given a partial assignment in which a certain subset $S = \{1, 2, ..., q\}$ of n departments is assigned to a subset $L = \{s_1, s_2, ..., s_q\}$ of n locations, it is obvious that the optimal objective function for a complete assignment is equal to the sum of the products of flow and distance computed for these three categories of departments:

1. Pairs of departments i, j such that $i, j \in S$
2. Pairs of departments i, j such that $i \in S, j \notin S$
3. Pairs of departments i, j such that $i, j \notin S$

Category 1 is a constant for a given partial assignment. We let k^* be the cost of the partial assignment, which is equal to the sum of the product of flow and distance computed only for those departments in S and locations in L.

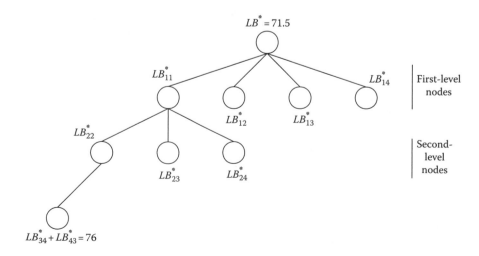

FIGURE 11.1 Illustration of the branch-and-bound algorithm.

The algorithm solves a LAP with a cost matrix to determine a lower bound with respect to categories (2) and (3). Solving the LAP for a given cost matrix is easy with an efficient optimal algorithm called the Hungarian method, which is available in most introductory operations research textbooks (e.g., Winston, 1994). What is difficult to explain, if not understand, however, is the computation of the cost matrix. The optimal algorithm forms vectors $[f_i]$ and $[d_j]$ exactly as mentioned earlier (i.e., removing entry f_{ii} from row i of the flow matrix $[f_{ij}]$ and rearranging the remaining entries in that row in a nonincreasing order and, similarly, removing entry d_{ii} from row i of the distance matrix $[d_{ij}]$ and rearranging the remaining entries in that row in a nondecreasing order). This is done only for those departments and locations that are *not* in S and L, respectively. As mentioned earlier, S and L indicate a partial assignment in which each department in S is assigned to a specific location in L and vice versa. Consider the matrix $[w'_{ij}]$ in which each element w'_{ij} is obtained using the following expression:

$$\sum_{k=1}^{q} f_{ik} d_{js_k} \quad \text{where } k \in S, i \notin S, s_k \in L, \text{ and } j \notin L \qquad (11.2)$$

Because i and j are not in S and L, respectively, it is obvious that the matrix $[w'_{ij}]$ has $n - q$ rows and $n - q$ columns. Now consider matrix $[w''_{ij}]$ in which each element w''_{ij} is equal to half the dot product of f_i and d_j (assuming symmetric flow and distance matrices), only for those i and j that are not in S and L, respectively. It is not difficult to see that this matrix also has $n - q$ rows and $n - q$ columns. If the elements w'_{ij} and w''_{ij} are added and placed in a new matrix $[w_{ij}]$ and a LAP solved with this new matrix as the cost matrix, then the resulting solution provides a lower bound for categories (2) and (3) mentioned earlier. Furthermore, if k^* is added to the OFV of the solution corresponding to the LAP, this provides a lower bound LB_{11}^* for all solutions in which department 1 is assigned to location 1. (To prove that this is a lower bound, see Francis and White (1974) and Gilmore (1962). If you can prove LB^* is a lower bound for all solutions, try proving that LB_{11}^* is a lower bound for all solutions in which department 1 is assigned to location 1.) Similarly, we can calculate the lower bounds $LB_{12}^*, LB_{13}^*, \ldots, LB_{1n}^*$, which are the lower bounds when department 1 is assigned to location 2, location 3, ..., location n, respectively. Each subset of solutions is represented as a node in Figure 11.1 and shown as the first level of nodes. In Figure 11.1, n is equal to 4. Again, if a "good" initial solution is known, the upper bound UB^* can be set equal to its OFV; if some LB_{1j}^* is greater than this UB^*, the tree can be "pruned or fathomed" from the node corresponding to that LB_{1j}^*. In other words, any complete assignment that is obtained from a partial assignment in which department 1 is assigned to location j need not be examined. All possible solutions are implicitly—but not explicitly—evaluated. This type of enumeration is referred to as controlled enumeration (Pierce and Crowston, 1971). If no bounds were considered for pruning the decision tree, then the procedure would have led to a complete enumeration algorithm that is computationally inefficient.

To facilitate the discussion of the branch-and-bound algorithm, we present two more definitions—*active* and *passive* nodes. A passive node is either a node from which the algorithm has branched to another or a node that has a lower bound greater than the upper bound. A node that does not fit this definition is an active node. At this stage, the root node is a passive one because the algorithm has branched into four first-level nodes in Figure 11.1.

The new lower bound LB^* is equal to the minimum of $\left\{ LB_{12}^*, LB_{13}^*, \ldots, LB_{1n}^* \right\}$. It should be evident from the discussion so far that the original LB^* can never be greater than any LB_{1j}^*. To obtain a good upper bound, we examine the subset of solutions (partial assignment) that has the least lower bound. A complete solution (or complete assignment) from this subset may then be grown. If the new lower bound equals the upper bound, the algorithm terminates. If not, additional work is necessary. The explanation provided so far covers step 2 of the algorithm.

In step 3, a partial assignment is constructed from each first level node in the tree in Figure 11.1. A lower bound is obtained as described earlier, and this provides the second level nodes. As soon as a second-level node is obtained, the first-level node from which branching took place can be made passive. The procedure for termination is the same as mentioned earlier. If termination is not possible, the algorithm proceeds far enough until a complete assignment is available. If the OFV of this assignment is less than or equal to the lower bound of all active nodes, then the algorithm terminates; an optimal solution has been found. If not, the new upper bound is set equal to the OFV of the best solution known so far, and those partial assignments (active nodes) with a smaller lower bound than the new upper bound are examined. This procedure is repeated until no active node exists; that is, no partial assignment has a lower bound less than that of the upper bound. When we have a partial assignment with $n-1$ departments assigned, the remaining department can be assigned to only one location, so the lower bound can be calculated by finding the OFV of the complete assignment. When the partial assignment has $n-2$ departments assigned, the remaining two departments i and j may be assigned in one of two ways to the remaining two locations k and l (i.e., i may be assigned to k and j to l or i may be assigned to l and j to k). For these two possibilities, it is easy to compute the sum of the product of flow and distance for category (2) mentioned earlier and choose the minimum of the two sums. Category (3) does not change because there are only two ways in which departments i and j can be assigned to locations k and l; the product of flow and distance computed for departments i and j is the same for the two assignments. The three steps of the branch-and-bound algorithm are presented in the following.

Specialized branch-and-bound algorithm for the QAP

Step 0: Calculate the OFV of a feasible solution in which department 1 is assigned to location 1, department 2 is assigned to location 2, and so on. Set UB = OFV of the feasible solution.

Step 1: Compute the lower bound (LB^*) of the root node by solving an LAP with a matrix $[w_{ij}]$. Matrix $[w_{ij}]$ is obtained by taking dot product of two vectors $[f_i]$ and $[d_i]$. Vectors $[f_i]$ and $[d_i]$ are obtained by removing f_{ii} and d_{ij} from each vector and arranging the remaining flow and distance values in nonincreasing and nondecreasing order, respectively. Set $p = 1$. Form n nodes with each node corresponding to the assignment of department p to location 1, 2, ..., n.

Step 2: Compute the lower bound for these nodes by solving a LAP with matrix $[w_{ij}]$ in which each element w_{ij} is obtained by adding w'_{ij} and w''_{ij}, which are obtained as specified next:

   a. Element w'_{ij} is obtained using expression (11.2).
   b. Element w''_{ij} is equal to half the dot product of vectors $[f_i]$ and $[d_j]$, set up for those departments i that have not yet been assigned, and locations j that are still available. Vectors $[f_i]$ and $[d_j]$ are obtained by arranging flow and distance values in nonincreasing and nondecreasing order, respectively.

Step 3: If the LB of a node calculated in step 2 is greater than UB, prune that node. If a node corresponds to a feasible solution, and OFV of that feasible solution is less than UB, set UB = OFV of the feasible solution. If there are one or more nodes with LB < UB, select one of them arbitrarily, form $n - p$ nodes with each node corresponding to the assignment of department $p + 1$ to each remaining, unassigned and open location, set $p = p + 1$ and go to step 2. Otherwise, *stop*. Return the best solution so far as the optimal solution.

Example 11.1

Solve the QAP formulation of the LonBank layout problem described in Example 10.3, using the specialized branch-and-bound algorithm. The flow and distance matrices for this problem are provided in Figure 11.2.

Solution

The first step of the branch-and-bound algorithm is to find a lower bound for the OFV by setting up matrix $[w_{ij}]$. For example, to find w_{11}, first arrange $[f_1]$ and $[d_1]$, which are row 1 of the flow

Office

	1	2	3	4
1	—	17	12	11
2	17	—	12	4
3	12	12	—	4
4	11	4	4	—

$[f_{ij}] =$ Office

Site

	1	2	3	4
1	—	1	1	2
2	1	—	2	1
3	1	2	—	1
4	2	1	1	—

$[d_{ij}] =$ Site

FIGURE 11.2 Flow and distance matrices for the LonBank layout problem.

$[w_{ij}] =$

	1	2	3	4
1	51	51	51	51
2	37	37	37	37
3	32	32	32	32
4	23	23	23	23

FIGURE 11.3 Calculation of w_{ij} values.

and distance matrices (without f_{11} and d_{11}) arranged in a nonincreasing and nondecreasing order, respectively. The dot product of $[f_1]$ and $[d_1]$ is then equal to w_{11}. The vectors $[f_i]$ and $[d_j]$ are given here; the matrix $[w_{ij}]$ is shown in Figure 11.3.

$$\begin{bmatrix} f_1 \end{bmatrix} = \begin{bmatrix} 17 & 12 & 11 \end{bmatrix}; \quad \begin{bmatrix} f_2 \end{bmatrix} = \begin{bmatrix} 17 & 12 & 4 \end{bmatrix}; \quad \begin{bmatrix} f_3 \end{bmatrix} = \begin{bmatrix} 12 & 12 & 4 \end{bmatrix}; \quad \begin{bmatrix} f_4 \end{bmatrix} = \begin{bmatrix} 11 & 4 & 4 \end{bmatrix}$$

$$\begin{bmatrix} d_1 \end{bmatrix} = \begin{bmatrix} 1 & 1 & 2 \end{bmatrix}; \quad \begin{bmatrix} d_2 \end{bmatrix} = \begin{bmatrix} 1 & 1 & 2 \end{bmatrix}; \quad \begin{bmatrix} d_3 \end{bmatrix} = \begin{bmatrix} 1 & 1 & 2 \end{bmatrix}; \quad \begin{bmatrix} d_4 \end{bmatrix} = \begin{bmatrix} 1 & 1 & 2 \end{bmatrix}$$

Note that only rows 2, 3, and 4 of the distance matrix had to be rearranged. Also, the row values of $[w_{ij}]$ are equal because the vectors $[d_1]$ through $[d_4]$ are identical. Solving the assignment problem for the distance matrix provides a lower bound of 143, so $LB^*/2$ is equal to 71.5. This is indicated above the root node in Figure 11.1. The upper bound, on the other hand, may be set to infinity or set equal to the OFV of the solution in which departments 1, 2, 3, and 4 are assigned to locations 1, 2, 3, and 4, respectively. Choosing the latter option yields a tighter upper bound of 83. Now consider the partial assignment in which department 1 is assigned to location 1. To calculate the lower bound, examine the sum of products of flow and distance for the three categories mentioned earlier. With respect to category (1), the cost of the partial assignment is zero because there is only one department in the partial assignment.

To calculate the cost with respect to category (2), set up the matrix $[w'_{ij}]$ using expression (11.2). This matrix is shown in Figure 11.4 along with the flow and distance matrices that correspond to the departments and locations *not* in the partial assignment, the f and d vectors, and the $[w''_{ij}]$ matrix. They are calculated as described earlier.

From the two matrices $[w'_{ij}]$ and $[w''_{ij}]$, it is easy to compute the $[w_{ij}]$ matrix, also shown in Figure 11.4. Solving the LAP with $[w_{ij}]$ as the cost matrix, we get the solution in which departments 1, 2, 3, and 4 are assigned to locations 1, 2, 3, and 4, respectively. The OFV of the solution to the LAP and hence the lower bound is equal to 75. Similarly, the other partial assignments (in which department 1 is assigned to locations 2, 3, and 4, respectively) all have a lower bound of 75. Thus, the new lower bound is now 75. Because the lower bound is not equal to the upper bound, additional work is necessary. The root node is made passive, and we proceed along the leftmost first level node in Figure 11.1.

$$[w'_{ij}] = \begin{array}{c} \\ 2 \\ 3 \\ 4 \end{array} \begin{array}{ccc} 2 & 3 & 4 \\ 17 & 17 & 34 \\ 12 & 12 & 24 \\ 11 & 11 & 22 \end{array} \qquad [f_{ij}] = \begin{array}{c} \\ 2 \\ 3 \\ 4 \end{array} \begin{array}{ccc} 2 & 3 & 4 \\ - & 12 & 4 \\ 12 & - & 4 \\ 4 & 4 & - \end{array} \qquad [d_{ij}] = \begin{array}{c} \\ 2 \\ 3 \\ 4 \end{array} \begin{array}{ccc} 2 & 3 & 4 \\ - & 2 & 1 \\ 2 & - & 1 \\ 1 & 1 & - \end{array}$$

$$[f_2] = [12\ 4] \quad [f_3] = [12\ 4] \quad [f_4] = [4\ 4]$$
$$[d_2] = [1\ 2] \quad [d_3] = [1\ 2] \quad [d_4] = [1\ 1]$$

$$[w''_{ij}] = \begin{array}{c} \\ 2 \\ 3 \\ 4 \end{array} \begin{array}{ccc} 2 & 3 & 4 \\ 10 & 10 & 8 \\ 10 & 10 & 8 \\ 6 & 6 & 4 \end{array} \qquad [w_{ij}] = \begin{array}{c} \\ 2 \\ 3 \\ 4 \end{array} \begin{array}{ccc} 2 & 3 & 4 \\ 27 & 27 & 42 \\ 22 & 22 & 32 \\ 17 & 17 & 26 \end{array}$$

FIGURE 11.4   Calculation of $[w_{ij}]$, $[w'_{ij}]$, and $[w''_{ij}]$ matrices assuming department i is assigned to location.

$$[w'_{ij}] = \begin{array}{c} \\ 3 \\ 4 \end{array} \begin{array}{cc} 3 & 4 \\ 36 & 36 \\ 19 & 26 \end{array} \qquad [f_{ij}] = \begin{array}{c} \\ 3 \\ 4 \end{array} \begin{array}{cc} 3 & 4 \\ - & 4 \\ 4 & - \end{array} \qquad [d_{ij}] = \begin{array}{c} \\ 3 \\ 4 \end{array} \begin{array}{cc} 3 & 4 \\ - & 1 \\ 1 & - \end{array}$$

$$[f_3] = [4] \quad [f_4] = [4] \quad [d_3] = [1] \quad [d_4] = [1]$$

$$[w''_{ij}] = \begin{array}{c} \\ 3 \\ 4 \end{array} \begin{array}{cc} 3 & 4 \\ 2 & 2 \\ 2 & 2 \end{array} \qquad [w_{ij}] = \begin{array}{c} \\ 3 \\ 4 \end{array} \begin{array}{cc} 3 & 4 \\ 38 & 38 \\ 21 & 28 \end{array}$$

FIGURE 11.5   Calculation of the $[w_{ij}]$ matrix assuming departments 1 and 2 are assigned to locations 1 and 2, respectively.

Consider the partial assignment in which departments 1 and 2 are assigned to locations 1 and 2. The cost of this partial assignment k^* is equal to 17. Notice that only two departments are to be assigned and only two locations remain. There are only two ways to assign these departments—to locations 3 and 4 or to locations 4 and 3. To calculate the cost with respect to category (2), we set up the matrix $[w'_{ij}]$ as before using expression (11.2). Similarly, the matrix $[w''_{ij}]$ can also be computed using the smaller flow and distance matrices. These are shown in Figure 11.5.

Solving the LAP with $[w_{ij}]$ as the cost matrix, we get a solution with an OFV of 59, in which departments 1, 2, 3, and 4 are assigned to locations 1, 2, 4, and 3, respectively. Thus, the new bound is equal to 17 + 59 = 76. It turns out that the actual cost of this assignment is also equal to 76. Thus, the new upper bound may be set to 76. If we proceed downward one more level along the other nodes that have a lower bound of 75, we obtain new bounds that are greater than the current bound of 76. Thus, the lower bound is also 76. Because the upper and lower bounds are equal, the branch-and-bound algorithm can be terminated, and an optimal solution has been found.

In addition to the optimal algorithm described here, Land (1963) and Gavett and Plyter (1966) developed two others. Instead of assigning one department to one location as discussed earlier, these algorithms assign pairs of departments to pairs of locations. The lower bound is also computed differently. The special-purpose optimal algorithm and others proceed on the basis of stage-by-stage assignment (or inclusion) of departments to locations. At each stage, backtracking may occur, certain partial or complete assignments (nodes) may be excluded, and the forward search process continues. Another algorithm that proceeds on the basis of stage-by-stage exclusion of pairs of assignments from a solution is discussed in Pierce and Crowston (1971).

In Chapter 5, we discussed hybrid algorithms. An algorithm that is a combination of two or more types of algorithms, e.g., construction and improvement, is considered a hybrid algorithm. An algorithm that has the characteristics of optimal and heuristic algorithms can also be considered a hybrid algorithm. This type of hybrid algorithm is essentially an optimal layout algorithm that is terminated before verifying that the best solution obtained is optimal. Some optimal algorithms have been modified to stop the search after a preset computer run time has been reached, and then the best solution obtained is improved further using an improvement algorithm. Such algorithms can be found in Burkard and Stratman (1978), Bazaraa and Sherali (1980), and Bazaraa and Kirca (1983).

11.2.2 Benders' Decomposition Algorithm

In this section, we discuss application of a specialized algorithm to solve one of the mixed-integer programming models (LMIP 1) from Chapter 10. Called Benders' decomposition algorithm, it works by separating the original mixed-integer program into a pure integer program and a pure linear program (Benders, 1962). Each is then solved separately in an iterative fashion until an optimal solution is reached. To understand how Benders' decomposition algorithm works, consider the following mixed-integer programming (MIP) model.

MIP

$$\text{Minimize } \mathbf{cx}$$

$$\text{Subject to } \mathbf{Ax} + \mathbf{By} \geq \mathbf{b}$$

$$\mathbf{x} \geq 0$$

$$\mathbf{y} = 0 \text{ or } 1$$

Now, consider a feasible *y* solution vector to MIP—say y^i. Then, MIP becomes the following linear programming (LP) model.

LP i

$$\text{Minimize } \mathbf{cx}$$

$$\text{Subject to } \mathbf{Ax} \geq \mathbf{b} - \mathbf{By}^i$$

$$\mathbf{x} \geq 0$$

The dual of LP *i* is the following dual linear programming (DLP) model. (The reader not familiar with dual LPs, duality theory, and complementary slackness property should consult an introductory operations research textbook before proceeding further.)

DLP i

$$\text{Maximize } \mathbf{u}(\mathbf{b} - \mathbf{By}^i)$$

$$\text{Subject to } \mathbf{uA} \leq \mathbf{c}$$

$$\mathbf{u} \geq 0$$

Let $\mathbf{u}^i$ be the optimal solution to DLP i. From duality theory, we know that $\mathbf{u}^i(\mathbf{b} - \mathbf{B}\mathbf{y}^i)$ is equal to the optimal OFV of LPi (because LP i and DLP i are both feasible). Hence, $\mathbf{u}^i(\mathbf{b} - \mathbf{B}\mathbf{y}^i)$ is equal to the OFV of some feasible solution to MIP (the one in which $\mathbf{y} = \mathbf{y}^i$). Because each variable y_{ij} in the vector $\mathbf{y}$ can take on a value of 0 or 1 only and because the number of such variables is finite, it is clear that the number of $\mathbf{y}$ vectors is also finite. In fact, if there are n y_{ij} variables, then the number of $\mathbf{y}$ vectors is equal to 2^n. Of course, not all of these may be feasible to the original problem. So, we assume that there are s feasible $\mathbf{y}$ solution vectors to MIP—$\{\mathbf{y}^1, \mathbf{y}^2, ..., \mathbf{y}^i, ..., \mathbf{y}^s\}$, arranged in any order.

Let DLP 1, DLP 2, ..., DLP i, ..., DLP s be the duals obtained by substituting $\mathbf{y}^1, \mathbf{y}^2, ..., \mathbf{y}^i, ...,$ $\mathbf{y}^s$ for $\mathbf{y}^i$ in DLP i. Let $\mathbf{u}^1, \mathbf{u}^2, ..., \mathbf{u}^i, ..., \mathbf{u}^s$ be the optimal solution vectors to DLP 1, DLP 2, ..., DLP i, ..., DLP s, respectively. The optimal OFV of each corresponds to the OFV of some feasible $\mathbf{y}$ solution vector to MIP. Because we have considered all feasible solution vectors, the dual with the least OFV among DLP 1, DLP 2, ..., DLP s provides the optimal OFV to MIP. Thus, the original problem MIP may be reduced to the following problem:

$$\text{Minimize } \mathbf{u}^i(\mathbf{b} - \mathbf{B}\mathbf{y})\} \; 1 \leq i \leq s$$

$$\text{Subject to } \mathbf{y} = 0 \text{ or } 1 \text{ and feasible to MIP}$$

This model can be restated as

$$\text{Minimize } z$$

$$\text{Subject to } z \geq \mathbf{u}^i(\mathbf{b} - \mathbf{B}\mathbf{y}) \quad i = 1, 2, ..., s$$

$$\mathbf{y}^s = 0 \text{ or } 1 \text{ and feasible to MIP}$$

Because z is a continuous variable, this problem is again a mixed-integer programming problem with one continuous variable and the remaining binary. If an integer upper bound on the value of the objective function z is known, however, then z can be written entirely in terms of new binary integer variables. For example, if it is known that an integer upper bound for z is UB, it can be written as $z = 2^0 z_0 + 2^1 z_1 + 2^2 z_2 + 2^3 z_3 + \cdots + 2^k z_k$, where k is the smallest integer satisfying $2^{k+1} - 1 > UB$. Thus, the model can be rewritten by substituting a vector $\mathbf{z}$ of binary variables for the continuous variable z. The revised model, called MP, is given here:

MP

$$\text{Minimize } \mathbf{z}$$

$$\text{Subject to } \mathbf{z} \geq \mathbf{u}^i(\mathbf{b} - \mathbf{B}\mathbf{y}) \quad i = 1, 2, ..., s$$

$$\mathbf{y} = 0 \text{ or } 1 \text{ and feasible to MIP}$$

$$\mathbf{z} = 0 \text{ or } 1$$

MP requires us to generate all the feasible $\mathbf{y}$ solution vectors and the corresponding s dual problems—DLP 1, DLP 2, ..., DLP s. This may not be computationally feasible because the number of dual problems, though finite, may be very large, as is true for the mixed-integer programming models in Chapter 10. For example, a problem with six departments has $6(5)/2 = 15$, y_{ij} binary integer variables, and hence, the number of possible $\mathbf{y}$ solution vectors we can have is $2^{15} = 32$, 7611. The dual associated with each of these has to be solved—a time-consuming task. When the number of departments is increased to 10, the increase in computational requirement is exponential. Because most of the constraints in MP are nonbinding in the optimal solution to it, however, we can overcome the computational problem by generating a subset of the constraints in MP and solving a restricted problem. In fact, this is how Benders' decomposition algorithm works. Because we are solving MP with only a small subset of constraints, its optimal solution will provide a lower bound on MIP. Thus, beginning with few or no constraints, we solve MP, obtain a new $\mathbf{y}$ vector, set up DLP i corresponding to this $\mathbf{y}$ vector, and obtain an upper bound. (Notice that the OFV of a feasible solution to MIP, provided by DLP i, yields an upper bound.) Using the optimal solution to DLP i, we add the corresponding constraint $[\mathbf{z} \geq \mathbf{u}^i(\mathbf{b} - \mathbf{By})]$ in the master problem MP and solve it. If the resulting lower bound is greater than or equal to the upper bound, we stop because the last solution to MP provides the optimal solution to MIP. Otherwise, we repeat the procedure until the termination criterion is met.

Here is a formal description of Benders' decomposition algorithm:

Benders' decomposition algorithm:

Step 0: Set $i = 1$, $\mathbf{y}^i = \{0, 0, \ldots, 0\}$, lower bound $LB = 0$ and upper bound $UB = \infty$.

Step 1: Solve DLP i. Let $\mathbf{u}^i$ be the optimal solution to DLP i. If $\mathbf{u}^i(\mathbf{b} - \mathbf{By}^i) < UB$, set $UB = \mathbf{u}^i(\mathbf{b} - \mathbf{By}^i)$.

Step 2: Update MP by adding the constraint $\mathbf{z} \geq \mathbf{u}^i(\mathbf{b} - \mathbf{By})$. Solve MP. Let $\mathbf{y}^*$ be the optimal solution and z be the optimal OFV of MP. Set $LB = z$. If $LB > UB$, stop. Otherwise, set $i = I + 1$, $\mathbf{y}^i = \mathbf{y}^*$ and return to step 1.

A more efficient, but less accurate version of the Benders' decomposition algorithm exists and is presented in Section 11.2.2.1 and also in Chapter 12. The next example shows how this algorithm can be applied to solve the single-row layout model LMIP 1 in Chapter 10.

Example 11.2

Consider Example 10.2. Assume that the furniture manufacturer has found a new machine that can do the same operations as machines 4 and 5. Although this machine costs more, the manufacturer feels that the savings in space and other factors are worth the extra cost and hence decides to purchase the new machine. The cost to move a unit load through unit distance is $1.00. The machine dimensions are given in Figure 11.6 along with the horizontal clearance required between each pair of machines and the revised frequency of trips estimated to be made per year for the foreseeable future. The machines are to be arranged so that their longer sides are parallel to the aisle. Formulate a linear programming model similar to LMIP 1 and solve it using Benders' decomposition algorithm.

Solution

LMIP 1 for this example can be set up as shown in the following computer printout. Variables x_i, z_{ij}, x_{ij}^+, and x_{ij}^- in model LMIP 1 are provided as Xi, Yij, XPij, and XNij, respectively.

Machine	Dimension	Horizontal clearance matrix					Flow matrix			
			1	2	3	4		1	2	3	4
1	25×20	1	—	3.5	5.0	5.0	1	—	25	35	50
2	35×20	2	3.5	—	5.0	3.0	2	25	—	10	15
3	30×30	3	5.0	5.0	—	5.0	3	35	10	—	50
4	40×20	4	5.0	3.0	5.0	—	4	50	15	50	—

FIGURE 11.6 Flow and clearance matrices and dimensions for four machines.

MIP

```
Data:
    N=4;
    M=999;
Enddata
Sets:
    Dept/1..N/: Length, Width, X;
    ObjFn(Dept,Dept)| &2 #GT# &1: Flow, HClear, XP, XN, Y;
Endsets

Data:
    Length=25 35 30 40;
    Width =20 20 20 20;
    HClear=3.5 5 5
            5 3
            5;
    Flow  =25 35 50
            10 15
            50;
Enddata
    ! Objective function;
    Min= @SUM(ObjFn(i,j)|j #GT# i: Flow(i,j)*(XP(i,j)+XN(i,j)));
    ! Constraints;
    @FOR (Dept(i)|i #LT# N:
        @FOR (Dept(j)|j #GT# i #AND# j #LE# N:
            X(i)-X(j)+M*Y(i,j)>= 0.5*(Length(i)+Length(j))+HClear
              (i,j);
            X(j)-X(i)+M*(1-Y(i,j))>= 0.5*(Length(i)+Length(j))
              +HClear(i,j);
            XP(i,j)-XN(i,j)=X(i)-X(j);
              @BIN(Y(i,j))
));
```

We now show how Benders' decomposition algorithm can be applied to solve the model.

Step 0: Set $i = 1$, $\mathbf{y}^i = \{0, 0, ..., 0\}$, lower bound $LB = 0$ and upper bound $UB = \infty$.
Step 1: Solve DLP i. The models LP 1, DLP 1 and their solutions are shown next.

LP 1

```
Data:
    N=4;
    M=999;
Enddata
Sets:
    Dept/1..N/: Length, Width, X;
    ObjFn(Dept,Dept)| &2 #GT# &1: Flow, HClear, XP, XN, Y;
Endsets

Data:
    Length = 25 35 30 40;
    Width  = 20 20 20 20;
    HClear = 3.5 5 5
              5 3
              5;
    Flow   = 25 35 50
              10 15
              50;
    Y      = 0 0 0
              0 0
              0;
Enddata
    ! Objective function;
```

```
        Min= @SUM(ObjFn(i,j)|j #GT# i: Flow(i,j)*(XP(i,j)+XN(i,j)));
        ! Constraints;
        @FOR (Dept(i)|i #LT# N:
            @FOR (Dept(j)|j #GT# i #AND# j #LE# N:
                X(i)-X(j)+M*Y(i,j)>= 0.5*(Length(i)+Length(j))+HClear
                 (i,j);
                X(j)-X(i)+M*(1-Y(i,j))>= 0.5*(Length(i)+Length(j))
                 +HClear(i,j);
                XP(i,j)-XN(i,j)=X(i)-X(j);
                @BIN(Y(i,j))
));
```

Solution to LP 1: Only variables with nonzero values are shown.

```
Global optimal solution found.
Objective value:                            12410.00
Total solver iterations:                         0

                    Variable           Value
                        X( 1)          111.0000
                        X( 2)          77.50000
                        X( 3)          40.00000
                        X( 4)          0.000000
                    XP( 1, 2)          33.50000
                    XP( 1, 3)          71.00000
                    XP( 1, 4)          111.0000
                    XP( 2, 3)          37.50000
                    XP( 2, 4)          77.50000
                    XP( 3, 4)          40.00000
```

DLP 1

```
Data:
    N=4;
    M=999;
Enddata
Sets:
    Dept/1..N/: Length, Width, X;
    Const(Dept,Dept)| &2 #NE# 1: WP, WN, U, V;
    ObjFn(Dept,Dept)| &2 #GT# &1: Flow, HClear, XP, XN, Y;
Endsets

Data:
    Length = 25 35 30 40;
    Width  = 20 20 20 20;
    HClear = 3.5 5 5
             5 3
             5;
    Flow   = 25 35 50
             10 15
             50;
    Y      = 0 0 0
             0 0
             0;
Enddata
    ! Objective function;
    Max = @SUM(ObjFn(i,j)|j #GT# i: (0.5*(Length(i)+Length(j))+HClear(i,j)
          -M*Y(i,j))*U(i,j)+
          (0.5*(Length(i)+Length(j))+HClear(i,j) -M*(1-Y(i,j)))*V(i,j));
    ! Constraints;
    @FOR (Dept(i)|i #LT# N:
        @FOR (Dept(j)|j #GT# i #AND# j #LE# N:
            WN(i,j)-WP(i,j)<= Flow(i,j);
            WP(i,j)-WN(i,j)<= Flow(i,j);
));
```

TABLE 11.1

Feasible Combinations of y Variables for a Triplet $\{i, j, k\}$

y_{ij}	y_{ik}	y_{jk}	Feasible?
0	0	0	Yes
0	0	1	Yes
0	1	1	Yes
1	0	0	Yes
1	1	0	Yes
1	1	1	Yes
1	0	1	No
0	1	0	No

```
@FOR (Dept(i)|i #LE# N:
    @SUM(Const(i,j)|j #NE# i #AND# j #GT# i:
      U(i,j)-V(i,j)+WP(i,j)-WN(i,j))+
@SUM(Const(j,i)|j #NE# i #AND# j #LT# i:
  -U(j,i)+V(j,i)-WP(j,i)+WN(j,i))<=0);
```

Solution to DLP 1: Only variables with nonzero values are shown.

```
Global optimal solution found.
Objective value:                          12410.00
Total solver iterations:                         7

            Variable          Value
            WN( 1, 2)       25.00000
            WN( 1, 3)       35.00000
            WN( 1, 4)       50.00000
            WN( 2, 3)       10.00000
            WN( 2, 4)       15.00000
            WN( 3, 4)       50.00000
             U( 1, 2)      110.0000
             U( 2, 3)      110.0000
             U( 3, 4)      115.0000
```

The OFV (12, 410) of LP 1 and its dual DLP 1 provide the OFV of a feasible solution and so the upper bound to the original problem is updated to 12,410. Due to the equality constraints in LP 1, the corresponding dual variables $W(i,j)$ are unrestricted in sign (urs). To provide these urs variables as the difference of two nonnegative variables, we introduce the variables $WP(i,j)$ and $WN(i,j)$ in the DLP 1 model.

Using the solution to DLP 1, we can construct the master problem MP 1. Notice that the first six constraints in MP 1 ensure that the values of the y_{ij} variables are feasible to MIP. To understand this, first it should be observed that the y variables corresponding to any triplet $\{i, j, k\}$ indicate whether or not the corresponding solution is feasible. Because $y_{ij} = 1(0)$ means that workstation i is to the left (right) of j, it is obvious that among the eight possible combinations shown in Table 11.1, the last two lead to infeasible combinations. We can avoid generating these two infeasible combinations by adding constraint (11.3) to MP, which does not prevent the six feasible combinations from being generated.

$$1 \geq y_{ij} + y_{jk} - y_{ik} \geq 0 \quad \forall \ i < n-1, \ i < j < n, \ j < k \leq n \qquad (11.3)$$

The seventh constraint in MP 1 is of the form $z \geq u^i(b - By)$.

MP 1

```
Data:
     N=4;
     M=999;
Enddata
Sets:
     Dept/1..N/: Length;
     Zees/1..15/: C,Z;
     ObjFn(Dept,Dept)| &2 #GT# &1: Flow, HClear, XP, XN, Y, U1, V1, U2, V2;
Endsets

Data:
     Length = 25 35 30 40;
     HClear = 3.5 5 5
                5 3
                5;
     C = 1 2 4 8 18 32 64 128 256 512 1024 2048 4096 8192 16384;
     U1 =110 0 0
          110 0
          115;
     V1 =0 0 0
         0 0
         0;
     Flow = 25 35 50
             10 15
             50;
Enddata
     ! Objective function;
     Min= @SUM(Zees(i): C(i)*Z(i));
     ! Constraints;
     Y(1,2)+Y(2,3)-Y(1,3) >= 0;
     Y(1,2)+Y(2,3)-Y(1,3) <= 1;
     Y(1,2)+Y(2,4)-Y(1,4) >= 0;
     Y(1,2)+Y(2,4)-Y(1,4) <= 1;
     Y(2,3)+Y(3,4)-Y(2,4) >= 0;
     Y(2,3)+Y(3,4)-Y(2,4) <= 1;
     @SUM(Zees(k): C(k)*Z(k)) >=
     @SUM(ObjFn(i,j)| j #GT# i: U1(i,j)*(0.5*(Length(i)+Length(j)) +
HClear(i,j)) + V1(i,j)*(-M+(0.5*(Length(i)+Length(j))+HClear (i,j))) -
M*U1(i,j)*Y(i,j) + M*V1(i,j)*Y(i,j));
     @FOR (ObjFn(i,j)| j #GT# i: @bin(Y(i,j)));
```

Solution to MP 1: Only variables with nonzero values are shown.

```
Global optimal solution found.
Objective value:                                    0.000000
Extended solver steps:                                     0
Total solver iterations:                                   0

                     Variable          Value
                     Y( 1, 2)       1.000000
                     Y( 3, 4)       1.000000
```

Step 2: The optimal solution y^* to MP 1 provides us a lower bound of 0, so more iterations are necessary. Because $LB < UB$, set $i = 1 + 1 = 2$, $y^2 = y^*$ and return to step 1. From the preceding set of y_{ij} values, it is easy to formulate the dual DLP 2 as shown next:
Step 1: Solve DLP 2.

FIGURE 11.7   Optimal machine layout in the new facility of the furniture manufacturing company.

DLP 2

```
Global optimal solution found.
Objective value:                          9600.000
Total solver iterations:                          6

                 Variable            Value
               WP( 1, 2)        25.00000
               WP( 3, 4)        50.00000
               WN( 1, 3)        35.00000
               WN( 1, 4)        50.00000
               WN( 2, 3)        10.00000
               WN( 2, 4)        15.00000
                U( 1, 4)        110.0000
                V( 1, 2)        50.00000
                V( 3, 4)        95.00000
```

The solution with a value of 9600 is now the new upper bound because it is lower than the previous value of 12,410. The master problem is updated with the new constraint provided by DLP 2, and MP 2 is shown next.

Step 2: Using the solution to DLP 2, we can update MP as shown in the following text. The optimal solution y^* to MP provides us a lower bound of 0, so more iterations are necessary. Because $LB < UB$, set $i = 2 + 1 = 3$, $y^3 = y^*$ and return to step 1.

   Solution to MP 2: Only variables with nonzero values are shown.

```
Global optimal solution found.
Objective value:                          0.000000
Extended solver steps:                            0
Total solver iterations:                          0

                 Variable            Value
                Y( 1, 2)        1.000000
```

Repeating the procedure of (1) solving the MP, (2) updating the lower bound *LB*, (3) obtaining the new **y** vector, (4) formulating the corresponding the dual, (5) updating the upper bound *UB*, if necessary, and (6) updating the master problem MP until the lower bound equals or exceeds the upper bound yields the models and solutions shown in http://sundere.okstate.edu/Book. It is obvious from the last solution that the lower and upper bounds are equal to 9600, so the MIP's optimal OFV must be 9600. The corresponding optimal solution is shown in Figure 11.7.

11.2.2.1   Making Benders' Decomposition More Efficient

The efficiency of the Benders' decomposition algorithm can be improved in a number of ways. We show one method by observing the following. The simplex algorithm is invoked as many times as there are constraints in the master problem MP. For large problems, the number of constraints could make the problem intractable. We therefore need to examine whether the dual problems DLP can be solved without invoking the simplex algorithm. Thus, we need to examine a given dual problem DLP *i* and see whether a more direct solution approach is possible.

   It turns out each dual problem DLP *i* has a unique solution that can be obtained by inspection. To see how, we first present two theorems.

Theorem 11.1: For a given feasible vector $\mathbf{y}^i$, where $\mathbf{y}^i = (y_{12}, y_{13}, \ldots, y_{n-1,n})$, the smallest value of M for adjacent workstations i and j in constraints (10.11) and (10.12) of model LMIP 1 is given by M_{ij}, where $M_{ij} = l_i + l_j$.*

Proof: Note from the definition of y_{ij} that a given feasible $\mathbf{y}^i$ will correspond to a unique arrangement of workstations. Each $y_{ij} = 1(0)$ specifies that workstation i is to the left (right) of workstation j. Consider any *adjacent* pair of workstations i and j. Constraints (10.11) and (10.12) from LMIP 1 are reproduced here (without d_{ij}) as constraints (11.4) and (11.5) for convenience.

$$x_i - x_j + Mz_{ij} \geq \frac{1}{2}\left(l_i + l_j\right) \quad i = 1, 2, \ldots, n-1, \ j = i+1, \ldots, n \tag{11.4}$$

$$-\left(x_i - x_j\right) + M(1 - z_{ij}) \geq \frac{1}{2}\left(l_i + l_j\right) \quad i = 1, 2, \ldots, n-1, \ j = i+1, \ldots, n \tag{11.5}$$

Note that either Equation 11.4 or 11.5 will be an equality for each pair of adjacent workstations.
Case 1 $(y_{ij} = 0)$:
 From Equation 11.4, we get $x_i - x_j = (1/2s)(l_i + l_j)$
 From Equation 11.5, we get $-(x_i - x_j) \geq (1/2)(l_i + l_j) - M_{ij}$
Substituting the value of $x_i - x_j$ in Equation 11.4, we get

$$-\left(\frac{1}{2}\right)(l_i + l_j) \geq \left(\frac{1}{2}\right)(l_i + l_j) - M_{ij}$$

$$M_{ij} \geq (l_i + l_j)$$

Case 2 $(y_{ij} = 1)$:
 From Equation 11.5, we get $-(x_i - x_j) = (1/2s)(l_i + l_j)$
 From Equation 11.4, we get $x_i - x_j + M_{ij} \geq (1/2)(l_i + l_j)$
 Again, substituting the value of $x_i - x_j$ in Equation 11.3, we get

$$-\left(\frac{1}{2}\right)(l_i + l_j) + M_{ij} \geq \left(\frac{1}{2}\right)(l_i + l_j)$$

$$M_{ij} \geq (l_i + l_j)$$

Hence, for both cases, the smallest value M_{ij} can take for any adjacent pair of workstations is $(l_i + l_j)$. For all other nonadjacent workstation pairs k, l, it is easy to show that the smallest value of M_{kl} is the sum of the lengths of all workstations between k and l, plus half the lengths of k and l. We set M_{ij} equal to $(l_i + l_j) + \varepsilon$ for adjacent pair of workstations and to $\Sigma_i l_i + \varepsilon$ for all others, however, to exploit the complementary slackness property exhibited by linear programming models and their duals.

Theorem 11.2: For a given feasible vector $\mathbf{y}^i$, where $\mathbf{y}^i = (y_{12}, y_{13}, \ldots, y_{n-1,n})$, there is a unique optimal solution to DLP i.

Proof: To prove the theorem, we use the complementary slackness conditions for a given primal-dual pair. DLP i for a feasible $\mathbf{y}^i$, written in the summation (nonmatrix) form, is provided in the following.

DLP i

$$\text{Maximize} \sum_{i=1}^{n-1} \sum_{j=i+1}^{n} u_{ij}\left[0.5\left(l_i + l_j\right) - My_{ij}\right] + v_{ij}\left[0.5\left(l_i + l_j\right) - M + My_{ij}\right] \tag{11.6}$$

---

* For the sake of clarity, we have not included here the clearance between each pair of machines, i.e., d_{ij} shown in constraints (10.11) and (10.12). All the results presented will hold even when the clearances are considered.

$$\text{Subject to} \sum_{j:i<j}\left(u_{ij}-v_{ij}+w_{ij}\right)-\sum_{j:i>j}\left(u_{ij}-v_{ij}+w_{ij}\right)\le 0 \quad i=1,2,...,n \tag{11.7}$$

$$-w_{ij}\le c_{ij}f_{ij} \quad i=1,2,...,n-1;\ j=i+1,...,n \tag{11.8}$$

$$w_{ij}\le c_{ij}f_{ij} \quad i=1,2,...,n-1;\ j=i+1,...,n \tag{11.9}$$

$$M_{ij}=\begin{cases}l_i+l_j+\varepsilon & \text{if facilities } i \text{ and } j \text{ are adjacent}\\[2mm]\displaystyle\sum_{i=1}^{n}l_i+\varepsilon & \text{otherwise } i=1,2,...,n-1;\ j=i+1,...,n\end{cases} \tag{11.10}$$

$$u_{ij},v_{ij}\ge 0, \quad w_{ij} \text{ unrestricted } i=1,2,...,n-1;\ j=i+1,...,n \tag{11.11}$$

As mentioned before, for a given feasible $\mathbf{y}^i$ it is easy to obtain the corresponding linear placement of the n workstations. Now let us examine DLP i for a feasible $\mathbf{y}^i$. By the complementary slackness theorem, constraints (11.7) have zero slack variables because the corresponding primal variables x_i are greater than 0 by construction. In fact, this is the reason for enforcing strict inequalities in constraint (10.15) in the LMIP 1 model. Thus, constraints (11.7) are equalities. Consider LP i for a feasible $\mathbf{y}^i$. It was shown in Theorem 11.1 that the smallest value of M_{ij} for adjacent pair of workstations i and j is (l_i+l_j). To ensure that only one of the constraints (11.4) and (11.5) is binding for each adjacent pair of workstations, we set it $(l_i+l_j)+\varepsilon$. For all other nonadjacent workstation pairs k and l, even though we could set M_{kl} to the sum of the lengths of workstations between k and l (including k and l), we set M_{kl} to $\Sigma_i l_i+\varepsilon$ to exploit the complementary slackness property. Because constraints (11.4) and (11.5) for each nonadjacent workstation pair will never be binding, the corresponding dual variables will take on zero values. Only the $n-1$ constraints corresponding to each adjacent workstation pair will be satisfied as an equation in the primal. Therefore, only the corresponding $n-1$ dual variables—u_{ij} or v_{ij} (but not both) for each adjacent workstation pair $\{i,j\}$—will take positive values. All the other u_{ij} and v_{ij} dual variables, including the $n(n-1)/2$ dual variables corresponding to each nonbinding primal constraint, are equal to zero (by the complementary slackness theorem). Thus, the number of such dual variables whose value is equal to 0 is

$$\left(n(n-1)/2-n+1\right)+n(n-1)/2=\left((n-1)(n/2-1+n/2)\right)=(n-1)^2$$

Moreover, because either x_{ij}^+ or x_{ij}^- is always nonzero, the corresponding dual constraint (which has only one variable) has a slack variable equal to zero, (see constraints (11.8) and (11.9)). Hence, the variable in these equations has a value equal to the constant on the right-hand side. The values of $n(n-1)/2$ more variables are therefore known.

To summarize, among the $3n(n-1)/2$ dual variables, the values of $(n-1)^2 + n(n-1)/2 = n(n-1)/2(3n-2)$ variables are known. The number of unknown variables is equal to $n-1$. The number of dual constraints with unknown variables is equal to n, and as shown earlier, these constraints are equalities. Thus, we have reduced the constraint set (11.7) through (11.11) in DLP i to a system of n equations with $n-1$ unknowns. Whenever an LP has fewer unknowns (variables) than equations, it must have a unique solution, so DLP i has a unique solution for a feasible $\mathbf{y}^i$.

We are now ready to discuss some refinements to the basic Benders' decomposition algorithm. First, as mentioned earlier, solution of MP with few or no constraints provides a lower bound on

4	3	2	1

FIGURE 11.8   Sample arrangement of our workstations.

the optimum OFV of MIP. To obtain a tight lower bound, the coefficient of y_{ij}'s in MP must be set as small as possible. Notice that these coefficients correspond to those of the y_{ij} variables in LP i and are equal to M_{ij}. We showed earlier that the coefficient of y_{ij}'s corresponding to adjacent workstations in LP i may be set to $(l_i + l_j) + c$, where $(1/2)(l_i + l_j)$ is the distance between the centers of adjacent workstations i and j. This will ensure that for a given $\mathbf{y}^i$ vector (or alternatively for a given linear arrangement of the workstations), M_{ij} is set to its minimum possible value plus ε. Because these values of M_{ij} are the coefficients of y_{ij}'s in MP, its optimal value and hence the lower bound to MIP will be tighter.

Second, as shown in Theorem 11.2, DLP i reduces to a set of n equations with $n - 1$ unknowns. Furthermore, in every equation, only those dual variables corresponding to adjacent workstation pairs will take positive values. Thus, each equation has at most two variables that are nonzero. Also, there are exactly two dual equations with only one variable—the equations corresponding to the workstations at the left and right extremes. Beginning with these variables, we can backtrack to find out the values of the other variables. We illustrate this using the linear arrangement of four workstations depicted in Figure 11.8.

Assume l_1, l_2, l_3, and l_4 are 4, 1, 3, and 2 units, respectively. Further assume that the coordinates of the centers of the four departments are $x_1 = 8$, $x_2 = 5.5$, $x_3 = 3.5$, and $x_4 = 1$. Because $x_i - x_j = (1/2)(l_i + l_j)$ for adjacent departments, the corresponding dual variables u_{ij} will take nonzero values. For all other primal constraints, the inequality strictly holds and hence the corresponding dual variables are zero. Because each dual constraint corresponds to the x_i (primal) variable, for the linear arrangement in Figure 11.8, we have only one dual variable u_{12} in the first dual constraint (corresponding to x_1). Similarly, we have exactly one variable u_{34} in the dual constraint corresponding to primal variable x_4. For the other two dual constraints, we have exactly two variables—u_{34} and u_{23}—in the constraint corresponding to x_3 and u_{12} and u_{23} in the constraint corresponding to x_2.

As specified in Theorem 11.2, these dual constraints (see constraint set [11.7]) are equalities, so we have the following equations:

$$
\begin{array}{rrrrrrrrcl}
u_{12} & & & +w_{12} & +w_{13} & +w_{14} & & & & = & 0 \\
-u_{12} & +u_{23} & & -w_{12} & & & +w_{23} & +w_{24} & & = & 0 \\
& -u_{23} & +u_{34} & & -w_{13} & & -w_{23} & & +w_{34} & = & 0 \\
& & -u_{34} & & & -w_{14} & & -w_{24} & -w_{34} & = & 0 \\
\end{array}
$$

In this example, the arrangement is such that $x_i > x_j$ for each $i < n$, and $i < j \le n$, so there are no v_{ij} variables in the preceding equations. Notice that w_{ij}'s are known, so there are at most two variables in any equation. This property can be exploited to design an efficient algorithm for solving DLP i.

Third, each time DLP i is solved, a corresponding Benders' cut is added and a pure integer program (MP) is solved. This step is the most time-consuming step of all. Obviously, if the termination criterion (lower bound greater than or equal to upper bound) is met early in the search process, few Benders' cuts will be generated and the pure integer program may be solved easily. On the other hand, if it takes a substantial number of Benders' cuts to verify optimality, this will have a significant impact on solution time.

Fourth, instead of solving MP to optimality at each stage, it is sufficient to find a feasible solution with an OFV equal to the upper bound minus a tolerance parameter. This requires the termination rules to be altered, as we discuss in the following algorithm, which is a more efficient version of the basic Benders' algorithm. It takes into consideration all the refinements we have discussed.

Modified Benders' decomposition algorithm

> *Step 0*: Set $i = 1$, $\mathbf{y}^i = \{0, 0, ..., 0\}$ and upper bound $UB = \infty$.
>
> *Step 1*: Because DLP i has a unique solution, find this using the technique discussed in the preceding paragraphs. Let $\mathbf{u}^i$ be the solution to DLP i. If $\mathbf{u}^i(\mathbf{b} - \mathbf{B}\mathbf{y}^i) < UB$, we set $UB = \mathbf{u}^i(\mathbf{b} - \mathbf{B}\mathbf{y}^i)$.
>
> *Step 2*: Update MP by adding these constraints:

$$\mathbf{z} \geq \mathbf{u}^i(\mathbf{b} - \mathbf{B}\mathbf{y})$$

$$\mathbf{z} \geq UB - \in$$

$$\mathbf{z}, \mathbf{y} = 0 \text{ or } 1 \text{ and } \mathbf{y} \text{ feasible to MIP.}$$

Find a feasible solution to MP. If it is infeasible, we have found an $\in$-optimal solution to MIP. Otherwise, let $\mathbf{y}^*$ be the feasible solution. Set $i = i + 1$, $\mathbf{y}^i = \mathbf{y}^*$ and return to step 1.

Details on this modified version are provided in Geoffrion and Graves (1974).

11.3   HEURISTIC ALGORITHMS

We discussed several heuristic algorithms in Chapters 5. The ones described in this section are versatile in that they can be adapted to solve numerous combinatorial optimization problems that are otherwise difficult to solve. They are also powerful because for many of these problems, they have yielded optimal or near-optimal solutions.

11.3.1   SIMULATED ANNEALING ALGORITHM

Simulated annealing is a technique based on ideas from statistical mechanics and is motivated by an analogy to the behavior of physical systems in the presence of a heat bath (Johnson et al., 1989). It has been used to solve many combinatorial problems such as the QAP, the TSP, and the graph partitioning problem with reasonable success. References on these and other applications, are listed in Johnson et al. (1989) and Collins et al. (1988).

The process of simulated annealing is analogous to the way in which crystals are formed. It is well known that if a liquid is heated to a high temperature and allowed to cool gradually, the final state reached or the final crystal produced is superior compared to that obtained by quenching or rapid cooling. Similarly, simulated annealing is an algorithm that obtains better final solutions by gradually going from one solution to the next.

To explain simulated annealing, we examine how it differs from simple improvement algorithms such as the 2-opt, 3-opt, or CRAFT algorithms discussed in Chapter 5. A primary distinction is that local optimization algorithms often restrict their search for the optimal solution in a downhill direction. In other words, the initial solution is changed only if it results in a decrease in the OFV for minimization problems; similarly, for maximization problems, a new solution is accepted only if it results in an increase in the OFV. Although this may help us find a local optimum solution, in many cases, that solution may be inferior to the global optimum solution. Moreover, the local optimum solution for 2-opt (or 3-opt) is dependent upon the local region in which the search takes place, which itself is determined by the initial solution provided to it. To overcome this entrapment in local optima, the local search algorithm is modified to allow it to back out of poor local optimal regions and explore other regions, so the chances of finding a better solution are greatly increased. This is the essence of the simulated annealing algorithm.

For every new solution, the simulated annealing algorithm determines the difference δ between the OFV of the previous (best) solution and the new solutions. If the difference is favorable (i.e., for a minimization problem, the OFV of the new solution is less than that of the previous best solution), the previous solution is discarded and the new one takes its place. (Note that the same thing is done in a local optimization algorithm.) If the difference is not favorable, then the new solution is accepted with a certain probability. Obviously, the probability of accepting a new worse solution depends upon the value of δ. Greater the value of δ, greater is the probability of a new solution being rejected. Thus, simulated annealing also searches for solutions in an uphill direction also and this is how it attempts to back out of a local optimum and search for better solutions in neighboring regions. Some implementations of the simulated annealing algorithm accept even superior solutions with a (high) probability. It is not clear whether there is any special advantage in doing so. In the version of the simulated annealing algorithm presented in this chapter, superior solutions are always accepted.

Our explanation provides an outline of the simulated annealing algorithm, but it does not answer two specific questions. (1) How many new solutions do we examine? (2) What must be done if we cannot improve the solutions?

The answer to the first question depends on whether a *frozen state* has been reached. If a frozen state (i.e., a state where the likelihood of finding new better solutions is relatively small) is reached, the cost of searching for new solutions is very high in comparison to the benefits that are likely to be obtained. Different implementations of the simulated annealing algorithm use different ways to determine whether a frozen state has been reached. In this chapter, we simply assume that if one of the following two conditions are met, a frozen state has been reached:

1. The number of solutions examined exceeds a predetermined value.
2. The number of new solutions accepted exceeds a predetermined value.

To answer the second question, the concept of *temperature* is introduced. Temperature refers to the state through which the simulated annealing algorithm passes in its search for a better solution. Starting with an initial temperature, we move to the next temperature only when we have reached a frozen state. When a frozen state is reached, the temperature is reduced by a cooling factor r ($0 < r < 1$), and the procedure is repeated until a certain predetermined number of temperature steps have been explored.

The notation used and the steps in the algorithm are provided next.

n	Number of departments in the layout problem
T	Initial temperature
r	Cooling factor
ITEMP	Number of times temperature T is decreased
NOVER	Maximum number of solutions evaluated at each temperature
NLIMIT	Maximum number of new solutions to be accepted at each temperature
δ	Difference in the OFVs of previous (best) solution and the current solution.

Simulated Annealing Algorithm

Step 0: Set: S = initial feasible solution; z = corresponding OFV; $T = 999.0$; $r = 0.9$; *ITEMP* = 0; *NLIMIT* = $10n$; *NOVER* = $100n$; p, q = maximum number of departments permitted in any row, column respectively.

Step 1: Repeat step 2 *NOVER* times or until the number of successful new solutions is equal to *NLIMIT*.

Step 2: Pick a pair of departments randomly and exchange the position of the two departments. If the exchange of the positions of the two departments results in the overlapping of some other pair(s) of departments, appropriately modify the coordinates of the centers of the concerned departments to ensure there is no overlapping. If the resulting solution S' has an OFV $\leq z$, set $S = S'$ and $z =$ corresponding OFV. Otherwise, compute $\delta =$ difference between z and the OFV of solution S', and set $S = S'$ with a probability $e^{-\delta/T}$.

Step 3: Set $T = rT$ and $ITEMP = ITEMP + 1$. If $ITEMP$ is ≤ 100, go to step 1; otherwise STOP.

Four aspects of this algorithm need further discussion. First, $e^{-\delta/T}$ in step 2 is called the acceptance function (Murty, 1995). Notice that the acceptance function is such that solutions with slightly higher OFVs are more likely to be accepted than those with OFVs much greater than the current solution's OFV. Also, the value of the *temperature* parameter T should be set so that more solutions will be accepted when T is high and fewer when it is low. Because the value of T is gradually reduced (see step 3), this implies that we want more worse solutions accepted in the beginning and fewer at the end. The rationale is that we want to avoid entrapment in a local optima in the early part of the search. Assigning too small a value to T will indeed make the simulated annealing algorithm behave like 2-opt (because very few uphill moves will be accepted) and trap it in an inferior local optimum.

Second, because T should be set so that more solutions will be accepted in the early part of the search process, this means it should be set so that $e^{-\delta/T}$ is relatively high (i.e., close to one). This in turn means that we must have a rough idea of the value of δ (difference in the OFVs of two successive solutions in the early part of the simulated annealing search). For example, if the OFV of an initial solution is around 5000 and the neighboring (inferior) solutions are approximately 400 units worse than the initial solution's OFV, a choice of 6000 for T will provide a value of 0.93 for the acceptance function, thereby allowing simulated annealing to accept solutions that have an OFV of 5400 units or more with a greater probability.

Third, the cooling schedule used in the simulated annealing algorithm is a simple yet powerful way of ensuring that the temperature is gradually reduced after a "frozen state or equilibrium" is reached at a certain state or temperature. Although we have chosen a cooling factor value of 0.9, any value from 0.8 to 0.99 could be used. Choosing a higher value for the cooling factor r will mean that the jump from one state to another is slow and gradual. A value close to 0.8 will mean that the drop in temperature is more significant, sometimes causing simulated annealing to terminate quickly. Once again, experimentation is needed to determine the "right" cooling factor value. Also, instead of using the simple cooling schedule $T_{t+1} = rT_t$ shown in step 3, sometimes the schedule $T_{t+1} = d/\log t$, where d is a positive constant, is also used (Murty, 1995).

Fourth, the efficiency of simulated annealing depends on the neighborhood structure used. Murty (1995) states that this can be improved substantially by designing neighborhoods that exploit the problem structure. Another important requirement in the neighborhood design is that we must be able to go from each feasible solution (layout) to every other. Clearly, the pairwise exchange routine used in step 2 allows us to go from one solution to any other.

In a general layout problem, we could be dealing with departments of unequal area, so we assume that the maximum number of departments permitted in each row and column and the department orientations are known *a priori*. This information enables us to apply the simulated annealing algorithm to layout problems with unequal area departments. Using information concerning the maximum number of departments in each row and column, we can easily construct a layout. If we have p, q rows, columns of departments, respectively, we can visualize the layout as a $p \times q$ grid in which the grid areas are different (because the departments are not all of equal area). Each grid defines a location and the area of the location depends upon the area of the department occupying it. For convenience, the locations can be numbered in increasing order from left to right and bottom to top. We assume that the departments are positioned so that the top edges of departments in a row are aligned horizontally, and the left edges are aligned vertically as in Figure 11.9. The locations in the figure are numbered from left to right and from bottom to top. Note that the departments are separated to maintain the minimum horizontal and vertical clearances.

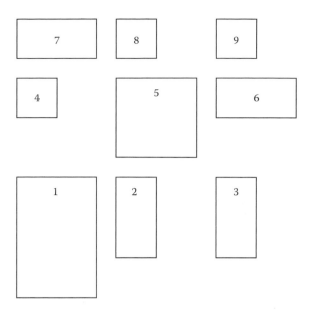

FIGURE 11.9 Department placement in which top edges are aligned horizontally and left edges are aligned vertically.

Step 1 of the simulated annealing algorithm is an initialization step in which an initial feasible solution; the corresponding OFV; parameters *ITEMP*, *NLIMIT*, and *NOVER*; and values for p and q are provided. Although any initial solution can be provided, the easiest is to assign department 1 to location 1, department 2 to location 2, and so on. In other words, the initial solution is such that department 1 is located in the extreme left position of the first (lowermost) row, department 2 is located immediately to the right of department 1, and so on.

The value of initial temperature T is to be set so that it is considerably larger than the largest δ normally encountered. This allows a greater number of inferior solutions to be accepted in the early part of the simulated search, thereby diversifying the search considerably. Our computational experience suggests that a value of 999.0 is appropriate, so we set T at 999.0. The cooling factor r is set to 0.90; *NOVER* to $100n$; and *NLIMIT* to $10n$; n is the number of departments in the layout problem. The reader is strongly encouraged to experiment with other values for these parameters.

In steps 1 and 2, the algorithm examines the random exchange of the positions of two departments. If the exchange results in a solution with a lower OFV, the new solution is accepted. Otherwise, δ, which is the difference between the OFV of the best solution obtained thus far and the current solution, is computed. The probability of acceptance of this solution is $e^{-\delta/T}$. In other words, greater the value of δ, greater is the probability of this solution being rejected. Step 2 is repeated *NOVER* times or until the number of new solutions accepted is equal to *NLIMIT*, whichever occurs first. Note that a layout rearrangement step is executed each time a random exchange is considered in step 2. This is because, in an unequal area layout problem, the exchange of the positions of two departments will require an adjustment in the position of some other departments, especially if the two departments that are exchanged do not have the same shape and area. For example, when the position of departments in locations 5 and 9 in Figure 11.9 is exchanged, the resulting solution has to be modified to ensure that (1) departments are separated just enough to avoid an overlap and the horizontal and vertical clearances are maintained and (2) the upper edges of departments in a row are aligned horizontally (see the modified solution in Figure 11.10). This is done each time an exchange is evaluated. Just as top edges are aligned horizontally, the left edges may also be aligned vertically so as to ensure that the material handling paths are rectilinear, as shown in Figures 11.9 and 11.10. However, the MULROW program available at http://sundere.okstate.edu/Book does not align the left edges vertically.

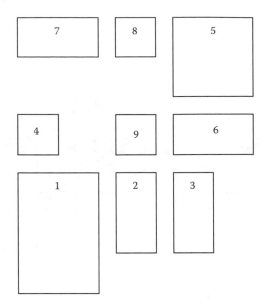

FIGURE 11.10 Modified solution after exchanging position of departments in locations 5 and 9.

Flow matrix (number of trips in thousands)

Machine	1	2	3	4	5	6	7	8	9	10	11	12
1	—	1	0	8	0	2	3	0	0	0	0	0
2	1	—	0	1	1	1	0	0	0	0	0	0
3	0	0	—	0	2	0	0	0	0	0	0	0
4	8	1	0	—	0	4	14	11	0	0	0	0
5	0	1	2	0	—	1	0	0	0	0	0	0
6	2	1	0	4	1	—	3	0	0	3	0	0
7	3	0	0	14	0	3	—	5	5	9	8	2
8	0	0	0	11	0	0	5	—	8	0	0	0
9	0	0	0	0	0	0	5	8	—	0	0	0
10	0	0	0	0	0	3	9	0	0	—	6	0
11	0	0	0	0	0	0	8	0	0	6	—	4
12	0	0	0	0	0	0	2	0	0	0	4	—

FIGURE 11.11 Flow matrix (thousands of trips).

Next, the algorithm decreases the value of the temperature by r and repeats steps 1 and 2. The process of examining NOVER solutions or accepting NLIMIT new solutions continues until 100 temperature steps (or annealing runs) have been considered (step 3). When 100 steps have been considered, the algorithm stops and provides the best layout obtained by it and the corresponding OFV. Instead of considering 100 temperature steps, one may also consider fewer steps, say 10 or 50. Generally, the larger the number of temperature steps, the better the final solution. Of course, the computation time requirement will also be significantly higher. A sample program input and output is available at http://sundere.okstate.edu/Book. In the following discussion, a numerical example is used to illustrate how simulated annealing may be used to solve real-world facility layout problems.

Example 11.3*

A manufacturer of fiber-optic and electro-optic medical equipment has a general machining section, and three manufacturing cells: one for manufacturing optical tube bases, another for otoscope base, and a third for studs. Nineteen parts are processed in the general machining section and traffic congestion is rather high. The production manager wants to develop a new layout for this section. Because the parts are small and lightweight, they are mounted on carts with hooks and transported manually from one machine to the next by machine operators. The transfer batch size is provided in Table 11.2. Also provided is a planned production volume for 3 years and part-routing information for each part. Using the data in Table 11.2 and the machine dimensions data in Table 11.3, construct a flow matrix (one that shows the flow *between* the 12 departments that process the 19 parts) and develop a layout using the simulated annealing

TABLE 11.2
Routing and Other Information for Each Part

Part Number	Part ID	Routing (Sequence of Machines Visited)	Planned Production Volume (in '000s)	Transfer Batch Size
1	031001-3	1–4–8–9	100,000	50
2	201052	1–4–7–4–8–7	240,000	80
3	270080-501	1–2–4–7–8–9	20,000	20
4	37019	1–4–7–9	30,000	10
5	380049-501	1–6–10–7–9	200,000	100
6	380049-502	6–10–7–8–9	100,000	100
7	400164-1	6–4–8–9	200,000	100
8	482111-501	3–5–2–6–4–8–9	50,000	50
9	490012-5	3–5–6–4–8–9	50,000	50
10	608115-1	4–7–4–8	80,000	40
11	635007-16	6	240,000	80
12	695426-503	11–7–12	80,000	80
13	710107-507	11–12	80,000	80
14	735001-502	11–7–10	300,000	100
15	760009-1	1–7–11–10–11–12	50,000	50
16	788014-1	1–7–11–10–11–12	100,000	50
17	999459	11–7–12	20,000	20
18	H3813	6–7–10	60,000	20
19	L999325-2	12	40,000	20

* Based on work done by the second author for a medical equipment manufacturer. The actual data have been modified and some additional data have been taken from Vakharia and Wemmerlov (1990).

TABLE 11.3
Machine Dimensions

Machine Number	Machine ID	Length (in Meters)	Width (in Meters)
1	GRIND-A	15	3
2	BROACH	10	10
3	AUTOCHUKR	6	4
4	SPCBENCH	8	8
5	VMC-MAT	9	11
6	SCREWMACH	4	2
7	PRESS-A	3	3
8	PRESS-B	5	5
9	INDUCTIONHEATR	4	6
10	CUTOFFSAW	10	9
11	MECHPRESS	4	2
12	HYDRAULICPRESS	3	3

```
9 9 8 8 8 8 4 4 4 4 3 3 3 3
9 9 8 8 8 8 4 4 4 4 3 3 3 3
9 9 8 8 8 8 4 4 4
9 9 8 8 8 8 4 4 4
12121111  7 7 1 1 1 1 1 1 1 1 1 1 1
12121111  7 7 1 1 1 1 1 1 1 1 1 1 1
101010101010 6 6 2 2 2 2 2 2 5 5 5 5 5 5
101010101010 6 6 2 2 2 2 2 2 5 5 5 5 5 5
101010101010   2 2 2 2 2 2 2 5 5 5 5 5 5
101010101010   2 2 2 2 2 2 2 5 5 5 5 5 5
101010101010   2 2 2 2 2 2 2 5 5 5 5 5 5
101010101010   2 2 2 2 2 2 2 5 5 5 5 5 5
```

FIGURE 11.12   Solution produced by simulated annealing algorithm for Example 11.3.

algorithm. Assume that the minimum clearance between the machines in the horizontal and vertical directions is 2 meters.

Table 11.2 enables us to develop the flow matrix shown in Figure 11.11. It is easy to see that the flow *between* a pair of machines depends on the parts processed consecutively on these machines, their production volume, and batch size. For example, parts 380049-501, 380049-502, 735001-502, and H3813 are the only ones that are processed consecutively on machines 7 and 10 (PRESS-A and CUTOFFSAW). Hence, the number of trips between these machines is equal to (200,000 + 100,000 + 300,000)/100 + (60,000/20) = 9,000.

We assume that in the initial layout, machine 1 is assigned to location 1, machine 2 to location 2, and so on. We also assume zero clearance between each pair of machines in both directions. The parameters T, r, $NLIMIT$, and $NOVER$ are set as shown in Step 0 of the algorithm. The computer program MULROW.EXE requests values for p, q (number of machines in a row and column) interactively and are provided as 4, 3, respectively. The final solution produced by the computer program is shown in Figure 11.12. It has an OFV of 887.

11.3.2   MODIFIED PENALTY ALGORITHM

The modified penalty (MP) algorithm involves transforming the constrained ABSMODELs discussed in chapter into an unconstrained one using the penalty method.* The square of each

---

* A model with constraints is called a constrained model, and one without any is called an unconstrained model.

constraint is multiplied by a penalty parameter and placed in the objective function. The reason for doing this is shown via the following constrained model:

$$\text{Minimize } c_{11}x_{11} + c_{12}x_{12} + \cdots + c_{3n}x_{3n} \tag{11.12}$$

$$\text{Subject to } a_{11}x_{11} + a_{12}x_{12} + \cdots + a_{1n}x_{1n} \geq b_1 \tag{11.13}$$

$$a_{21}x_{21} + a_{22}x_{22} + \cdots + a_{2n}x_{2n} \leq b_2 \tag{11.14}$$

$$a_{31}x_{31} + a_{32}x_{32} + \cdots + a_{3n}x_{3n} = b_3 \tag{11.15}$$

$$x_{21}, x_{22}, \ldots, x_{3n} \geq 0 \tag{11.16}$$

The constrained model is transformed into an unconstrained model using penalty parameters β_1, β_2, and β_3:

$$\text{Minimize } c_{11}x_{11} + c_{12}x_{12} + \cdots + c_{3n}x_{3n} + \beta_1 \left[\max\{0, b_1 - a_{11}x_{11} - a_{12}x_{12} - \cdots - a_{1n}x_{1n}\} \right]^2$$

$$+ \beta_2 \left[\max\{0, a_{21}x_{21} + a_{22}x_{22} + \cdots + a_{2n}x_{2n} - b_2\} \right]^2$$

$$+ \beta_3 \left[\max\{0, a_{21}x_{21} + a_{22}x_{22} + \cdots + a_{2n}x_{2n} - b_3\} \right]^2 \tag{11.17}$$

In the transformed model presented in expression (11.17), there is a penalty if any of the constraints (11.13) through (11.15) in the original (constrained) model is violated. Notice that the third term in the unconstrained model must be squared, but the first and second terms need not.

To understand the transformation, assume that $a_{11}x_{11} + a_{12}x_{12} + \cdots + a_{1n}x_{1n} < b_1$; then $\max\{0, b_1 - a_{11}x_1 - a_{12}x_{12} - \cdots - a_{1n}x_{1n}\}$ is positive and the objective function will not be minimized. In order to minimize the OFV, theoretically, the solution will be such that all the constraints in the original model are satisfied. Care has to be exercised in setting values for the penalty parameters β_1, β_2, and β_3, however. If these values are large compared with the objective function coefficients in the original unconstrained model, it indicates that satisfying the constraints is given more importance than minimizing the objective function of the unconstrained model. On the other hand, if the value of the penalty parameters is small, the objective function (of the constrained model) will be minimized, but the constraints may be violated. Moreover, as the problem size increases, the number of constraints in the ABSMODEL increases, and determining the penalty parameter values becomes even more difficult. The penalty parameter values must be set depending on the properties of the unconstrained model—whether it is convex, whether it is continuously differentiable, and whether certain other conditions, such as the consistency condition, are satisfied. It is a research topic by itself and is outside the scope of this textbook. We suggest that the reader perform trial runs with various values for the penalty parameters, and use the one that minimizes the objective function. For more details on the penalty algorithm and setting values for the penalty parameters, the reader is referred to Bazaraa and Shetty (1979) or Avriel (1976).

Here are the steps of the MP algorithm. A computer program* for the algorithm is available at http://sundere.okstate.edu/Book.

Step 0: Obtain values for β_1, β_2, and β_3 from the user. Set S = initial solution vector and z = corresponding OFV.

Step 1: Transform the constrained model into an unconstrained one.

---

* We are grateful to Numerical Recipes Software Inc. for allowing us to use five of their subroutines in the MP algorithm.

Step 2: Solve the unconstrained minimization model using the Powell algorithm. If the OFV of the resulting solution is less than or equal to z, set $S^* =$ new solution vector and $z^* = $ OFV corresponding to S^*.

Step 3: Modify solution vector S^* so that a feasible solution is obtained.

Step 4: Improve the solution using greedy 2-opt. Stop.

Notice that the solution obtained is improved, if possible, using the greedy 2-opt algorithm in step 4. An advantage of the algorithm given here is that the user can provide a feasible or infeasible solution. For example, the value of all the variables x_i in the initial solution to ABSMODEL 1 may be set to zero. There is not much difference in the quality of final solutions produced with feasible or infeasible starting solutions. (More computational details are provided in Heragu and Kusiak (1991).) Even though any feasible solution may be provided, it is convenient to provide a simple and standard initial solution in which department 1 is assigned to location 1, department 2 is assigned to location 2, and so on. Of course, this can be done only if the locations are known *a priori*. If not, one has to use other means to generate a standard initial solution. For example, for the single-row layout problem, department 1 can be assigned to the left extreme position, department 2 to the immediate right of department 1, department 3 to the immediate right of department 2, and so on. The Powell algorithm used in step 2 is a general algorithm for unconstrained optimization (Powell, 1964). Any other suitable algorithm for unconstrained optimization could be used in place of the Powell algorithm, such as the more recent algorithms in software packages like LINGO and CPLEX.

Step 3 is necessary because the solution produced in step 2 by the Powell algorithm is usually not feasible. The value of the variables may be such that two or more departments overlap. To make the solution feasible, we need an additional step that modifies the values of the variables corresponding to the overlapping departments while maintaining the relative positions of the departments provided by the Powell algorithm in step 2. This modification is necessary for each pair of overlapping departments and is done in step 3 of the MP algorithm. For example, for the single-row layout problem, assume that the values x_i, x_j, and x_k for three departments i, j, and k are such that $x_i + x_j \leq 0.5 * (l_i + l_j) + d_{ij}$, $x_j + x_k \leq 0.5 * (l_j + l_k) + d_{jk}$, $x_i + x_k \geq 0.5 * (l_i + l_k) + d_{ik}$, $x_i > x_j > x_k$. This solution indicates that

- Departments i and j are to be located to the left of department k.
- Department i is to be located to the left of department j.

Due to the nature of the penalty method, the constraints are not fully satisfied, so departments i, j, and k overlap. Therefore, the values x_i, y_i and x_j, y_j are to be modified so that $x_i + x_j \geq 0.5 * (l_i + l_j) + d_{ij}$, $x_j + x_k \geq 0.5 * (l_j + l_k) + d_{jk}$, $x_i + x_k \geq 0.5 * (l_i + l_k) + d_{ik}$. This modification is necessary for each pair of overlapping departments and is done in step 3 of the MP algorithm. For the single-row layout problem, this modification step can be thought of as *stretching* the solution provided by the Powell algorithm to eliminate overlap. Notice that there is only one constraint for ABSMODEL 1. Constraint (10.3) was included in ABSMODEL 1 to familiarize with the department-within-building constraint, but it need not be considered in the MP algorithm. Because the objective is to minimize the distance traveled by the material handling carrier, the layout generated will be compact. In other words, there will be no empty space between the facilities unless the user specifies that clearance is required between them. Hence, while solving the single-row layout problem, the user has to provide value only for β_1 parameters because ABSMODEL 1 has only greater than or equal to constraints. A modified version of the model in Section 10.6.3 could also be used to develop a feasible final layout from the layout produced by the MP algorithm.

Example 11.4

A facility analyst has been asked to develop a layout of 11 workstations in a renovated part of an assembly facility. Due to restrictions imposed by the shape of the renovated area, the analyst has determined that only a single-row layout is feasible. The engineering department has used product flow information to generate the frequency of trips matrix in Figure 11.13. The entries specify the number of trips that are anticipated between each pair of workstations for the next planning period (3 years). The lengths of the workstations are provided next to the matrix in the figure. All the workstations must be oriented so that their longer sides are parallel to a horizontal line. Notice that the other dimension of the facilities is not considered because the analyst is developing a single-row layout.

Solution

The unconstrained model for this problem is presented in expression (11.18).

$$\text{Minimize} \sum_{i=1}^{10} \sum_{j=i+1}^{11} f_{ij}|x_i - x_j| + \beta_{ij}\left(\max\left\{0, 0.5(l_i + l_j) - |x_i - x_j|\right\}\right) \tag{11.18}$$

The f_{ij} and l_i values in the aforementioned model are already known (see the matrix in Figure 11.13) and the values for β_{ij} are provided by the user. (The input file prepared for solving this example using the enclosed SINROW.EXE program as well as the output file generated are available at http://sundere.okstate.edu/Book. The reader may want to refer to them now.) For this example, all the β_{ij} values were set at one. Solving the preceding model using the Powell algorithm (step 2 of the MP algorithm) results in a particular solution. The values of the x_i variables are given in the aforementioned website. With these x_i values as the coordinates of the workstation centers, the solution shown in Figure 11.14 is easily developed. Note that a number of facilities in this solution overlap. If the solution in Figure 11.14 is stretched so that the relative positions of the machines is maintained, we get the solution shown in Figure 11.15.

	1	2	3	4	5	6	7	8	9	10	11	Workstation lengths (meter)
1	0	20	2	8	0	9	5	7	0	20	3	3
2	20	0	8	9	13	17	16	1	8	6	7	9
3	2	8	0	18	0	10	4	18	5	8	0	3
4	8	9	18	0	6	16	10	4	2	14	6	7
5	0	13	0	6	0	6	0	11	0	8	2	3
6	9	17	10	16	6	0	6	13	2	7	18	7
7	5	16	4	10	0	6	0	1	11	15	7	5
8	7	1	18	4	11	13	1	0	1	7	2	9
9	0	8	5	2	0	2	11	1	0	12	0	6
10	20	6	8	14	8	7	15	7	12	0	3	5
11	3	7	0	6	2	18	7	2	0	3	0	10

$[f_{ij}] =$ Workstation

FIGURE 11.13 Frequency of trips matrix and lengths for 11 workstations.

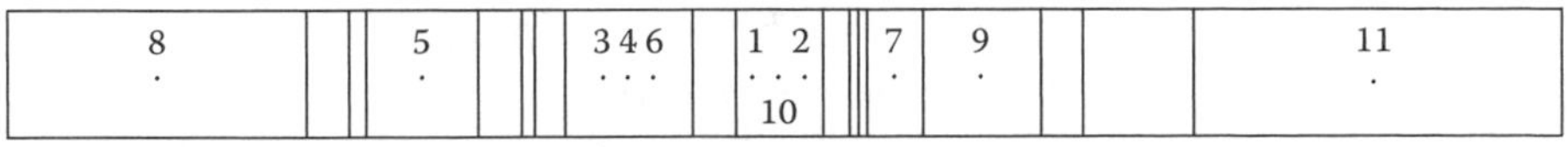

8		5		3 4 6	1 2		7	9			11
.		.		. . .	. . .		.	.			.
					10						

FIGURE 11.14 Solution of unconstrained model using Powell algorithm.

8	5	3	4	6	1	10	2	7	9	11

FIGURE 11.15   Feasible layout for Example 11.4.

If necessary, this solution can be improved using the 2-opt greedy exchange algorithm. The MP algorithm combined with the 2-opt greedy exchange algorithm has been found to produce solutions very close to the optimal ones, especially for the single-row layout problems. More computational details are provided in Heragu and Alfa (1992).

11.3.3   Hybrid Simulated Annealing Algorithm

In this section, we discuss a hybrid simulated annealing (HSA) algorithm. It first attempts to improve an initial user-provided solution using the MP algorithm discussed in the Section 11.3.2 and greedy 2-opt algorithms described in Chapter 5. The solution is improved further using the simulated annealing algorithm (see Figure 11.16). The notation is similar to the one used in the previous section.

Step 1: Set S = initial feasible solution; z = corresponding OFV; $T = 999.0$; $r = 0.9$; $ITEMP = 0$; $NOVER = 100n$; $NLIMIT = 10n$; and p, q = maximum number of departments permitted in any row, column respectively.

Step 2: Apply the MP algorithm to the initial feasible layout. If the departments overlap, modify the coordinates of the departments to eliminate overlapping. If z' (OFV of the resulting solution S') is $\leq z$, set $z = z'$; $S = S'$. Set $i = 1$; $j = i + 1$.

Step 3: If $i \leq n - 1$, exchange the positions of departments i and j; otherwise, go to step 4. If the exchange of the positions of departments i, j results in the overlap of some other pair(s) of departments, appropriately modify the coordinates of the centers of the concerned departments to ensure there is no overlap. If the resulting solution has an OFV $z' \leq z$, set $S = S'$; $z = z'$; $i = 1$; $j = i + 1$ and repeat step 3. Otherwise, set $j = j + 1$. If $j \geq n$, set $i = i + 1$, $j = i + 1$ and repeat step 3.

Step 4: Repeat step 5 *NOVER* times or until the number of successful new solutions is equal to *NLIMIT.*

Step 5: Pick a pair of departments randomly and exchange the position of the two departments. If the exchange of the positions of the two departments results in the overlap of

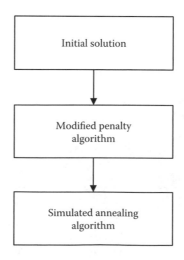

FIGURE 11.16   Flowchart illustrating the solution procedure of HSA.

some other pair(s) of departments, appropriately modify the coordinates of the centers of the concerned departments to ensure there is no overlap. If the resulting solution S' has an OFV $\leq z$, set $S = S'$ and z = corresponding OFV. Otherwise, compute δ = difference between z and the OFV of solution S', and set $S = S'$ with a probability $1 - e^{\delta/T}$.

Step 6: Set $T = rT$ and *ITEMP* = *ITEMP* + 1. If *ITEMP* ≤ 100, go to step 4; otherwise STOP.

Step 1 of the algorithm is an initialization step in which an initial feasible solution, corresponding OFV, other parameters, and the maximum number of departments permitted in each row and column are provided. As mentioned previously, it is convenient to provide an initial solution such that department 1 is in location 1, department 2 is in location 2, and so on.

Next, the MP algorithm discussed in Section 11.3.2 is applied to the initial solution provided in step 1. Because this usually results in a layout in which departments overlap, the layout modification step is executed. In the layout modification step, the x coordinates of the departments determined by the MP algorithm are arranged in increasing order. Because the maximum number of departments permitted in a row is p, the first p departments with the least x coordinate values are placed in the lower most row of the layout. To determine the left to right arrangement of these p departments, their y coordinates are examined. Among these p departments, the one with the least y coordinate value is placed in the leftmost position of the lowermost row, the department with the next higher y coordinate value is placed immediately to the right, and so on. The positioning of the next row of departments is determined in a similar manner. If this leads to a solution with a lower OFV, the improved solution and its corresponding OFV are retained.

In step 3, the greedy 2-opt exchange algorithm is applied. Every time the algorithm examines the exchange of the positions of two departments, it ensures that there is no overlap as a result of the exchange. If there is some overlap, the coordinates of the overlapping departments are modified accordingly, i.e., the layout rearrangement step is executed. If the resulting solution has a lower OFV, the algorithm performs the exchange, sets $i = 1$, $j = i + 1$, and repeats step 3. Otherwise, it examines the next available pair of departments. This is repeated until the exchange of the positions of two departments does not lead to a solution with a lower OFV.

In steps 4 and 5, the algorithm examines the random exchange of the positions of two departments. If the exchange results in a solution with a lower OFV, the new solution is accepted. Otherwise, δ, which is the difference between the OFV of the best solution obtained thus far and the current solution, is computed. The probability of acceptance of this solution is $1 - e^{\delta/T}$. In other words, greater the value of δ, greater is the probability of this solution being rejected. Step 4 is repeated *NOVER* times or until the number of new solutions accepted is equal to *NLIMIT*, whichever occurs first. As in the case of greedy 2-opt, the layout rearrangement step is executed each time a random exchange is considered in step 5. Next, the algorithm decreases the value of the temperature by r and repeats steps 4 and 5. The process of examining *NOVER* solutions or accepting *NLIMIT* new solutions continues until 100 temperature steps have been considered (step 6). When 100 steps have been considered, the algorithm stops and provides the best layout obtained by it and the corresponding OFV.

Example 11.5

Solve the problem described in Example 11.3 using HSA.

Solution

As before, the flow matrix is constructed and the parameters are initialized as detailed in step 1. The input data file is prepared and provided to HSA. The solution produced by the algorithm is provided in Figure 11.17. The OFV of the solution is 922. Although the OFV is greater than the one determined by the simulated annealing algorithm, it should be noted that for many test problems in the literature, the opposite was true.

```
9 9 8 8 8 8 4 4 4 4 3 3 3 3
9 9 8 8 8 8 4 4 4 4 3 3 3 3
9 9 8 8 8 8 4 4 4 4
9 9 8 8 8 8 4 4 4 4
1111  7 71212 1 1 1 1 1 1 1 1 1 1 1
1111  7 71212 1 1 1 1 1 1 1 1 1 1 1
101010101010 6 6 2 2 2 2 2 2 5 5 5 5 5 5
101010101010 6 6 2 2 2 2 2 2 5 5 5 5 5 5
101010101010     2 2 2 2 2 2 5 5 5 5 5 5
101010101010     2 2 2 2 2 2 5 5 5 5 5 5
101010101010     2 2 2 2 2 2 5 5 5 5 5 5
101010101010     2 2 2 2 2 2 5 5 5 5 5 5
```

FIGURE 11.17   Solution produced by HSA.

11.3.4 TABU SEARCH

Tabu search (TS) is a higher-level or "meta" heuristic conceived by Glover (1990). Like SA, this algorithm has also been used to solve many combinatorial problems with reasonable success. Malek et al. (1989) applied TS to solve the TSP, Barnes and Laguna (1990) applied it to machine scheduling, and Skorin-Kapov (1990) used a modified version (called tabu navigation algorithm) to solve the QAP. In the following discussion, we provide a brief discussion of the basic ideas in TS. A tutorial on TS is provided in Glover (1989) and Glover (1990), and the reader is strongly encouraged to consult these two sources for detailed information about TS.

TS is a procedure that can be superimposed on other search techniques, such as 2-opt and simulated annealing, to make the search efficient and avoid entrapment in local optima. Although SA also has a mechanism that allows the search procedure to escape out of local optima, because it is a memoryless procedure, we could be repeatedly examining solutions or solution spaces that were previously explored and thus waste precious computation time. This disadvantage known as cycling can occur in deterministic approaches (for an example, see Skorin-Kapov, 1990) but is generally not a problem for probabilistic procedures such as SA. However, computation time is wasted in memoryless procedures. TS overcomes both drawbacks by maintaining, in short-term memory, a tabu list of solutions (or attributes of solutions) that must not be considered. These solutions must not be considered in the short run (unless they satisfy some aspiration criteria) because they may have been recently explored or because they tend to lead us toward suboptimal regions. The short-term memory is updated as the search progresses. A solution (or solution attribute) that is tabu at one time may not be a tabu later on and could reappear as a tabu as the search progresses even further. TS admits tabu solutions provided they satisfy some specified aspiration criterion. While deciding whether to make an exchange, the TS algorithm first looks up the tabu list. If the solution resulting from the exchange is in the tabu list, the algorithm checks to see whether the aspiration criterion is satisfied. A simple aspiration criterion may be the following. If the exchange leads to a solution that has an OFV better than the best we currently have, the aspiration criterion is said to be satisfied. Thus, TS guides the search and makes it efficient by keeping track of recently explored solutions and determining whether a recently explored solution may be admitted. This method of search is continued until one of several termination criteria is met. While the use of short-term memory may by itself produce good solutions, Glover (1990) suggests the use of intermediate and long-term memory structures to intensify and diversify the search. Intensification can be achieved by keeping track of solution features that have been found to be good in the long run and exploring the neighborhood of such solutions. Diversification is achieved by driving the algorithm into new unexplored regions. Intensification and diversification strategies employing long-term memory (LTM) have been found to lead to the best solutions for hard combinatorial problems. We now briefly discuss a TS algorithm for solving the QAP. This version of the TS algorithm, called tabu navigation, uses short-term as well as long-term memory structures and was developed by Skorin-Kapov (1990).

Tabu Navigation Algorithm

Step 1: Read the flow (**F**) and distance (**D**) matrices. Construct the zero **LTM** matrix of size $n \times n$, where n is the number of departments in the problem.

Step 2: Construct an initial solution using any construction algorithm. Obtain values for the following two short-term memory parameters—size of tabu list (t) and maximum number of iterations (v). Construct the zero tabu list (**TL**) vector and set iteration counter $k = 1$.

Step 3: For iteration k, examine all possible pairwise exchanges to the current solution and make the exchange $\{i, j\}$ that leads to the greatest reduction in the OFV *and* satisfies one of the following two conditions:

    a.   Exchange $\{i, j\}$ is not contained in the tabu list.

    b.   If exchange $\{i, j\}$ is in the tabu list, it satisfies the aspiration criteria.

       Update tabu list vector **TL** by including the pair $\{i, j\}$ as the first element in **TL**. If the number of elements in **TL** is greater than v, drop the last element.

       Update **LTM** matrix by setting $\mathbf{LTM}_{ij} = \mathbf{LTM}_{ij} + 1$.

Step 4: Set $k = k + 1$. If $k > v$, invoke **LTM** by replacing the original distance matrix **D** with **D** + **LTM** and go to step 2. Otherwise, *stop*.

One could modify step 4 by restarting the improvement algorithm with the solution given by the construction algorithm using new values for t and v. Another option is to restart the improvement algorithm with the best solution obtained thus far using new values for t and v.

In step 1, the algorithm reads the flow and distance matrices. Because the problem considered is the QAP, the locations and the distances between them are known. The **LTM** matrix (which is a zero matrix of size $n \times n$) is constructed to keep track of how many times a particular exchange has been considered. Because the **D** matrix is updated by **D** + **LTM** in step 4, the algorithm has an incentive to explore exchanges (or regions) not previously explored sufficiently or not at all. A penalty is incurred if previously well-explored exchanges are made, and therefore, the algorithm has a tendency to stay away from these exchanges. Thus, the *LTM* permits search to be diversified.

The tabu list size is an important parameter and must be carefully selected. Skorin-Kapov (1990) mentions that it is better to set it as an increasing function of the problem size. In other words, the larger the problem size, the larger the tabu list size. For a given problem size, if the tabu list size is too small, cycling will occur. On the other hand, if the tabu list is too big, the algorithm may not intensify search in local optimum regions and hence may not identify the local optimum solutions, some of which may be globally optimal or near-optimal. From empirical tests, it appears that a tabu list size value of $0.33n - 0.6n$ (n is the number of departments) yields good solutions (Skorin-Kapov, 1990). Also, a good choice for the maximum number of iterations is $7n - 10n$. Of course, the more the maximum number of iterations, the greater is the chance of finding a better solution, but more computation time is required. Thus, a trade-off decision has to be made.

Notice that if we were to use SA as the search procedure in step 3, we may sometimes be accepting inferior solutions. In the tabu search algorithm however, we use the 2-opt search strategy and hence accept an exchange only if it leads to an improved one. Whether we use SA or 2-opt, an exchange is made only if the corresponding department pair

1. Is not contained in the tabu list.
2. Or, if the pair is contained in the tabu list, it must satisfy the aspiration criterion (for example, the exchange should lead to a solution that has an OFV better than the best we currently have).

The tabu list (which is initially a zero vector of size t) is updated to include the exchange just made. If the number of elements in the tabu list is greater than the tabu list size, the oldest element in the list is dropped to make room for the newest one. Thus, the tabu list size is always equal to *t*. In addition to updating the tabu list, the algorithm updates the **LTM** matrix by setting element **LTM**$_{ij}$ = **LTM**$_{ij}$ + 1. The more times a particular exchange {*i, j*} is considered, the less likely it will be explored when **LTM** is invoked (in step 4) for reasons discussed earlier.

There are several variations possible in the TS algorithm and the algorithms available in the literature use different short-term and long-term memory structures. In addition, the aspiration criteria are different as are the intensification and diversification strategies. This discussion is meant to introduce the reader to TS and we encourage the interested reader to consult other sources mentioned for a detailed discussion of TS.

Heragu and Alfa (1992) conducted a performance comparison using 15 test problems to test the efficiency of the aforementioned algorithms. The problems ranged in size from 5 to 90 and are available in the literature. It was found that the HSA algorithm yielded solutions of good quality in reasonable computation time. For seven out of eight single-row test problems, the algorithm produced optimal or best known solutions and in some cases solutions better than those previously known. For multirow layout problems also, it produced the best (or best-known) solutions. Based on the performance comparison of the algorithms presented in this section, Heragu and Alfa (1992) concluded the following. First, 3-opt performs better than 2-opt and MP algorithms, but requires very high CPU time, especially as the problem size increases. MP algorithm performs better than 2-opt. The version of simulated annealing and HSA presented in this chapter appears to produce better-quality solutions than the version of TS presented in Skorin-Kapov (1990) and 2-opt, 3-opt, and MP algorithms. Also, the greater the number of annealing runs (temperature steps), the better the OFV and the greater the CPU time requirement for the SA and HSA algorithms. Moreover, the solutions produced by simulated annealing and HSA algorithms converge as the number of annealing runs is increased. Figure 11.18 is a graph showing the OFVs of solutions generated by the 2-opt, 3-opt, MP, simulated annealing, HSA, and TS algorithms for some large layout problems.

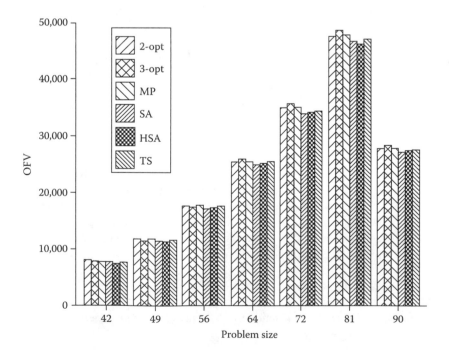

FIGURE 11.18   OFV of solutions produced by several algorithms for some test problems.

11.3.5 Genetic Algorithm

GA is yet another general-purpose algorithm that has been successfully applied to the layout problem. What we mean by general purpose is that this algorithm can be applied to a variety of combinatorial optimization problems such as the TSP, graph partitioning problem, graph coloring problem, jobshop scheduling, and others. This method was developed by Holland (1975) and was inspired by biological systems that produce organisms that not only adapt to the environment successfully but also thrive. The survival of the fittest principle seen in natural biological systems has been applied to the layout and other combinatorial optimization problems with remarkable success. In the following discussion, we provide a "bare-bones" explanation of how the basic GA works. For more details about the GA itself, there are several excellent sources, among which the books by Davis (1991), Goldberg (1989), and Holland (1975) are very popular.

Biological systems reproduce as follows. Fit individuals from the population mate to produce offspring. The offspring are made up of genetic material similar to those found in the parents. As we move from one generation to the next, we find that generally, the fittest survive and the offspring are able to adapt to their environment successfully and thrive. The GA works along similar principles.

Step 0: Obtain the maximum number of individuals in the population N and the maximum number of generations G from the user, generate N solutions for the first generation's population randomly and represent each solution as a string. Set generation counter $N_{gen} = 1$.

Step 1: Determine the fitness of each solution in the current generation's population and record the string with the best fitness.

Step 2: Generate solutions for the next generation's population as follows.
  a. Retain $0.1N$ of the solutions with the best fitness in the previous population.
  b. Generate $0.89N$ solutions via mating.
  c. Select $0.01N$ solutions from the previous population randomly and mutate them.

Step 3: Update $N_{gen} = N_{gen} + 1$. If $N_{gen} \leq G$, go to step 1. Otherwise, *stop*.

In step 0, the GA obtains information concerning the population size and the number of generations. The number of individuals in the population is generally fixed at a certain number, say N. The GA first generates a random population of feasible solutions of size N (using any available method) and represents each solution as a string. The representation of solutions may be trivial for some problems but require careful analysis for others. For example, for the single-row layout problem, each layout can be represented as a string made up of n alphanumerics (corresponding to *genes*), where n is the number of departments in the layout problem. The ordering of the alphanumerics determines the ordering of the departments. For example, if we assign a unique number 1 through n to each of the n departments, then the string $\{1, 2, 3, ..., n\}$ corresponds to a layout in which department 1 is in the left extreme position, department 2 immediately to the right of department 1, 3 to the right of 2, and so on. To ensure the initial population has only feasible solutions, each string must be a permutation of $\{1, 2, ..., n\}$ and therefore cannot have the same integer (or department) appear more than once.

Similarly, for the unequal area layout problem, a valid string may be determined by first dividing each department into unit square blocks and recording the department numbers in a left-to-right and top-to-bottom sequence. Of course, the number of blocks for each department depends upon its area. For example, if the area of departments 1 and 6 is 3 square units, 2 and 4 is 2 square units, and 3 and 5 is 1 square unit, using unit square blocks, we can represent departments 1 and 6 using 3 blocks, 2 and 4 using 2 blocks, and 3 and 5 using 1 block, as shown in Figure 11.19. The string representation for the layout shown in Figure 11.19 is $\{1, 1, 6, 1, 6, 6, 5, 2, 2, 3, 4, 4\}$.

$$
\begin{array}{ccc}
1 & 1 & 6 \\
\\
1 & 6 & 6 \\
\\
5 & 2 & 2 \\
\\
3 & 4 & 4
\end{array}
$$

FIGURE 11.19   Unit square blocks representing departments with unequal areas.

Now that an initial population of solutions is available, the GA mimics the natural reproduction process based on the survival of the fittest principle. To do so, it determines the fitness of the individuals (solutions) so that individuals with high fitness are selected from the population and used to generate offspring for the next generation. To determine the fitness of the individuals, we must first observe that, in minimization problems, highly fit individuals are those that have lower OFVs. Thus, we must transform the OFV into an appropriate fitness measure so that there is an inverse relationship between the two. In other words, a solution with lower OFV must have a corresponding higher fitness when compared to another solution with a higher OFV. For single-row layout problems, we may represent the fitness function (F) shown in Equation 11.19:

$$
F = \frac{1}{\left[\sum_{i=1}^{n-1} \sum_{j=i+1}^{n} c_{ij} f_{ij} \left| x_i - x_j \right| \right]}
\tag{11.19}
$$

Notice that the denominator in the right-hand side of the fitness function is the same as expression (10.1) and c_{ij}, f_{ij}, x_i, x_j, are as defined in that chapter. Although the preceding fitness function is rather simple, a problem is that the difference between two solutions may be pronounced when we consider OFVs, but may appear insignificant when we look at the fitness function. Thus, the solution selection mechanism must be sensitive to this aspect. For the unequal area layout problem, the fitness function in the following expression may be used (Kochhar et al., 1998).

$$
F = -k \left[\sum_{i=1}^{N} \sum_{j=1}^{N} w_{ij} \right] \quad \text{where } w_{ij} = \frac{g_{ij} \left[\left| x_i - x_j \right| + \left| y_i - y_j \right| \right]}{n_i n_j}
\tag{11.20}
$$

N is the total number if blocks occupied by all the n departments, n_i, n_j are the number of blocks occupied by departments i and j, respectively, x_i, y_i are the coordinates of block i, and k is a nonnegative scaling factor. In expression (11.20), the total interaction between departments i and j (g_{ij}) is prorated among the pairs of blocks belonging to the "two" departments by dividing g_{ij} by $n_i n_j$. If blocks i and j belong to the same department, then w_{ij} is set to an arbitrarily large positive number M to ensure the blocks are located adjacently. Kochhar et al. (1996) use a more complex fitness function that appears to work well. In addition, they add a penalty component to the fitness function to minimize the generation of infeasible solutions. There are several ways of ensuring feasibility and we refer the reader to Kochhar et al. (1998) for one such method. Thus, in addition to capturing the inverse relationship between fitness function and OFVs, if necessary, the former must have a penalty term to minimize the generation of infeasible solutions.

After the fitness of solutions in a population is determined, the GA selects individuals for the mating procedure using a probability function. Individuals with a higher fitness have a greater probability of being selected than individuals with a lower fitness value. In addition to employing the mating principle to produce offspring for the next generation's population, the GA

1. Retaining a small percentage (typically in the range of 10%–30%) of individuals from the previous generation.
2. Mutating, i.e., randomly altering a randomly selected chromosome (or individual), from the previous population. The percentage of individuals in a generation produced through mutation is typically very small (0.1%–1%).

Thus, 70%–90% of the individuals in a population are obtained via mating. As indicated in Tate and Smith (1995), there is a great deal of flexibility in the choice of parameters in the GA. Although we have used 0.89 and 0.01 probabilities for mating and mutation, these should be treated only as a guideline and the user must test other values to see which works best for a given problem. Two methods for mutation are:

1. *Reproduction method*, in which a prespecified percentage of individuals is retained based on probabilities that are inversely proportional to their OFVs
2. *Clonal propagation method*, in which xN individuals with the best fitness are retained (x is the prespecified proportion of individuals that are to be retained from the previous generation and N is the population size)

Whereas the clonal propagation method is deterministic, the former is probabilistic. To perform Several methods are available to perform mating. For details on how available methods could be adapted to the layout problem, see Kochhar et al. (1998). Two common method are:

1. Two-point crossover method
2. Partially matched crossover method

In the two-point crossover method, given two parent chromosomes $\{x_1, x_2, ..., x_n\}$ and $\{y_1, y_2, ..., y_n\}$, two integers r, s, such that $1 \le r < s \le n$ are randomly selected and the genes in positions r to s of one parent are swapped (as one complete substring without disturbing the order) with that of the other to get two offspring as follows:

$$\{x_1, x_2, ..., x_{r-1}, y_r, y_{r+1}, ..., y_s, x_{s+1}, x_{x+2}, ..., x_n\}$$

$$\{y_1, y_2, ..., y_{r-1}, x_r, x_{r+1}, ..., x_2, y_{s+1}, y_{s+2}, ..., y_n\}$$

Either both or the better of the two offspring are then included in the new population. This mutation procedure is repeated until the required number of offspring is generated. A variant of the two-point crossover method, called the one-point crossover method, is sometimes used in place of the former. In the one-point crossover method, a number r, $1 \le r \le n$ is randomly selected and the GA swaps the entire substring to the right of the gene in the $r - 1$ position of the first parent with that of the second. An obvious disadvantage with both methods is that for some problems such as the single-row layout problem, infeasible offspring or solutions are often produced. For example, suppose we have a 5-department single-row layout problem and have identified two parents as follows: $\{3, 2, 1, 4, 5\}$ and $\{2, 1, 4, 3, 5\}$. If the GA randomly picks $r = 3$, $s = 5$, the two offspring will be $\{3, 2, 4, 3, 5\}$ and $\{2, 1, 1, 4, 5\}$, which are not permutations of $\{1, 2, 3, 4, 5\}$. To avoid generating infeasible solutions, for such problems, we often use the partially matched crossover method. In this as before, two integers r and s are randomly selected. The GA then selects the first parent $x = \{x_1, x_2, ..., x_{r-1}, x_r, x_{r+1}, ..., x_{s-1}, x_s, x_{s+1}, ..., x_n\}$, sets $t = r$, and swaps genes x_t and y_t in x only if

$x_t \neq y_r$. This swapping is done for $t = r$ through s, one by one. When completed, we have a feasible offspring. The same is done for $y = \{y_1, y_2, ..., y_{r-1}, y_r, y_{r+1}, ..., y_{s-1}, y_s, y_{s+1}, ..., y_n\}$. Because the partially matched crossover method swaps genes within a parent, feasible offspring are obtained. This method is illustrated for the two parents considered earlier $\{3, 2, 1, 4, 5\}$ and $\{2, 1, 4, 3, 5\}$ and $r = 3, s = 5$. Set $t = r = 3$. Because $x_3 \neq y_3$, we swap genes $x_3 = 1$ and $y_3 = 4$ in x to get $\{3, 2, 4, 1, 5\}$. Because $x_4 \neq y_4$, we swap genes $x_4 = 4$ and $y_4 = 3$ (again in x) to get $\{4, 2, 3, 1, 5\}$. Because $x_5 = y_5$ and $t = s = 5$, we are done with the swapping. Repeating the same procedure for y yields the feasible offspring $\{2, 3, 1, 4, 5\}$.

As mentioned previously, the GA also uses mutation to generate a population in addition to reproduction, clonal propagation, and mating. Once again, mutation may be done in several ways. The easiest method is perhaps to randomly select an individual from the previous generation and mutate (alter) it by swapping any two randomly selected genes.

Now that we have a new population, the procedure of determining fitness of the individuals and generating the next generation's population via reproduction/clonal propagation, mating, and mutation is continued until a prespecified number of populations have been generated or there is no appreciable improvement in the fitness of the best individual for several generations. The best solution obtained is then provided to the user.

11.4 MULTICRITERIA LAYOUT PROBLEMS

In this chapter and Chapters 4, 5, and 10, we described quantitative and qualitative approaches to solve the facility layout problem. For example, the QAP and other mixed-integer formulations in Chapter 10 and the 2-opt, 3-opt and CRAFT algorithms presented in Chapter 5 were quantitative approaches. They are typically concerned with the minimization of material handling cost. SLP and the graph theoretic method are, for the most part, qualitative approaches and are concerned with maximizing closeness rating measures. It has been argued that both of these approaches have some drawbacks. For example, if two machines must be placed as far apart as possible, due to noise, dust, or safety reasons, the quantitative approaches cannot effectively capture such a constraint. On the other hand, there is too much subjectivity in the qualitative approaches, and a layout is developed based on preassigned numeric values for the various closeness ratings. It has been suggested that to arrive at a good layout decision, the facility analyst must develop an efficient frontier of layouts, from which the *best* layout may be selected (Rosenblatt, 1979). Using the notation in Chapter 10, we present the following models.

Model M1

$$\text{Minimize } C = \sum_{i=1}^{n} \sum_{j=1}^{n} \sum_{\substack{k=1 \\ i \neq k}}^{n} \sum_{\substack{l=1 \\ j \neq l}}^{n} f_{ik} c_{jl} x_{ij} x_{kl} \tag{11.21}$$

$$\text{Subject to } \sum_{j=1}^{n} x_{ij} = 1 \quad i = 1, 2, ..., n \tag{11.22}$$

$$\sum_{i=1}^{n} x_{ij} = 1 \quad j = 1, 2, ..., n \tag{11.23}$$

$$x_{ij} = 0 \text{ or } 1 \quad i, j = 1, 2, ..., n \tag{11.24}$$

Now, consider the model for maximizing the qualitative, closeness ratings.

Model M2

$$\text{Maximize } R = \sum_{i=1}^{n}\sum_{j=1}^{n}\sum_{\substack{k=1\\i\neq k}}^{n}\sum_{\substack{l=1\\j\neq l}}^{n} r_{ijkl}x_{ij}x_{kl} \tag{11.25}$$

and the constraints (11.22) through (11.25). Note that

$$r_{ijkl} = \begin{cases} t_{ij} & \text{if locations } i \text{ and } j \text{ are adjacent} \\ 0 & \text{otherwise} \end{cases}$$

and t_{ij} is the numeric value assigned to the relationship code between departments i and j.

Models M1 and M2 can be combined to form a multiobjective problem because they both have the same feasible region. Model M3 shown next minimizes the material handling cost and maximizes the total closeness ratings. Note that maximizing R is the same as minimizing $-R$.

Model M3

$$\text{Minimize } w_1 C - w_2 R \tag{11.26}$$

$$\text{Subject to constraints (11.22) through (11.25)}$$

We have already seen several models for the layout. Clearly any one of them could be used to model the quantitative aspects. Once the models are developed, the next step is to solve them optimally using any available algorithm. We can then develop an efficient frontier. To develop the efficient frontier, we use the following procedure: We draw a graph with vertical and horizontal axes to represent C and R values, (i.e., the objective functions of models M1 and M2), respectively. For any available layout—say layout i—we can compute the values C_i and R_i easily and thus can represent it as a point in the graph (see point A in Figure 11.20). A point (layout) is said to be efficient if it dominates all other layouts; i.e., there is no other point j such that (1) $C_i \le C_j$ and $R_i \ge R_j$, and (2) $C_j < C_i$ or $R_j > R_i$. For nonnegative weights w_1 and w_2, it can be seen that any point in the shaded area

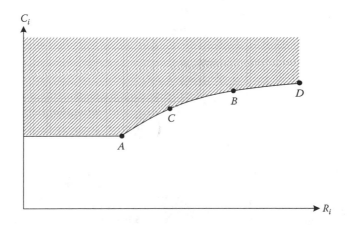

FIGURE 11.20 Graph showing efficient frontier.

is a dominated point. The set of nondominated solutions is called the efficient frontier.* Once the efficient frontier is generated (see the curve in Figure 11.20), the decision maker can make trade-offs to choose layouts. Heuristics for generating the efficient frontier can be found in Rosenblatt (1979) and Dutta and Sahu (1982).

11.5   OPTIMAL APPROACH TO SOLVING CELLULAR MANUFACTURING SYSTEM (CMS) DESIGN PROBLEMS[†]

In Chapter 6, we were primarily concerned with identifying the block diagonal structure from a given part-machine processing indicator matrix so that we minimize the number of nonzero elements outside the final block diagonal. In other words, we were trying to minimize the need for a part to visit a secondary or tertiary cell for processing. For example, given the initial matrix in Figure 11.21a, we reach the block diagonal solution shown in Figure 11.21b.

In addition to minimizing the number of nonzero elements outside the block diagonal, it sometimes is desirable to minimize the number of zero elements inside the blocks. This is important if we want to maximize the utilization of resources (machines) in the cell visited by each part. In this section, we present a mathematical model that explicitly attempts to minimize the number of nonzero elements outside the block diagonal as well as the number of zero elements inside it. It also considers two critical aspects of the CMS design problem. The first is alternate routings. Unlike traditional manufacturing systems in which each part typically has only one routing or process plan, in advanced manufacturing systems, the consideration of alternate routings provides planning flexibility and allows us to reduce production lead times. For example, if a part has multiple process plans, and it turns out that one of the machines in the primary process plan is busy or down for repair, we are able to route the part using an alternate plan. The second aspect involves consideration of many of the practical constraints listed in Section 6.4.

11.5.1   MODEL DESCRIPTION

The model presented in this section is a modified version of the nonlinear mixed-integer programming model in Adil et al. (1997). Its objective is to minimize the intercellular material handling cost and the cost related to resource utilization of parts in each cell. The model makes some practical assumptions: (1) Product mix and demand information is known as *a priori*, (2) operation requirements for each part

	Machine							Machine				
	1	2	3	4	5			1	3	5	2	4
1	0	1	0	1	0		2	1	1	1	0	0
2	1	0	1	0	1		4	1	1	0	0	0
Part 3	0	1	1	1	0		Part 5	1	0	1	0	0
4	1	0	1	0	0		1	0	0	0	1	1
5	1	0	0	0	1		3	0	1	0	1	1
(a)							(b)					

FIGURE 11.21   Sample part-machine processing (a) indicator matrix and (b) identification of clusters.

---

* Because the number of possible layouts is too many ($n!$ for equal area layout problems), it is practically impossible to evaluate all the points (layouts) and generate a complete set of efficient frontier points. We therefore generate a subset of the nondominated solutions and use it as the efficient frontier.
† This section includes material from Dr. Ja-Shen Chen's PhD dissertation (Chen, 1995). The author is grateful to Dr. Ja-Shen Chen for granting permission to use the material in this section.

are known, and (3) the required types and number of machines are available. When the model is solved, the part families and machine cells are identified simultaneously. The following notation is used.

Parameters

i, j, k Part, machine, and cell indices, respectively

a_{ij} $\begin{cases} = 0 & \text{if machine } j \text{ is not required for part } i \\ \leq 1 & \text{otherwise} \end{cases}$

c_i Intercellular movement cost per unit for part i
v_i Number of units of part i
u_{ij} Cost of part i not utilizing machine j
o_{ij} Number of times each part i requires operation on machine j
M_{max} Maximum number of machines permitted in a cell
M_{min} Minimum number of machines permitted in a cell
C_u Maximum number of cells permitted
S_1 Sets of machine pairs that cannot be located in the same cell
S_2 Sets of machine pairs that must be located in the same cell
np Total number of part types
nm Total number of machines

Decision variables

$$x_{ik} \begin{cases} = 0 & \text{if part } i \text{ is not processed in cell } k \\ \leq 1 & \text{otherwise} \end{cases}$$

$$y_{jk} \begin{cases} = 0 & \text{if machine } j \text{ is not in cell } k \\ \leq 1 & \text{otherwise} \end{cases}$$

Before we present the model, we describe six departures from the assumptions made in Chapter 6. First, the elements of the part-machine processing indicator matrix, a_{ij}'s, are considered to be real variables between 0 and 1. This allows us to include alternate routings in the model. For example, if part 1 has two routes with the first through machines 1, 3, 5 and the second through machines 1, 4, 5, and the probability of using the two routes is 0.8 and 0.2, respectively, then in the part-machine processing indicator matrix, we assign $a_{11} = 1$, $a_{13} = 0.8$, $a_{14} = 0.2$, and $a_{15} = 1$. Second, it is common practice to have multiple machines of a given type. By strategically placing copies of a machine in various cells, we can identify machine cells and part families to minimize the number of exceptional elements. In fact, some research papers even suggest that we purchase additional machines of a given type and duplicate them in appropriate cells to minimize the exceptional elements and thereby reduce intercellular material handling costs (for example, see Seifoddini and Wolfe, 1986). Such an approach, which calls for high capital expenditures to minimize small operating costs, may not be practical, however, and the main reason to have multiple machines is to ensure that the system has adequate capacity to process all the parts. In our analysis of the CMS design problem, we first determine the number of machines of each type required to satisfy the capacity constraint and work with these numbers to design a CMS.

Third, we do not assume that the number of cells is predetermined. In practice, designers usually do not know the number of cells that would yield the best CMS design. Thus, it is unrealistic to set the number of cells to a given value. In our approach, the optimal number of cells is determined by the model and is equal to the initial assigned number of cells minus the number of empty cells, which will be known from the solution. The initially assigned number of cells is any number less than or equal to the maximum number of machines in the problem because there can never be more cells than machines.

Fourth, our model assumes that the cost of intercellular movement is the same between any pair of cells for a given part. This cost need not necessarily be the same for all parts, however, because it depends upon the material handling devices used and may be different for different parts. Although our model ignores the distance between cells, this underestimation does not cause serious problems because there is a large fixed cost involved in moving material between cells and the variable (distance) component is typically small. If this is not true in a problem, we can reduce the weight given to intercellular movement in the objective function. Also, the number of intercellular moves is generally small. Furthermore, because the layout of cells is not known at this stage, it is impossible to determine the exact cost of intercellular movement.

Fifth, our model allows some units of a part type to be processed in one cell and the remaining units in others. In Section 6.2, we attempted to develop solutions so that a part type is assigned to only one cell. This may sometimes be too restrictive and not enable us to exploit the availability of alternate process plans. As mentioned before, it may sometimes be advantageous to route some units of a part type via a secondary cell, if a machine in its primary cell is busy or down for repair. Hence, decision variable x_{ik} is allowed to take any value between 0 and 1 to indicate the probability that part i is processed in cell k. A major advantage of our model is that it eliminates integer restriction on a large number of variables and hence makes the model much easier to solve. Thus, the number of integer variables in our problem does not depend upon the number of parts, but rather on the number of machines and cells.

Sixth, although u_{ij} is defined for every part-machine pair, the corresponding cost is incurred only when part i and machine j are assigned to the same cell and part i does *not* require processing on machine j. A nonlinear CMS design model is presented next.

Model M4

$$\text{Minimize} \sum_{i=1}^{np}\sum_{j=1}^{nm}\sum_{k=1}^{C_u} c_i v_i o_{ij} a_{ij} x_{ik}(1 - y_{jk}) + \sum_{i=1}^{np}\sum_{j=1}^{nm}\sum_{k=1}^{C_u} u_{ij}(1 - a_{ij}) x_{ik} y_{jk} \tag{11.27}$$

$$\text{Subject to} \sum_{k=1}^{C_u} x_{ik} = 1 \quad i = 1,2,...,np \tag{11.28}$$

$$\sum_{k=1}^{C_u} y_{jk} = 1 \quad j = 1,2,...,nm \tag{11.29}$$

$$y_{sk} + y_{tk} \leq 1 \quad k = 1,2,...,C_u, \{s,t\} \in S_1 \tag{11.30}$$

$$y_{sk} - y_{tk} = 0 \quad k = 1,2,...,C_u, \{s,t\} \in S_2 \tag{11.31}$$

$$M_{\min} \leq \sum_{j=1}^{nm} y_{jk} \leq M_{\max} \quad k = 1,2,...,C_u \tag{11.32}$$

$$0 \leq x_{ik} \leq 1 \quad i = 1,2,...,np, k = 1,2,...,C_u \tag{11.33}$$

$$y_{jk} = 0 \text{ or } 1 \quad j = 1,2,...,nm, k = 1,2,...,C_u \tag{11.34}$$

The objective function (11.27) minimizes the total cost of intercellular movement as well as the cost of resource underutilization. Constraint (11.28) allows a part to be processed in multiple cells. Constraint (11.29) states that each machine can only be assigned to one cell. Constraint (11.30) ensures that the machine pairs included in S_1 cannot be placed in the same cell. Similarly, constraint (11.31) ensures machine pairs in S_2 are placed in the same cell. Constraint (11.32) specifies the minimum and maximum number of machines allowed in any cell. Unfortunately, model 1 is nonlinear, and nonlinear models are usually much harder to solve optimally than linear models. We reformulate the model as a mixed-integer linear programming model by introducing a new set of variables z_{ijk} to replace the $x_{ik}y_{jk}$ product.

Model M5

$$\text{Minimize} \sum_{i=1}^{np}\sum_{j=1}^{nm}\sum_{k=1}^{C_u} c_i v_i o_{ij} a_{ij} x_{ik} + \sum_{i=1}^{np}\sum_{j=1}^{nm}\sum_{k=1}^{C_u} \left[u_{ij}(1-a_{ij}) - c_i v_i o_{ij} a_{ij} \right] z_{ijk} \tag{11.35}$$

Subject to constraints (11.28) through (11.34)

$$z_{ijk} \leq x_{ik} \quad i = 1,2,\dots,np,\ j = 1,2,\dots,nm,\ k = 1,2,\dots,C_u \tag{11.36}$$

$$z_{ijk} \leq y_{jk} \quad i = 1,2,\dots,np,\ j = 1,2,\dots,nm,\ k = 1,2,\dots,C_u \tag{11.37}$$

$$x_{ik} + y_{jk} - z_{ijk} \leq 1 \quad i = 1,2,\dots,np,\ j = 1,2,\dots,nm,\ k = 1,2,\dots,C_u \tag{11.38}$$

$$0 \leq z_{ijk} \leq 1 \quad i = 1,2,\dots,np,\ j = 1,2,\dots,nm,\ k = 1,2,\dots,C_u \tag{11.39}$$

Here, x_{ik} and z_{ijk} are real variables, whereas y_{jk} are integer variables. A major advantage of model M5 is that the number of integer variables is a function of the product of the total number of machines and the number of cells. For many problems, this number is not large, so the general-purpose branch-and-bound algorithm or slightly more sophisticated ones found in LINGO may be used. When the product of the number of machines and the number of cells is large, however, as in some large real-world problems, the available branch-and-bound algorithms may not be practical, because the computation time required may be enormous. Hence, in the next section, we discuss the application of Benders' decomposition for solving model M5. Due to the combinatorial nature of the cell formation problem, more heuristic than optimal algorithms have been developed. We use Benders' decomposition approach to obtain optimal solutions of large, real-world cell formation problems.

11.5.2 SOLUTION ALGORITHM: BENDERS' DECOMPOSITION

As discussed in Section 11.2.2, the basic idea in Benders' decomposition approach is to iterate between two problems: a subproblem and a master problem—both derived from the original mixed-integer problem. If the OFV of the optimal solution to the master problem is greater than or equal to that of the subproblem, we terminate the iteration and obtain the optimal solution of original mixed-integer problem. Otherwise, we add Benders' cuts, one at a time to the master problem, and solve it until the termination criterion is met.

To apply Benders' decomposition, we need to derive a primal problem from model M5 by fixing the values of integer variables y_{jk}'s to yield a feasible solution and thus ignoring the constraints with only y_{jk}, i.e., Equations 11.11 through 11.13. Here is the reformulated primal problem:

Model P (primal problem)

$$\text{Minimize} \sum_{i=1}^{np} \sum_{j=1}^{nm} \sum_{k=1}^{C_u} c_i v_i o_{ij} a_{ij} x_{ik} + \sum_{i=1}^{np} \sum_{j=1}^{nm} \sum_{k=1}^{C_u} \left[u_{ij}(1-a_{ij}) - c_i v_i o_{ij} a_{ij} \right] z_{ijk} \qquad (11.35)$$

Subject to constraints (11.28), (11.33), and (11.36) through (11.39)

Note that the y_{jk}'s in model P are known. Thus, the primal problem is a linear problem with two sets of real variables x_{ik} and z_{ijk}. The total number of variables in model P is $np \times C_u$ and the total number of constraints is $3 \times np \times nm \times C_u + np$. Next, we derive the following dual model from the primal problem:

Model D (Dual problem)

$$\text{Maximize} \sum_{i=1}^{np} \sum_{j=1}^{nm} \sum_{k=1}^{C_u} 0 \cdot l_{ijk} + \sum_{i=1}^{np} \sum_{j=1}^{nm} \sum_{k=1}^{C_u} (-y_{jk}) \cdot m_{ijk} + \sum_{i=1}^{np} \sum_{j=1}^{nm} \sum_{k=1}^{C_u} (y_{jk} - 1) \cdot n_{ijk} + \sum_{i=1}^{np} p_i \qquad (11.40)$$

$$\text{Subject to} \sum_{j=1}^{nm} l_{ijk} + 0 \cdot m_{ijk} + (-1) \sum_{k=1}^{C_u} n_{ijk} + p_i \le \sum_{j=1}^{nm} c_i v_i a_{ij} o_{ij} \quad i = 1,2,...,np; k = 1,2,...,np \qquad (11.41)$$

$$-l_{ijk} - m_{ijk} + n_{ijk} + 0 \left(p_i \right) \le u_{ij}(1-a_{ij}) - c_i v_i a_{ij} o_{ij} \quad i = 1,2,...,np, j = 1,2,...,nm, k = 1,2,...,C_u \qquad (11.42)$$

$$l_{ijk}, m_{ijk}, n_{ijk} \ge 0 \quad i = 1,2,...,np, j = 1,2,...,nm, k = 1,2,...,C_u \qquad (11.43)$$

$$p_i \text{ free} \quad i = 1,2,...,np \qquad (11.44)$$

The dual problem is linear with four sets of real variables l_{ijk}, m_{ijk}, n_{ijk}, and p_i corresponding to constraint sets (11.28), (11.36) through (11.38), respectively. The l_{ijk}, m_{ijk}, and n_{ijk} are nonnegative variables and p_i are free variables. Last, we formulate the master problem:

Model M (master problem)

$$\text{Minimize } Z \qquad (11.45)$$

Subject to (11.29) through (11.32), (11.34), and

$$Z + \sum_{i=1}^{np} \sum_{j=1}^{nm} \left[\sum_{k=1}^{C_u} \left(m_{ijk} - n_{ijk} \right) \right] (y_{jk}) \ge \sum_{i=1}^{np} p_i - \sum_{i=1}^{np} \sum_{j=1}^{nm} \sum_{k=1}^{C_u} n_{ijk} \qquad (11.46)$$

$$Z \ge 0 \qquad (11.47)$$

Although the master problem is a mixed-integer problem (due to variable Z), it can be easily converted to a pure 0–1 integer problem, as discussed in Chapter 10. Only the constraints with y_{jk} terms in model M5 are considered in the model M; constraint (11.46) is the Benders' cut, which is derived from the objective function of the dual problem. The number of constraints in the master problem is $nm + 3 \times C_u +$ number of iterations. The number of integer variables is $nm \times C_u$.

Here is the formal procedure for Benders' decomposition algorithm:

Step 0: Set $LB = -\infty$, $UB = +\infty$ and the initial y_{jk}'s $= 0$ for every j, k in model D.

Step 1: Solve model D; set the current optimal solution $= z^*$ and optimal variable values as l_{ijk}, m_{ijk}, n_{ijk}, and p_i. Set $UB = \min\{UB, z^*\}$. If $LB \geq z^*$, stop; we have the optimal solution of the original mixed-integer problem. Otherwise, go to step 2.

Step 2: Update model M by adding a new constraint (11.46) to it, solve the model and obtain optimal solution Z^* and optimal values for integer variables y_{jk}. Set $LB = Z^*$, and update y_{jk}'s in the dual problem. Go to step 1.

We now elaborate on the algorithm. First, by adding constraints (11.29) through (11.32) in the master problem, we guarantee that the solution to it will always be feasible to the original problem. Second, OFV of the solution to the master problem is a lower bound for the original problem. We add a stronger cut to the master problem to improve the lower bound in each iteration. If the lower bound is greater than or equal to the optimal solution in model D, we get an optimal solution to the original problem. Of course, the master problem is always feasible because the original problem is feasible. Third, the computational efficiency of the algorithm depends mainly on the number of iterations required and time needed to solve the master problem in each iteration. Although the master problem is a mixed-integer problem, the only continuous variable Z in the master problem can be rewritten entirely in terms of additional binary variables to make it a pure integer problem. Even after this conversion to a pure integer problem, it still takes more effort to solve it optimally in each iteration. Therefore, the variant of Benders' decomposition presented in Geoffrion and Graves (1974) and discussed in Section 11.2.2 is used. Instead of solving the master problem optimally, we solve only until the first feasible solution is found with value below $UB - \varepsilon$, $\varepsilon > 0$, where UB is the best solution of model D among the iterations, and ε is an allowed error margin. The master problem is modified as follows:

Model M′ (modified master problem)

$$\text{Minimize} \sum_{i=1}^{np} \sum_{j=1}^{nm} \left[\sum_{k=1}^{C_u} \left(m_{ijk} - n_{ijk} \right) \right] \left(y_{jk} \right) \tag{11.48}$$

Subject to (11.11) through (11.14), (11.16) and

$$Z + \sum_{i=1}^{np} \sum_{j=1}^{nm} \left[\sum_{k=1}^{C_u} \left(m_{ijk} - n_{ijk} \right) \right] \left(y_{jk} \right) \geq \sum_{i=1}^{np} p_i - \sum_{i=1}^{np} \sum_{j=1}^{nm} \sum_{k=1}^{C_u} n_{ijk} \tag{11.49}$$

Model M' is a pure integer problem and the objective function can be arbitrarily assigned. If $\varepsilon > 0$, constraint (11.49) will ensure that the same integer solution will not be regenerated. Here is the modified algorithm:

Step 0: Set $UB = +\infty$ and $y_{jk} = 0$ for every j, k.

Step 1: Solve model D; set the current optimal solution $= z^*$ and optimal variable values as l_{ijk}, m_{ijk}, n_{ijk}, and p_i. Set $UB = \min\{UB, z^*\}$.

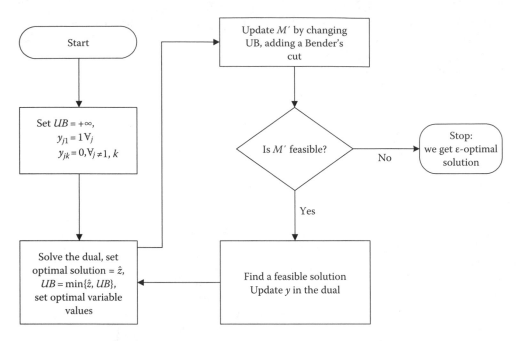

FIGURE 11.22   Flow chart for the modified Benders' decomposition algorithm.

Step 2: Update model M' by adding a new constraint (11.49) to it. Obtain a feasible solution to M' and update y_{jk} values. Go to step 1. If M' is infeasible, stop—we have an ε-optimal solution.

The flowchart for the modified Benders' decomposition algorithm is provided in Figure 11.22.

Example 11.6

In this section, we provided an example to show how Benders' decomposition approach may be used for solving practical cell formation problems. A company manufactures 20 part types on 6 machines and a maximum number of 3 cells is allowed. The related costs, information of alternate routings, multiple identical machines, and practical constraints are all considered and assumed. These required data are provided:

1. The part-machine processing indicator matrix (**A**) in Figure 11.23 includes the part operation requirement information and alternate routings. This matrix has two additional columns that show the fixed cost of handling a unit load between cells and the number of parts manufactured and transported per batch.
2. Machine relationship matrix (**R**) is in Figure 11.24.
3. The resource underutilization cost for each part-machine pair is 0.2.
4. The minimum and maximum number of machines in a cell are 1 and 5, respectively.

The first row of matrix **A** indicates that 80% of part 1 is processed on machines 1, 2, and 4 and 20% is processed on machines 1, 2, and 5. A similar interpretation can be derived from all the other rows of the **A** matrix. In the machine relationship indicator matrix **R**, an O indicates no restriction is applied in the relative placement of the corresponding two machines, an A indicates that the two machines have to be grouped in the same cell, and an X indicates the two machines cannot be grouped in the same cell. For this problem, the **R** matrix shows that machines 2 and 4 must be in the same cell, machines 3 and 6 cannot be in the same cell, and no restrictions hold for any other machine pairs.

	Machine						Intercellular movement cost	Batch size
	1	2	3	4	5	6		
1	1	1	0	0.8	0.2	0	1	1
2	0	0	0	0	0	1	0.5	2
3	0	0	0	0.85	0.15	0	1	1
4	1	1	0	1	0	1	0.3	1
5	1	0	0	1	0	0	1	1
6	0	0	1	0	0	1	1	1
7	0	1	0	1	0	0	1	1
8	0	0	0	0	1	1	1	2
9	0.1	0.9	1	1	0	0	1	1
10	1	1	0	0	0	0	1.2	1
11	0	0	0	0	0	1	0.2	1
12	0	0	0	0	1	1	1	1
13	0	0	1	0	0	0	1	3
14	0	0	0	0	1	0	1	1
15	1	0	0	0	1	1	1	1
16	1	0	0	0	1	1	1	1
17	0	0	0	0	1	1	1	1
18	0.8	0.2	0	1	0	0	1	1
19	1	1	0	0	0	0	1	2
20	0	0	1	0	0	0	1	1

$A =$ Part number

FIGURE 11.23 Part–machine processing indicator matrix.

	Machine					
	1	2	3	4	5	6
1	O	O	O	O	O	O
2	O	O	O	A	O	O
3	O	O	O	O	O	X
4	O	A	O	O	O	O
5	O	O	O	O	O	O
6	O	O	X	O	O	O

$R =$ Machine

FIGURE 11.24 Machine relationship indicator matrix.

		Machine							Intercellular movement cost	Batch size
		1	2	3	4	5	6	1'		
	1	0.5	1	0	0.8	0.2	0	0.5	1	1
	2	0	0	0	0	0	1	0	0.5	2
	3	0	0	0	0.85	0.15	0	0	1	1
	4	0.5	1	0	1	0	1	0.5	0.3	1
	5	0.5	0	0	1	0	0	0.5	1	1
	6	0	0	1	0	0	1	0	1	1
	7	0	1	0	1	0	0	0	1	1
	8	0	0	0	0	1	1	0	1	2
	9	0.05	0.9	1	1	0	0	0.05	1	1
$\mathbf{A} =$	10	0.5	1	0	0	0	0	0.5	1.2	1
	11	0	0	0	0	0	1	0	0.2	1
	12	0	0	0	0	1	1	0	1	1
	13	0	0	1	0	0	0	0	1	3
	14	0	0	0	0	1	0	0	1	1
	15	0.5	0	0	0	1	1	0.5	1	1
	16	0.5	0	0	0	1	1	0.5	1	1
	17	0	0	0	0	1	1	0	1	1
	18	0.4	0.2	0	1	0	0	0.4	1	1
	19	0.5	1	0	0	0	0	0.5	1	2
	20	0	0	1	0	0	0	0	1	1

Part number (row label)

FIGURE 11.25   Revised part–machine processing indicator matrix.

Before we solve this problem, we need to determine whether we have adequate production capacity. Using a linear programming model, we find the minimum number of each machine type required to satisfy the machine capacity constraint. Two units of machine type 1 and one unit of the remaining are required, so the total number of machines in the system is 7. We assign the additional unit of machine 1 a number 1', and then, in the part machine matrix, we equally divide the probability of parts processed on machine 1 into two identical machines 1 and 1'. This information is then used to develop model M4. The revised part-machine processing indicator matrix is provided in Figure 11.25.

The next step is to solve model M5, which is a mixed-integer problem with 480 real variables and 21 integer variables. The modified Benders' decomposition is used to ensure quick termination. The error margin, ε, is set at one, so the possible largest difference between the solution value we find and the optimal solution is 1. After 65 iterations, a near-optimal solution with an OFV of 11.72 is found. From the values of x_{ik}'s and y_{jk}'s, we can identify the part families and the corresponding machine cells. The solution is illustrated in Table 11.4.

TABLE 11.4
Final Solution for Example 11.6

Cell 1	Part family 1	Part 1, 3, 4, 5, 7, 9, 18, 19
	Machine group 1	Machine 1, 1′, 2, 4
Cell 2	Part family 2	Part 10, 13, 20
	Machine group 2	Machine 3
Cell 3	Part family 3	Part 2, 6, 8, 11, 12, 14, 15, 16, 17
	Machine group 3	Machine 5, 6

11.6 NEXT-GENERATION FACTORY LAYOUTS

All the layout models presented in Chapter 10 and solution techniques presented in Chapters 5 and 11 treat the layout problem in a deterministic manner. For example, a major assumption made by traditional layout models and algorithms is that the material flow between each pair of departments is known well into the future (say for a period of 5 years or more) and that these values do not change. Current realities simply do not allow us to look at these values as being static. The dynamic nature of today's manufacturing and service systems does not allow us to predict material flows that far into the future. Often, we are fortunate if we can predict material flows for the upcoming planning period. Moreover, two changes taking place in the manufacturing design and manufacturing process technologies enable us to envision layouts that can be changed relatively frequently for relatively low cost (Heragu and Kochhar, 1994):

1. Because we are making more and more products with lightweight composites that can be engineered to have all the desired tensile and mechanical strength–related properties, machines processing these composite products or material handling systems or transporting them need not be heavy or require elaborate foundations.
2. Noncontact manufacturing processes such as electron beam welding and laser cutting also suggest that we do not need heavy machines with elaborate foundations. Layouts in which lightweight machines are mounted on tracks using wheels attached to their base are highly flexible because the machines can be clamped in any desired location in a grid of tracks on a factory floor. Such a layout has been proposed as the next-generation factory layout (Heragu and Kochhar, 1994). This design allows machines to be easily moved from one location to another as frequently as production changes warrant.

Composites are the primary choice for many discrete manufactured components. Aluminum composites, for instance, can now replace cast iron parts. Phenolics can replace aluminum parts (Arimond and Ayles, 1993; Fujine et al., 1993). Not only are these light, but they can be engineered to have excellent mechanical properties such as hardness, heat resistance, tensile strength, and vibration absorption. The last property permits machine tool designers to design functionally equivalent, but lighter tools that do not require an elaborate foundation, making them easily movable. Nonabrasive manufacturing process technology such as laser cutting, electron beam hardening, and molecular nanotechnology also support machine tool designers' quest for making lightweight machining equipment (Asari, 1993). Permanent magnetic chucks that facilitate quick mounting and dismounting of tools have been developed (American Machinist, 1993). In fact, these chucks do not magnetize the cutting tool, carry their own energy source, and do not obstruct machining. These features by themselves support rapid equipment reconfiguration. This trend is likely to continue well into the next two decades. In fact, through a workshop and a delphi survey, the committee on Visionary Manufacturing Challenges for 2020 has identified adaptable processes and equipment and reconfiguration of manufacturing operations as two key enabling technologies that will help companies overcome two of the six grand challenges or fundamental goals to remain productive and profitable in the year 2020 (National Research Council, 1998). These grand challenges are to "achieve concurrency in all operations" and to "reconfigure manufacturing enterprises rapidly in response to changing needs and opportunities."

Numerous examples where facility layouts are modified on a frequent basis, sometimes every few months, are cited in Benjafaar et al. (2002). For example, Northern Telecom, in one of its manufacturing facilities facing constant product design changes, employs conveyor-mounted work cells that can be readily relocated just before a scheduled production/assembly change (Assembly Magazine, 1996c). A primary advantage of reconfiguring a layout when warranted by changes in product mix and volume is that material handling cost can be minimized because equipment can be reconfigured to suit the new production mix and volume. Of course, this cost must more than offset the cost of moving equipment from its current location to a new one.

A layout problem becomes more difficult to solve when multiple layout contexts must be considered and the problem needs to be solved frequently in a real-time mode. A typical current-day manufacturing company faces constantly changing product volumes and mix, which makes it necessary to update the layout accordingly in order to operate efficiently. Simultaneously, the rapid advances in materials engineering and manufacturing technology have made it practical and economical to switch layout when needed.

11.6.1 The Need for Stochastic Analysis of Layout Problems

Given the new reality that layouts are likely to change very frequently, possibly every few months rather than years, and that at best, we only have knowledge of the production activities during the upcoming planning period, we need to develop a layout only for the next planning period. In addition, due to the short-term life of a given layout and availability of production data for this time period, it is possible to consider optimizing operational performance measures such as minimizing part cycle times and work-in-process (WIP) inventory. Notice that it is relatively easy to get detailed data on material flows, machine setup, process and transfer batching, and others relative to the production activities in the next period compared to manufacturing activities for the next 5 years. Although the required data are available, it is well known that these data are not static. For example, the time to transport a load from one machine to another is a function of where the material handling device (assigned to transfer that load) is located at the time the material transfer request comes in. Even if we were to assume that the device is always at the same known location, the travel time is determined by congestion and other factors and is highly variable.

Thus, it makes sense to develop stochastic models for layout analysis. Furthermore, because many factors—deterministic as well as stochastic—have an impact on determining the suitability of a layout, we need tools that can design and analyze a layout with respect to static design criteria and stochastic operational performance criteria. Developing layouts on the basis of static design criteria alone is inadequate and, in some cases, could lead to undesirable consequences. It has been shown that a layout that minimizes material transfer costs when considering only loaded trips (static design criteria) may have much higher WIP inventory (stochastic operational performance measure) than another layout that may not minimize (loaded) material transfer costs (Benjafaar, 2002). In fact, it is quite possible that a layout that minimizes loaded material transfer costs might be infeasible in the sense that it leads to an infinite WIP accumulation at one or more machines. We present an example to highlight this point. Consider the layout shown in Figures 11.26 and 11.27. Assume the only product manufactured in this facility visits machines in this sequence—{1, 2, 3, 4, 3, 4, 5, 6, 7, 8, 9, 10, 11, 10, 11, and 12}. Assuming material handling is accomplished by a pool of identical devices, realistic values for setup times, processing and transfer times, and processing and transfer batch sizes, we can use the Manufacturing Performance Analyzer (MPA) software available at http://sundere.okstate.edu/downloadable-software-programs-and-data-files, to determine if the average WIP inventory in the layout in Figure 11.26 is much higher than that in the

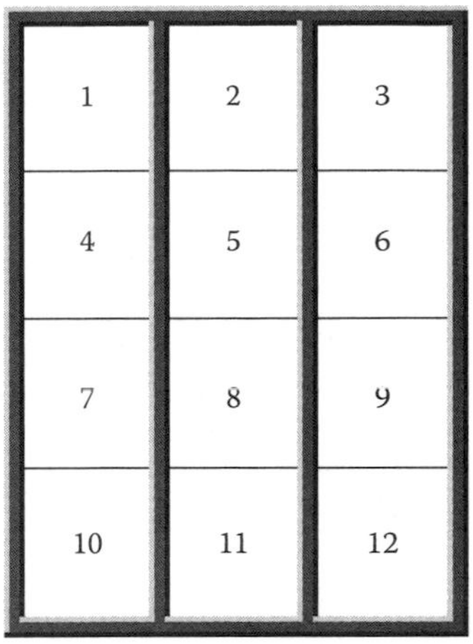

FIGURE 11.26   One possible layout for a facility manufacturing a single product.

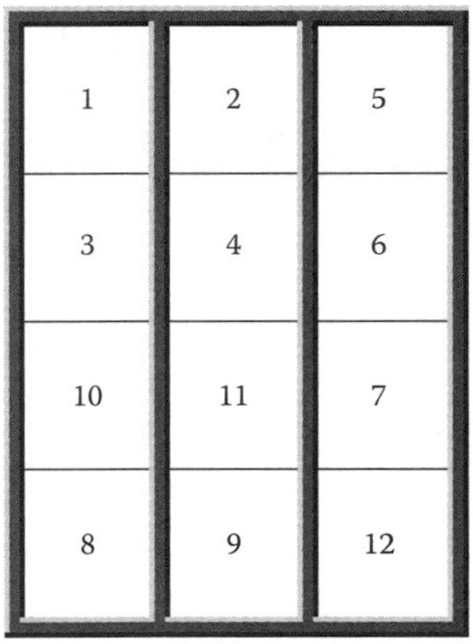

FIGURE 11.27   Alternate layout for the facility illustrated in Figure 11.26.

layout of Figure 11.27 (see Exercise 29). Although the former layout minimizes the static design criteria of minimizing loaded material handling costs, it performs so poorly relative to stochastic operational performance measures that it is completely unsuitable due to significant operational inefficiencies.

This example illustrates the fact that placing machines or sets of machines that may have very few loaded trips between them is sometimes advantageous because it minimizes empty material handling device travel, thereby increasing device utilization and decreasing product wait times as well as WIP inventory. Traditional methods that focus on minimizing static design criteria (e.g., minimize loaded material handling costs) completely ignore the effects of idle travel and thus will prefer layout in Figure 11.26 instead of the layout in Figure 11.27. In some cases, the WIP inventories of some layouts may be so significant that it could render a particular layout infeasible (Benjafaar et al., 2002). We refer the interested reader to Benjaafar (2002) and Benjafaar et al. (2002) for more details on this topic.

To summarize, the potential to frequently alter layouts, therefore, in a sense, transforms the modern layout problem from a strategic problem in which only long-term material handling costs are considered to a tactical problem in which operational performance measures such as reduction of product flow times, WIP inventories, and maximizing throughput rate are considered in addition to material handling and machine relocation costs when changing from one layout configuration to the next.

11.6.2   THREE-PHASE APPROACH FOR NEXT-GENERATION FACTORY LAYOUTS

Consider a three-phase procedure for the design and analysis of a next-generation factory layout. In the first phase, multiple, alternate layouts that perform well with respect to static design criteria are generated using available algorithms for facility layout. In the second phase, each alternate layout is evaluated with respect to several stochastic operational performance measures using the MPA software. MPA is described in more detail in Chapter 13. The analyses from phases 1 and 2 are combined to rank a set of three or so layouts that perform well with respect to the user-weighted design and operational performance criteria.

We apply the three-phase procedure described in the preceding paragraph to a small machine shop in this section. Table 11.5 provides the sequence of machines visited by 19 parts. It also shows part arrival rates. The following are assumed in the machine shop:

1. The machines have dimensions of 1 meter × 1 meter.
2. The shop floor space has the fixed dimension of 6 meters × 4 meters.
3. Work is assigned to specific machine within a machine type using "shortest travel time from previous operation machine" rule and each machine adopts a first-come, first-served service priority.

Table 11.5 shows the routings of products in the machine shop. Note that the letters in the routings represent machine types, not a specific machine. As shown in the current layout (see Figure 11.28),

TABLE 11.5
Operation Sequences of Products Produced in the Facility

Product #	Sequence	Arrival Rate (per Hour)
1	A → D → H → I	0.2
2	A → D → G → D → H → G	0.3
3	A → B → D → G → H → I	0.1
4	A → D → G → I	0.3
5	A → F → J → G → I	0.2
6	F → J → G → H → I	0.1
7	F → D → H → I	0.2
8	C → E → B → F → D → H → I	0.1
9	C → E → F → D → H → I	0.1
10	D → G → D → H	0.2
11	F	0.3
12	K → G → L	0.1
13	K → L	0.1
14	K → G → J	0.3
15	A → G → K → J → K → L	0.1
16	A → G → K → J → K → L	0.2
17	K → G → L	0.1
18	F → G → J	0.3
19	L	0.2

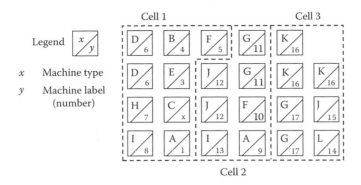

FIGURE 11.28 L0: Current cellular layout.

there may be multiple of a machine type. Note that duplicate machines are dispersed among multiple cells. For example, machine type J is dispersed in cells 2 and 3, labeled as machines 12 and 15, respectively. When several machines of the same type are placed together within a cell, they are labeled using one number. For example, the two type D machines in cell 1 are labeled together as machine number 6. The assumption here is that machines of same type placed together share the same incoming queue, thus working as a single station with multiple servers.

11.6.2.1 Generate Candidate Layouts

Candidate layouts can be generated using an existing technique such as the ones described in this chapter and Chapter 5, or by varying an existing layout. Because the current layout is cellular, we build a functional layout from scratch and generate several alternate cellular layouts. In all the layouts, we assume that machine **G** (#17 in Figure 11.28) cannot be moved due to hard constraints.

A pure functional layout is shown in Figure 11.29. Machines of the same type are grouped together into a work center. The relative positions of work centers are determined so as to minimize the material handling cost.

Layout L2 in Figure 11.30 is a slight variation of the current cellular layout L0 in Figure 11.28. There is no change in cell 1, but cells 2 and 3 are rotated clockwise by 90°. The idea is to put the two copies of machine A physically adjacent to their own machine types in the adjacent cells. Besides reorientation, layout L3 in Figure 11.31 further reshapes cells to achieve the maximum physical adjacency between machines of the same type. The cells are allowed to have L or S shape, compared with the U shape (or rectangular shape) in Figure 11.31. Layout L4 in Figure 11.32 tries to solve this problem by setting aside an extra cell and placing the machine types shared by different cells into that cell. The distance matrix of a layout is determined by the Manhattan distance metric between the mass center of the two work centers.

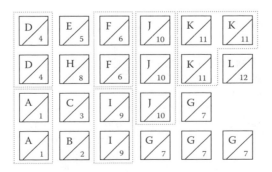

FIGURE 11.29 L1: Functional layout.

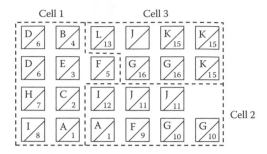

FIGURE 11.30   L2: Cellular layout with reorientation of cell 2 and cell 3.

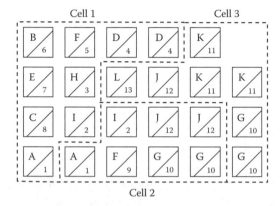

FIGURE 11.31   L3: Cellular layout with reorientation and reshaping of cells.

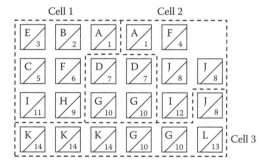

FIGURE 11.32   L4: Cellular layout with a remainder cell.

11.6.2.2 Choosing between Existing and Candidate Layouts

With the preceding processing and layout data, we can estimate the performance measures relative to each layout. These measures include both deterministic measures such as material handling cost and stochastic measures such as average waiting time in queue and average queue length. The stochastic measures are determined using MPA. To facilitate the decision-making process, some of the measures need to be aggregated into more informative measures. Product lead time is the sum of all processing times spent on the machine, waiting time in queue before each machine, and transfer time between machines. Average work-in-process (WIP) inventory level of the shop floor is the sum of the average queue length, over all the machines on the shop floor. Table 11.6 shows the WIP inventory level of the shop floor for each of the five layouts. Table 11.7 shows the material handling cost and lead time for each product for each of the five layouts. Product-specific material handling cost is

TABLE 11.6
Overall WIP Level of Different Layouts

			WIP Inventory		
M	L0	L1	L2	L3	L4
1	3.2	4.6	4.6	4.6	2.3
2	1.5	1.5	1.5	2.6	1.5
3	1.0	1.5	1.0	2.8	1.0
4	1.5	2.5	1.5	2.5	44.8
5	5.1	1.0	11.3	1.6	1.5
6	2.5	4.2	2.5	1.5	1.3
7	2.8	3.4	2.8	1.0	2.5
8	1.7	2.8	0.8	1.5	5.0
9	4.7	2.6	1.5	14.6	2.8
10	2.2	5.0	2.3	3.4	3.4
11	1.3	2.7	4.4	2.7	1.7
12	5.4	1.6	5.4	5.0	9.9
13	1.5		35.6	1.6	1.6
14	1.6		0.7		2.7
15	2.1		2.7		
16	2.7		2.9		
17	4.4				
Sum	45.1	33.4	81.4	45.4	82.1

TABLE 11.7
Material Handling Cost and Product Lead Times Corresponding to Different Layouts

		Material Handling Cost (Distance)						Lead Time (Hours)				
P	λ	L0	L1	L2	L3	L4	Due Date	L0	L1	L2	L3	L4
1	0.2	6.0	6.0	6.5	9.0	4.5	7.6	11.3	7.9	7.3	7.9	6.5
2	0.3	15.5	20.0	25.0	20.0	11.0	9.0	9.3	9.1	9.8	9.1	7.8
3	0.1	13.0	111.0	111.5	16.0	9.5	17.0	17.6	17.3	17.1	17.3	15.9
4	0.3	11.0	11.0	15.5	11.0	6.5	6.4	6.9	6.6	6.4	6.6	5.3
5	0.2	11.0	11.0	11.5	9.0	11.0	42.7	211.7	16.2	26.3	31.2	111.9
6	0.1	9.0	11.0	11.0	11.5	11.0	31.5	15.4	11.9	311.6	27.0	64.6
7	0.2	5.0	6.0	6.0	7.5	6.0	30.1	17.9	9.5	36.3	24.5	62.1
8	0.1	11.0	14.5	9.0	11.5	11.0	35.5	311.8	30.3	411.4	30.2	29.7
9	0.1	11.0	9.5	7.0	11.5	11.0	25.0	211.3	19.8	37.9	19.7	19.2
10	0.2	7.5	13.5	9.5	11.5	6.5	6.2	5.8	6.0	7.1	6.0	6.0
11	0.3	0.0	0.0	0.0	0.0	0.0	13.2	2.5	2.9	2.1	2.9	55.6
12	0.1	4.0	5.0	4.0	6.0	5.0	13.1	9.4	6.2	37.5	6.2	6.2
13	0.1	3.0	1.0	3.0	3.0	4.0	11.4	4.4	4.4	34.6	4.4	4.4
14	0.3	4.0	5.0	4.0	4.0	5.0	14.7	11.4	11.0	311.3	11.0	11.0
15	0.1	11.0	11.5	13.5	13.5	19.5	24.2	25.0	16.0	49.1	16.0	14.7
16	0.2	11.0	11.5	11.5	13.5	19.5	23.4	25.0	16.0	45.4	16.0	14.7
17	0.1	4.0	5.0	4.0	6.0	5.0	13.7	11.0	6.8	311.1	6.8	6.8
18	0.3	3.5	6.5	3.5	4.0	6.0	11.8	11.2	11.7	11.3	23.7	11.2
19	0.2	0.0	0.0	0.0	0.0	0.0	11.0	2.0	2.0	32.2	2.0	2.0
	MHD_Cost	132.6	164.0	160.0	161.5	139.0	Over due	11.2	0.9	2011.7	11.8	175.8

added to get the overall material handling cost of the system, using the formula $\sum_i \lambda_i T mhd_i$, where λ_i is the arrival rate of product type i and T is the length of the planning horizon, mhd_i is the material handling cost incurred while handling product i. To relate product lead time with cost, we calculate the delay (estimated completion time minus product due date), then sum over all product types to get the overall lateness of the products. The formula used is $\sum_i T\lambda_i$ overdue$_i$. The results are shown in the last row in Table 11.7.

Choosing among existing and candidate layouts is a multiple objective decision problem. Different companies might be concerned with different sets of cost terms. While most companies use deterministic terms such as material handling and relocation costs as well as stochastic terms such as WIP inventory cost and lead time, some companies might want to include unused space, machine utilization, or cell/machine center shape for consideration.

We have now aggregated the product- and machine-specific cost or performance measures (see Tables 11.6 and 11.7) into four system-wide cost measures: material handling cost, WIP inventory cost, product lateness penalty cost, and relocation cost (Tables 11.8 and 11.9). We use these cost terms to select the final layout. The next step is to combine the four cost measures into a single one. A layout can then be selected based on this aggregate cost measure.

To help us in choosing among the available layouts, we assume a set of cost measures with a corresponding unit cost vector. For example, let us assume that every distance unit traveled by the material handling device costs $5, one unit of shop floor inventory space costs $1 per hour, every hour of lateness of product delivery incurs a penalty of $10, and that the unit distance cost of relocating a machine is $0.1. Thus, for this set of cost measures—{WIP, material handling, Overdue, relocation}—the unit cost vector is {5, 1, 10, 0.1}. The overall cost of the manufacturing system is the weighted sum of all cost measures. Table 11.8 shows the overall cost for the five layouts with the unit cost vector of {5, 1, 10, 0.1}, and Table 11.9 shows the cost with unit cost vector of {1, 10, 1, 0.1}.

TABLE 11.8

Overall Cost with Unit Cost Vector of {5, 1, 10, 0.1}

Criteria	Unit Cost	L0	L1	L2	L3	L4
WIP	5	45.1	33.4	81.4	45.4	82.1
Material handling	1	132.6	164.0	160.0	161.5	139.0
Overdue	10	11.2	0.9	2011.7	11.8	175.8
Relocation	0.1	0.0	32.0	20.0	36.0	56.0
Overall cost		470.0	**342.6**	2656.0	509.6	2311.9

TABLE 11.9

Overall Cost with Unit Cost Vector of {1, 10, 1, 0.1}

Criteria	Unit Cost	L0	L1	L2	L3	L4
WIP	1	45.1	33.4	81.4	45.4	82.1
Material handling	10	132.6	164.0	160.0	161.5	139.0
Overdue	1	11.2	0.9	2011.7	11.8	175.8
Relocation	0.1	0.0	32.0	20.0	36.0	56.0
Overall cost		**1382.5**	1677.4	1892.1	1675.7	1653.5

The cost measures vector is generic, in the sense that any discrete manufacturing system can have the same set of cost measures. But the unit costs are typically company specific, reflecting a manufacturing system's resources, production control policy, and even management strategy. A low WIP inventory unit cost may be due to the fact that the company has relatively more shop floor space (i.e., shop floor space premium is not high) and a high overdue unit cost that reflect the emphasis of the company's eagerness to be responsive to market demand (i.e., penalize production delays). Note that the unit cost of relocation in the two examples is relatively small. There are two reasons. First, for reasons mentioned in Section 11.6.1, we are assume that relocation cost in a reconfigurable manufacturing system is relatively small and is a one-time cost, whereas the other costs (WIP, material handling, overdue penalty) accumulate over time and depend upon production volume.

The unit cost vector can also carry information about the user's solution approach to the layout problem. A unit cost vector of {1, 10, 1, 0.1} emphasizes the importance of material handling cost over WIP inventory and due-date-related penalty costs with a ratio of 10:1, which is close to the scenario of traditional layout problem where only deterministic measures are considered. The unit cost vector of {5, 1, 10, 0.1} emphasizes the importance of WIP and cycle time–related penalty costs over the relocation or material handling costs. Obviously, the unit cost vector determines the candidate layout that is finally chosen. As shown in Tables 11.8 and 11.9, different unit costs lead to different choices of layout. When the unit cost vector is {5, 1, 10, 0.1}, the pure functional layout L1 has the minimum overall cost. But when the unit cost vector is {1, 10, 1, 0.1}, cellular layout L0 has the minimum cost, which suggests keeping the current layout.

11.6.2.3 Refinement of Selected Layout

The last step is to refine the selected layout before actually applying it to the manufacturing system. One intuitive way is to combine the good feature of other competing candidate layouts into the selected one, without jeopardizing benefits of the current layout. With respect to layout L1, one refinement might pertain to the position of machine **D**. Because there are significant amount of appearances of **D** → **G** and **G** → **D** in product routings, switching the position of two type **G** machines with those of machine type **B** and **C** will not change the routing of the product, but it reduces material handling cost of those products that transfer between machine types **D** and **G** (Figure 11.33). Other refinements are possible and the designer must explore appropriate ones before settling on a layout for the next period.

11.7 SUMMARY

In this chapter, we discussed several algorithms for solving the layout problem. The first two were optimal algorithms and the others were heuristics. Although the optimal algorithms can only be used for small-sized problems due to their high computational requirements, they provide useful

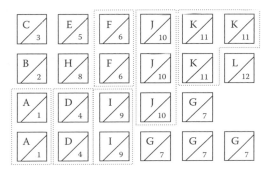

FIGURE 11.33 Final layout: Switch the position of D with B&C.

insight into the layout problem and allow us to devise efficient heuristics. Among the heuristic algorithms discussed in this chapter, simulated annealing, TS and GA are relatively new and very efficient. They provide near-optimal solutions to the layout problem and have also been successfully used in other applications.

The reconfigurable layout problem assumes that the production data are available only for the next planning period and are thus more realistic. While choosing between candidate layouts, reconfigurable layout considers not only deterministic material handling and relocation costs, but also the stochastic performance measures such as WIP inventory level and product lead time, making it a more comprehensive decision model. As a performance evaluation tool, MPA fits well in reconfigurable layout framework. It takes arrival, routing, processing, and facility data as input and yields stochastic cost (WIP level, lead time) of the manufacturing system.

11.8   REVIEW QUESTIONS AND EXERCISES

1. Why are heuristics preferred to optimal algorithms for solving layout problems?
2. What does the following statement mean: "The QAP is NP-complete."
3. Discuss and compare the systematic layout planning method (see Chapter 4) and the layout algorithms presented in this chapter.
4. Explain in your own words the terms assignment, partial assignment, and complete assignment in the context of a QAP. Provide examples.
5. Given the flow and distance matrices in Figure 11.34, compute a lower bound (see Section 11.1) on the optimal OFV, assuming that the flow matrix is
   (a) A *trips between* matrix
   (b) A *trips from–to* matrix
6. Find the optimal assignment of four departments to the four locations shown in Figure 10.19, using the flow matrix in Figure 11.35. Use the branch-and-bound technique to solve the problem.
7. Repeat Exercise 6 for the flow matrix in Figure 11.36.
8. Explain the differences between the general-purpose branch-and-bound algorithm (see Hillier and Lieberman, 1990) and the special-purpose branch-and-bound algorithm discussed in this chapter.
9. Consider the data provided in the Example 11.2. Set up DLP and MP for these data. Perform six iterations of Benders' decomposition algorithm. Is the solution you get optimal? Explain.

Flow matrix							Distance matrix					
	1	2	3	4	5	6		1	2	3	4	5	6
1	—	6	2	3	6	2	1	—	1	2	1	2	3
2	6	—	3	4	6	8	2	1	—	1	2	1	2
3	2	3	—	0	2	0	3	2	1	—	3	2	1
4	3	4	0	—	3	3	4	1	2	3	—	2	1
5	6	6	2	3	—	1	5	2	1	2	2	—	1
6	2	8	0	3	1	—	6	3	2	1	1	1	—

FIGURE 11.34   Flow and distance matrices for Exercise 5.

Flow matrix

	1	2	3	4
1	—	6	2	3
2	3	—	8	0
3	2	0	—	2
4	4	2	2	—

FIGURE 11.35 Flow matrix for Exercise 6.

Flow matrix

	1	2	3	4
1	—	1	1	2
2	2	—	3	4
3	1	2	—	4
4	2	1	0	—

FIGURE 11.36 Flow matrix for Exercise 7.

10. Solve the problem described in Example 2 using the modified Benders' decomposition algorithm and an ε value of 4.
11. Write a computer program in any language (C++, TurboPascal, FORTRAN, etc.) for solving the single-row layout problem using the simulated annealing algorithm. The program should read the following data: (a) number of departments, (b) frequency-of-trips-between-departments matrix, (c) horizontal dimension of each department, (d) arrangement of departments in the initial solution of corresponding objective function value, and (e) simulated annealing parameters T, r, *NLIMIT*, and *NOVER*. It should provide the following output:
    (i) Each improved solution and corresponding OFV
    (ii) The best solution of OFV obtained at the end of the run
        Test your program using data from Example 5.1.
12. Modify the program in Exercise 11 so that 2-opt is used instead of simulated annealing. Test your program using data from Example 5.1.
13. Modify the program in Exercise 11 so that greedy 2-opt is used instead of simulated annealing. Test your program using data from Example 5.1.
14. Write a computer program to solve the single-row layout problem using the hybrid simulated annealing algorithm. Test your program using data from Example 5.1.
15. Consider Exercise 6 in Chapter 5. Solve it using the modified penalty algorithm available at http://sundere.okstate.edu/downloadable-software-programs-and-data-files. Generate a layout using the sequence produced by the program.
16. Repeat the aforementioned question for Exercise 7 in Chapter 5.
17. Repeat the aforementioned question for Exercise 8 in Chapter 5.
18. Repeat the aforementioned question for Exercise 9 in Chapter 5.

19. Repeat Exercise 11 for a multirow layout problem. Test your program using data from Example 11.3.

20. Repeat Exercise 12 for a multirow layout problem. Test your program using data from Example 11.3.

21. Repeat Exercise 13 for a multirow layout problem. Test your program using data from Example 11.3.

22. Repeat Exercise 14 for a multirow layout problem. Test your program using data from Example 11.3.

23. *Project*: Collect relevant literature on TS and the single-row layout problem from your library. Write a computer program in a language of your choice to solve the single-row layout problem via the TS algorithm. Use the parameters specified in Skorin-Kapov (1990). Test your algorithm's performance on the test problems provided in Heragu and Alfa (1992) and compare your results with the best results provided in Heragu and Alfa (1992).

24. *Project*: Collect relevant literature on GA and the single-row, QAP, and multirow unequal area layout problems from your library. Write a computer program in a language of your choice to solve the single-row layout problem via the GA. Use the parameters specified in Tate and Smith (1995). Test your algorithm's performance on the test problems provided in Heragu and Alfa (1992) and compare your results with the best results provided in Heragu and Alfa (1992).

25. Solve the CMS design problem for the data in Exercise 7 of Chapter 6 using the Bender's decomposition algorithm. Assume that (a) the intercellular movement costs and resource utilization costs are 0.4 and 0.2, respectively, (b) there is no special adjacency relationship for the machines, and (c) the minimum and maximum number of machines in a cell are 1 and 5, respectively.

26. Repeat Exercise 25 assuming that the intercellular movement costs and resource utilization costs are 0.2 and 0.4, respectively. Comment on the difference between the two answers.

27. Solve the CMS design problem for the data in Exercise 8 of Chapter 6 using the Bender's decomposition algorithm. Assume that (a) the intercellular movement costs and resource utilization costs are 0.2 and 0.4, respectively, (b) there is no special adjacency relationship for the machines, and (c) the minimum and maximum number of machines in a cell are 1 and 5, respectively.

28. Repeat Exercise 27 assuming that the intercellular movement costs and resource utilization costs are 0.4 and 0.2, respectively. Comment on the difference between the two answers.

29. Consider the layout shown in Figures 11.26 and 11.27. Assume the only product manufactured in this facility visits machines in this sequence—{1, 2, 3, 4, 3, 4, 5, 6, 7, 8, 9, 10, 11, 10, 11, and 12}. Assuming material handling is accomplished by a pool of identical devices, realistic values for setup times, processing and transfer times, and processing and transfer batch sizes, use the MPA software to show that the average work-in-process inventory in the layout in Figure 11.26 is much higher than that in the layout of Figure 11.27.

30. Run the MPA software for the small data set that is provided with the software and determine the queuing related performance measures for all the parts, machines, and material handling devices.

31. Run the MPA software for the large data set that is provided with the software and determine the queuing related performance measures for all the parts, machines, and material handling devices. Run the automatically generated Promodel simulation output using Promodel simulation program and determine whether the analytical results are the same as simulation results. Explain any differences in the two solutions.

12 Advanced Location and Routing Models

12.1 MOTIVATING CASE STUDY

The Nike distribution center (DC) in Laakdal, Belgium (see Figure 12.1), which employs 800 people and has an annual turnover of 10.5 million units of footwear and apparel, covers 25 acres and cost $139 million to build. A location and design study was done in 1992 and the building was completed in two phases, the last in 1995. Let us examine why Nike selected Laakdal from several available locations in Europe. First, one of Nike's main business objectives was to serve 75% of its customers in less than 24 hours. Because of its proximity to major customer markets, Laakdal was a natural choice. The proximity to ports of entry for footwear and apparel manufactured overseas, the road network in and around Laakdal, and access to major highways were contributing factors. Because its citizens are required to go to school at least until the age of 18, Belgium has an educated workforce. Other factors also favored Laakdal.

12.2 INTRODUCTION

Much of this chapter focuses on the more realistic, multifacility location problem. As discussed in Chapter 9, logistics problems can be classified as

- Location problems
- Allocation problems
- Location–allocation problems

The typical questions that must be addressed in the more general, location–allocation problem are as follows:

1. How many new facilities are to be located in the distribution network that consists of previously established facilities and customers?
2. Where should the new facilities be located?
3. How large should each new facility be? In other words, what is the capacity of the new facility?
4. How should customers be assigned to the new and existing facilities? More specifically, which facilities should be serving each customer?
5. Can more than one facility serve a customer?

A model that can answer all or most of these questions would be desirable, but the more features we add to a model, the more difficult it is to solve. For multifacility location, however, we do have a model that captures a wide variety of factors and yet is relatively easy to solve. Moreover, this model has been used by companies (e.g., Hunt–Wesson Foods, Inc.) to make strategic logistical decisions covering large geographical areas. The algorithm to solve the model, however, is quite involved. It is based on Benders' decomposition approach introduced in Chapter 11, and is discussed in Section 12.5.3. We first cover models for the location and allocation problems that are rather easy to solve.

FIGURE 12.1 Nike distribution center in Laakdal, Belgium.

12.3 LOCATION MODELS

Problems in which the new facilities have *no* interaction among themselves can be looked at as several *independent* single-facility location problems. For example, if we have to introduce three new facilities into an existing distribution network and there is no interaction between the three new facilities, then we can set up three independent single-facility location problems with the appropriate distance measure (rectilinear, squared Euclidean, or Euclidean), solve each using the techniques discussed in Chapter 9, and simply combine the results to get a solution to the original problem. Although we can solve such special multifacility problems easily, the reader should be cautioned that, if the location of one or more new facilities coincides with that of an existing one, finding alternative optimal feasible locations using the contour line method is extremely difficult for all but trivial two-facility problems. Finding alternative suboptimal solutions is relatively easy however. Location problems in which there is interaction among new facilities and existing facilities and customers is more representative of the real world, so we turn our attention to such problems.

12.3.1 Multiple-Facility Problems with Rectilinear Distances

Consider a distribution network with m facilities. It is desired to add n new facilities to the network. Coordinates of the ith existing facility are (a_i, b_i). The problem is to find coordinates of the n new facilities, (x_i, y_i), $i = 1, 2, \ldots, n$ that minimizes the total distribution cost. The flow from a new facility i to an existing facility j is denoted by g_{ij}, and that between new facilities i and j is f_{ij}. The cost per unit distance of travel between new facilities i and j is denoted as c_{ij}, whereas that between new facility i and existing facility j is denoted as d_{ij}. This is the location problem:

Model 1

$$\text{Minimize} \sum_{i=1}^{n}\sum_{j=1}^{n} c_{ij}f_{ij}\left[\left|x_i - x_j\right| + \left|y_i - y_j\right|\right] + \sum_{i=1}^{n}\sum_{j=1}^{m} d_{ij}g_{ij}\left[\left|x_i - a_j\right| + \left|y_i - b_j\right|\right] \quad (12.1)$$

This nonlinear unconstrained model can be transformed easily into an equivalent linear, constrained model using the techniques in Chapters 10. For example, we define

$$x_{ij}^{+} = \begin{cases} x_i - x_j & \text{if } (x_i - x_j) > 0 \\ 0 & \text{otherwise} \end{cases} \quad (12.2)$$

$$x_{ij}^- = \begin{cases} x_j - x_i & \text{if } (x_i - x_j) \le 0 \\ 0 & \text{otherwise} \end{cases} \qquad (12.3)$$

We can observe that

$$\left| x_i - x_j \right| = x_{ij}^+ + x_{ij}^- \qquad (12.4)$$

$$x_i - x_j = x_{ij}^+ - x_{ij}^- \qquad (12.5)$$

A similar definition of $y_{ij}^+, y_{ij}^-, xa_{ij}^+, xa_{ij}^-, yb_{ij}^+,$ and yb_{ij}^- yields

$$\left| y_i - y_j \right| = y_{ij}^+ + y_{ij}^- \qquad (12.6)$$

$$y_i - y_j = y_{ij}^+ - y_{ij}^- \qquad (12.7)$$

$$\left| x_i - a_j \right| = xa_{ij}^+ + xa_{ij}^- \qquad (12.8)$$

$$x_i - a_j = xa_{ij}^+ - xa_{ij}^- \qquad (12.9)$$

$$\left| y_i - b_j \right| = yb_{ij}^+ + yb_{ij}^- \qquad (12.10)$$

$$y_i - b_j = yb_{ij}^+ - yb_{ij}^- \qquad (12.11)$$

Thus, the transformed linear model is

Model 2

$$\text{Minimize} \sum_{i=1}^{n} \sum_{j=1}^{n} c_{ij} f_{ij} \left(x_{ij}^+ + x_{ij}^- + y_{ij}^+ + y_{ij}^- \right) + \sum_{i=1}^{n} \sum_{j=1}^{m} d_{ij} g_{ij} \left(xa_{ij}^+ + xa_{ij}^- + yb_{ij}^+ + yb_{ij}^- \right) \qquad (12.12)$$

Subject to constraints (12.5), (12.7), (12.9), and (12.11).

$$x_{ij}^+, x_{ij}^-, y_{ij}^+, y_{ij}^- \ge 0 \quad i, j = 1, 2, \dots, n \qquad (12.13)$$

$$xa_{ij}^+, xa_{ij}^-, yb_{ij}^+, yb_{ij}^- \ge 0 \quad i = 1, 2, \dots, n; \ j = 1, 2, \dots, m \qquad (12.14)$$

$$x_i, y_i \text{ unrestricted in sign} \quad i = 1, 2, \dots, n; \ j = 1, 2, \dots, m \qquad (12.15)$$

As discussed in Chapter 10, for the transformed linear model 2 to be equivalent to expression (12.1), the solution must be such that one of the two new variables introduced, x_{ij}^+ or x_{ij}^-, but not both, is greater than zero. (If both are, then the values of x_{ij}^+ and x_{ij}^- do not satisfy their definition in Equations 12.2 and 12.3.) Similarly, only one of these pairs, y_{ij}^+, y_{ij}^-, and xa_{ij}^+, xa_{ij}^-, and yb_{ij}^+, yb_{ij}^-, must be greater than zero. Recall that this condition had to be satisfied for the LMIP 1 model in Chapter 10 as well as the median location model in Chapter 9. Fortunately, they are automatically satisfied in the linear model, just as they were for the median location model. Verification is left as an Exercise to the reader (see Exercise 2).

It turns out that the optimum x coordinate of each new facility is the same as that of an existing facility. The same is true for the y coordinates. If it turns out that the x and y coordinates of a new facility coincide with the x and y coordinates of a *single* existing facility, we must find alternate feasible locations heuristically using rules of thumb—for example, locate a new facility in a feasible location that is within 5 miles of the optimal one. It is rather difficult to use the contour line methods that worked so well for the single-facility case.

Model 1 can be simplified further by noting that x_i can be substituted as $a_j + xa_{ij}^+ - xa_{ij}^-$ due to Equation 12.9 and the fact that x_i is unrestricted in sign (urs). Similarly, y_i can also be substituted, resulting in a model with $2n$ fewer constraints and variables than model 1 (see Exercises 10 and 11).

Example 12.1

Tires and Brakes, Inc., is an automobile service company specializing in tire and brake replacement. It has four service centers and a warehouse in a large metropolitan area. The warehouse supplies tires, brakes, and other components to the service centers. The company manager has determined that he needs to add two more warehouses to improve the component delivery service. At the same time, he wants the location of the two new warehouses to minimize the cost of delivering components from the new warehouses to the existing facilities (four service centers and the existing warehouse) as well as between the new warehouses. The four service centers and warehouse have these coordinate locations—(8, 20), (8, 10), (10, 20), (16, 30), and (35, 20). It is anticipated that there will be one trip per day between the new warehouses. The number of trips between the new warehouses (W_1, W_2) and four service centers (SC_1–SC_4) as well as the existing warehouse (SC_5) is provided in the matrix in Figure 12.2. Develop a model similar to the transformed model 2 to minimize the distribution cost and solve it using LINGO.

Solution
Because the cost per unit of distance traveled is not given, we assume that the same type of vehicle is used for distribution and that the cost per unit distance traveled between any two facilities is 1. Here are the LINGO model for the problem and a partial solution.

	SC_1	SC_2	SC_3	SC_4	W_1
W_2	7	7	5	4	2
W_3	3	2	4	5	2

FIGURE 12.2   Number of trips between warehouses and service centers in Example 12.1.

```
Data:
    N = 2;
    M = 5;
Enddata
Sets:
    NewFac/1..N/: X, Y;
    ExistFac/1..M/: A, B;
    NewLinks(NewFac,NewFac): F, XP, XN, YP, YN;
    ExistLinks(NewFac,ExistFac): G, XAP, XAN, YBP, YBN;
Endsets
Data:
    F = 0 1
        1 0;
    G = 7 7 5 4 2
        3 2 4 5 2;
    A = 8 8 10 16 35;
    B = 20 10 20 30 20;
Enddata
! Objective function;
Min= @SUM(NewLinks(i,j): F(i,j)*(XP(i,j)+XN(i,j)+YP(i,j)+YN(i,j)))+
@SUM(ExistLinks(i,j): G(i,j)*(XAP(i,j)+XAN(i,j)+YBP(i,j)+YBN(i,j)));
! Constraints;
    @FOR (NewLinks(i,j): X(i)-X(j)=XP(i,j)-XN(i,j);
                    Y(i)-Y(j)=YP(i,j)-YN(i,j));
    @FOR (ExistLinks(i,j): X(i)-A(j)=XAP(i,j)-XAN(i,j);
                    Y(i)-B(j)=YBP(i,j)-YBN(i,j));
End
```

```
    Global optimal solution found.
    Objective value:                        370.0000
    Total solver iterations:                      16
```

Variable	Value
X(1)	8.000000
X(2)	10.00000
Y(1)	20.00000
Y(2)	20.00000
XP(2, 1)	2.000000
XN(1, 2)	2.000000
XAP(2, 1)	2.000000
XAP(2, 2)	2.000000
XAN(1, 3)	2.000000
XAN(1, 4)	8.000000
XAN(1, 5)	27.00000
XAN(2, 4)	6.000000
XAN(2, 5)	25.00000
YBP(1, 2)	10.00000
YBP(2, 2)	10.00000
YBN(1, 4)	10.00000
YBN(2, 4)	10.00000

As mentioned, we could have reduced the problem size by substituting values for the urs variables using some of the equality constraints. This would have eliminated not only the corresponding row that was used for the substitution but also the urs variable, thus reducing problem size. In this example, we made no substitution for any of the urs variables and, instead, explicitly declared them.

In the solution to the model, notice that the location of each new facility coincides with that of an existing one. We find alternate feasible locations heuristically by choosing an available location close to the optimal one for both new warehouses. Thus, coordinate locations of (8.6, 20) and (9.3, 20) could be used for the two new warehouses. In fact, because the warehouses are so close together, the manager may even consider locating just one larger warehouse at coordinate location (9, 20) or reformulate the preceding model under the assumption that only one new warehouse will be built and solve the resulting model to obtain the new location (Exercise 12).

12.3.2 MULTIPLE-FACILITY PROBLEMS WITH EUCLIDEAN DISTANCES

Consider this objective for the Euclidean distance problem. (Recall that the notation was introduced for the rectilinear distance problem).

Model 3

$$\text{Minimize} \sum_{i=1}^{n} \sum_{j=1}^{n} c_{ij} f_{ij} \left[\sqrt{\left(x_i - x_j\right)^2 + \left(y_i - y_j\right)^2} \right] + \sum_{i=1}^{n} \sum_{j=1}^{m} d_{ij} g_{ij} \left[\sqrt{\left(x_i - a_j\right)^2 + \left(y_i - b_j\right)^2} \right]$$

(12.16)

As in the single-facility model, we can take the partial derivative of expression (12.16) with respect to the variables x_i and y_i, set the equations to zero, and solve for the variables, because expression (12.16) can be shown to be a convex function. Taking the partial derivatives, we get

$$\sum_{j=1}^{n} \frac{c_{ij} f_{ij} \left(x_i - x_j\right)}{\sqrt{\left(x_i - x_j\right)^2 + \left(y_i - y_j\right)^2}} + \sum_{j=1}^{m} \frac{d_{ij} g_{ij} \left(x_i - a_j\right)}{\sqrt{\left(x_i - a_j\right)^2 + \left(y_i - b_j\right)^2}} \quad i = 1, 2, ..., n$$

(12.17)

$$\sum_{j=1}^{n} \frac{c_{ij} f_{ij} \left(y_i - y_j\right)}{\sqrt{\left(x_i - x_j\right)^2 + \left(y_i - y_j\right)^2}} + \sum_{j=1}^{m} \frac{d_{ij} g_{ij} \left(y_i - b_j\right)}{\sqrt{\left(x_i - a_j\right)^2 + \left(y_i - b_j\right)^2}} \quad i = 1, 2, ..., n$$

(12.18)

Because we have $2n$ variables and an equal number of constraints, we can solve Equations 12.17 and 12.18 to get optimal (x, y) coordinates for all the n new facilities. As noted in the single-facility Euclidean distance model, however, we must be able to guarantee that the optimal location of any new facility does not coincide with that of any existing facility. Because we cannot guarantee, we can develop an iterative heuristic procedure by adding a small quantity ε to the denominator in each term on the left-hand side of Equations 12.17 and 12.18. This is done in the following equations:

$$x_i' = \frac{\displaystyle\sum_{j=1}^{n} \frac{c_{ij} f_{ij} x_j}{\sqrt{\left(x_i - x_j\right)^2 + \left(y_i - y_j\right)^2} + \varepsilon} + \sum_{j=1}^{m} \frac{d_{ij} g_{ij} a_j}{\sqrt{\left(x_i - a_j\right)^2 + \left(y_i - b_j\right)^2} + \varepsilon}}{\displaystyle\sum_{j=1}^{n} \frac{c_{ij} f_{ij}}{\sqrt{\left(x_i - x_j\right)^2 + \left(y_i - y_j\right)^2} + \varepsilon} + \sum_{j=1}^{m} \frac{d_{ij} g_{ij}}{\sqrt{\left(x_i - a_j\right)^2 + \left(y_i - b_j\right)^2} + \varepsilon}} \quad i = 1, 2, ..., n$$

(12.19)

$$y_i' = \frac{\displaystyle\sum_{j=1}^{n} \frac{c_{ij} f_{ij} y_j}{\sqrt{\left(x_i - x_j\right)^2 + \left(y_i - y_j\right)^2} + \varepsilon} + \sum_{j=1}^{m} \frac{d_{ij} g_{ij} b_j}{\sqrt{\left(x_i - a_j\right)^2 + \left(y_i - b_j\right)^2} + \varepsilon}}{\displaystyle\sum_{j=1}^{n} \frac{c_{ij} f_{ij}}{\sqrt{\left(x_i - x_j\right)^2 + \left(y_i - y_j\right)^2} + \varepsilon} + \sum_{j=1}^{m} \frac{d_{ij} g_{ij}}{\sqrt{\left(x_i - a_j\right)^2 + \left(y_i - b_j\right)^2} + \varepsilon}} \quad i = 1, 2, ..., n$$

(12.20)

Because Equations 12.19 and 12.20 are now defined even when the optimal location of a new facility coincides with that of an existing one, we can begin with an initial value for x_i, y_i for each new facility i, substitute these values in Equations 12.19 and 12.20 to get the new values of x_i, y_i (denoted as x_i', y_i', respectively).

The new values of x_i, y_i are substituted in the right-hand side of Equations 12.19 and 12.20 to get the next set of values. This procedure is continued until two successive x_i, y_i values or the objective function values (OFVs) (obtained by substituting x_i, y_i values in expression (12.16)) are nearly equal. Although it cannot be proved, we assume convergence has occurred at this point and stop. Upper and lower bounds on the optimal OFV for the Euclidean distance problem can be found by examining the rectilinear distance solution (see Francis and White (1974) and Pritsker and Ghare (1970) for more details). Based on these bounds, we can tell how far off a given Euclidean solution is for a particular problem. For many practical problems, it has been found that the x_i, y_i values for the new facilities determined via the iterative procedure are very close to optimal. The iterative procedure is rather easy to set up in a spreadsheet. Note that large values of ε will ensure faster convergence, but the quality of the final solution is inferior compared with that obtained using a relatively smaller ε value. Thus, the user has to trade off quick convergence and solution quality and choose an appropriate value.

Example 12.2

Consider Example 12.1. Assuming the Euclidean distance metric is more appropriate and that Tire and Brakes, Inc., does not currently have a warehouse, determine where the two new warehouses are to be located.

Solution
Because there is no existing warehouse, we disregard information corresponding to that warehouse provided in Example 12.1. A partial spreadsheet setup to iteratively calculate the x_i and y_i values for the resulting problem is shown in Table 12.1. Also, shown in the spreadsheet are the flow and ε values as well as the coordinate locations of the existing service centers. The columns labeled C_1 through C_4 give the values of the following part of Equation 12.19 calculated for each service center's coordinate location (a_j, b_j).

$$\frac{d_{ij}g_{ij}}{\sqrt{\left(x_i - a_j\right)^2 + \left(y_i - b_j\right)^2} + \varepsilon}$$

Because this part of the expression does not change for Equation 12.20, we do not show the values again in the y_i rows. The column labeled C_5 in Table 12.1 shows the values for the following part of Equations 12.19 and 12.20:

$$\frac{c_{ij}f_{ij}}{\sqrt{\left(x_i - x_j\right)^2 + \left(y_i - y_j\right)^2} + \varepsilon}$$

Once again, because it is the same in both expressions, it is not shown in the y_i rows. Notice that in each iteration, this value is the same for each x_i row, because we only have two new warehouses to be located.

The column labeled C_6 gives the sum of the values in columns C_1 through C_5 and is the denominator of Equations 12.19 and 12.20. Using an initial seed of (8, 10) and (9, 10) for the two facilities, we begin the iterative procedure. To determine the coordinates of the two new warehouses for the kth iteration, we use the ε, flow, (a_j, b_j) values, values in columns C_1 through C_6 for the previous ($k - 1$)th iteration, and Equations 12.19 and 12.20. This procedure is repeated until two successive x_i, y_i values are equal. This occurs in the 13th iteration, and we therefore

TABLE 12.1

Warehouse Location Determination for Example 12.2 Using Spreadsheet

			SC_1	SC_2	SC_3	SC_4			
	W_1		7	7	5	4			
	W_2		3	2	4	5			
			a_1	a_2	a_3	a_4			
	x coordinate		8	8	10	16			
			b_1	b_2	b_3	b_4			
	y coordinate		20	10	20	30			
	ε		0.02						

Iteration		Coordinates	C_1	C_2	C_3	C_4	C_5	C_6	TC
1	x_1	8	0.6999	49.5	0.4902	0.183	0.99	52	441
	y_1	10							
	x_2	9	0.2985	1.98	0.398	0.236	0.99	3.9	
	y_2	10							
2	x_1	8.047	0.7216	20.89	0.5053	0.183	0.35	22.7	411
	y_1	10.3							
	x_2	8.941	0.4243	0.637	0.5644	0.272	0.35	2.25	
	y_2	12.994							
			.	.	.				
11	x_1	8.841	7.7071	0.734	3.8643	0.197	0.078	12.7	218
	y_1	19.497							
	x_2	9.967	1.5198	0.194	23.146	0.431	0.78	26.1	
	y_2								
12	x_1	8.794	7.5662	0.731	3.8568	0.197	0.77	12.1	218
	y_1	19.546							
	x_2	9.967	1.5198	0.194	23.149	0.431	0.77	26.1	
	y_2	20.094							
13	x_1	8.754	7.9405	0.729	3.7693	0.197	0.75	12.4	218
	y_1	19.566							
	x_2	9.967	1.5198	0.194	23.15	0.431	0.75	26	
	y_2	20.094							

stop the procedure. If we had used the total cost, shown in the last column as TC, to check for convergence, we would have stopped at the 12th iteration, because solutions in this and the 11th iteration yield the same total cost of 218. Had we used large values of ε, convergence would have occurred much sooner, but then we might have obtained a solution inferior to the current one. We encourage the reader to solve the example with different values of ε and with different initial seeds (see Exercises 8 and 9). Notice that this solution (8.754, 19.566) and (9.967, 20.094) is close to the one obtained for Example 12.1 and that both warehouses are located very close to existing service centers. If these locations are infeasible, we must determine alternate feasible solutions heuristically.

12.4 ALLOCATION MODEL

Service and manufacturing companies often find it necessary to maintain proximity to their markets and also to input sources. For service organizations, the input sources may be a skilled labor pool. For manufacturing companies, the input sources may be raw materials, power, water, and so on. The allocation problem is then to find the quantity of raw material that each supply source should be supplying to each plant, as well as the quantity of finished goods each plant should be supplying each customer. For the single product case, this problem can be set up as a transportation model and solved rather easily using the transportation model (Das and Heragu, 1988). A subproblem also involves finding how much material to transport through available routes so that we do not exceed route capacity restrictions while ensuring the material reaches its destination as required. We now discuss three classes of models—network flow problem, two-stage transportation problem, and vehicle-routing problem (VRP).

12.4.1 NETWORK FLOW MODEL

The model considered here—the minimum cost network flow problem—is a representation of the general network flow problem and it can be modified to model several special classes of models including the transportation model (discussed in Chapter 9), shortest path problem, and the maximal flow problem. We have already seen application of the transportation model for the location problem. In Section 12.3.2, we will use it to model an allocation problem. The more general network flow problem is can be used to model many allocation and location–allocation problems. This problem assumes we have n nodes, with an arc between each pair of nodes, and there is at least one source node where all the flow originates and a sink node to absorb that flow (Figure 12.3). If flow is to be prohibited along any arc, we can always set the cost per unit flow along that node to be very large. We will first present notation used in the model, then the model, and solve an allocation problem using a numerical example.

Model 4
Consider this notation:

c_{ij} Cost of sending one unit of flow on arc (i, j)
U_{ij} Upper bound on the units of flow that can be sent on arc (i, j), i.e., capacity of arc (i, j)
L_{ij} Lower bound on the units of flow that can be sent on arc (i, j)
D_i Net flow generated at node i
x_{ij} Number of units of flow on arc (i, j)

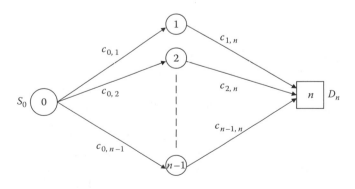

FIGURE 12.3 Graphical representation of a network flow problem.

$$\text{Minimize} \sum_{i=1}^{n} \sum_{j=1}^{n} c_{ij} x_{ij} \tag{12.21}$$

$$\text{Subject to} \sum_{j=1}^{n} x_{ij} - \sum_{i=1}^{n} x_{ji} = D_i \quad i = 1, 2, ..., n \tag{12.22}$$

$$L_{ij} \le x_{ij} \le U_{ij} \quad i, j = 1, 2, ..., n \tag{12.23}$$

The objective function (12.21) minimizes the total cost of sending the flow through the arcs. Constraint (12.22) is a flow conservation equation. Note that D_i is negative for the source node, positive for the sink node and 0 for transshipment nodes. Constraint (12.23) ensures that the flow on each arc does not exceed that arc's capacity and that it is above a certain minimum. The maximum and minimum could be set to infinity and zero, respectively, unless there are hard lower bound and upper bound constraints on the capacity of an arc. This model can be modified to formulate the transportation, assignment, shortest path, and maximum flow problems (Exercises 19 through 21). We now present an optimal algorithm, called the network simplex algorithm (NSA), to solve the network flow problem.

NSA

Step 1: Construct a spanning tree for the n nodes. The variables x_{ij} corresponding to the arcs (i, j) in the spanning tree are basic variables and the remaining are nonbasic. Find a basic feasible solution to the problem so that
1. The basic variables satisfy $L_{ij} < x_{ij} < U_{ij}$
2. The nonbasic variables take on a value of L_{ij} or U_{ij} to satisfy constraint (12.22)

Step 2: Set $u_1 = 0$ and find $u_j, j = 2, ..., n$ using the formula $u_i - u_j = c_{ij}$ for all basic variables.

Step 3: If $u_i - u_j - c_{ij} \le 0$ for all nonbasic variables x_{ij} with a value of L_{ij}, and $u_i - u_j - c_{ij} \ge 0$ for all nonbasic variables x_{ij} with a value of U_{ij}, then the current basic feasible solution is optimal; *stop*. Otherwise, go to step 4.

Step 4: Select the variable x_{i*j*} that violates the optimality condition (in step 3) the most, i.e., the largest of the $u_i - u_j - c_{ij}$ for those nonbasic variables with $x_{ij} = L_{ij}$, and the smallest of the $u_i - u_j - c_{ij}$ for those nonbasic variables with $x_{ij} = U_{ij}$. Make the arc $(i*, j*)$ a basic variable and add arc $(i*, j*)$ to the spanning tree. Make one of the other basic variables in the loop of basic variables (formed by including arc $(i*, j*)$) a nonbasic variable such that
1. x_{i*j*} takes on the largest possible value
2. Constraint (12.21) is satisfied for all the n nodes
3. Constraint (12.22) is satisfied for all the arcs in the loop

Remove the arc corresponding to the nonbasic variable just identified so that we have a spanning tree once again. Go to step 2.

A spanning tree is a set of arcs connecting all the nodes without forming a loop. Problems in which basic variables take on a value equal to L_{ij} or U_{ij} are degenerate problems, and we proceed by treating these variables as basic. The reason we construct a spanning tree is because it is well known that a basic solution always corresponds to a spanning tree. It can be shown that the NSA is an optimal algorithm. Consult introductory operations research textbooks (for example, Winston,

1994) for additional details. We now present an example to illustrate how the network flow model can be used to model an allocation problem.

Example 12.3

The Fast Shipping Company manages the distribution of lawnmowers from a company that has two factories (F_1 and F_2) in the Northeast to two large customer bases (C_1 and C_2) in the Southwest. For cost and freight consolidation reasons, Fast Shipping would like to route the shipments via three intermediate nodes ($T_1 - T_3$) located in the Midwest. The relevant data are provided in Tables 12.2 and 12.3. Set up a model to determine how the shipment is to take place from the two factories to the two destinations via the three intermediate shipment points.

Solution

A linear programming (LP) formulation and its LINGO solution (by the simplex algorithm in LINDO) is shown next. The NSA is a more efficient algorithm than simplex and the reader is encouraged to solve the problem using NSA (Exercise 22).

```
Data:
     N = 2;
     M = 3;
Enddata
Sets:
     OriginNodes/1..N/: DemandO;
     DestinationNodes/1..N/: DemandD;
     IntermediateNodes/1..M/: DemandI;
     OutboundArcs(OriginNodes,IntermediateNodes): C, X, LO, UO;
     InboundArcs(IntermediateNodes,DestinationNodes): D, Y, LD, UD;
Endsets
```

TABLE 12.2

Supply and Demand Information for Two Factories and Two Customers Bases in Example 12.3

Facility	Supply	Demand
F_1	900	—
F_2	600	—
C_1	—	750
C_2	—	750

TABLE 12.3

Inbound and Outbound Transportation Costs (Arc Capacities) for Example 12.3

	T_1	T_2	T_3
F_1	8(500)	11(1500)	5(350)
F_2	12(1200)	8(750)	5(450)
C_1	6(1000)	12(750)	9(1000)
C_2	3(150)	1(200)	19(1500)

```
Data:
    C = 8 11 5
        12 8 5;
    LO = 0 0 0
         0 0 0;
    UO = 500 500 350
         1200 750 450;
    D = 6 3
        12 1
        9 19;
    LD = 0 0
         0 0
         0 0;
    UD = 1000 150
          750 200
         1000 1500;
    DemandO=900 600;
    DemandD=750 750;
    DemandI=0 0 0;
Enddata
! Objective function;
Min= @SUM(OutboundArcs(i,j): C(i,j)*X(i,j))+
     @SUM(InboundArcs(i,j): D(i,j)*Y(i,j));
! Constraints;
     @FOR (OriginNodes(i): @SUM(IntermediateNodes(j): X(i,j))=DemandO(i));
     @FOR (DestinationNodes(j): @SUM(IntermediateNodes(i):
-1*Y(i,j))=-DemandD(j));
     @FOR (IntermediateNodes(i): @SUM(OriginNodes(j): X(j,i))=
@SUM(DestinationNodes(j): Y(i,j)));
     @FOR (OutboundArcs(i,j): X(i,j)<=UO(i,j));
     @FOR (InboundArcs(j,i): Y(j,i)<=UD(j,i));
End


    Global optimal solution found.
    Objective value:                    23700.00
    Total solver iterations:                   8


                    Variable           Value
                    X( 1, 1)        500.0000
                    X( 1, 2)         50.00000
                    X( 1, 3)        350.0000
                    X( 2, 2)        150.0000
                    X( 2, 3)        450.0000
                    Y( 1, 1)        350.0000
                    Y( 1, 2)        150.0000
                    Y( 2, 2)        200.0000
                    Y( 3, 1)        400.0000
                    Y( 3, 2)        400.0000
```

12.4.2 Two-Stage Transportation Model

Here, we consider an allocation model with two stages of distribution. We formulate an LP model for this problem, and show how a corresponding transportation tableau can be set up. The ideas are subsequently illustrated in a numerical example.

Consider this notation:

S_i Capacity of supply source i, $i = 1, 2, ..., p$
P_j Capacity of plant j, $j = 1, 2, ..., q$
D_k Demand at customer k, $k = 1, 2, ..., q$

c_{ij} Cost of transporting one unit from supply source i to plant j

d_{jk} Cost of transporting one unit from plant j to customer k

x_{ij} Number of units of raw material shipped from supply source i to plant j

y_{jk} Number of units of product shipped from plant j to customer k

The LP model is as follows:

Model 5

$$\text{Minimize} \sum_{i=1}^{p} \sum_{j=1}^{q} c_{ij} x_{ij} + \sum_{j=1}^{q} \sum_{k=1}^{r} d_{jk} y_{jk} \tag{12.24}$$

$$\text{Subject to} \sum_{j=1}^{q} x_{ij} \le S_i \quad i = 1, 2, ..., p \tag{12.25}$$

$$\sum_{i=1}^{p} x_{ij} \le P_j \quad j = 1, 2, ..., q \tag{12.26}$$

$$\sum_{j=1}^{q} y_{jk} \ge D_k \quad k = 1, 2, ..., r \tag{12.27}$$

$$\sum_{i=1}^{p} x_{ij} = \sum_{k=1}^{r} y_{jk} \quad j = 1, 2, ..., q \tag{12.28}$$

$$x_{ij}, y_{jk} \quad i = 1, 2, ..., p; \; j = 1, 2, ..., q; \; k = 1, 2, ..., r \tag{12.29}$$

The objective function (12.24) minimizes the cost of inbound as well as outbound shipments. Constraint (12.25) ensures the raw material shipped out from each supply source does not exceed its capacity limits. Constraint (12.26) ensures the raw material shipment received from all the supply sources at each plant does not exceed the capacity limits. Constraint (12.27) requires that the total amount of finished products shipped from the plants to each customer be sufficient to cover the demand. Constraint (12.28) is a material balance equation ensuring all the raw material that comes into each plant is shipped out as finished product to customers. Notice that we are implicitly assuming a unit of finished product requires one unit of raw material. If this is not the case, we can adjust the model easily as discussed in Das and Heragu (1988).

For model 5 to be transformed into an equivalent transportation model, it is required that either the plants or the raw material supply sources—but not both—have limited capacity. If the plants and raw material sources both have limited capacity, the problem cannot be set up as a transportation model and hence, we cannot use the transportation simplex algorithm (TSA) discussed in Chapter 11. However, we solve the model 5 using the simplex algorithm.

We now discuss the special case where either the plants or the raw material sources have limited capacity. Depending on whether supply sources or plants have limited capacities and whether supply exceeds demand, the following four cases arise:

1. Supply source capacity is unlimited, plant capacity is limited, and total plant capacity exceeds total demand.

2. Supply source capacity is unlimited, plant capacity is limited, and total demand exceeds total plant capacity.
3. Plant capacity is unlimited, supply source capacity is limited, and total supply source capacity exceeds total demand.
4. Plant capacity is unlimited, supply source capacity is limited, and total demand exceeds total supply source capacity.

In our discussion here, the supply sources are assumed to have unlimited capacities and the total plant capacity is more than the total demand (case 1). (Model 5 can be transformed rather easily into an equivalent transportation model. This is left to the reader as Exercise 24.) The transportation tableau is set up as shown in Table 12.4. The third case is discussed in Example 12.4. For the other two cases also, the transportation tableaus are rather easy to set up and are shown in Das and Heragu (1988); also, see Exercises 25 and 26.

In the transportation tableau, there are p rows corresponding to each plant even though only one plant can be set up at each location j, where $j = 1, 2, \ldots, q$. This accounts for the possibility that each plant may receive raw material from any supply source i. Because $p - 1$ excess rows have been introduced for each plant j with a capacity of P_j, we remove this excess by introducing dummy plants $1, 2, \ldots, q$ to absorb the excess plant capacity. The "demand' in these q columns is therefore $(p - 1)$ $P_j, j = 1, 2, \ldots, q$. The unit transportation costs for each of these columns are zero in the corresponding rows and are large (denoted as M) in others. In other words, the dummy plant column j has zero transportation costs in the jth row for each supply source i, $i = 1, 2, \ldots, p$. It is M in all other rows for column j. The last demand column is introduced to absorb the excess of total plant capacity over the total demand and has zero cost in all the rows.

Now that the transportation tableau is set up, it may be efficiently solved using the TSA. Although the discussion thus far has pertained to problems with unlimited supply source capacities, in Example 12.4, we assume that plant capacities are unlimited in order to show the versatility of the approach.

TABLE 12.4

Unit Transportation Costs for Case 1 of the Two-Stage Transportation Model

Supply Source	Plant	Customer				Dummy Plant				Excess Plant Capacity	Capacity
		1	2	...	r	1	2	...	q		
1	1	c_{111}	c_{112}	...	c_{11r}	0	M	...	M	0	P_1
	2	c_{121}	c_{122}	...	c_{12r}	M	0	...	M	0	P_2
	$\vdots$	$\vdots$	$\vdots$	$\ddots$	$\vdots$	$\vdots$	$\vdots$	$\ddots$	$\vdots$	$\vdots$	$\vdots$
	q	c_{1q1}	c_{1q2}	...	c_{1qr}	M	M	...	0	0	P_q
2	1	c_{211}	c_{212}	...	c_{21r}	0	M	...	M	0	P_1
	2	c_{221}	c_{222}	...	c_{22r}	M	0	...	M	0	P_2
	$\vdots$	$\vdots$	$\vdots$	$\ddots$	$\vdots$	$\vdots$	$\vdots$	$\ddots$	$\vdots$	$\vdots$	$\vdots$
	q	c_{2q1}	c_{2q2}	...	c_{2qr}	M	M	...	0	0	P_q
$\ddots$											
	1	c_{p11}		...	c_{p1r}	0	M	...	M	0	P_1
p	2	c_{p21}	c_{p22}	...	c_{p2r}	M	0	...	M	0	P_2
	$\vdots$	$\vdots$	$\vdots$	$\ddots$	$\vdots$	$\vdots$	$\vdots$	$\ddots$	$\vdots$	$\vdots$	$\vdots$
	q	c_{pq1}	c_{pq2}	...	c_{pqr}	M	M	...	0	0	P_q
Demand		D_1	D_2	...	D_r	$(p-1)P_1$	$(p-1)P_2$	...	$(p-1)P_q$	$\sum_{j=1}^{q} P_j - \sum_{k=1}^{r} D_k$	$p\sum_{j=1}^{q} P_j$

Example 12.4

Two-Stage Distribution Problem: RIFIN Company has recently developed a new method of manufacturing a type of chemical. The method involves refining a certain raw material that can be obtained from four overseas suppliers A, B, C, and D who have access to the four ports at Vancouver, Boston, Miami, and San Francisco, respectively. RIFIN wants to determine the location for the plants that will refine the material. Once refined, the chemical will be transported via trucks to five outlets located in Dallas, Phoenix, Oregon, Montreal, and Orlando. After an initial study, the choice of location for RIFIN's refineries has been narrowed down to Denver, Atlanta, and Pittsburgh. Assume that one unit of the raw material is required to make one unit of the chemical. The amount of raw material that can be obtained from suppliers A, B, C, and D and the amount of chemical required at the five outlets are known and are given in Table 12.5. The cost of transporting the raw material from each port to each potential refinery and the cost of trucking the chemical to outlets are also provided in Tables 12.6 and 12.7, respectively. Determine the locations of RIFIN's refining plants, their capacities, and the distribution pattern for the raw material and processed chemical.

Solution

Figure 12.4 is a pictorial representation of the problem. We can reasonably assume that there is no practical limit on the capacity of the refineries at any of the three locations, Atlanta, Denver, and Pittsburgh, because the refineries have not been built yet. This assumption allows us to use the two-stage transportation method.

TABLE 12.5
Supply and Demand Information for Four Sources and Five Outlets

Raw Material Source	Supply	Outlet	Demand
A	1000	Dallas	900
B	800	Phoenix	800
C	800	Portland	600
D	700	Montreal	500
		Orlando	500

TABLE 12.6
Inland Raw Material Transportation Cost

To From	Denver	Atlanta	Pittsburgh
Vancouver	4	13	9
Boston	8	8	5
Miami	12	2	9
San Francisco	11	11	12

TABLE 12.7
Chemical Trucking Cost

To From	Dallas	Phoenix	Portland	Montreal	Orlando
Denver	28	26	12	30	30
Atlanta	10	22	23	29	8
Pittsburgh	18	21	23	18	21

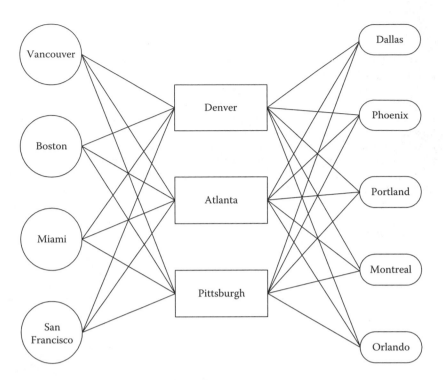

FIGURE 12.4 Pictorial representation of the RIFIN example.

The transportation setup is shown in Table 12.8. Because we assumed that a refinery capable of handling the total raw material supply could be built at each location, the supply rows of the transportation tableau are bounded by the capacity of each supply source. Also, because we are introducing more supply than is actually available at each source, we have to remove these excess units via the "dummy source" columns. Note that the cells that lie at the intersection of rows and column corresponding to a specific supply source have zero costs in them. The remaining cells in that column are assigned a large positive value, *M*, to prohibit use of these cells in the solution.

TABLE 12.8
Transportation Tableau Setup for Example 12.4

Supply Source	Refinery	Dal	Ph	Port	Mont	Orl	Van	Bos	Miami	SanFr	Capacity
		Customer					**Dummy Source**				
Vancouver (A)	Denver	32	30	16	34	34	0	M	M	M	1000
	Atlanta	23	35	36	42	41	0	M	M	M	1000
	Pittsburgh	47	30	32	27	30	0	M	M	M	1000
Boston (B)	Denver	36	34	20	38	38	M	0	M	M	800
	Atlanta	18	30	31	37	16	M	0	M	M	800
	Pittsburgh	23	26	28	23	26	M	0	M	M	800
Miami (C)	Denver	40	38	24	42	42	M	M	0	M	800
	Atlanta	12	24	25	31	10	M	M	0	M	800
	Pittsburgh	27	30	32	27	30	M	M	0	M	800
San Francisco (D)	Denver	39	37	23	41	41	M	M	M	0	700
	Atlanta	21	33	34	40	19	M	M	M	0	700
	Pittsburgh	30	33	35	30	33	M	M	M	0	700
Demand		900	800	600	500	500	2000	1600	1600	1400	9900

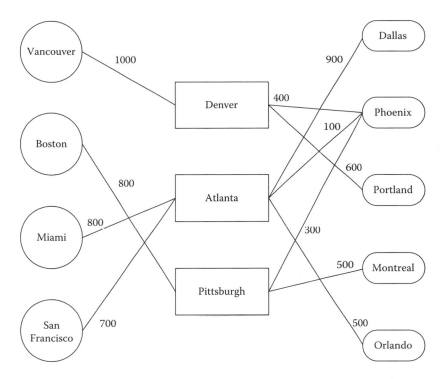

FIGURE 12.5 Solution to the RIFIN allocation problem.

For this problem, the total actual supply is equal to the total demand, and hence, the excess source capacity column or the excess demand row is not needed. The transportation problem may be solved to yield the solution (with a total cost of $65,400) indicated in Figure 12.5. Notice that the solution indicates that refineries be built at all locations.

12.4.3 Vehicle-Routing Problem

Consider the vehicle-routing problem (VRP) faced by a parcel delivery company in its hub location. It wants to determine the number of vehicles required to (1) serve its customers (pick up or deliver parcels) so that each customer is visited once and only once per day, (2) the vehicle capacity is not exceeded, and (3) the total travel time is minimized. Let,

T_{ij} Time to travel from customer i to customer j, $i, j = 1, 2, ..., n$
D_i Demand at customer i, $i = 1, 2, ..., n$
C_k Capacity of vehicle k, $k = 1, 2, ..., p$

$$x_{ijk} = \begin{cases} 1 & \text{if truck } k \text{ visits customer } j \text{ immediately after visiting customer } i \\ 0 & \text{otherwise} \end{cases}$$

The VRP model is

Model 6

$$\text{Minimize} \sum_{i=1}^{n} \sum_{j=1}^{n} \sum_{k=1}^{p} T_{ij} x_{ijk} \qquad (12.30)$$

$$\text{Subject to} \sum_{i=1}^{n} \sum_{k=1}^{p} x_{ijk} = 1 \quad j = 1, 2, ..., n \tag{12.31}$$

$$\sum_{j=1}^{n} \sum_{k=1}^{p} x_{ijk} = 1 \quad i = 1, 2, ..., n \tag{12.32}$$

$$\sum_{i=1}^{n} D_i \sum_{j=1}^{n} x_{ijk} \leq C_k \quad k = 1, 2, ..., p \tag{12.33}$$

$$\sum_{i=1}^{n} x_{ilk} - \sum_{j=1}^{n} x_{ljk} = 0 \quad l = 1, 2, ..., n; \, k = 1, 2, ..., p \tag{12.34}$$

$$\sum_{j=2}^{n} x_{jk} \leq 1 \quad k = 1, 2, ..., p \tag{12.35}$$

$$\sum_{i=2}^{n} x_{ik} \leq 1 \quad k = 1, 2, ..., p \tag{12.36}$$

$$u_i - u_j + n \sum_{k=1}^{p} x_{ijk} \leq n - 1 \quad 2 \leq i \neq j \leq n \tag{12.37}$$

$$x_{ijk} = 0 \text{ or } 1 \quad i, j = 1, 2, ..., n; \, k = 1, 2, ..., p \tag{12.38}$$

The objective function (12.30) minimizes the total travel time. Constraint (12.31) ensures that we arrive at each customer location from only one other location. Constraint (12.32) ensures that we leave each customer location for only one other. Constraint (12.33) ensures that vehicle capacity limits are not exceeded. Constraint (12.34) ensures that if a vehicle arrives at a customer location, then it must leave that location. Constraints (12.35) and (12.36) ensure that trucks leave and arrive at the hub. Customer 1 is considered the hub location. Constraint (12.37) prevents the formation of subtours. Also, $T_{ii} = \infty$ for each i. Constraint (12.38) imposes (binary) integer restrictions on the decision variables.

Numerous heuristics are available to solve the VRP. See Bodin et al. (1983) for a survey of the various algorithms and models for this problem. Alfa et al. (1991) have proposed a combined 3-opt and simulated annealing algorithm for this problem. We will not provide any algorithm here, but recommend the reader to consult the previous two sources for ways of solving the VRP.

12.5 LOCATION–ALLOCATION MODELS

In Chapter 8, we formulated the product assignment problem in a warehouse as a generalized assignment problem (see Section 8.6.7.1). Clearly, this model can be used to formulate location–allocation problems in which the objective is to determine the location of facilities to minimize the cost of assigning facilities to customers subject to the constraint that each facility be assigned to a prespecified number of customers. Similarly, the quadratic assignment model discussed in the context of a layout problem in Chapter 10 can be used at a macrolevel to determine the location of facilities given that these facilities have flow (interaction) among themselves. We consider three other location–allocation models, each with specific applications:

1. Set covering model
2. Uncapacitated location–allocation model
3. Comprehensive location–allocation model

The models are discussed in order of the difficulty in solving them. For all the models, we present good heuristic or optimal solution procedures. The models determine the number of facilities to be located, where these are to be located as well as the interaction between the facilities and customers. The first two are rather simple. The first considers only the cost of covering each customer with a facility. The second considers a single product, one stage of distribution, facilities with unlimited capacity, and allows a customer to be served from several facilities. The last model relaxes several of these assumptions and therefore better represents the real-world location–allocation problem. To facilitate understanding of the last model, to provide a sound introduction, and to illustrate the use of efficient branch-and-bound algorithms, we begin our discussion of location–allocation problems with two simple models.

12.5.1 SET COVERING MODEL

The set covering problem arises in situations where it is necessary to ensure that each customer is covered by at least one service facility. For example, fire stations and other emergency facilities, libraries, community colleges, or state university campuses have to be located so that each population area or "customer" is within a certain range of distance from at least one facility. If a customer is within the desired range, we say the customer is covered. Defining

c_j Cost of locating facility at site j

$$a_{ij} = \begin{cases} 1 & \text{if customer } i \text{ can be served by facility located at site } j \\ 0 & \text{otherwise} \end{cases}$$

$$x_j = \begin{cases} 1 & \text{if facility is located at site } j \\ 0 & \text{otherwise} \end{cases}$$

The set covering problem is to

Model 7

$$\text{Minimize} \sum_{j=1}^{n} c_j x_j \tag{12.39}$$

$$\text{Subject to} \sum_{j=1}^{n} a_{ij} x_j \geq 1 \quad i = 1, 2, \ldots, m \tag{12.40}$$

$$x_j = 0 \text{ or } 1 \quad j = 1, 2, \ldots, n \tag{12.41}$$

In this 0–1 integer programming model, the number of customers and facilities are m and n, respectively. Constraint (12.40) ensures that each customer is covered by at least one facility. The objective function (12.39) minimizes the cost of covering each customer. The model may be solved optimally using a general-purpose branch-and-bound technique, but that may be too time consuming for large problems. Hence, the following greedy algorithm is used to obtain suboptimal solutions efficiently. It assumes that $c_j \geq 0, j = 1, 2, \ldots, n$.

Greedy Heuristic for Set Covering Problem

Step 1: If $c_j = 0$, for any $j = 1, 2, \ldots, n$, set $x_j = 1$ and remove all constraints in which x_j appears with a coefficient of +1.

Step 2: If $c_j > 0$, for any $j = 1, 2, \ldots, n$ and x_j does not appear with a +1 coefficient in any of the remaining constraints, set $x_j = 0$.

Step 3: If for any $i = 1, 2, \ldots, n$, $a_{ik} = 1$ for some k and $a_{ij} = 0$ for all $j \neq k$, set $x_j = 1$.

Step 4: For each of the remaining variables, determine c_j/d_j, where d_j is the number of constraints in which x_j appears with a +1 coefficient. Select the variable k for which c_k/d_k is minimum, set $x_k = 1$, and remove all constraints in which x_j appears with a +1 coefficient. Examine the resulting model.

Step 5: If there are no more constraints, set all the remaining variables to 0 and *stop*. Otherwise, go to step 1.

Example 12.5 illustrates the greedy heuristic.

Example 12.5

A rural county administration wants to locate several medical emergency response units so that it can respond to any call within the county within 8 minutes. The county is divided into seven population zones. The distances between the centers of each pair of zones are known and are given in the matrix in Figure 12.6. The response units can be located in the center of population zones 1–7 at a cost (in \$10,000s) of 100, 80, 120, 110, 90, 90, and 110, respectively. Assuming the average travel speed during an emergency is 60 miles per hour, formulate an appropriate set covering model to determine where the units are to be located and how the population zones are to be covered. Solve the model using the greedy heuristic.

Solution

We define

$$
a_{ij} = \begin{cases} 1 & \text{if zone } i\text{'s center can be reached from the center of zone } j \text{ within 8 minutes} \\ 0 & \text{otherwise} \end{cases}
$$

and note that $d_{ij} > 8$ and $d_{ij} \leq 8$ yield a_{ij} values of 0 and 1, respectively. We can then set up the $[a_{ij}]$ matrix in Figure 12.7.

		1	2	3	4	5	6	7
	1	0	4	12	6	15	10	8
	2	8	0	15	60	7	2	3
	3	50	13	0	8	6	5	9
$[d_{ij}] =$	4	9	11	8	0	9	10	3
	5	50	8	4	10	0	2	50
	6	30	5	7	9	3	0	27
	7	8	5	9	7	25	27	0

FIGURE 12.6 Distance between seven zones in Example 12.5.

$$
[a_{ij}] = \begin{array}{c|ccccccc|}
 & 1 & 2 & 3 & 4 & 5 & 6 & 7 \\
\hline
1 & 1 & 1 & 0 & 1 & 0 & 0 & 1 \\
2 & 1 & 1 & 0 & 0 & 1 & 1 & 1 \\
3 & 0 & 0 & 1 & 1 & 1 & 1 & 0 \\
4 & 0 & 0 & 1 & 1 & 0 & 0 & 1 \\
5 & 0 & 1 & 1 & 0 & 1 & 1 & 0 \\
6 & 0 & 1 & 1 & 0 & 1 & 1 & 0 \\
7 & 1 & 1 & 0 & 1 & 0 & 0 & 1 \\
\end{array}
$$

FIGURE 12.7 Revised, binary distance matrix for Example 12.5.

The corresponding set covering model is

$$
\text{Minimize } 100x_1 + 80x_2 + 120x_3 + 110x_4 + 90x_5 + 90x_6 + 110x_7
$$

$$
\begin{array}{llllllll}
\text{Subject to} & x_1 & +x_2 & & +x_4 & & & +x_7 & \geq & 1 \\
 & x_1 & +x_2 & & & +x_5 & +x_6 & +x_7 & \geq & 1 \\
 & & & x_3 & +x_4 & +x_5 & +x_6 & & \geq & 1 \\
 & & & x_3 & & & & +x_7 & \geq & 1 \\
 & & x_2 & & & +x_5 & +x_6 & & \geq & 1 \\
 & & x_2 & & & +x_5 & +x_6 & & \geq & 1 \\
 & x_1 & +x_2 & & +x_4 & & & +x_7 & \geq & 1 \\
 & x_1, & x_2, & x_3, & x_4, & x_5, & x_6, & x_7 & = & 0 \text{ or } 1
\end{array}
$$

Greedy Heuristic

Step 1: Because each $c_j > 0$, $j = 1, 2, \ldots, 7$, go to step 2.
Step 2: Because x_j appears in each constraint with a +1 coefficient, go to step 3.
Step 3: Because there is no constraint such that only one a_{ij} for that constraint is 1, go to step 4.
Step 4: $\dfrac{c_1}{d_1} = \dfrac{100}{3} = 33.3$; $\dfrac{c_2}{d_2} = \dfrac{80}{5} = 16$; $\dfrac{c_3}{d_3} = \dfrac{120}{4} = 30$; $\dfrac{c_4}{d_4} = \dfrac{110}{4} = 27.5$;

$\dfrac{c_5}{d_5} = \dfrac{90}{4} = 22.5$; $\dfrac{c_6}{d_6} = \dfrac{90}{4} = 22.5$; $\dfrac{c_7}{d_7} = \dfrac{110}{4} = 27.5$

Because the minimum c_k/d_k occurs for $k = 2$, set $x_2 = 1$ and remove the first two and the last thee constraints. Here is the resulting model:

$$
\text{Minimize } 100x_1 + 120x_3 + 110x_4 + 90x_5 + 90x_6 + 110x_7
$$

$$
\begin{array}{llllllll}
\text{Subject to} & & x_3 & +x_4 & +x_5 & +x_6 & & \geq & 1 \\
 & & x_3 & +x_4 & & & +x_7 & \geq & 1 \\
 & x_1, & x_3, & x_4, & x_5, & x_6, & x_7 & = & 0 \text{ or } 1
\end{array}
$$

Step 5: Because we have two constraints, go to step 1.

Step 1: Because $c_j > 0$, $j = 1, 3, 4, ..., 7$, go to step 2.

Step 2: Because $c_j > 0$ and x_1 does not appear in any of the constraints with a +1 coefficient, set $x_1 = 0$.

Step 3: Because there is no constraint such that only one a_{ij} for that constraint is 1, go to step 4.

Step 4: $\dfrac{c_3}{d_3} = \dfrac{120}{2} = 60$; $\dfrac{c_4}{d_4} = \dfrac{110}{2} = 55$; $\dfrac{c_5}{d_5} = \dfrac{90}{1} = 90$; $\dfrac{c_6}{d_6} = \dfrac{90}{1} = 90$; $\dfrac{c_7}{d_7} = \dfrac{110}{1} = 110$

Because the minimum c_k/d_k occur for $k = 4$, set $x_4 = 1$ and remove both constraints in the preceding model, because x_4 has a +1 coefficient in each. The resulting model is

$$\text{Minimize } 120x_3 + 90x_5 + 90x_6 + 110x_7$$
$$x_3, \quad x_5, \quad x_6, \quad x_7 \quad \geq \quad 0$$

Step 5: Because there are no constraints in the preceding model, set $x_3 = x_5 = x_6 = x_7 = 0$ and stop.

The solution is $x_2 = x_4 = 1$; $x_1 = x_3 = x_5 = x_6 = x_7 = 0$. The cost of locating emergency response units to meet the 8 minute response service level is therefore \$800,000 + \$1,100,000 = \$1,900,000.

12.5.2 Uncapacitated Location–Allocation Model

Consider this notation:

m Number of potential facilities

n Number of customers

c_{ij} Cost of transporting one unit of product from facility i to customer j

F_i Fixed cost of opening and operating facility j

D_j Number of units demanded at customer j

x_{ij} Number of units shipped from facility i to customer j

$$y_i = \begin{cases} 1 & \text{if facility } i \text{ is opened} \\ 0 & \text{otherwise} \end{cases}$$

The basic location–allocation model is then to

Model 8

$$\text{Minimize } \sum_{i=1}^{m} F_i y_i + \sum_{i=1}^{m}\sum_{j=1}^{n} c_{ij} x_{ij} \tag{12.42}$$

$$\text{Subject to } \sum_{i=1}^{m} x_{ij} = D_j \quad j = 1, 2, ..., n \tag{12.43}$$

$$\sum_{j=1}^{n} x_{ij} \leq y_i \sum_{j=1}^{n} D_j \quad i = 1, 2, ..., m \tag{12.44}$$

$$x_{ij} \geq 0 \quad i = 1, 2, ..., m; \ j = 1, 2, ..., n \tag{12.45}$$

$$y_i = 0 \text{ or } 1 \quad i = 1, 2, ..., m \tag{12.46}$$

The objective function (12.42) minimizes the variable transportation cost as well as the fixed cost of opening and operating the facilities needed to support the distribution activities. Constraint (12.43) ensures that each of the n customers' demand is met fully by one or more of m facilities. The objective function (12.42) and constraints (12.44) and (12.46) ensure that if a facility i ships goods to one or more customers, a corresponding fixed cost is incurred, and that the total number of units shipped does not exceed the total demand at all the customers. On the other hand, if a facility does not ship goods to any customer, no fixed cost is incurred. Constraint (12.45) is a nonnegativity constraint.

We now modify model 8 by making the following transformation of the x_{ij} variables and the c_{ij} parameter:

$$x'_{ij} = \frac{x_{ij}}{D_j} \quad c'_{ij} = c_{ij}D_j \quad i = 1, 2, \ldots, m; \; j = 1, 2, \ldots, n \tag{12.47}$$

Then model 8 can be rewritten as

Model 9

$$\text{Minimize} \sum_{i=1}^{m} F_i y_i + \sum_{i=1}^{m} \sum_{j=1}^{n} c'_{ij} x'_{ij} \tag{12.48}$$

$$\text{Subject to} \sum_{i=1}^{m} x'_{ij} = 1 \quad j = 1, 2, \ldots, n \tag{12.49}$$

$$\sum_{j=1}^{n} x'_{ij} \le n y_i \quad i = 1, 2, \ldots, m \tag{12.50}$$

$$x'_{ij} \ge 0 \quad i = 1, 2, \ldots, m; \; j = 1, 2, \ldots, n \tag{12.51}$$

$$y_i = 0 \text{ or } 1 \quad i = 1, 2, \ldots, m \tag{12.52}$$

Notice that x'_{ij} is the fraction of customer j's demand that is met by facility i. It can be seen that Expressions (12.48) and (12.49) are obtained by substituting $x'_{ij} = x_{ij}/D_j$, $i = 1, 2, \ldots, m$, $j = 1, 2, \ldots, n$. Substituting $x'_{ij} = x_{ij}/D_j$ in constraint (12.44), we get

$$\sum_{j=1}^{n} x'_{ij} D_j \le y_i \sum_{j=1}^{n} D_j \quad i = 1, 2, \ldots, m \tag{12.53}$$

Then dividing left- and right-hand sides of Equation 12.53 by ΣD_j, we get

$$\frac{1}{\sum_{j=1}^{n} D_j} \left[\sum_{j=1}^{n} x'_{ij} D_j \right] \le y_i \quad i = 1, 2, \ldots, m \tag{12.54}$$

Because the sum of the terms in the left-hand side of expression (12.54) is less than or equal to y_i, each term must also be less than or equal to y_i, because x'_{ij} and D_j are all greater than or equal to zero, for $i = 1, 2, ..., m, j = 1, 2, ..., n$. This gives us

$$\frac{x'_{ij}D_j}{\sum_{j=1}^{n} D_j} \leq y_i \quad i = 1, 2, ..., m \tag{12.55}$$

Because $D_j/\Sigma D_j$ is a positive fraction for each j, it follows that

$$x'_{ij} \leq y_i, \quad j = 1, 2, ..., n \tag{12.56}$$

Adding the preceding n equations, we get

$$\sum_{j=1}^{n} x'_{ij} \leq n y_i \quad i = 1, 2, ..., m \tag{12.50}$$

Thus, constraint (12.50) is equivalent to constraint (12.44) in the following sense. Like expression (12.44), constraint (12.50) together with (12.52) ensures that if facility i serves any customer, a corresponding fixed cost is incurred. Otherwise, it is not. Thus, model 9 is equivalent to model 8. Model 9 may be solved using the general-purpose branch-and-bound technique found in most introductory Operations Research textbooks, e.g., Winston (1994) and Hillier and Lieberman (1995). This entails setting up a root node (i.e., a subproblem with model 9 without the integer restriction on the y_i variables, solving this subproblem using the simplex algorithm, selecting a y variable with a fractional value, say y_i, branching on this variable, setting up two subproblems (nodes), one with subproblem at the root node plus the constraint $y_i = 0$ and another with $y_i = 1$, solving the two subproblems (again using simplex), and deciding whether or not to prune a node based on the following two tests:

1. The bound at the node is greater than or equal to the OFV of the best known feasible solution. (If no feasible solution has been identified yet, we proceed to test 2.)
2. The solution to the subproblem at the node is an all-integer (binary) solution.

If a node passes either of the two tests, it is pruned and we update the best known OFV if necessary. Otherwise, we determine (arbitrarily or using specialized branching rules) the fractional y_i variable on which to branch, set up two additional subproblems (nodes), solve, and make pruning decisions as before. This procedure is repeated until all the nodes are pruned. At this point, we have the optimal solution to the problem.

Although the general-purpose branch-and-bound technique can be applied to solve model 9, it is not very efficient, because we have to solve several subproblems—one at each node—using the simplex algorithm. We now present a very efficient way of solving the subproblems that does not use the simplex algorithm. To facilitate its discussion, it is convenient to refer to x'_{ij}, the fraction of customer j's demand met by facility i in model 9, as simply x_{ij}. Thus, x_{ij} in the remainder of this section does not refer to the number of units rather a fraction. Similarly, c_{ij} now refers to c'_{ij}.

The central idea of the branch-and-bound algorithm is based on the following result. Suppose, at some stage of the branch-and-bound solution process, we are at a node where some facilities are closed (corresponding $y_i = 0$), some are open ($y_i = 1$), and the remaining are free, i.e., a decision whether to open or close has not yet been taken ($0 < y_i < 1$). Let us define

S_0 as the set of facilities whose y_i value is equal to 0; $\{i: y_i = 0\}$
S_1 as the set of facilities whose y_i value is equal to 1; $\{i: y_i = 1)$
S_2 as the set of facilities whose y_i value is greater than 0 but less than 1; $\{i: 0 < y_i < 1\}$

Now examine the location–allocation model (12.48) through (12.51) for this node. It can be rewritten as

Model 10

$$\text{Minimize} \sum_{i \in S_1} F_i + \sum_{i \in S_1} \sum_{j=1}^{n} c_{ij} x_{ij} + \sum_{i \in S_2} F_i y_i + \sum_{i \in S_2} \sum_{j=1}^{n} c_{ij} x_{ij} \qquad (12.57)$$

$$\text{Subject to} \sum_{i=1}^{m} x_{ij} = 1 \quad j = 1, 2, ..., n \qquad (12.58)$$

$$\sum_{j=1}^{n} x_{ij} \le n y_i \quad i = 1, 2, ..., m \qquad (12.59)$$

$$x_{ij} \ge 0 \quad i = 1, 2, ..., m; \ j = 1, 2, ..., n \qquad (12.60)$$

Because x_{ij} is a fraction, it can be proved by contradiction that equality of Equation 12.59 holds at optimality. From this, expression (12.59) can be written as

$$y_i = \frac{1}{n} \sum_{j=1}^{n} x_{ij} \quad i = 1, 2, ..., m \qquad (12.61)$$

Because the maximum value each x_{ij} can take is 1 (due to constraint (12.58)) and the right-hand side of expression (12.61) is the sum of n x_{ij}'s divided by n, it is obvious that the maximum value that y_i can take is also 1. Substituting the value of y_i from Equation 12.61 for $i \in S_2$ in Equation 12.55, we get

Model 11

$$\text{Minimize} \sum_{i \in S_1} F_i + \left\{ \sum_{i \in S_1} \sum_{j=1}^{n} c_{ij} x_{ij} + \left[\sum_{i \in S_2} F_i \sum_{j=1}^{n} \frac{x_{ij}}{n} + \sum_{i \in S_2} \sum_{j=1}^{n} c_{ij} x_{ij} \right] \right\}$$

$$= \sum_{i \in S_1} F_i + \text{Minimize} \left\{ \sum_{i \in S_1} \sum_{j=1}^{n} c_{ij} x_{ij} + \left[\sum_{i \in S_2} F_i \sum_{j=1}^{n} \left(c_{ij} + \frac{F_i}{n} \right) x_{ij} \right] \right\} \qquad (12.62)$$

subject to constraints (12.58) and (12.61).

Model 11, which is equivalent to model 9 without the integer restrictions on the y variables, is a half assignment problem. It can be proved (again, by contradiction) that for each $j = 1, 2, ..., n$, only one of $x_{1j}, x_{2j}, ..., x_{mj}$ will take on a value of 1, due to constraint (12.58). In fact, for each j, the x_{ij} taking on a value of 1 will be the one that has the smallest coefficient in expression (12.62). Thus, in order to solve model 11, we only need to find for a specific j, the smallest coefficient of x_{ij} in

Equation 12.62, $i = 1, 2, \ldots, m$, set the corresponding x_{ij} equal to 1 and all other x_{ij}'s to 0. Of course, this is to be done for each j as shown next. We list the coefficients for each j as follows:

$$
\begin{aligned}
& c_{ij} && \text{if } i \in S_1 \\
& c_{ij} + \frac{F_i}{n} && \text{if } i \in S_2
\end{aligned}
\tag{12.63}
$$

Select the smallest coefficient, set the corresponding $x_{ij} = 1$ and all other x_{ij}'s to zero. This method of determining x_{ij}'s is called as the *minimum coefficient rule*. Note that expression (12.61) does not include facility $i \in S_0$, because these are closed. Because the x_{ij}'s are known, y_i values for $i \in S_2$ can be determined from Equation 12.59. Moreover, a lower bound on the partial solution of the node under consideration can be obtained via Equation 12.60 or simply by adding $\sum_{i \in S_1} F_i$ to the sum of the coefficients of the x_{ij} variables, which have taken a value of 1 (because all the other x_{ij}'s are equal to zero per the minimum coefficient rule). If it turns out that all the y_i values ($i \in S_2$) obtained from Equation 12.59 are binary, we have a feasible solution and the lower bound obtained for the node from Equation 12.60 is also an upper bound for the original location–allocation problem. The node can therefore be pruned. If, on the other hand, one or more y_i variables take on fractional values, then we need to branch on *one* of these variables, first by setting it equal to 0 (and then to 1), create two corresponding nodes, update S_0 or S_1 as appropriate, set up model 10 for the nodes, and obtain a solution and a lower bound via the minimum coefficient rule, and Equations 12.59 and 12.60. If the solution at a node has a lower bound greater than or equal to the best upper bound determined so far for the overall location–allocation problem, it can be pruned as branching further on this node can only lead to worse solutions. We repeat the preceding procedure of branching on nodes, solving the problem at each newly created node, determining the lower bound, and making pruning decisions until all the nodes are pruned. When this occurs, we have an optimal solution to the location–allocation model, which is given by the node, that has a feasible solution with the smallest cost among all the nodes. A formal listing of the branch-and-bound algorithm steps follows.

Branch-and-bound algorithm for uncapacitated location–allocation model:

Step 1: Set the best known upper bound $UB = \infty$, the node counter $p = 1$, $S_0 = S_1 = \{\Phi\}$, and $S_2 = \{1, 2, \ldots, m\}$.

Step 2: Construct a subproblem (node) p with the current values of the y variables.

Step 3: Solve the subproblem corresponding to the node under consideration using the minimum coefficient rule and Equation 12.61.

Step 4: If all the y variables in the solution take an integer (0 or 1) value, go to step 7. Otherwise, go to step 5.

Step 5: Determine the lower bound of node p using Equation 12.62. Arbitrarily select one of the facilities—say, k—which has taken on a fractional value for y_k ($0 < y_k < 1$) and create two subproblems (nodes) $p + 1$ and $p + 2$ as follows.

Subproblem $p + 1$: Include facility k and others with a y_k value of 0 in S_0, facilities with y value of 1 in S_1, and all other facilities in S_2.

Subproblem $p + 2$: Include facility k and others with a y_k value of 1 in S_1, facilities with y value of 0 in S_0, and all other facilities in S_2. If, $x_{kj} = 1$ for $j = 1, 2, \ldots, n$, in the solution to subproblem p, remove each such customer j from consideration in subproblem $p + 2$, and reduce n by the number of j's for which $x_{kj} = 1$.

Step 6: Solve subproblem $p + 1$ using the minimum coefficient rule and Equation 12.61. Set $p = p + 2$. Go to step 4.

Step 7: Determine the lower bound of node p using Equation 12.54. If it is less than UB, set UB = lower bound of node p. Prune node p as well as any other node whose bound is greater than or equal to UB. If there are no more nodes to be pruned, stop. Otherwise, consider any unpruned node and go to step 3.

Notice that although we could have set up the root node by relaxing the integer restrictions as the y variables and solving it via the simplex algorithm (as is done in the general-purpose branch-and-bound algorithm), we find it convenient to set up the root node in which all the y_i variables are assumed to be fractional. Given that it has not yet been decided whether to open or close any facility, we seek to determine via the minimum coefficient rule and Equation 12.61 which, if any, should be open or closed. Also, because the subproblem at each node is solved by the minimum coefficient rule, there is really no need to construct the subproblem as indicated in step 2. When we implement the branch-and-bound algorithm in Example 12.6, we therefore skip step 2.

Step 5 calls for an arbitrary selection of the branching variable. However, by carefully selecting the variable on which to branch, we can improve the efficiency of the branch-and-bound algorithm. For example, we may select the variable that is close to 1 or 0 and branch on it. Other more sophisticated branching rules that depend upon fixed costs, maximum or minimum bounds on cost reduction for opening a facility, and other factors are discussed in detail in Khumawala (1972). For ease of discussion, our selection of the variable to branch on is made arbitrarily.

In subproblem $p + 2$ in step 5, notice that we remove any customer who in the solution to subproblem p was served by facility k. This is done because of the following simple observation. If it is cheapest to ship goods from facility k to customer j (among all facilities) when fixed costs at facility k are included in the cost calculation, it must be even cheaper to ship goods from facility k to customer j when fixed costs at facility k are excluded in the cost calculation. Thus, because customer j will continue to be served from warehouse k, we may eliminate customer j from consideration in the $p + 2$ node as well as other nodes emanating from it. Consequently, the number of available customers must also be reduced.

Example 12.6

The nation's leading retailer Sam-Mart wants to establish its presence in the Northeast by opening five department stores. The store locations have already been determined. To serve these stores, the retailer wants to have a maximum of three distribution warehouses. The potential locations for these warehouses have already been selected, and there are no practical limits on the size of the warehouses. The fixed cost (in $1,000,000s) of building and operating the warehouse at each location are 6, 5, and 3, respectively. The variable cost of serving each warehouse from each of the potential warehouse locations is given in Figure 12.8 (again in $1,000,000s). Determine

	1	2	3	4	5
1	20	12	14	12	10
2	15	10	20	8	15
3	12	16	25	11	10

FIGURE 12.8 Variable cost of serving each warehouse from each site in Example 12.6.

how many warehouses are to be built and in what locations. Also determine how the customers (department stores) are to be served.

Solution

Step 1: Set $UB = \infty$, $p = 1$, $S_0 = S_1 = \{\Phi\}$, and $S_2 = \{1, 2, 3\}$.

Step 3: *Solution of subproblem 1 using the minimum coefficient rule*: Determine the x_{ij} coefficients as follows:

$$c_{11} = 20 + \frac{6}{5} = 21\frac{1}{5}; \quad c_{21} = 15 + \frac{5}{5} = 16; \quad c_{31} = 12 + \frac{3}{5} = 12\frac{3}{5}$$

Because the minimum occurs for $c_{ij} = c_{31}$, set $x_{31} = 1$ and $x_{11} = x_{21} = 0$. Then,

$$c_{12} = 12 + \frac{6}{5} = 13\frac{1}{5}; \quad c_{22} = 10 + \frac{5}{5} = 11; \quad c_{32} = 16 + \frac{3}{5} = 16\frac{3}{5}$$

Because the minimum occurs for $c_{ij} = c_{22}$, set $x_{22} = 1$ and $x_{12} = x_{32} = 0$. Then,

$$c_{13} = 14 + \frac{6}{5} = 15\frac{1}{5}; \quad c_{23} = 20 + \frac{5}{5} = 21; \quad c_{33} = 25 + \frac{3}{5} = 25\frac{3}{5}$$

Because the minimum occurs for $c_{ij} = c_{13}$, set $x_{13} = 1$; $x_{23} = x_{33} = 0$. Then,

$$c_{14} = 12 + \frac{6}{5} = 13\frac{1}{5}; \quad c_{24} = 8 + \frac{5}{5} = 9; \quad c_{34} = 11 + \frac{3}{5} = 11\frac{3}{5}$$

Because the minimum occurs for $c_{ij} = c_{24}$, set $x_{24} = 1$ and $x_{14} = x_{34} = 0$. Then,

$$c_{15} = 10 + \frac{6}{5} = 11\frac{1}{5}; \quad c_{25} = 15 + \frac{5}{5} = 16; \quad c_{35} = 10 + \frac{3}{5} = 10\frac{3}{5}$$

Because the minimum occurs for $c_{ij} = c_{35}$, set $x_{35} = 1$ and $x_{15} = x_{25} = 0$. We have

$$y_1 = \frac{1}{5}\left(x_{11} + x_{12} + x_{13} + x_{14} + x_{15}\right) = \frac{1}{5}\left(0 + 0 + 1 + 0 + 0\right) = \frac{1}{5}$$

$$y_2 = \frac{1}{5}\left(x_{21} + x_{22} + x_{23} + x_{24} + x_{25}\right) = \frac{1}{5}\left(0 + 1 + 0 + 1 + 0\right) = \frac{2}{5}$$

$$y_3 = \frac{1}{5}\left(x_{31} + x_{32} + x_{33} + x_{34} + x_{35}\right) = \frac{1}{5}\left(1 + 0 + 0 + 0 + 1\right) = \frac{2}{5}$$

Step 4: Because all three y variables have fractional values, go to step 5.

Step 5: The lower bound of node 1 is

$$0 + 12\frac{3}{5} + 11 + 15\frac{1}{5} + 9 + 10\frac{3}{5} = 58\frac{2}{5}$$

Arbitrarily select variable y_1 to branch on, and create subproblems 2 and 3 as follows:

Subproblem 2: $S_0 = \{1\}$, $S_1 = \{\Phi\}$, and $S_2 = \{2, 3\}$.
Subproblem 3: $S_1 = \{1\}$, $S_0 = \{\Phi\}$, and $S_2 = \{2, 3\}$.

Because $x_{13} = 1$ in the solution to subproblem 1, remove customer 3 from consideration in node 3. Reduce n by 1, or $n = 5-1 = 4$.

Step 6: *Solution of subproblem 2 using minimum coefficient rule*: Determine x_{ij} coefficients as follows:

$$c_{21} = 15 + \frac{5}{5} = 16; \quad c_{31} = 12 + \frac{3}{5} = 12\frac{3}{5}$$

Because the minimum occurs for $c_{ij} = c_{31}$, set $x_{31} = 1$ and $x_{21} = 0$. Then,

$$c_{22} = 10 + \frac{5}{5} = 11; \quad c_{32} = 16 + \frac{3}{5} = 16\frac{3}{5}$$

Because the minimum occurs for $c_{ij} = c_{22}$, set $x_{22} = 1$ and $x_{32} = 0$. Then,

$$c_{23} = 20 + \frac{5}{5} = 21; \quad c_{33} = 25 + \frac{3}{5} = 25\frac{3}{5}$$

Because the minimum occurs for $c_{ij} = c_{23}$, set $x_{23} = 1$ and $x_{33} = 0$. Then,

$$c_{24} = 8 + \frac{5}{5} = 9; \quad c_{34} = 11 + \frac{3}{5} = 11\frac{3}{5}$$

Because the minimum occurs for $c_{ij} = c_{24}$, set $x_{24} = 1$ and $x_{34} = 0$. Then,

$$c_{25} = 15 + \frac{5}{5} = 16; \quad c_{35} = 10 + \frac{3}{5} = 10\frac{3}{5}$$

Because the minimum occurs for $c_{ij} = c_{35}$, set $x_{35} = 1$ and $x_{25} = 0$. We have

$$y_2 = \frac{1}{5}\left(x_{21} + x_{22} + x_{23} + x_{24} + x_{25}\right) = \frac{1}{5}\left(0 + 1 + 1 + 1 + 0\right) = \frac{3}{5}$$

$$y_3 = \frac{1}{5}\left(x_{31} + x_{32} + x_{33} + x_{34} + x_{35}\right) = \frac{1}{5}\left(1 + 0 + 0 + 0 + 1\right) = \frac{2}{5}$$

Solution of subproblem 3 using minimum coefficient rule: Determine x_{ij} coefficients as follows:

$$c_{11} = 20; \quad c_{21} = 15 + \frac{5}{4} = 16\frac{1}{4}; \quad c_{31} = 12 + \frac{3}{4} = 12\frac{3}{4}$$

Because the minimum occurs for $c_{ij} = c_{31}$, set $x_{31} = 1$ and $x_{11} = x_{21} = 0$. Then,

$$c_{12} = 12; \quad c_{22} = 10 + \frac{5}{4} = 11\frac{1}{4}; \quad c_{32} = 16 + \frac{3}{4} = 16\frac{3}{4}$$

$$c_{14} = 12; \quad c_{24} = 8 + \frac{5}{4} = 9\frac{1}{4}; \quad c_{34} = 11 + \frac{3}{4} = 11\frac{3}{4}$$

Because the minimum occurs for $c_{ij} = c_{22}$, set $x_{22} = 1$ and $x_{12} = x_{32} = 0$. Then,

$$c_{15} = 10; \quad c_{25} = 15 + \frac{5}{4} = 16\frac{1}{4}; \quad c_{35} = 10 + \frac{3}{4} = 10\frac{3}{4}$$

Because the minimum occurs for $c_{ij} = c_{24}$, set $x_{24} = 1$ and $x_{14} = x_{34} = 0$. Then,

$$c_{11} = 20; \quad c_{21} = 15 + \frac{5}{4} = 16\frac{1}{4}; \quad c_{31} = 12 + \frac{3}{4} = 12\frac{3}{4}$$

Because the minimum occurs for $c_{ij} = c_{15}$, set $x_{15} = 1$ and $x_{25} = x_{35} = 0$. We have

$$y_2 = \frac{1}{4}(x_{21} + x_{22} + x_{24} + x_{25}) = \frac{1}{4}(0 + 1 + 1 + 0) = \frac{1}{2}$$

$$y_3 = \frac{1}{4}(x_{31} + x_{32} + x_{34} + x_{35}) = \frac{1}{4}(1 + 0 + 0 + 0) = \frac{1}{4}$$

Set $p = 1 + 2 = 3$.

Step 4: Because the solution for subproblem 2 is not all integer, go to step 5.

We repeat steps 3–7 until all the nodes are pruned. We then have an optimal solution. These steps are summarized in Table 12.9.

Figure 12.9 illustrates the solution and lower bound at each of the nodes. In Table 12.9 and Figure 12.9, variable values of zero are not shown. As can be observed, the problem has multiple optimal solutions with an optimal OFV of 68. Choosing the solution at node 5, we see that warehouses are to be opened at sites 2 and 3. From the seventh line in Table 12.9, we find that warehouse 2 should serve department stores 2, 3, and 4, and warehouse 1 should serve customers 1 and 5.

12.5.3 COMPREHENSIVE LOCATION–ALLOCATION MODEL

In all the models we studied so far in this chapter and Chapter 9, we did not explicitly consider multiple commodities. We now present a comprehensive model that considers real-world factors and constraints. Consider this problem. Different types of products are produced at several plants with known production capacities. The demand for each product type at each of several customer areas is also known. The products are shipped from plants to customer areas via intermediate warehouses with the restriction that each customer area be serviced by only one warehouse. The latter is done to improve customer service. Upper and lower bounds on the capacity of each warehouse, potential locations for these, inbound and outbound transportation costs at each of these warehouses, and the fixed cost of opening and operating a warehouse at each potential location are known.

The problem is to find the locations for the warehouses, the corresponding capacities, the customers served by each warehouse, and how products are to be shipped from each plant, to minimize the fixed and variable costs of opening and operating warehouses as well as the distribution costs. We use the following notation:

S_{ij} Production capacity of product i at plant j
D_{il} Demand for product i at customer zone l
F_i Fixed cost of operating warehouse i
V_{ij} Unit variable cost of handling product i at warehouse j
c_{ijkl} Average unit cost of producing and transporting product i from plant j via warehouse k to customer area l
UC_i Upper bound on the capacity of warehouse i
LC_i Lower bound on the capacity of warehouse i
X_{ijkl} Number of units of product i transported from plant j via warehouse k to customer area l

$$y_{kl} = \begin{cases} 1 & \text{if warehouse } i \text{ serves customer area } l \\ 0 & \text{otherwise} \end{cases}$$

$$z_k = \begin{cases} 1 & \text{if warehouse is opened at location } k \\ 0 & \text{otherwise} \end{cases}$$

TABLE 12.9

Details of the Solution Steps for the Problem in Example 12.6

Step	Node	Subproblem	Solution	Remarks	Lower Bound	Branching Variable
5	2				64.2	y_2
	4	$S_0 = \{1,2\}, S_2 = \{3\}$				
	5	$S_0 = \{1\}, S_1 = \{2\}, S_3 = \{3\}$		$n = 2$.		
6	4		$x_{3j} = 1, j = 1,2, \ldots, 5,$ $y_3 = 1$	Go to step 4.		
4	4			Go to step 7.		
7	4			Pruned; select node 5 and go to step 3.	77	
3	5		$x_{22} = x_{23} = x_{24} = x_{31} =$ $x_{31} = y_2 = y_3 = 1$			
4	5			Go to step 7.		
7	5			Pruned; select node 3 and go to step 3.	68	
3	3		$x_{31} = x_{22} = x_{24} = x_{15} = 1$ $y_1 = 1, y_2 = 0.5,$ $y_3 = 0.25$			
4	3			Go to step 5.		
5	3				63.25	y_2
	6	$S_0 = \{2\}, S_1 = \{1\}, S_2 = \{3\}$				
	7	$S_1 = \{1, 2\} \, S_2 = \{3\}$		$n = 1$.		
6	6		$x_{31} = x_{34} = x_{12} = x_{15} =$ $y_1 = 1$ $y_3 = 0.5$	Go to step 4.		
4	6			Go to step 5.		
5	6				66.5	y_3
	8	$S_0 = \{2, 3\}, S_1 = \{1\}$				
	9	$S_0 = \{2\}, S_1 = \{1, 3\}$		$n = 2$.		
6	8		$x_{1j} = 1, j = 1,2, \ldots, 5,$ $y_1 = 1$	Go to step 4.		
4	8			Go to step 7.		
7	8			Pruned; select node 9 and go to step 3.	74	
3	9		$x_{31} = x_{12} = x_{13} = x_{34} =$ $x_{15} = y_1 = y_3 = 1$			
4	9			Go to step 7.		
7	9			Pruned; select node 7 and go to step 3.	68	
3	7		$x_{31} = y_1 = y_2 = y_3 = 1$			
4	7			Go to step 7.		
7	7			Pruned; stop.	68	

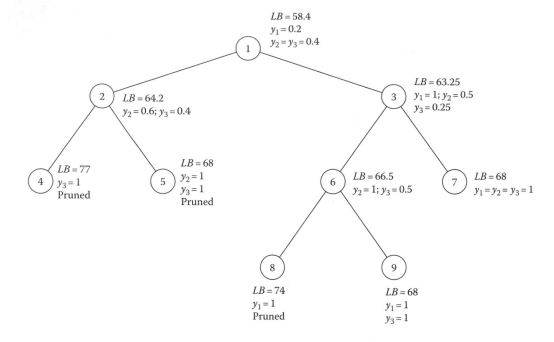

FIGURE 12.9   Branch-and-bound tree for Example 12.6.

The model for location–allocation follows:

Model 12

$$\text{Minimize} \sum_{i=1}^{p}\sum_{j=1}^{q}\sum_{k=1}^{r}\sum_{l=1}^{s} c_{ijkl}x_{ijkl} + \sum_{i=1}^{p}\sum_{l=1}^{s} D_{il}\sum_{k=1}^{r} V_{ik}Y_{kl} + \sum_{k=1}^{r} F_k z_k \tag{12.64}$$

$$\text{Subject to} \sum_{k=1}^{r}\sum_{l=1}^{s} x_{ijkl} \le S_{ij} \quad i=1,2,...,p; \; j=1,2,...,q \tag{12.65}$$

$$\sum_{j=1}^{q} x_{ijkl} \ge D_{il}y_{kl} \quad i=1,2,...,p; \; k=1,2,...,r; \; l=1,2,...,s \tag{12.66}$$

$$\sum_{k=1}^{r} y_{kl} = 1 \quad l=1,2,...,s \tag{12.67}$$

$$\sum_{i=1}^{p}\sum_{l=1}^{s} D_{il}y_{kl} \ge LC_k z_k \quad k=1,2,...,r \tag{12.68}$$

$$\sum_{i=1}^{p}\sum_{l=1}^{s} D_{il}y_{kl} \le UC_k z_k \quad k=1,2,...,r \tag{12.69}$$

$$x_{ijkl} \ge 0 \quad i=1,2,...,p; \; j=1,2,...,q; \; k=1,2,...,r; \; l=1,2,...,s \tag{12.70}$$

$$y_{kl}, z_k = 0 \text{ or } 1 \quad k=1,2,...,r; \; l=1,2,...,s \tag{12.71}$$

The objective function (12.64) minimizes the inbound and outbound transportation costs as well as production costs for each product at each warehouse. It also minimizes the fixed and variable costs of opening and operating the required number of warehouses. Constraint (12.65) ensures that the capacity constraints for making each product at each plant are not violated. Constraint (12.66) ensures the demand of each product at each customer zone is met. Constraint (12.67) requires that each customer area be serviced by a single warehouse. Constraints (12.68) and (12.69) have a dual purpose. Not only do they enforce the upper and lower bound on the warehouse capacity, they also "connect" the y_{kl} and z_k variables. Because a warehouse can serve a customer only if it is opened, we must have $y_{kl} = 0$, when $x_k = 0$ and $y_{kl} = 1$, when $z_k = 1$, for each warehouse–customer area $\{k, l\}$ pair. These two conditions are satisfied by constraints (12.68) and (12.69), respectively.

We can easily add more linear constraints not involving x_{ijkl} variables to model 12 to

- Impose upper and lower limit on the number of warehouses that can be opened.
- Enforce precedence relations among warehouses (e.g., open warehouse at location 1 only if another is opened at location 3).
- Enforce service constraints (e.g., if it is decided to open a certain warehouse, then a specific customer area must be served by it)

Other constraints that can be added are discussed further in Geoffrion and Graves (1974). Such constraints reduce the solution space, so they allow quicker solution of the model while giving the modeler much flexibility.

Model 12 may be solved using available mixed-integer programming software, but due to the presence of binary integer variables y_{kl} and x_k, only small-sized problems can be solved. Real-world problems such as the Hunt–Wesson Foods, Inc., location–allocation problem considered in Geoffrion and Graves (1974), which had more than 11,000 constraints, 23,000 x_{ijkl} variables, and 700 y_{kl} and x_k binary variables, cannot be solved via general mixed-integer programming algorithms. Interestingly, such large problems have been rather easily solved using the modified Benders' decomposition algorithm discussed in Chapter 11. Before we formally present the algorithm for the comprehensive location–allocation model, we must present the linear program obtained by fixing the binary integer variables along with its dual and the master problem.

Suppose we fix the values of binary variables y_{kl} and z_k so that constraints (12.67) through (12.69) are satisfied. Of course, because there are many feasible solutions to model 12, we have multiple ways of fixing the binary variables. Select any of them. Then, model 12 reduces to the following linear program, which we will refer to as model transportation problem (TP).

Model TP

$$\text{Minimize} \sum_{i-1}^{p}\sum_{j=1}^{q}\sum_{l-1}^{s} c_{ijk(l)l} x_{ijk(l)l} + K \tag{12.72}$$

$$\text{Subject to} \sum_{l=1}^{s} x_{ijk(l)l} \le S_{ij} \quad i = 1,2,...,p; \; j = 1,2,...,q \tag{12.73}$$

$$\sum_{j=1}^{q} x_{ijk(l)l} \ge D_{il} y_{k(l)l} \quad i = 1,2,...,p; \; l = 1,2,...,s \tag{12.74}$$

$$x_{ijk(l)l} \ge 0 \quad i = 1,2,...,p; \; j = 1,2,...,q; \; l = 1,2,...,s \tag{12.75}$$

where

$$K = \sum_{i=1}^{p}\sum_{l=1}^{s} D_{il}V_{ik(l)}y_{k(l)l} + \sum_{k=1}^{r} F_k z_k \tag{12.76}$$

In model TP, $k(l)$ is defined as the k-index for which $y_{kl} = 1$. (Recall that we have temporarily fixed each y_{kl} and z_k to zero or one.) For each l, there can be only k index for which $y_{kl} = 1$, due to constraint (12.67). This means that for every l, there is only one $k(l)$, and hence, there is no need for summation over $k(l)$ in model TP. Also, the right-hand side of expression (12.74) can be written simply as D_{il}, because $y_{k(l)l} = 1$. However, for ease in setting up the master problem, we explicitly keep the $y_{k(l)l}$ variables that are fixed at one. The second term of expression (12.72) is represented as K, because it is a constant quantity when the y_{kl} and z_k variables are temporarily fixed at zero or one.

Model TP can be decomposed into i separate transportation problems, model TP_i, because the variables pertaining to a specific product appear only in the rows (constraints) corresponding to that product and not elsewhere. (Notice that we have temporarily eliminated K from TP_i.)

Model TP_i

$$\text{Minimize } \sum_{j=1}^{q}\sum_{l=1}^{s} c_{ijk(l)l} x_{ijk(l)l} \tag{12.77}$$

$$\text{Subject to } \sum_{l=1}^{s} x_{ijk(l)l} \leq S_{ij} \quad j = 1,2,...,q \tag{12.78}$$

$$\sum_{j=1}^{q} x_{ijk(l)l} \geq D_{il}y_{k(l)l} \quad l = 1,2,...,s \tag{12.79}$$

$$x_{ijk(l)l} \geq 0 \quad j = 1,2,...,q; l = 1,2,...,s \tag{12.80}$$

The dual of model TP_i designated as model DTP_i, is

Model DTP_i

$$\text{Maximize } -\sum_{j=1}^{q} u_{ij}S_{ij} + \sum_{l=1}^{s} v_{il}D_{il}y_{k(l)l} \tag{12.81}$$

$$\text{Subject to } -u_{ij} + v_{il} \leq c_{ijk(l)l} \quad j = 1,2,...,q; l = 1,2,...,s \tag{12.82}$$

$$u_{ij} \geq 0 \quad i = 1,2,...,p; j = 1,2,...,q \tag{12.83}$$

$$v_{il} \geq 0 \quad i = 1,2,...,p; l = 1,2,...,s \tag{12.84}$$

Because we have the duals DTP_i, $i = 1, 2, ..., p$, we can now combine the duals and set up the master problem MP as we did in Chapter 11.

Model MP

Minimize T

Subject to $T \geq -\sum_{i=1}^{p}\sum_{j=1}^{q} u_{ij}S_{ij} + \sum_{i=1}^{p}\sum_{k=1}^{r}\sum_{l=1}^{s} v_{il}D_{il}y_{kl} + \sum_{i=1}^{p}\sum_{l=1}^{s} D_{il}\sum_{k=1}^{r} V_{ik}y_{kl} + \sum_{k=1}^{r} F_k z_k$

and constraints (12.64) through (12.67) and (12.69).

Using the following modified Benders' decomposition algorithm, which is similar to the one discussed in Chapter 11, we can solve model 12 efficiently.

Modified Benders Decomposition Algorithm

Step 0: Set upper bound $UB = \infty$, and convergence tolerance parameter ε to a desired small, positive value. Set y_{kl}, $z_k = 0$ or 1, $k = 1, 2, \ldots, r$, $l = 1, 2, \ldots, s$ such that the resulting values satisfy Equations 12.67 through 12.69.

Step 1: Set up TP$_i$, $i = 1, 2, \ldots, p$ and determine K using expression (12.76) for the current values of y_{kl}, z_k, $k = 1, 2, \ldots, r$, $l = 1, 2, \ldots, s$. Set up the corresponding dual model DTP$_i$ for each i. Solve each DTP$_i$ and add K to the sum of the optimal OFV of each DTP$_i$. If this sum is less than or equal to UB, set $UB = K +$ sum of original OFVs of each DTP$_i$.

Step 2: Set up model MP for the current values of u_{ij}, v_{il}, $i = 1, 2, \ldots, p$; $j = 1, 2, \ldots, q$, $l = 1$, 2, $\ldots$, s. Find a feasible solution to MP such that $T \leq UB - \varepsilon$. If there is no such feasible solution to the current MP, *stop*. We have an ε-optimal solution. Otherwise, go to step 1 with the current values of the y_{kl} and z_k variables.

Several strategies to improve the efficiency of the algorithm, for example, by solving the DTP$_i$ models by inspection (as done in Chapter 11) rather than using the simplex algorithm, are discussed in Geoffrion and Graves (1974). We urge the interested reader to consult this source for details.

12.6   SUMMARY

In this chapter, we discussed models for the multifacility location, allocation, and location–allocation problem with numerical examples. The field of facility location is very mature, and there are several models and corresponding algorithms. The reader interested in additional models available for the location problem is referred to Daskin (2013) and Love et al. (1988). The reader interested in design issues in location problems and the broader topic of logistics management is referred to Ballou (1988) and Ghiani et al. (2004).

12.7   REVIEW QUESTIONS AND EXERCISES

1. Give ten specific questions that arise in a real-world facility location problem.
2. Consider the following pairs of variables in the model defined by expressions (12.12) through (12.15). Show that only one of each pair of variables can take on a strictly positive value.
   (a) x_{ij}^+, x_{ij}^-
   (b) y_{ij}^+, y_{ij}^-
   (c) xa_{ij}^+, xa_{ij}^-
   (d) yb_{ij}^+, yb_{ij}^-
3. If there were two training facilities to be located in Exercise 29 of Chapter 9, how would you solve the problem if (1) there were no interaction between the two new facilities and if (2) there was significant interaction between the two. Do not attempt to solve it. Simply state the method you would use for each in 1 or 2 sentences.

4. Consider Exercise 27 in Chapter 9. Suppose another new machine is to be introduced into the flexible manufacturing cell and that the second new machine has an interaction of 8, 10, 4, and 12, respectively, with the four existing machines. Determine the optimal location of the two new machines assuming travel takes place along horizontal and vertical aisles.

5. A grocery store has entered a densely populated area of a city in an aggressive manner and has decided to open two new distribution centers (DCs) in order to improve service and reduce inventories at its existing three stores. The interaction between the first DC and store 1 is expected to be approximately 10 deliveries per week; to store 3, it is 7 deliveries per week. The second DC is expected to make 11, 3 deliveries per week to stores 2 and 3, respectively. Assume that the unit cost of deliveries is the same for all stores and that the deliveries made between the two DCs are two per week. Further assume that the distance metric is rectilinear and that the coordinates (x_i, y_i) for each of the three stores are (18, 2), (11, 5), and (2, 18), respectively.

   (a) Determine the optimal location of the two DCs, by setting up a linear programming model for the problem and solve it using any available software.

   (b) Are the locations obtained in (a) feasible? If not, determine alternate near-optimal, feasible locations.

6. A dry-cleaning service company wants to have two "factories" (F_1 and F_2) to clean and iron clothes that are collected from five drop-off locations (DL_1–DL_5). It also wants to add another drop-off location (DL_6) in a suburban area. This new drop-off location serves as a feeder to the existing drop-off locations. It does not interact directly with the factories. The two factories have two trips between them per day. The daily interaction between the new factories and existing as well as new drop-off locations is given in Figure 12.10. Assume that the unit cost of deliveries is the same for all locations. Further assume that the distance metric is rectilinear and that the coordinates (x_i, y_i) for the five existing drop-off locations are (12, 12), (5, 8), (2, 6), (10, 5), and (2, 18), respectively.

   (a) Determine the optimal location of the two factories as well as the sixth drop-off location by setting up a linear programming model for the problem and solve it using any available software.

   (b) Are the locations obtained in (a) feasible? If not, determine alternate near-optimal, feasible locations.

7. Example 12.2 shows only the first two and the last three iterations. Using a spreadsheet, show all the intermediate iterations and verify that the final answer you get is the same as that in Example 12.2.

8. Selecting an ε value of 0.2, solve the problem described in Example 12.2 with a different initial seed. Is the solution you get different from that in Example 12.2? Explain.

9. Repeat Exercise 8 with an ε value of 0.002.

	DL_1	DL_2	DL_3	DL_4	DL_5
F_1	3	7	1	2	9
F_2	2	8	4	6	5
DL_6	1	0	1	0	1

FIGURE 12.10   Interaction between drop-off locations and two dry-cleaning factories in Exercise 6.

10. Eliminate the x_i and y_i variables in Exercise 5 using Equations 12.9 and 12.11, set up a linear model for the same problem, and resolve the model using LINGO. Verify that the answer you get is the same as that in Exercise 5.

11. Eliminate the x_i and y_i variables in Exercise 6 using Equations 12.9 and 12.11, set up a linear model for the same problem, and resolve the model using LINGO. Verify that the answer you get is the same as that in Exercise 6.

12. Consider Example 12.1. Assume that only one warehouse is to be built. The interaction of this warehouse to each existing and new service center is equal to 10, 9, 9, 9, and 4, respectively. Set up a linear model for this problem and solve it using LINGO or any other available LP software. Is the answer different from what you obtained in Example 12.1? Explain.

13. Suppose the flexible manufacturing cell described in Exercise 4 is to be served by overhead material-handling equipment and the Euclidean metric is more appropriate than rectilinear. Determine the optimal coordinates of the two new facilities. Are they different from that obtained in Exercise 4? Explain. (Hint: You may find it easier to do the computations on a spreadsheet.)

14. Experiment with large, medium, and small values of ε for Exercise 8. Analyze the results obtained and discuss the advantages/disadvantages of using large and small values of ε.

15. Consider Exercise 5.
    (a) Assuming the distance metric in Exercise 5 is not rectilinear but Euclidean, determine the coordinates of the optimal location using the technique discussed in Section 12.2.2. Is it different from what you obtained in Exercise 5. Why or why not? Explain. (Note: It may be helpful to develop results for the model in a spreadsheet. Try using a low value for ε, so that convergence takes place gradually and therefore better results can be obtained.)
    (b) Change the value of ε to a high value so that convergence takes place rather quickly. Are the answers in (a) and (b) too far off? Explain.

16. Repeat the questions in Exercise 15 for the data in Exercise 6.

17. If there were no interaction between the two distribution centers in Exercise 5, what would be their new optimal locations (assuming rectilinear distances)? Is it different from what you obtained in Exercise 5 or 15a? Why or why not? Explain.

18. If there were no interaction between the two distribution centers in Exercise 6, what would be their new optimal locations (assuming rectilinear distances)? Is it different from what you obtained in Exercise 6 or 16a? Why or why not? Explain.

19. Given a set of nodes including a source node and a sink node and the distance between each pair of nodes, the shortest path problem finds the shortest path from the source to the sink. Modify the network flow problem in Section 12.3.1 to model the shortest path problem.

20. Given a set of nodes, including a set of source nodes, a set of sink nodes, and no transshipment nodes, the cost of sending a unit of flow from each source node to each sink node, no capacity restrictions on the arcs, the transportation problem is concerned with minimizing the cost of meeting the demand at sink nodes using capacity from the source nodes. Modify the network flow problem in Section 12.3.1 to model the transportation problem. Is it identical to the one you saw in Chapter 9? Explain.

21. Given a set of nodes, including one source node, one sink node, a set of transshipment nodes, capacity restrictions on the arcs, the maximal flow problem is concerned with determining the maximum flow that can be sent from the source node to the sink node. Modify the network flow problem in Section 12.3.1 to model the maximal flow problem.

22. Solve Example 12.3 using the Network Simplex Algorithm. Is the answer you get the same as that shown in Example 12.2? Why or why not?

23. Consider model 5. Suppose that two units of raw material yield one unit of finished product. Modify the model to reflect this.

24. Consider model 5. Reformulate it as a transportation model with a set of supply constraints and a set of demand constraints.

25. Consider the transportation tableau setup in Table 12.4. Suppose the supply source capacity is unlimited, plant capacity is limited, and total demand exceeds total plant capacity. Modify Table 12.4 to reflect the aforementioned changes.

26. Repeat Exercise 25 for the case where plant capacity is unlimited, supply source capacity is limited, and total demand exceeds total supply source capacity.

27. Set up a transportation (LP) model for the situation described in Exercise 25.

28. Set up an LP model for the situation described in Exercise 26.

29. Solve the transportation model in Example 12.4 using TSA. Verify the results you get are the same as that shown in Figure 12.3.

30. Consider Example 12.4. Suppose Orlando has a demand of 400 units instead of 500. Set up a transportation tableau for the modified situation and solve it using LINDO. How is the solution different from that obtained in Example 12.4? Explain.

31. Solve the transportation model in Exercise 30 using the TSA. Verify the results you get are the same as that in Exercise 30.

32. Repeat Exercise 30 assuming Orlando has a demand of 600 units instead of 500 units.

33. Solve the transportation model in Exercise 32 using the TSA. Verify the results you get are the same as that in Exercise 32.

34. Consider a two-stage transportation problem with three supply sources, four demand centers, and three plants. The supply, demand, and unit transportation cost data are provided in Table 12.10. Determine how the goods are to be shipped from supply sources to demand centers.

35. Consider a two-stage transportation problem with two supply sources, three demand centers, and two plants. The supply, demand, and unit transportation cost data are provided in Table 12.11.

    (a) Set up a transportation (LP) model to determine how the goods are to be shipped from supply sources to demand centers via the plants.
    (b) Solve the model you set up in (a).

TABLE 12.10

Supply, Demand, and Transportation Cost Data for Exercise 34

Supply Source	Supply Capacity	Demand Center	Demand
1	450	1	200
2	380	2	355
3	220	3	190
		4	455

Inbound Transportation Cost			
To-From	1	2	3
1	70	300	60
2	50	500	80
3	60	400	70

Outbound Transportation Cost				
To-From	1	2	3	4
1	180	240	140	200
2	140	180	60	100
3	320	340	260	220

TABLE 12.11

Supply, Demand, and Transportation Cost Data for Exercise 35

Plant	Plant Capacity	Demand Center	Demand
1	400	1	200
2	300	2	300
		3	100

Inbound Transportation Cost		
To-From	1	2
1	50	30
2	40	60

Outbound Transportation Cost			
To-From	1	2	3
1	100	120	100

(c) Set up a transportation tableau similar to the one in Table 12.4 and determine a feasible solution for the distribution problem. Find an improved solution, if possible. (Hint: You are not required to use any sophisticated algorithm, but only to develop a feasible solution and find an improved one.)

(d) Solve the transportation problem you set up in (c) using the TSA. (You may use a software for this purpose.) Compare the solution you get with that obtained in (b) and (c). Are they different? Why or why not?

36. The department of defense (DoD) can purchase nine key technologies required in a critical defense project from seven aerospace companies. Each of these companies can supply several of the technologies, but not all nine. The fixed charge (in $ millions) of doing business with a company is rather high and is given in the matrix in Figure 12.11. Also shown in the matrix are

Company\technology	1	2	3	4	5	6	7	8	9	Fixed cost
1	1	1	0	0	0	0	0	0	1	20
2	1	1	0	1	0	0	0	0	0	34
3	0	0	1	0	0	0	0	1	0	28
4	0	0	0	1	0	1	0	0	0	92
5	1	0	0	1	1	0	0	0	0	59
6	0	0	0	1	0	0	1	1	0	51
7	0	0	0	0	0	0	0	0	1	12

FIGURE 12.11 Binary matrix showing nine technologies that can be supplied by seven companies in Exercise 3.

binary values that tell us whether or not a company can supply a technology. The DoD wants to select companies so that all the nine technologies are supplied at minimum cost.
  (a) Set up an integer programming model for this problem and solve using LINGO.
  (b) Solve the model by hand using the greedy algorithm.
  (c) Are the answers you obtained in (a) and (b) the same? Why or why not? Explain why it may be advantageous to solve heuristically using greedy algorithm instead of the optimal branch-and-bound algorithm in LINGO.
37. There are three potential plant sites and five customers in a distribution network. The per unit variable cost of serving each customer from each potential plant site, fixed cost of opening and operating each plant as well as the demand at each customer are provided in Figure 12.12. All figures are in hundreds of thousands of dollars.
  (a) Set up a mixed integer programming model similar to model 6 and solve it using LINGO.
  (b) Use the specialized branch-and-bound algorithm and solve the problem by hand. Verify that the answer you get by hand is the same as that obtained by computer.
38. A national discount store is considering the location of new warehouse(s) to serve its four outlets in the Northeast. The location of warehouses has been narrowed to three sites. The cost of serving each outlet from the three sites (in $ millions) is given in the matrix in Figure 12.13. The costs of opening a warehouse at sites 1, 2, and 3 are 3, 4, and 3 million dollars, respectively.
  (a) Using the branch-and-bound technique discussed in Section 12.4.2, find the solution and lower bound at the root node. In step 5 of the algorithm, select a variable to branch on using the "largest fractional y value rule," and find the optimal solution as well as the lower or upper bound at each node. (Be sure to indicate whether the bound you obtain at each node is a lower or upper bound.)
  (b) If all the fixed costs were equal to 0, what is the most efficient algorithm to solve the problem? Can you develop the solution by inspection? If so, what is it?

	1	2	3	4	5	Fixed cost
1	4	3	10	12	8	50
2	6	4	5	8	4	40
3	8	7	7	6	6	60
Demand	10	20	16	22	5	

FIGURE 12.12   Fixed and variable cost of serving five customers from three plants in Exercise 37.

	1	2	3	4
1	4	2	8	3
$[c_{ij}] =$ 2	3	1	1	3
3	2	1	2	3

FIGURE 12.13   Variable cost of serving four retail outlets from three warehouses in Exercise 38.

TABLE 12.12

Variable Cost of Serving Customers in Exercise 39

From-To	1	2	3	4	5	Fixed Cost
1	4	3	10	12	8	5
2	6	4	5	8	4	4
3	8	7	7	6	6	6

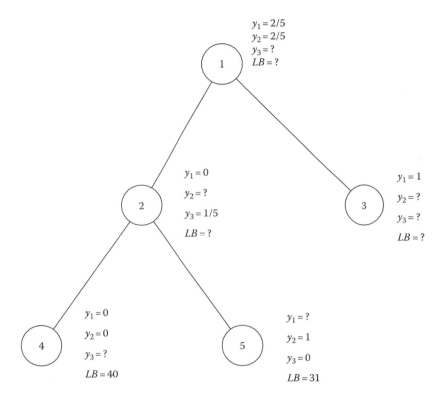

FIGURE 12.14 Branch-and-bound tree for Exercise 39.

39. Consider the following uncapacitated location–allocation problem in which there are three facilities and five customers. The variable cost of meeting each customer's demand using each facility is shown in Table 12.12. The fixed cost of opening each facility is also given. Assume all figures are in hundreds of thousands of dollars. At some stage of solution of the problem via the specialized branch-and-bound algorithm, the following partial solution is obtained.

(a) Facility 1 is open, whereas no decision has been made regarding facilities 2 and 3. Customers 1 and 2 are to be served by the open facility 1. Solve this subproblem (node) only using the minimum coefficient rule discussed in Section 12.4.2. What is the solution? Show your calculations.

(b) Suppose you are given a partial branch-and-bound tree for this problem (see Figure 12.14). What can you say about the optimal solution to the problem? Justify your answer.

(c) Values of some binary variables or lower/upper bounds are intentionally left blank in Figure 12.14. Provide suitable values. Do not calculate, but only state what values they could possibly take.

40. *Project*: Consult the literature and find an efficient algorithm for the VRP. Write a computer program for the algorithm and compare its efficiency to other algorithms using datasets available in the literature. Write a report on your work and present your findings to your class.

41. *Project*: Write a computer program for the Benders' decomposition and modified Benders' decomposition algorithms. Generate large datasets with up to 20 plants, 40 warehouses, and 25 customers. Compare the efficiencies of the two other algorithms. Write a report on your work and present your findings to your class.

13 Introduction to Queuing, Queuing Network, and Simulation Modeling*

13.1 INTRODUCTION

As the name implies, queuing theory involves the study of queues or waiting lines. Unlike many other modeling methods, it is not process specific and can be used to model any dynamic system in which discrete events alter the state of the system. Queuing theory can be used to study any manufacturing or service system where a queue buildup occurs over time. Thus, the system can be an airport, a walk-in medical clinic, fast-food restaurants, or machine shop. In an airport, we see queue buildup occurring when departing airplanes wait for clearance from air traffic control to take off. Similarly, in a walk-in medical clinic, customers wait their turn for consultation with medical staff, and in a fast-food restaurant, customers wait to place an order and also to pick up their order. Jobs wait to be machined on an automated lathe in a machine shop. In this chapter, we will focus on the application of queuing theory to manufacturing environments. Queuing theory can be used to answer the following questions with respect to a given system:

- What is the expected number of parts waiting in a queue?
- What is the expected time a part spends waiting in a queue?
- What is the probability that a machine will be idle?
- What is the probability of a queue being filled to capacity?

In a queuing system, customers arrive by some arrival process and wait in a queue for the next available server. In the manufacturing context, customers could be parts and servers could be machines or material handling carriers or pick and place robots. When a server becomes available, a customer is selected by some queue discipline for service. Service is completed by an output process and the customer leaves the system. The simple queuing system illustrated in Figure 10.2 is reproduced again as Figure 13.1.

The arrival process tells us how customers arrive at the queue. Generally, it is assumed that no more than one customer can arrive at any given instant of time. However, in some situations, especially in manufacturing, parts typically arrive in batches. Such arrivals are called bulk arrivals. Unless otherwise stated, the arrival process is assumed to be independent of the number of customers already in the system. The most common exception to this assumption occurs when the arrival rate decreases due to overcrowding, i.e., customers arrive at the system but do not enter it because the queue is too long. For example, a customer may not wait at a restaurant, if there are many people already waiting. This phenomenon is referred to as balking. Another situation in which the arrival process depends upon the number of customers already in queue or service occurs when there is a finite calling population. Assume a machine repairman is assigned to the

---

* Parts of this chapter are reproduced from Heragu, S.S. and Lucarelli, C.M., Automated manufacturing system design using analytical techniques, in J.D. Irwin (Ed.), *The Industrial Electronics Handbook*, CRC Press Inc., Boca Raton, FL, 1997, pp. 677–693. The author is grateful to C.M. Lucarelli for permission to use that material.

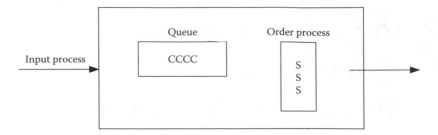

FIGURE 13.1   A queuing system (C, customers; S, serves).

repair of five machines. If all of them have broken down and have arrived for repair or are being repaired, no more arrivals can take place. On the other hand, if only one machine is currently being repaired, there is greater probability that another will break down and enter the queue soon. The calling source or calling population from which arrivals take place is said to be finite for such situations. In many models discussed in this chapter, the infinite calling population assumption is more relevant.

An arrival process is usually described by the probability distribution of the number of arrivals in a fixed interval of time. The most commonly used distribution is the Poisson distribution. As will be discussed later, if the number of arrivals at any interval of time follows a Poisson distribution, then the time between consecutive arrivals (known as the interarrival time) follows an exponential distribution.

The queue discipline is the method by which a part or customer is selected from a queue for service. When customers are served in the order in which they arrive, the queue discipline is first come, first served (FCFS). Other queue disciplines include last come, first served (LCFS), service in random order (SIRO), or priority-based selection.

The service process is typically described by a probability distribution. The most commonly used distribution is the exponential distribution. As with the arrival process, the service process is assumed to be independent of the number of parts already in the system. For example, the service rate is not assumed to increase to accommodate a backlog. The service process may have one or more servers in series (in which case each customer has to go through a sequence of servers before service is completed) or in parallel (each customer visits one of many servers depending upon who is available at the time the customer is ready to depart the queue for service). The service rate is the number of customers served per unit time. Similarly, the arrival rate of a queuing system is the terms of number of customers arriving per unit time.

13.1.1   Modeling of the Arrivals and Service Process

Let us define T_i as the ith interarrival time (or service time for the ith customer) and assume it to be an independent, continuous random variable F. Let the probability density function (pdf), expected value, and variance of F be $f(t)$, $E(F)$, and $V(F)$, respectively. Then,

$$P(F \leq c) = \int_0^c f(t)dt \quad \text{and} \quad P(F > c) = \int_c^\infty f(t)dt$$

Suppose that F follows an exponential distribution with parameter λ. The exponential distribution has several characteristics that enable us to analyze queuing models rather easily.

The pdf $f(t)$ for an exponential distribution is a strictly decreasing function of t ($t \geq 0$). So,

$$P\{0 \leq F \leq \Delta t\} > P\{t \leq F \leq t + \Delta t\} \quad \text{for any } t \text{ and } \Delta t > 0$$

It is not difficult to infer from this property that F is likely to take on a very small value, perhaps near zero. See Hillier and Lieberman (1990) for further explanation on this aspect.

Suppose that F represents service time and the time to serve a customer is typically small, but an occasional customer requires extensive service. For such a situation, clearly the service time distribution can be approximated as an exponential distribution. For example, jobs arriving at a machining center may require essentially the same machining process (and hence the setup and machining time), but every now and then, a job requiring a large setup or machining time or both may arrive at the queue for processing. The reader should be cautioned that not all systems exhibit such a characteristic. In many systems, the type of service may be such that it is almost identical for every customer. In such a case, the actual service time will be close to the expected service time and hence, the exponential distribution cannot be used. However, in most models discussed in this chapter, we will assume that the interarrival times and service times follow an exponential distribution. Exceptions are discussed in a separate section.

It is well known that the pdf of an exponential distribution is given by

$$f(t) = \begin{cases} \lambda e^{-\lambda t} & \text{for } t > 0 \\ 0 & \text{for } t \leq 0 \end{cases} \tag{13.1}$$

The cumulative probabilities for an exponential distribution are given by

$$P(F \leq c) = \int_0^c f(t)dt = \int_0^c \lambda e^{-\lambda t}\, dt = 1 - e^{-\lambda c} \tag{13.2}$$

The preceding result can be verified by substituting $x = -\lambda t$. Then, $dt = -(1/\lambda)dx$. Therefore,

$$\int_0^c \lambda e^{-\lambda t}\, dt = \int_0^c -e^x\, dx = -e^x \Big|_0^c = -e^{\lambda t}\Big|_0^c = 1 - e^{\lambda c} \tag{13.3}$$

Also,

$$P(F > c) = \int_c^\infty \lambda e^{-\lambda t}\, dt = -e^{-\lambda t}\Big|_c^\infty = e^{-\lambda c} \tag{13.4}$$

The expected value of F, which is given by

$$E(F) = \int_c^\infty t f(t)dt$$

for a general distribution, can be shown to be $1/\lambda$ for an exponential distribution. Similarly, the variance of F, $V(F)$, which is equal to

$$V(F) = \int_{c}^{\infty} \left[t - E(t) \right]^{2} f(t) dt$$

for a general distribution, can be shown to be $1/\lambda^{2}$ for an exponential (see Exercise 6). We are now ready to discuss two fundamental properties of the exponential distribution.

Property 13.1: The exponential distribution is *memoryless*. If the interarrival time distribution is exponential, it means that the time until the next arrival is independent of how long it has been since the last arrival. Thus, it is completely random. This property of the exponential distribution is also referred to as the Markovian, lack of memory, no-memory, or forgetfulness property. Mathematically, we have

$$P(F > t + \Delta t | F > t) = P(F > \Delta t), \quad \text{provided } t, \Delta t \geq 0$$

Proof of this result is shown in the following. From the conditional probability rule, we know that

$$P(F > t + \Delta t | F > t) = \frac{P(F > t + \Delta t \cap F > t)}{P(F > t)} = \frac{P(F > t + \Delta t)}{P(F > t)}$$

From Equation 13.4, we also know that

$$P(F > t) = e^{-\lambda t} \quad \text{and} \quad P(F > t + \Delta t) = e^{-\lambda(t + \Delta t)}$$

Hence,

$$P(F > t + \Delta t | F > t) = \frac{P(F > t + \Delta t)}{P(F > t)} = \frac{e^{-\lambda(t + \Delta t)}}{e^{-\lambda t}} = P(F > \Delta t)$$

The no-memory property implies that to predict future arrivals, we do not have to remember how long it has been since the last. If a customer's service in an S-server system has just begun, the probability that his or her service will be completed is equal to the probability of the remaining customers' service being completed, although they may have begun to receive service much earlier. Thus, in a machine shop with two lathes (servers), if a job has just been loaded on the second lathe while the first has been processing a job for the past 10 or more minutes, the no-memory property implies that the service time of the job on the second lathe has the same distribution as the remaining service time of the job on the first lathe. It is perhaps obvious to the reader now why the no-memory property makes queuing models relatively easy to solve. If it were not for this property, it would be very difficult to compute the remaining service time distribution of the job on the first lathe. It has been shown that the exponential distribution is the only continuous pdf having the no-memory property in Feller (1957). If the interarrival times are not exponential, we would have to use some other distribution, e.g., Erlang.

Property 13.2: If the interarrival time has an exponential distribution with parameter λ, then the number of arrivals occurring over a time period t, denoted by N_{t}, has a Poisson distribution with parameter λt. In such a case, the arrival process is said to be Poisson.

To understand the aforementioned result, consider an arrival process $\{N_t=n\}$, where $N_t=n$ is the number of arrivals occurring during time interval t ($t \geq 0$). Assume that $N_0=0$ and the arrival process satisfies the following three conditions:

1. The probability of an arrival occurring between times t and $t+\Delta t$ depends only upon the length Δt and does not depend upon either the number of arrivals occurring until time t or the specific value of t. The probability is equal to $\lambda \Delta t + o(\Delta t)$, where $o(\Delta t)$ is a small quantity in comparison to Δt, especially as Δt tends to zero. In other words,

$$\lim_{\Delta t \to 0} \frac{o(\Delta t)}{\Delta t} = 0 \qquad (13.5)$$

2. The probability of more than one arrival occurring during a very small time interval Δt is equal to $o(\Delta t)$.
3. The number of arrivals occurring in nonoverlapping intervals is independent. In other words, the number of arrivals between the first 20 time units has no bearing on the number of arrivals between the next 20, 30, or 40 time units.
   We define $P\{N_t=n\}$ as the probability of $N_t=n$ arrivals occurring during time interval t. With the understanding that probability of negative arrivals is zero, it can be shown that

$$\lim_{\Delta t \to 0} \frac{P\{N_{t+\Delta t} = n\} - P\{N_t = n\}}{\Delta t} = -\lambda \left[P\{N_t = n\} - P\{N_t = n-1\} \right] \quad \text{for } n = 0,1,2,\dots \quad (13.6)$$

The left-hand side of Equation 13.6 is just the differentiation of $P\{N_t=n\}$ with respect to t. These infinite linear, first-order differential equations can be solved and $P\{N_t=n\}$ can be shown to be $(1/n!)(\lambda t)^n e^{-\lambda t}$ for $n = 0, 1, 2, \dots$ (Gross and Harris, 1985). Given the arrival process described by $P(N_t=n)$, assume F is the random variable for the time between successive arrivals—the interarrival time. We wish to show that F follows an exponential distribution. To do so, note that $P\{t \geq T\} = P\{N_T = 0\}$. This equation states that the probability of the next arrival exceeding any specific value T is equivalent to having no arrival in time T:

$$P\{t \geq T\} = \int_0^\infty f(t)dt \quad \text{and} \quad P\{N(t) = 0\} = e^{-\lambda t}$$

Taking the derivative of both sides, we get $f(t) = \lambda e^{-\lambda T}$, which is an exponential distribution. Hence, if the number of arrivals occurring over a time period t—denoted by N_t—has a Poisson distribution with parameter λt, the random variable F for the interarrival time follows an exponential distribution with parameter λ. The reverse is also shown to be true (Gross and Harris, 1985).

It is well known that the mean of the Poisson distribution given by $E\{N_t\} = \lambda t$ is equal to the variance. In fact, this is the only distribution to have such a property. Hence, if λt is the mean, then λ must be the mean rate at which arrivals occur. Using the relationship between the exponential and Poisson distributions and Property 13.1, it is easy to see that the Poisson arrival process is completely random.

We conclude this section by showing that much of our discussion also holds for the service process. As before, if we assume service times are independent random variables F coming from a pdf $f(t)$, then $E(F) = \int_0^\infty tf(t)dt = 1/\mu$, where μ is the service rate and $1/\mu$ is the mean service time. The two properties discussed in the context of an arrival process also hold for a service process also. To verify this, replace the terms interarrival time and arrival with service time and service completion, respectively. Thus, Property 13.2 says that the service time follows an exponential distribution if the service completion process is Poisson with parameter μ. Note that the notation used for the arrival process parameter is λ, whereas for the service process, it is μ.

13.1.2 TERMINOLOGY AND NOTATION

Depending upon the assumptions made regarding the arrival and service rates, the number of servers, the queue capacity, discipline, and the calling population size, we can have several models. It is therefore convenient to use a simple notation that will enable us to refer to each model concisely. The Kendall–Lee notation is widely used in the queuing literature. It was first devised by Kendall (1953) and modified later by Lee (1966). It clearly and concisely specifies the assumptions made about the system under study. The notation takes the following general form:

$$A/B/C/D/E/F$$

where
> A denotes the nature of the arrival process. An M in the A position denotes that the arrival process is Poisson (or Markovian). (As mentioned in Property 13.2, this means that the interarrival time follows an exponential distribution. Similarly, a D, E_k, GI specify that the interarrival time is constant, follows an Erlang distribution with shape parameter k, and general distribution, respectively. The GI symbol stands for general time distribution, meaning it represents any arbitrary, but specified distribution representing independent and identically distributed [siid] random variables.)
> B denotes the service time distribution. (Again, an M, D, E_k, and GI in the B position are used to denote the nature of the service time distribution.)
> C denotes the number of parallel servers ($S \geq 1$).
> D denotes the queue discipline (first-come, first-served (FCFS), last-come, first-served (LCFS), service in random order (SIRO), general discipline (GD), etc.).
> E denotes the maximum number of customers allowed in the system (system capacity) ($C \geq 1$).
> F denotes the size of the calling population (finite or infinite).

For example, a queuing model with exponentially distributed interarrival and service times, four servers, an FCFS queue discipline, and no limits on the queue capacity or calling population would have the notation M/M/4/FCFS/∞/∞. Because D, E, and F are typically assumed to be FCFS, infinite, and infinite, respectively, we sometimes use the notation $A/B/C$ instead of $A/B/C/D/E/F$. The assumption here is that the queue discipline is FCFS, with no limits on queue and calling population sizes.

Generally, models for which analytical solutions can be easily found have the following characteristics:

1. The calling population is infinite. Results under this assumption are much more easily calculated than under the finite assumption. Simple analytical results are also available for some finite population models, for example, the M/M/1/GD/K/K model—typically called the machine repairman model. In this model, the calling population is finite (e.g., a limited number of machines come to a technician for maintenance and service, or a finite number of ships come to a shipyard for repair and service).
2. The queue capacity is infinite. If the queue can accommodate only a finite number of customers (e.g., due to space limitations, a workstation can have only four pallets in its input buffer), appropriate models are available and can be solved analytically rather easily. For example, the M/M/S/GD/C/∞ model, in which we have a Poisson input process, exponential service times, S servers, customers selected according to GD, a queue with a maximum

capacity of C customers, and the customers coming from an infinite calling population, has a simple solution as discussed later.

3. For many of the models discussed in this section, analytical results are independent of the queue discipline, so the service discipline is assumed to be general. This means that the manner in which customers are selected for service could be FCFS, LCFS, based on jobs having the largest or shortest processing time, priority rules, or any other basis. For a few special models, analytic solution is possible for only FCFS and priority disciplines. The priorities are assumed to be nonpreemptive, meaning the service to a customer interrupted will never be even if a high-priority job comes in after the current customer has been selected for service.

4. The probability distribution of the number of arrivals in a fixed interval of time is Poisson or, equivalently, the distribution of the interarrival time is exponential. Similarly, the probability distribution of the service times is exponentials.

13.2 BASIC QUEUING MODELS

In this section, we discuss several basic queuing models. The first one is called the birth–death model.

13.2.1 BIRTH–DEATH MODEL

The birth–death process plays a significant role in queuing theory. It is a continuous time Markov chain or stochastic process in which the state of the system at any continuous time (rather than discrete points in time) is a nonnegative integer. The role of the birth–death process in automated manufacturing is discussed in detail in many sources including Viswanadham and Narahari (1992).

Assume that a system has $N_t = n$ customers at time t, the pdf of the remaining time until the next arrival (birth) and the pdf of the remaining time until the next service completion (death or departure) are exponential with parameters λ_n, μ_n, respectively, for $n = 0, 1, 2, \ldots$. Further, assume that only one event (either a birth or a death) can occur in any period of time Δt. Although the birth and death rates are independent of each other, they are state dependent, i.e., the rate at which birth occurs is dependent upon the state (number of customers) of the system. However, because of the exponential assumption, birth and death occur randomly. While the birth–death process may be finite or infinite, Figure 13.2 shows an infinite birth–death process in which birth and death rates are state dependent. Such a diagram that shows the possible transitions into and out of each state is called a rate diagram. Analysis of most birth–death processes is computationally feasible only when the system has reached steady state. Until steady state is reached, the system is said to be in a transient state and its analysis is difficult. Therefore, this section only deals with steady-state analysis.

To analyze a birth–death process that has reached steady state, we make use of the following simple, but important result. The mean rate of entering any state n is equal to the mean rate of leaving that state. The equation describing this result is called a balance equation or flow conservation equation. Deriving this result is easy. We use the approach presented in Hillier and Lieberman (1990).

FIGURE 13.2 An infinite birth–death process.

Denote $E_n(t)$, $L_n(t)$ as the number of times the process enters and leaves state n by time t, respectively. Because the number of times we have entered any state must differ from the number of times we have left that state by 1, $|E_n(t) - L_n(t)| \leq 1$. Dividing the left- and right-hand sides by t and letting $t \to \infty$, we get

$$\lim_{t \to \infty} \left| \frac{E_n(t)}{t} - \frac{L_n(t)}{t} \right| = 0$$

Hence,

$$\lim_{t \to \infty} \left| \frac{E_n(t)}{t} \right| = \lim_{t \to \infty} \left| \frac{L_n(t)}{t} \right| \tag{13.7}$$

The left- and right-hand sides of the Equation 13.7 correspond to the mean rate at which the process enters and leaves state n, respectively. Because the preceding result has been derived for any general state n, it is clear that the mean rate of entering any state n is equal to the mean rate of leaving that state.

To analyze steady-state performance of a birth–death process, we solve a system of equations obtained by setting up flow conservation equations for each state. It can be easily verified that we will always have one more variable than the number of equations, so we need another equation to get the value of each variable. This is rather easy. Let P_j be the steady-state probability of being in state j. For state 0 in the rate diagram (Figure 13.2), the flow conservation equation requires

$$\mu_1 P_1 = \lambda_0 P_0$$

Therefore,

$$P_1 = (\lambda_0/\mu_1) P_0$$

Similarly, for state 1,

$$\mu_2 P_2 + \lambda_0 P_0 = \lambda_1 P_1 + \mu_1 P_1$$

Therefore,

$$P_2 = (\lambda_1/\mu_2) P_1 + (1/\mu_2)(\mu_1 P_1 - \lambda_0 P_0) = (\lambda_1/\mu_2)(\lambda_0/\mu_1) P_0 + (\lambda_0/\mu_2)(P_0 - P_0) = (\lambda_1/\mu_2)(\lambda_0/\mu_1) P_0$$

For state 2,

$$\mu_3 P_3 + \lambda_1 P_1 = \lambda_2 P_2 + \mu_2 P_2$$

Therefore,

$$P_3 = (\lambda_2/\mu_3) P_2 + (1/\mu_3)(\mu_2 P_2 - \lambda_1 P_1) = (\lambda_2/\mu_3)(\lambda_1/\mu_2)(\lambda_0/\mu_1) P_0, \text{ and so on.}$$

Let $c_n = (\lambda_{n-1}/\mu_n)(\lambda_{n-2}/\mu_{n-1})\ldots(\lambda_0/\mu_1)$ for $n = 1, 2, \ldots$. Then, $P_n = c_n P_0$, for $n = 1, 2, \ldots$.

Also, $\sum_{n=0}^{\infty} P_n = 1$. Hence, $P_n + \sum_{n=1}^{\infty} P_n = 1$. So,

$$\left(1 + \sum_{n=1}^{\infty} c_n\right) P_0 = 1 \quad \text{and} \quad P_0 = \frac{1}{\left(1 + \sum_{n=1}^{\infty} c_n\right)}$$

We now define these variables:

L Average number of customers in the system
L_q Average number of customers in the queue
L_s Average number of customers in service
W Average time a customer spends in the system
W_q Average time a customer spends in the queue
W_s Average time a customer spends in service
λ Mean arrival rate
μ Mean service rate

One of the most fundamental results in queuing theory that allows us to determine the performance measures of a system is known as Little's law. It defines the relationship between the operational characteristics of a queuing system. It is one of the simplest, yet most powerful formulae in queuing theory. It basically states that under steady-state conditions, the average number of customers in any queuing system is equal to the product of the mean arrival rate and the average time spent in the system by a customer. This result holds for any interarrival and service time distributions, any service discipline, and for any number of servers. The only requirement is that the queuing system must be in steady state.

Mathematically, Little's law is given by $L = \lambda W$. It has also been proved that $L_q = \lambda W_q$ and $L_s = \lambda W_s$. For a rigorous proof of this result, see Little (1961), Stidham (1974), and Ross (1970). However, it can be proved in an intuitive manner rather easily as discussed. Assume a queuing system has Poisson arrivals with a mean arrival rate of λ. Then, there will be one arrival every $1/\lambda$ time units. Also assume that the current queue length is L customers and a new customer has just arrived. Then, the average time spent by the customer in the system is L/λ, which is equal to W, assuming W is the average waiting time in the system (waiting time in queue plus service time) for any customer, including the customer in question. Hence, $L = \lambda W$.

If we know the mean service time, then it is easy to see that $W = W_q + 1/\mu$. If any of the six key variables (L, L_q, L_s, W, W_q, W_s) is known, the remaining can be immediately determined using the equations $L = \lambda W$, $L_q = \lambda W_q$, and $L_s = \lambda W_s$. For the birth–death process discussed thus far, notice that we can analytically find the value of L, L_q, and L_s using these formulae:

$$L = \sum_{n=0}^{\infty} n P_n, \quad L_q = \sum_{n=S}^{\infty} (n - S) P_n, \quad L_s = L - L_q$$

Because P_n can be determined entirely in terms of λ_n and μ_n, where $n = 1, 2,\ldots, \infty$ (which are parameters with known values), we can obtain values for L and L_q. Furthermore, because the mean arrival rate for the birth–death process is given by

$$\lambda = \sum_{n=0}^{\infty} \lambda_n P_n$$

we can obtain values for W, W_q, and W_s using Little's law. The P_n formula includes a geometric series, and its computation is relatively easy for many special cases of the birth–death model. These special cases are described in the next two sections.

The remainder of this chapter, including the numeric examples, focuses on applying queuing theory to automated manufacturing and logistical system design problems. In an automated manufacturing system, parts are transported to a workstation for processing. If the workstation is busy, the part will have to wait in a queue until the workstation is available. Otherwise, there is no waiting time. Thus, if there are S servers at a workstation and $S - 1$ or fewer jobs in the system, an incoming job will not have to wait for service. After processing, the part enters an output queue where it waits to be transported to the next operation. The part processing time includes any setup or dismantle time involved in the operation. A workstation may have one or more servers that can be a worker, a combination of machine and worker, or any other resource.

13.2.2 M/M/1/GD/∞/∞ Queuing Systems

The first manufacturing system discussed is one with a single server. Assume a birth–death process with constant part arrival (birth) and service completion (death) rates. Specifically, let λ and μ be the arrival and service rates of parts, per unit time. If the arrival rate is greater than the service rate, the queue will grow infinitely. The ratio of λ to μ is called the traffic intensity or utilization factor of the server and is denoted as $\rho = \lambda/\mu$. Thus, one can intuitively see that for the system to be in steady state, the traffic intensity must be less than 1.

For this system, the steady-state conditions for the various states can be written exactly as done for the birth–death model. Of course, $\lambda_j = \lambda$; $\mu_j = \mu$, for $i = 1, 2, \ldots$ and $\mu_0 = 0$. From the earlier results, it is easy to see that $P_1 = (\lambda/\mu)P_0$, $P_2 = (\lambda/\mu)^2 P_0$, $\ldots$, $P_n = (\lambda/\mu)^n P_0$. Substituting $\rho = (\lambda/\mu)$ and assuming $0 \le \rho < 1$, we get $P_1 = \rho P_0$, $P_2 = \rho^2 P_0$, $\ldots$, $P_n = \rho^n P_0$, $\ldots$

We know that $P_0 + P_1 + \cdots + P_n + \cdots = 1$. Hence, $P_0(1 + \rho + \rho^2 + \cdots + \rho^n + \cdots) = 1$.

To evaluate $(1 + \rho + \rho^2 + \cdots + \rho^n + \cdots)$, substitute $S = (1 + \rho + \rho^2 + \cdots + \rho^n + \cdots)$. Then $\rho S = (\rho + \rho^2 + \cdots + \rho^n + \cdots)$. Clearly, $S - \rho S = 1$. Hence, $S = 1/(1 - \rho)$.

Thus, $P_0(1 + \rho + \rho^2 + \cdots + \rho^n + \cdots) = P_0/(1 - \rho) = 1$. Therefore, $P_0 = 1 - \rho$. In a similar manner, $P_1 = \rho(1 - \rho)$, $P_2 = \rho^2(1 - \rho)$, $\ldots$, $P_n = \rho^n(1 - \rho)$, $\ldots$.

Now that the state probabilities are known, the other operating characteristics of the queuing system can be calculated. The expected number of parts at a workstation is simply the probability of a given state n multiplied by the probability of its occurrence:

$$L = \sum_{n=0}^{\infty} nP_n = \sum_{n=0}^{\infty} n\rho^n (1-\rho) = (1-\rho)\sum_{n=0}^{\infty} n\rho^n$$

Once again, the summation can be determined by substituting

$$S' = \sum_{n=0}^{\infty} n\rho^n = \rho + 2\rho^2 + 3\rho^3 + \cdots$$

Then $S' - \rho S' = \rho + \rho^2 + \cdots + \rho^n + \cdots$

The right-hand side of this equation is equal to ρS, S was shown previously to be equal to $1/(1 - \rho)$. Hence, we get $S' - \rho S' = \rho/(1 - \rho)$ and $S' = \rho/(1 - \rho)^2$. Therefore,

$$L = \frac{(1-\rho)\rho}{(1-\rho)^2} = \frac{\rho}{(1-\rho)} = \frac{(\lambda/\mu)}{(1-\lambda/\mu)} = \frac{\lambda}{(\mu-\lambda)}$$

L_q can also be determined in this way:

$$L_q = \sum_{n=1}^{\infty}(n-1)P_n = \sum_{n=1}^{\infty}nP_n - \sum_{n=1}^{\infty}P_n = L-(1-P_0) = L-1+1-\rho = L-\rho$$

Therefore,

$$L_q = \frac{\lambda}{(\mu-\lambda)} - \rho = \frac{\lambda}{(\mu-\lambda)} - \frac{\lambda}{\mu} = \frac{\mu\lambda - \mu\lambda + \lambda^2}{\mu(\mu-\lambda)} = \frac{\lambda^2}{\mu(\mu-\lambda)}$$

$$L_s = L - L_q = \frac{\lambda}{(\mu-\lambda)} - \frac{\lambda^2}{\mu(\mu-\lambda)} = \lambda\left[\frac{1}{(\mu-\lambda)} - \frac{\lambda}{\mu(\mu-\lambda)}\right] = \frac{\lambda}{\mu} = \rho$$

Using Little's formula, we can then determine W, W_q, and W_s as

$$W = \frac{L}{\lambda} = \frac{1}{(\mu-\lambda)} \quad W_q = \frac{\lambda}{\mu(\mu-\lambda)} \quad W_s = \frac{1}{\mu}$$

An example illustrating how the M/M/1 model can be applied to evaluate a system design is provided.

Example 13.1

Consider the forge press operation shown in Figure 13.3. After heating, parts arrive at the press with at the rate of ten parts per hour. The pressing process takes an average of 5 minutes. Find the following values assuming the service times are exponentially distributed and the arrival rate is Poisson.

1. The percentage of the time that the press is idle
2. The average number of parts in the queuing system
3. The average queue length
4. The throughput time of the system
5. The amount of time spent in the queue

Solution

The forge press operation can be modeled as an M/M/1 queuing system. The rate diagram for this problem is shown in Figure 13.4. The arrival and departure rates are given as $\lambda = 10$ per hour and $\mu = 12$ per hour, respectively. The utilization factor is $\rho = \lambda/\mu = 0.833$.

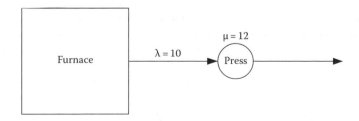

FIGURE 13.3 Furnace and press system.

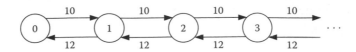

FIGURE 13.4   Rate diagram for Example 13.1.

1. The probability that the press is idle is the probability that there are no jobs being served or waiting. $P_0 = 1 - \rho = 0.167$ or the press is idle 16.7% of the time.
2. The average number of parts in the queuing system is $L = \rho/(1 - \rho) = 5$ parts. This is also the average work-in-process (WIP) inventory.
3. The average queue length is $L_q = \rho^2/(1 - \rho) = 4.167$ parts.
4. The throughput time (or the average time spent by a part in the system) is $W = 1/(\mu(1 - \rho)) = 0.5$ hours. Because λ and μ were given in parts per hour, the throughput time is also in hours.
5. The amount of time spent in the queue is $W_q = \rho/(\mu(1 - \rho)) = 0.4167$ hours $= 25$ minutes.

13.2.3   M/M/S/GD/∞/∞ Queuing Systems

Consider a queuing system in which there are S servers and the ith server, $i = 1, 2, ..., S$, has an exponentially distributed service time with parameter μ_i. Then, it can be shown that the multiple-server system performs just like a single-server system where the service time has an exponential distribution with parameter $\sum_{i=1}^{S} \mu_i$. To prove this result, we use the following property of the exponential distribution.

Property 13.3: The minimum of independent exponential random variables is also exponential. Let $F_1, F_2, ..., F_n$ be independent exponential random variables with parameters $\mu_1, \mu_2, ..., \mu_n$. Let $F_{\min} = \min \{F_1, F_2, ..., F_n\}$. Then for any $t \geq 0$,

$$P(F_{\min} > t) = P\{F_1 > t\} P\{F_2 > t\} \cdots P\{F_n > t\} \quad = e^{\mu_1 t} e^{\mu_2 t} \cdots e^{\mu_n t} = e^{[\mu_1 + \mu_2 + \cdots + \mu_n]t}$$

An interesting implication for this property to interarrival times is discussed in Hillier and Lieberman (1990). Suppose n types of customers, with the ith type of customer having an exponential interarrival time distribution with parameter λ_i, arrive at a queuing system. Let us assume that an arrival has just taken place. Then, from the no-memory property, it follows that the time remaining until the next arrival is also exponential. Using Property 13.3, it is easy to see that the interarrival time for the entire queuing system (which is the minimum amongst all interarrival times) has an exponential distribution with parameter $\sum_{i=1}^{S} \lambda_i$.

We now derive the results for the multiple-server model. Assume λ is the arrival rate and μ is the service rate for each of the S servers. Clearly, if $n \leq S$ customers are present in the system, all of them are in service. If $n > S$ customers are present, all S servers are busy and $n - S$ customers are waiting. From Property 13.3, service rate for the entire queuing system is $\sum_{i=1}^{S} \mu_i$. Thus, for an M/M/2 system with each server having a service rate of 5 parts per hour, the departure rate is 5 parts per hour if only one server is working. If both servers are working, the departure rate is 10 parts per hour. The rate diagram for this system is shown in Figure 13.5.

The utilization factor for an M/M/S system is given by $\rho = \lambda/S\mu$. We can derive the operating characteristics of a multiple-server system from the state probabilities in a similar manner as we did for a single-server system. Clearly, $\mu_n = \mu\{\min\{n, S\}\}$. Hence, $\mu_n = n\mu$ for $n = 1, 2, ..., S$ and $\mu_n = S\mu$ for $n = S + 1, S + 2, ..., \infty$. Also, $\lambda_n = \lambda$ for $n = 0, 1, 2, ..., \infty$. Hence, $P_n = P_0 c_n$, for $n = 1, 2, ..., S$. We get

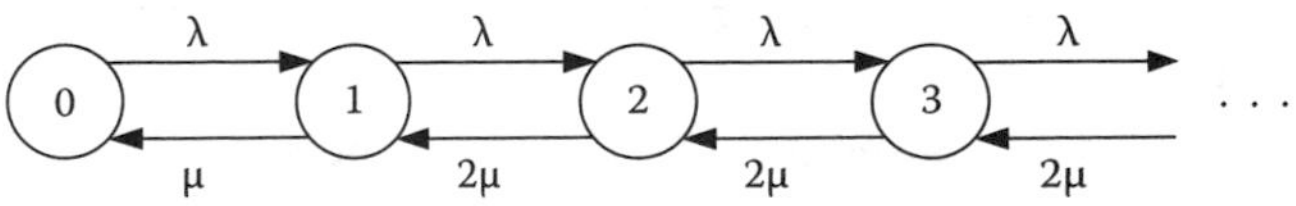

FIGURE 13.5 M/M/2 rate diagram.

$$P_n = P_0 \left[\frac{\lambda}{1\mu} \cdot \frac{\lambda}{2\mu} \cdots \frac{\lambda}{n\mu} \right] = \frac{P_n}{n!} \left[\frac{\lambda}{\mu} \right]^n \quad \text{for } n = 1,2,\ldots,S \tag{13.8}$$

$$P_n = P_0 \left[\frac{\lambda}{1\mu} \cdot \frac{\lambda}{2\mu} \cdots \frac{\lambda}{S\mu} \cdot \frac{\lambda}{S\mu} \cdot \frac{\lambda}{S\mu} \cdots \right] = \frac{P_0}{S!} \left[\frac{\lambda}{\mu} \right]^S \left[\frac{\lambda}{S\mu} \right]^{n-S}$$

$$= \left[\frac{P_0}{S! S^{n-S}} \right] \left[\frac{\lambda}{\mu} \right]^n \quad \text{for } n = S+1, S+2,\ldots,\infty \tag{13.9}$$

$$P_0 = \frac{1}{\displaystyle\sum_{n=0}^{\infty} c_n} = \frac{1}{\displaystyle\sum_{n=0}^{S} c_n + \sum_{n=S}^{\infty} c_n} = \frac{1}{\displaystyle\sum_{n=0}^{S-1} \frac{(\lambda/\mu)^n}{n!} + \frac{(\lambda/\mu)^S}{S!} \sum_{n=S}^{\infty} \left(\frac{\lambda}{S\mu} \right)^{n-S}} \tag{13.10}$$

Note that the second summation in the denominator of the Equation 13.10 is of the form $1+x+x^2+x^3+\cdots$ where $x = \lambda/(S\mu)$. As seen in derivation of the M/M/1 model, this summation is equal to $1/(1-x)$ or $1/(1-(\lambda/S\mu))$. Hence,

$$P_0 = \frac{1}{\displaystyle\sum_{n=0}^{S-1} \frac{(\lambda/\mu)^n}{n!} + \frac{(\lambda/\mu)^S}{S!} \left(\frac{1}{1-\lambda/(S\mu)} \right)}$$

Substituting $\rho S = (\lambda/\mu)$, we get

$$P_0 = \frac{1}{\displaystyle\sum_{n=0}^{S-1} \frac{(\rho S)^n}{n!} + \frac{(\rho S)^S}{S!} \left(\frac{1}{1-\rho} \right)} \tag{13.11}$$

$$P_0 = \begin{cases} \dfrac{P_0(\rho S)^n}{n!} & \text{for } n = 1,2,\ldots,S \\[4mm] \dfrac{P_0(\rho S)^n}{S! S^{n-S}} & \text{for } n = S+1, S+2,\ldots \end{cases} \tag{13.12}$$

We now find L_q:

$$L_q = \sum_{n=S}^{\infty} (n-S)P_n = \sum_{i=0}^{\infty} iP_{S+i} = \sum_{i=0}^{\infty} i \frac{(\rho S)^S \rho^i P_0}{S!} = \frac{(\rho S)^S P_0}{S!} \sum_{i=0}^{\infty} i\rho^i = \frac{(\rho S)^S P_0 \rho}{S!(1-\rho)^2} \tag{13.13}$$

TABLE 13.1
M/M/1/GD/∞/∞ and M/M/S/GD/∞/∞ Operating Characteristics

Model	M/M/1/GD/∞/∞	M/M/S/GD/∞/∞
P_0	$1 - \rho$	$1\big/\left[\left(\rho S\right)^s\big/\left(S!(1-\rho)\right) + \sum_{n=0}^{S-1}\left(\left(\rho S\right)^n/n!\right)\right]$
L	$\lambda/(\mu - \lambda)$	$L_q + \lambda/\mu$
L_q	$L - \rho$	$\left[\left(\rho S\right)^s P_{0\rho}\right]\big/\left[S!(1-\rho)^2\right]$
W	$1/(\mu(1 - \rho))$	$W_q + 1/\mu$
W_q	$\lambda/(\mu(\mu - \lambda))$	L_q/λ

Note that the last summation in Equation 13.13 denoted as S' in the M/M/1 model was previously shown to be equal to $\rho/(1 - \rho)^2$. Now that we have an expression for L_q, we can determine the other variables as $W_q = L_q/\lambda$, $W = W_q + 1/\mu$, $L = \lambda W = \lambda(W_q + 1/\mu) = L_q + \lambda/\mu$. Because the results for the basic M/M/1 and M/M/S models will be used again in later sections, they are summarized in Table 13.1. It should be noted that $\rho = \lambda/\mu$ for the single-server model, but $\rho = \lambda/(\mu S)$ for the multiple-server case. In both models, $P_n = c_n P_0$. Formulae for c_n are summarized in the following.

Single-server model

$$c_n = \rho^n, \quad \text{for } n = 1, 2, \ldots, \infty \tag{13.14}$$

$$P_0 = 1 - \rho \tag{13.15}$$

Multiple-server model

$$c_n = \begin{cases} \dfrac{1}{n!}\left(\dfrac{\lambda}{\mu}\right)^n & \text{for } n = 1, 2, \ldots, S \\[3mm] \dfrac{1}{S!S^{n-S}}\left(\dfrac{\lambda}{\mu}\right)^n & \text{for } n = S+1, S+2, \ldots, \infty \end{cases} \tag{13.16}$$

$$P_0 = \frac{1}{\displaystyle\sum_{n=0}^{S-1}\frac{(\lambda/\mu)^n}{n!} + \frac{(\lambda/\mu)^S}{S!}\left(\frac{1}{1-\lambda/(S\mu)}\right)} \tag{13.17}$$

Note that the P_0 calculation, which itself is required to calculate L_q and W_q, is somewhat tedious. Sakasegawa (1977) came up with an approximate formula to calculate W_q from which L_q can be obtained using Little's law. The W_q formula is shown next:

$$W_q = \frac{(\lambda/\mu S)^{\sqrt{2(S+1)}-1}}{S\mu(1-\lambda/\mu S)}$$

We conclude this section with the discussion of another important property of the exponential distribution that is used later in this section.

Property 13.4: If we have multiple arrivals in a queuing system and each interarrival time has an exponential distribution, the interarrival time for the system as a whole is also exponential with parameter equal to the sum of the parameters for each interarrival.

From Property 13.2, we know that a Poisson arrival rate results in an exponential interarrival the Property 13.4 thus also applies to the Poisson distribution. We will therefore discuss it now in the context of a Poisson distribution. Basically, Property 13.4 states that for a queuing system with multiple arrivals, each occurring according to a Poisson input process, the sum of these arrivals is also Poisson. Let n different customers arrive according to a Poisson input process with parameter λ_i, $i = 1, 2, \ldots, n$. Then, the arrival of all the customers (i.e., aggregate arrival) also follows a Poisson input process with parameter $\lambda = \lambda_1 + \lambda_2 + \cdots + \lambda_n$. The reverse is also true. If we have n customer types and it is known that the arrival of all the customers follows a Poisson distribution with parameter λ, then each type of customer also arrives according to a Poisson input process with parameter λ_i where $\lambda_i = P_i \lambda$, P_i is the probability that an arriving customer is of type i and $P_1 + P_2 + \cdots + P_n = 1$.

Example 13.2 illustrates the effects of multiple servers on a manufacturing system design.

Example 13.2

In an effort to reduce WIP inventory, the press in Example 13.1 is modified to have two servers. Find the following:

1. The average queue length
2. The average number of parts in the queuing system (WIP)
3. The expected amount of time spent in the queue
4. The throughput of the system

Solution

This new queuing system is an M/M/2 model. The rate diagram for this problem is shown in Figure 13.6. The arrival and service rates are the same as in Example 13.1; $\lambda = 10$ per hour and $\mu = 12$ per hour. The two servers result in a utilization factor of $\rho = \lambda/S\mu = 0.417$.

The new idle time of the press is $P_0 = 1/\left[(\rho S)^S / S!(1-\rho) + \sum_{n=0}^{1}\left((\rho S)^n / n!\right)\right] = 7/17 = 0.412$, or 41.2% of the time.

1. The expected queue length in the two server system is $L_q = [\rho(\rho S)^S P_0]/[S!(1-\rho)^2] = 7/40 = 0.175$ parts.
2. The expected WIP is $L = L_q + \lambda/\mu = 121/120 = 1.0083$ parts.
3. The expected time spent by a part in the queue is $W_q = [(\rho S)^S P_0]/[S! S\mu(1-\rho)^2] = 7/400 = 0.0175$ hours $= 1.05$ minutes.
4. The throughput time in the system is $W = W_q + 1/\mu = 0.10083$ hours $= 6.05$ minutes.

Multiple servers are just one factor affecting the characteristics of a queuing system. The physical design of the queue can affect the operating characteristics. The following example illustrates how different system designs can yield different results.

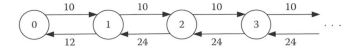

FIGURE 13.6   Rate diagram for Example 13.2.

Example 13.3

A workstation in a company receives parts from two sources—a subcontractor and a machining center within the company. The arrival rate of each is Poisson with parameter $\lambda/2$. The company has purchased another identical workstation in order to increase throughput. Each has a service rate of μ. The company has the following three ways of designing the system.

1. Have all the parts from the subcontractor visit the first workstation, and all the parts from the internal machining center visit the recently purchased workstation for processing.
2. Because the parts coming from the subcontractor are identical to those from the internal machining center, and the two workstations can be combined into one so as to have a service rate of 2μ, the company can have parts from both sources join a single queue and visit the combined workstation.
3. Combine the two arrivals into a single queue but keep the workstations separate.

Making the usual assumptions that the service times are random variables following an exponential distribution,

1. Determine the effective arrival and service rates of the aforementioned systems.
2. Determine which system would result in the most waiting time for the parts. Which system would result in the least waiting time?
3. If reducing WIP becomes a more important criterion than waiting time in the system, show that system (c) is preferred to system (b). Assume that WIP is measured as the mean number of parts in the queue.

Solution

System (a) is an M/M/1 queue with parameters $\lambda/2$ and μ. System (b) is an M/M/1 queue with parameters λ and 2μ. System (c) is an M/M/2 queue with parameters λ and μ. The three systems are depicted in Figure 13.7.

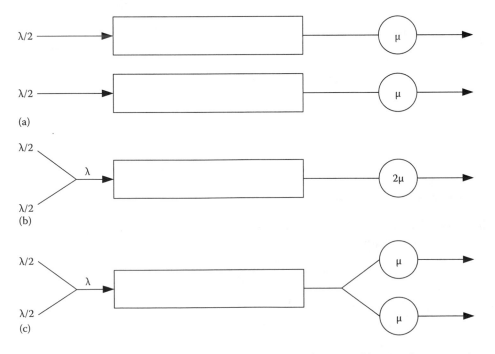

FIGURE 13.7 Three-system designs. (a) Two separate queues and servers, (b) merged queues and servers, and (c) merged queues and separate servers.

1. As discussed previously, the sum of two Poisson variables yields a Poisson variable. Because two Poisson arrivals are superimposed for systems (b) and (c), the arrival rate is $\lambda/2 + \lambda/2 = \lambda$. The effective arrival and service rates are λ and 2μ for all three systems.
2. For an M/M/1 system with parameters λ and μ, the waiting time in a queue (W) and the length of the queue (L_q) are

$$W = \frac{1}{\mu(1-\rho)} = \frac{1}{\mu - \lambda}$$

$$L_q = \frac{\lambda^2}{\mu(\mu - \lambda)}$$

For an M/M/2 system,

$$P_0 = \frac{1}{\frac{(2\rho)^2}{2!(1-\rho)} + \sum_{n=0}^{1} \frac{(2\rho)^n}{n!}} = \frac{1}{\frac{(2\rho^2)}{2!(1-\rho)} + \frac{(2\rho)^0}{0!} + \frac{(2\rho)^1}{1!}}$$

$$= \frac{2(1-\rho)}{2 + 4\rho - 2\rho - 4\rho^2 + 4\rho^2} = \frac{1-\rho}{1+\rho}$$

$$L_q = \frac{(2\rho)^2 P_0 \rho}{2!(1-\rho)^2} = \frac{(2\rho)^2(1-\rho)\rho}{2!(1+\rho)(1-\rho)^2}$$

$$= \frac{2\rho^3}{1-\rho^2} = \frac{\lambda^3}{\mu(4\mu^2 - \lambda^2)}$$

$$W_q = \frac{L_q}{\lambda} = \frac{\lambda^2}{\mu(4\mu^2 - \lambda^2)}$$

$$W = W_q + \frac{1}{\mu} = \frac{\lambda^2}{\mu(4\mu^2 - \lambda^2)} + \frac{1}{\mu} = \frac{4\mu}{4\mu^2 - \lambda^2}$$

For system (a), therefore, $W^{(a)} = 1/(\mu - \lambda/2) = 2/(2\mu - \lambda)$. $L_q^{(a)} = \lambda^2/4\mu(\mu - \lambda/2) = \lambda^2/2\mu(2\mu - \lambda)$. For system (b), $W^{(b)} = 1/(2\mu - \lambda)$. $L_q^{(b)} = \lambda^2/2\mu(2\mu - \lambda)$. For system (c), $W^{(c)} = 4\mu/(4\mu^2 - \lambda^2)$. $L_q^{(c)} = \lambda^3/[\mu(4\mu^2 - \lambda^2)]$. To prevent an infinite queue, λ must be less than 2μ (i.e., $\rho < 1$). Therefore, $W^{(c)} = 4\mu/(4\mu^2 - \lambda^2) = [4\mu(2\mu + \lambda)][1/(2\mu - \lambda)] > 1/(2\mu - \lambda) = W^{(b)}$. Thus, $W^{(a)} > W^{(c)} > W^{(b)}$. Hence, system (a) would result in the most waiting time and system (b) the least.
3. The WIP in the queue of system (c) is given by $L_q^{(c)} = [\lambda^2/(2\mu(2\mu - \lambda))][2\lambda/(2\mu + \lambda)] < [\lambda^2]/[2\mu(2\mu - \lambda)] = L_q^{(b)}$. When WIP in the queue is the criterion, system (c) is preferred to system (b), but when manufacturing lead time is considered, (b) is better than (c).

13.3 OTHER VARIATIONS OF THE BASIC QUEUING MODELS FOR WHICH ANALYTICAL SOLUTION IS AVAILABLE

Excluding the birth–death model, we studied two basic queuing models in the previous two sections. Analytical results for the two models were easily obtained. These models serve as building blocks for all other models in queuing theory. In this section, we will relax many of the assumptions made in the last two, one at a time, and where available, provide exact analytical results for each.

Although derivations of the results are not provided, it is rather easy for the reader to prove the results for many of the models on his or her own. If necessary, the reader may refer to any standard queuing text such as Gross and Harris (1985) or Kleinrock (1975).

13.3.1   FINITE QUEUE SIZE MODELS

Thus far, we have assumed that the capacity of the queue is infinite. However, in some applications, this may not be true. For example, each machining center in a manufacturing system has a finite input buffer (waiting area for parts that need processing). Suppose that the queue can accommodate a maximum of C parts (customers). Then, the maximum number of servers that can ever be used is also C, i.e., $S \leq C$. Also, the arrival rate λ_n for states $n \geq C$ is equal to 0. Using this condition, we can determine c_n for the single-server and multiple-server models as shown in the following. Of course, because P_n, $c_n P_0$, and P_0 can be determined (see Table 13.2), we can find the steady-state probabilities for each possible state. As in Table 13.1, ρ is λ/μ, $\lambda/(\mu S)$ for the single- and multiple-server models, respectively.

Single-server model

$$c_n = \begin{cases} \rho^n & \text{for } n = 1, 2, ..., C \\ 0 & \text{for } n > C \end{cases} \tag{13.18}$$

Multiple-server model

$$c_n = \begin{cases} (\rho S)^n / n! & \text{for } n = 1, 2, ..., S \\ [(\rho S)^n] / [S! S^{n-S}] & \text{for } n = S+1, S+2, ..., C \\ 0 & \text{for } n > C \end{cases} \tag{13.19}$$

Using the preceding formulae, we can develop results shown in Table 13.2 for the M/M/1/GD/C/∞ and M/M/S/GD/C/∞ models. For all the previous models, we made the assumption that the ratio of the arrival rate to the service rate must be less than 1; otherwise, the queue size would have grown infinitely. We need not worry about this problem here because due to the finite queue capacity, the queue size can never increase beyond its capacity. Hence, the results in Table 13.2 hold as long as $\rho \neq 1$. When $\rho = 1$, it can be shown that $L = C/2$ for the single-server model and $L_q = P_0 (\rho S)^S (C - S)(C - S + 1)/2S!$

TABLE 13.2
Results for Finite Queue Models

Model	M/M/1/GD/C/∞	M/M/S/GD/C/∞
P_o	$(1 - \rho)/(1 - \rho^{C+1})$	$1\Big/\Big[\sum_{n=1}^{S}(\rho S)^n/n! + \big((\rho S)^S/S!\big)\Big/\big(\sum_{n=S+1}^{C}\rho^{n-S}\big)\Big]$
L	$\rho/(1 - \rho) - (C+1)\rho^{C+1}/(1 - \rho^{C+1})$	$L_q + \lambda(1 - P_C)/\mu$
L_q	$L + P_0 - 1$	$(\rho S)^S P_0 \rho[1 - \rho^{C-S} - (C-S)\rho^{C-S}(1-\rho)]/[S!(1-\rho)^2]$
W	$L/(\lambda(1 - P_C))$	$L/(\lambda(1 - P_C))$
W_q	$L_q/(\lambda(1 - P_C))$	$L_q/(\lambda(1 - P_C))$

Because the finite queue size ensures that the arrival rate after the queue has reached its capacity is zero, the effective arrival rate is to be calculated using the following formula:

$$\lambda_{eff} = \sum_{n=0}^{C-1} \lambda P_n = \lambda(1 - P_C)$$

Now the other variables shown in Table 13.2 can be determined.

13.3.2 Finite Source Models

All the models discussed so far assumed that the source or the calling population was infinite. We now relax this assumption and present some results. Of course, if the calling population is finite, it does not make sense to have an infinite queue. So, we assume that the queue size is equal to the size of the calling population. The finite source model is also known as the machine repairman problem because a machine repairman is usually assigned to a finite number of machines. Thus, the calling population is finite for such types of models. Let us assume the size of the calling population 15 N. As mentioned previously, the arrival rate for this model is state dependent. Note that λ_0 is equal to 0 for this model. The effective arrival rate is given by the product of the mean arrival rate and the average number of machines working. Because $N - L$ machines (customers) are working and do not need repair (servicing) on average, the effective arrival rate to be used in the calculation of W and W_q is $\lambda(N - L)$.

Single-server model

$$c_n = \begin{cases} \dfrac{N! \rho^n}{(N-n)!} & \text{for } n = 0, 1, 2, ..., N \\ 0 & \text{for } n > N \end{cases} \tag{13.20}$$

Multiple-server model

$$c_n = \begin{cases} \dfrac{N! \lambda^n}{(N-n)! n! \mu^n} & \text{for } n = 0, 1, 2, ..., S \\ \dfrac{N! \lambda^n}{(N-n)! S! S^{n-S} \mu^n} & \text{for } n = S+1, S+2, ..., N \\ 0 & \text{for } n > N \end{cases} \tag{13.21}$$

Results are shown in Table 13.3 for the M/M/1/GD/N/N and M/M/S/GD/N/N models. If a machine is not in repair (service), it must be in production. Bunday and Scraton (1980) have shown that as long as the production time is independent with mean $1/\lambda$, the results in Table 13.3 will hold, regardless of the distribution of the production (nonrepair or interarrival) time. However, when the repair (service) time is nonexponential, analytical results have been obtained only for the single-server case (Takacs, 1962).

13.3.3 Nonexponential Models

Thus far, we have assumed that the interarrival and service times have an exponential distribution. While it enables us to find simple analytical results, it cannot be used in some real-world

TABLE 13.3
Results for Finite Source Models

Model	M/M/1/GD/N/N	M/M/S/GD/N/N
P_0	$1/\left[\sum_{n=0}^{N}\lambda^n N!/\left(\mu^n(N-n)!\right)\right]$	$1/\left[\sum_{n=0}^{S-1}\left(\lambda^n N!\right)/\left(\mu^n n!(N-n)!\right)+\sum_{n=S}^{N}\left(\lambda^n N!\right)/\left(\mu^n S! S^{n-S}(N-n)!\right)\right]$
L	$N-\mu(1-P_0)/\lambda$	$L_q+\sum_{n=0}^{S-1}nP_n+S\left(1-\sum_{n=0}^{S-1}P_n\right)$
L_q	$N-(\lambda+\mu)(1-P_0)/\lambda$	$\sum_{n=S}^{N}(n-S)P_n$
W	$L/(\lambda(N-L))$	$L/(\lambda(N-L))$
W_q	$L_q/(\lambda(N-L))$	$L_q/(\lambda(N-L))$

situations, especially in many automated manufacturing systems. For example, in a manufacturing system, the arrival of parts may not be random but scheduled to occur at regular intervals. The arrival process is no longer Poisson and the appropriate model is therefore D/M/S, assuming we have an exponential service time distribution. Results for this and the E_k/M/S models are available in tabular form in Hillier et al. (1981). Exact results for the more general G/M/1 and G/M/S models are available in Gross and Harris (1985). Unfortunately, although we can get exact results for the exponential service time and general interarrival time distribution models, we cannot do so for the general service time and exponential interarrival time distribution models. The latter are more often encountered in practice.

Consider, for example, a manufacturing system in which we have sophisticated equipment with a high degree of precision and accuracy. In such systems, the processing (service) times are identical from one part (customer) to the next. Similarly, in an automated assembly line, the variance in service times is near zero.

We discussed earlier that the mean and the standard deviation of the exponential distribution are equal. Hence, using the models discussed in the previous sections could lead to incorrect determination of some of the key variables. In fact, the error could be as high as 100 percent! (see Suri et al., 1993). We therefore turn our attention now to models in which either the interarrival or service times are not required to be exponential and for which exact (optimal) or approximate (heuristic) results are available.

Suppose that the service process in a queuing system is made up of k phases in which each phase has a service time with an exponential distribution. Further, let the mean service time in each phase be $1/k\mu$. Then, it can be shown that the *total* service time has an Erlang distribution with shape parameter k and rate parameter $k\mu$.

It is well known that the pdf of an Erlang distribution is $f(t)=R(Rt)^{k-1}e^{-Rt}/(k-1)!$ for all $t \geq 0$ and R, k are called the rate parameter and shape parameter, respectively. k must be a positive integer. The mean and variance are given by $E(T)=k/R$ and $V(T)=k/R^2$, respectively. When $k=1$, $E(T)=V(T)$ and we get an exponential distribution. When k is very large, $V(T) \simeq 0$ and the Erlang distribution becomes deterministic (i.e., approaches a random variable with zero variance). Thus, one extreme of the Erlang is deterministic, while the other is exponential. For values of k between 1 and ∞, we get other Erlang distributions with different shapes, hence the name shape parameter for k.

Not all nonexponential systems will fit the k (exponential service time) phase pattern that we discussed earlier. However, it may be known that the service time distribution is Erlang. In order to derive steady-state results for queuing models in which the service time has an Erlang distribution, we use a formula called Pollaczek–Khintchine (PK) formula:

$$L = \rho + \frac{\lambda^2 \sigma^2 + \rho^2}{2(1-\rho)}$$

where

$\rho = \lambda/\mu$ and $1/\mu$ and σ^2 are the mean and variances of the service time distribution

λ is the mean of the (exponential) interarrival time distribution

Using Little's formula, we can therefore determine L_q as $(\lambda^2 \sigma^2 + \rho^2)/[2(1 - \rho)]$ and W_q as L_q/λ.

For a proof of the PK formula, see Gross and Harris (1985) or Viswanadham and Narahari (1992). A remarkable feature of this formula is that it holds for *any* service time distribution, so long as we are able to estimate its mean and variance. The arrival process, however, must be Poisson. The formula holds only for the single-server case. Results for general service time distributions for the multiple-server case unfortunately are extremely difficult to calculate. Although some results are available for a handful of special cases, no general results like the one for the single-server case are available.

Turning our attention back to the Erlang service time distribution case, we can at least obtain results for the single-server case using the PK formula, because the latter is applicable to any service time distribution, including the Erlang. Because $1/\mu = k/R$, we get $R = k\mu$ and $\sigma^2 = k/R^2 = 1/(k\mu^2)$. Substituting $\sigma^2 = 1/(k\mu^2)$ in the PK formula, we get

$$L_q = \frac{\lambda^2 \sigma^2 + \rho^2}{2(1-\rho)} = \frac{\lambda^2/(k\mu^2) + \lambda^2/\mu^2}{2(1-\lambda/\mu)}$$

$$= \frac{\lambda^2 + \lambda^2 k}{2k(\mu-\lambda)} = \left(\frac{1+k}{2k}\right)\left(\frac{\lambda^2}{\mu(\mu-\lambda)}\right) \tag{13.22}$$

$$W_q = \frac{L_q}{\lambda} = \left(\frac{1+k}{2k}\right)\left(\frac{\lambda}{\mu(\mu-\lambda)}\right) \tag{13.23}$$

From Equations 13.22 and 13.23, we can determine that $W = W_q + 1/\mu$ and $L = \lambda W$.

The reader may verify that with $k = 1$, steady-state results for exponential service time distribution models (in which there is only one server and the arrivals process is Poisson) obtained via the previous formulae are identical to those in Table 13.1. Further, by setting $k = \infty$, we can also obtain steady-state results for the deterministic service time model. Gross and Harris (1985) present results for two variations of the earlier model, the first in which the queue size is finite and the second in which bulk arrivals are allowed to occur. They also provided some additional results for the M/G/S model.

The ratio of the variance of a random variable to the square of its mean (σ^2/μ^2) is called the squared coefficient of variation (SCV) of the random variable. SCV is typically denoted as c^2. The SCV of a random variable with an exponential distribution is equal to 1, because its mean and standard deviation are the same. It can be verified easily that the SCV of any Erlang distribution is less than one if the shape parameter $k > 1$. In fact, when $k = \infty$, SCV approaches zero. Thus, depending upon the value of k, the Erlang distribution can model situations with $c^2 \leq 1$. In contrast, the hyperexponential distribution always has $c^2 \geq 1$. Such a distribution has many applications in the manufacturing world. Consider, for example, a highly automated system in which the processing (service) times are identical (due to the high level of accuracy inherent in the manufacturing system), but then an occasional major breakdown may occur dramatically increasing the processing time of one part. The standard deviation for such systems may be more than the mean, in which case we have c^2 values greater than 1.

Perhaps the most general model for which analytical solution is desired would be the one in which we have general service time and general interarrival time distributions and multiple servers. Such a model has applications in many manufacturing systems. Although the exact analysis of such models is not possible, we do have some heuristic results for the single-server and multiple-server cases of the earlier model.

Assume the interarrival and service (or interdeparture) times in a system have a general distribution with SCV given by c_a^2 and c_s^2, respectively. For the single-server case, the following result for the average waiting time in queue is available:

$$W_q \approx \left(\frac{c_a^2 + c_s^2}{2} \right) \left(\frac{\lambda}{\mu(\mu - \lambda)} \right) \qquad (13.24)$$

As before, approximate values for the other three variables can be determined from Equation 13.24 as $L_q = \lambda W_q$; $W_q + 1/\mu$; and $L = \lambda W$. From the expression 13.24, it is a simple matter to obtain results for the M/M/1 and M/G/1 special cases (by noting that $c^2 = 1$ for a Poisson arrival process and $c^2 = 1$ for exponential service times) and verify them to be exactly the same as those determined earlier. Thus, although the preceding result is an approximation for the GI/GI/1 case, it is exact for the M/M/1 and M/GI/1 special cases.

For the multiple-server case of the preceding general model (i.e., interarrival and service times with general distributions), another simple heuristic result is available for the average waiting time of parts (customers) in queue, as shown in the following:

$$W_q \approx \left(\frac{c_a^2 + c_s^2}{2} \right) \left(W_q^{\text{M/M/S}} \right) \qquad (13.25)$$

where $W_q^{\text{m/m/s}}$ is the average waiting time of parts in queue obtained for the M/M/S model and shown in Table 13.1.

13.3.4 PRIORITY-BASED AND OTHER QUEUING MODELS

Consider a manufacturing system in which parts are processed not on a FCFS basis, but on the priority attached to them. Suppose that we have m types of jobs ranked so that job type 1 has the highest priority and job type m has the lowest. The criteria for assigning priorities may be the importance of the job, which itself may be a function of the job's "value" to the company. Assume that jobs are selected for service in order of their priority. Thus, job type 1 is selected first; if the queue does not have any job of type 1, then a job of type 2 is selected, and so on. Within a priority level, however, jobs are selected for service on a FCFS basis. In other words, if there are no type 1 or 2 jobs in the queue, and we have 5 jobs of type 3, one of these is selected in the order in which they entered the queue. Further, once a job is selected for service, it will not exit the server (and the system) until its service is completed. In other words, its service will not be interrupted if a job with a higher priority arrives after the service has begun. Such a priority rule is called the nonpreemptive rule. In contrast, the preemptive rule allows for a job to be "bumped" from service if a higher priority job arrives at the queue, even if the arrival occurred after the lower-priority job began receiving its service.

For the single-server nonpreemptive case, we have results for models with Poisson arrivals and general service time distributions. For the multiple-server case, we have results for models with Poisson arrivals and exponential service time distributions (see, for example, Taha, 1992). Variations of these models are treated in Jaiswal (1968).

It should be noted that in addition to the models discussed so far, there are others for which results have been obtained. For many of these, results are available in the form of tables for various values of key variables and parameters. For example, results for models with certain state-dependent arrival and departure rates are available in table format in Conway and Maxwell (1961a) and Hillier et al. (1964). Results for the M/M/S/GD/N/N model and M/D/S/GD/∞/∞ are available in Peck and Hazelwood (1958) and Hillier et al. (1981). In addition, analytical results for the "self-service" models and bulk arrival models are provided in basic queuing textbooks, including Gross and Harris (1985) and Kleinrock (1975) and in other elementary textbooks in Operations Research, for example, Winston (1994), Taha (1992), and Hillier and Lieberman (1990). The self-service and bulk arrival models as well as models in which a queue is not allowed to form are of interest in some situations. Consider, for example, a system in which there is no server and the customer serves himself or herself. Examples may be students taking the SAT, or television viewers tuning in to a station. In such a model, the number of "servers" is infinite. Similarly, we can have models in which bulk arrivals occur or the number of servers in a finite queue capacity model may be equal to the queue size. A queue is not allowed to form in the latter. Also, in some applications, it may be necessary to determine the probability distribution of the waiting time in system for a random arrival. This can be done for the basic models discussed in the previous sections provided it is assumed that the queue discipline is FCFS. Once again, results for these are provided in Gross and Harris (1985) and Hillier and Lieberman (1990), amongst others. The interested reader is encouraged to consult these sources. We now turn our attention to networks of queues.

13.4　QUEUING NETWORKS

The models we studied in the previous sections assumed that the product undergoes one stage of service and then departs the system. While single stage service models demonstrate the value of queuing theory in the design of manufacturing systems, multiple stage service models represent more realistic scenarios. These multiple stages can be modeled as networks of queues. Throughout the remainder of this section, we therefore consider networks of queues and derive analytical results for two types of networks. Before we do so, however, we define a queuing network. Consider a network of i machine centers in which each center has S_i identical parallel servers (machines). In the most general case, parts arrive at each center i from an external source (e.g., a supplier) or another internal machine center, visit a series of other centers (perhaps including center i), and depart from some machine center. Different parts may enter and exit the system from different machine centers and may visit different sets of machine centers in different sequences to complete processing. Let us refer to machine centers as nodes. A network with the following properties is referred to as Jackson network (Jackson, 1957, 1963):

1. The arrival process from the external node to node i is Poisson with mean rate of γ_i.
2. The service time at node i follows an exponential distribution with parameter μ_i.
3. p_{i0} is the probability that a part will exit the network after completion of processing at node i and p_{ij} is the probability that a part will visit node j after completion of processing at node i. p_{i0} and p_{ij} are assumed to be known and independent of the state of the system.

Two types of Jackson networks are possible—open networks and closed networks. For each type of queuing network, we have two classes of models depending upon whether the system considered has exponentially distributed interarrival and service times or generally distributed times. Open queuing network models with exponential and general service time distributions and closed queuing network models with exponential service times are discussed next.

13.4.1　Open Queuing Network Model for Exponential Systems

In the open queuing network model, jobs can arrive from an external source at one or more of the machine centers. As will be seen later, an open queuing network is simply an expansion of the single machine system previously discussed. Although results are provided for the general

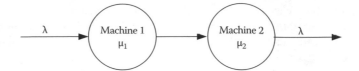

FIGURE 13.8 Two-machine system.

m-node open queuing network later, it is instructive to develop results for the two-node network first. Consequently, we will consider an open network system with two single-server machines in tandem (or series) as shown in Figure 13.8. Assume there is infinite capacity in front of each node (machine). Parts arrive at the network according to a Poisson process with parameter λ, i.e., the interarrival time follows an exponential distribution with mean $1/\lambda$ and each is served by machine 1 first and then machine 2, which have independent exponential service time distributions of parameters μ_1 and μ_2, respectively.

Let N_{t1}, N_{t2} be the number of customers at nodes 1, 2 respectively, at time t. The stochastic process defining the number of parts at each of the nodes at time t, $t \geq 0$, is a Markov chain. Assuming we have reached steady state, let n_1 and n_2 represent the average number of parts at nodes 1 and 2, respectively. Because steady state has been reached, we will denote $P(n_1, n_2)$ as the steady-state joint probability of having n_1 and n_2 parts at machines 1 and 2, respectively. Whenever a service completion or arrival occurs, the state changes. In Figure 13.9, a partial steady-state diagram is shown for the general state in which we have n_1 and n_2 parts at nodes 1 and 2, respectively. Recall, we constructed the state diagram for a single machine system previously and found the steady-state probability to be $P_n = (1 - \rho)\rho^n$. Using the same principle, the steady-state probabilities can be calculated from the multiple machine state diagram. Steady state implies that the rate out of state n is equal to the rate into state n. Of course, when defining the state of the system for the multiple machine system, both machines must be considered. The state of the system is the number of jobs at machine

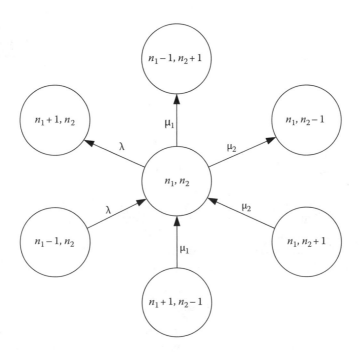

FIGURE 13.9 State diagram of two-machine network.

1 and machine 2. It is easy to write down the flow balance (rate in equals rate out) (Equation 13.26) for any $\{n_1, n_2\}$ pair greater than 0, from the diagram:

$$(\lambda + \mu_1 + \mu_2)P(n_1, n_2) = \lambda P(n_1 - 1, n_2) + \mu_2 P(n_1, n_2 + 1) + \mu_1 P(n_1 + 1, n_2 - 1) \tag{13.26}$$

For the remaining states $(0, 0)$, $(n_1, 0)$, $(0, n_2)$, the following flow equations are also easy to derive:

$$\lambda P(0, 0) = \mu_2 P(0, 1) \tag{13.27}$$

$$(\lambda + \mu_1)P(n_1, 0) = \mu_2 P(n_1, 1) + \lambda P(n_1 - 1, 0) \tag{13.28}$$

$$(\lambda + \mu_2)P(0, n_2) = \mu_1 P(1, n_2 - 1) + \mu_2 P(0, n_2 + 1) \tag{13.29}$$

As before, we need another equation to solve for the various steady-state probabilities in the Equations 13.26 through 13.29. This is provided by the fact that the sum of all possible steady-state probabilities is equal to 1. This equation is $\Sigma\Sigma P(n_1, n_2) = 1$. From Equations 13.27 through 13.29, it is easy to derive the following solution for the general steady-state probability (see Exercise 25):

$$P(n_1, n_2) = \rho_1^{n_1}(1 - \rho_1)\rho_2^{n_2}(1 - \rho_2) = P_{n_1} P_{n_2} \tag{13.30}$$

$$P(0, 0) = (1 - \rho_1)(1 - \rho_2) = P_0 P_0 \tag{13.31}$$

As before, $\rho_1 = \lambda/\mu_1$ and $\rho_2 = \lambda/\mu_2$ in the equations. Also, ρ_1 and ρ_2 must be less than 1 so that the queue will not grow without bound. The solution given by Equations 13.30 and 13.31 is generally referred to as the product form solution, because it is a product of the individual steady-state probabilities derived earlier for the M/M/1 model. This remarkable result says that, in steady state, the joint probability of having n_1 parts at node 1 and n_2 in node 2 is simply the product of the individual probabilities of having n_1 parts at node 1 and n_2 parts at node 2. Hence, n_1 and n_2 are independent of each other. Notice that the right-hand sides of the preceding Equations 13.30 and 13.31 are exactly the steady-state probabilities derived for the M/M/1 model previously. It is obvious that the first node is an M/M/1 queue. Moreover, it has been shown that the departure process from an M/M/1 queue is Poisson with parameter λ provided the arrival process is also Poisson with parameter λ (see Burke (1956) for the original proof and also Gross and Harris (1985) and Viswanadham and Narahari (1992)). Because the output stream from machine 1 goes directly to machine 2, the interarrival rate at machine 2 is also exponential (see Example 13.4). Hence, the second node is also an M/M/1 queue. Thus, the aforementioned product form solution implies that the network behaves "as if" each node is an independent M/M/1 queue with parameters λ and μ_i for the two nodes 1 and 2. It should be cautioned that the network does "not" break up into two independent M/M/1 queues with the arrival process being a true Poisson process with parameter λ. In fact, it has been shown that this is not necessarily true (Disney, 1981). However, what the result given by Equations 13.30 and 13.31 states is that despite the fact that the arrival process into each node is not a true Poisson process, the network can be treated as if it is made up of two independent M/M/1 nodes. Therefore, each machine in the network can be treated independently. This allows for a divide and conquer strategy of analyzing each machine in a two-machine network as two single machine queues. From the preceding product form solution, it is easy to see that the mean number of parts in the system L is equal to $\rho_1/(1 - \rho_1) + \rho_2/(1 - \rho_2)$ and average time spent in the system W is equal to L/λ.

Although we demonstrated Jackson's result using a two-machine center network, it can be expanded to any number of machine centers. Moreover, the preceding results hold even when the number of servers in each of these centers is greater than 1. Of course, all other conditions, including infinite buffers in front of each machine, independent service time distributions at each node, identical machines at each machine center (node), and Poisson arrival process, must hold. In fact, even when there is a feedback loop, i.e., parts may reenter the system for rework, Jackson (1963) has showed that the product form solution will hold. However, when the infinite buffer assumption is violated, the product form result does not.

We conclude this section by providing steady-state results for the m-node Jackson network. Under the assumptions stated, the analysis of an m-node Jackson network is a three-step process. The first step is to calculate the effective arrival rate. Recall that each machine center in a network has two sources of arrivals. The first arrival source is from the external environment. The second arrival source is from other machines in the system. For each node i, $i = 1, 2, \ldots, m$, let

γ_i be the mean rate of the Poisson arrival process from the external world to node i
$1/\mu_i$ be the mean (exponentially distributed) service time at node i
S_i be the number of identical servers at node i
p_{i0} be the probability that a part will exit the network after completion of processing at node i
p_{ij} be the probability that a part will visit node j after completion of processing at node i

Because the arrival at a node is made up of external and internal arrivals—the latter occurring as a result of service completions at other internal nodes—the arrival rate at node i, $i = 1, 2, \ldots, m$, is

$$\lambda_i = \gamma_i + \sum_{j=1}^{m} p_{ji} \lambda_j \tag{13.32}$$

If p_{i0} and γ_j are greater than 0 for some i, j, and ρ_i given by $\lambda_i/(S_i\mu_i)$, for $i = 1, 2, \ldots, m$, is less than 1 (so that the queue does not grow infinitely at any node), then it can be shown that the product form solution will hold. Specifically, it can be shown that the joint probability of having $n_1, n_2, \ldots, n_m$ parts at nodes 1, 2, \ldots, m is simply the product of the following probabilities—having n_1 parts at node 1, n_2 parts at node 2, \ldots, and n_m parts at node m. Of course, the latter probabilities are obtained by treating each node as an independent M/M/S_i queue and $n_1, n_2, \ldots, n_m$ are independent of each other. Because we are able to treat the m-node Jackson network as m independent M/M/S_i queuing systems (for the same reasons mentioned in the two-node case), we can obtain the performance measures for the entire network using equations from Table 13.1.

Thus, the second step is to analyze each machine in the m-machine system independently. The number of servers in each center will determine whether the M/M/1 or M/M/S results are to be used. The final step is to combine the results from each machine center to analyze the performance of the entire system.

In order to keep the presentation simple, we only present results for the M/M/1 case. The reader can get results similarly for the M/M/S_i case using Table 13.1. The average number of parts in the network and the average waiting time for each part in the network, i.e., flow time, are given by

$$L = \sum_{i=1}^{m} L_i = \sum_{i=1}^{m} \frac{\lambda_i}{(\mu_i - \lambda_i)}, \quad W = \frac{L}{\lambda_{eff}}$$

where λ_{eff}, the effective arrival rate into the network from the outside, is given by

$$\sum_{i=1}^{m} \gamma_i$$

In addition to the overall performance measures, we can obtain some specific performance measures at the individual nodes, e.g., the sojourn times, i.e., the time spent by the part in traversing each node of the network. This can be done in a manner similar to the one we used to write down the arrival rate into each node.

We now present three examples illustrating some variations of the open Jackson network model. The first example shows the effect of a *rework* loop in a manufacturing system. It utilizes Jackson's result to calculate effective arrival rates from multiple arrival sources and illustrates the cost of defects in manufacturing.

Example 13.4

Consider a system consisting of a preheat furnace, a forge press, and an inspection station. An ingot arrives at the furnace to be heated and then sent to the press. After pressing, the ingot is transferred to inspection. Ingots are delivered to the furnace at a rate of 12 per hour. The furnace heats the ingots at a rate of 15 per hour. The press and inspection each have capacities of 18 and 21 ingots per hour. All interarrival and service times are assumed to follow an exponential distribution. The three-machine network is shown in Figure 13.10.

1. Determine the expected number of jobs in the furnace (*F*), press (*P*), and inspection (*I*) as well as the WIP and processing time of the system.
2. Suppose the capacity of the queue in front of the press is 5 ingots, is the current design sufficient?

Solution

1. We can analyze each machine in the three-machine system as three single machines. From Figure 13.10, we can see that jobs are conserved. All jobs eventually leave the system. Therefore, the arrival rate for both *P* and *I* is $\lambda = 12$ ingots per hour.

$$\lambda_F = 12 \text{ per hour} \qquad \mu_F = 15 \text{ per hour} \qquad \rho_F = 0.80$$
$$\lambda_P = 12 \text{ per hour} \qquad \mu_P = 18 \text{ per hour} \qquad \rho_P = 0.67$$
$$\lambda_I = 12 \text{ per hour} \qquad \mu_I = 21 \text{ per hour} \qquad \rho_I = 0.57$$

The WIP and processing time are the expected number of jobs in the queuing system that are waiting or being served and the expected time that a job spends in the queuing system waiting and being served. Using the equations in Table 13.1, we find the operating characteristics as follows:

$$W_F = 1/(\mu_F(1 - \rho_F)) = 1/(15(1 - 0.8)) = 0.33 \text{ hours} = 20 \text{ minutes}$$

$$L_F = \rho_F/(1 - \rho_F) = 0.8/(1 - 0.8) = 4 \text{ ingots}$$

$$W_P = 1/(\mu_P(1 - \rho_P)) = 1/(18(1 - 0.67)) = 0.167 \text{ hours} = 10 \text{ minutes}$$

$$L_P = \rho_P/(1 - \rho_P) = 0.67/(1 - 0.67) = 2 \text{ ingots}$$

$$W_I = 1/(\mu_I(1 - \rho_I)) = 1/(21(1 - 0.57)) = 0.11 \text{ hours} = 6.67 \text{ minutes}$$

$$L_I = \rho_I/(1 - \rho_I) = 0.57/(1 - 0.57) = 1.33 \text{ ingots}$$

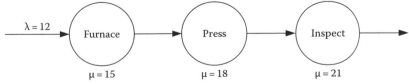

FIGURE 13.10   Three-machine network.

The system WIP and processing time is the sum of the individual values.

$$L = L_F + L_P + L_I = 4 + 2 + 1.33 = 7.33 \text{ ingots}$$

$$W = W_F + W_P + W_I = 20 + 10 + 6.67 = 36.67 \text{ minutes}$$

2. To find out if a design with a maximum of 5 ingots is sufficient, we must find the probability that more than 5 ingots are in the queue. Recall that P_n, the probability of n jobs being at a machine in steady state, is given by the equation $P_n = (1 - \rho)\rho^n$, $n \geq 1$. Therefore,

$$P_{n>5} = 1 - P_{n\leq5} = 1 - (1 - 0.67)(1.00 + 0.67 + 0.67^2 + 0.67^3 + 0.67^4 + 0.675^5) = 0.09046$$

This design will result in the queue in front of the press being at full capacity 9.05% of the time. When the queue is full, there will be a blocking problem in the queuing system and the steady-state results in part 1 will not hold because they were derived under the assumption that each machine has an infinite queue in front of it.

Example 13.5

Consider the open network queuing system in Example 13.4. Suppose that the rejection rate at inspection is 10%. Rejected parts are sent back to the furnace and are reworked through the entire system. This new network is shown in Figure 13.11. Determine the expected number of jobs in the furnace (F), press (P), and inspection (I) as well as the WIP and processing time of the system.

Solution

Assume that the queuing discipline is FCFS and that all the machines have only one server. Therefore, we will use the operating characteristics of an M/M/1 queue from Table 13.1.

Step 1: Calculate the arrival rate at each center. From Figure 13.11, we can see that jobs are conserved (all jobs eventually leave the system). Therefore, the arrival rate for both P and I is equal to the effective arrival rate at the furnace. Hence, this is the only effective arrival rate that must be calculated:

$$\lambda_F = \gamma_F + \lambda_R$$

where
  λ_F is the effective arrival rate to the furnace
  γ_F is the external arrival rate
  λ_R is the internal arrival rate (rejected parts) $= P(R)\lambda_F$

We also have the following:
  $P(R)$ is the probability of rejection
  $P(A)$ is the probability of acceptance $= 1 - P(R)$

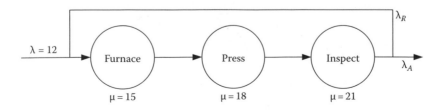

FIGURE 13.11   Three-machine network with rework loop.

Therefore,

$$\lambda_F = \gamma_F/(1 - P(R)) = \gamma_F/P(A)$$

The arrival and service rates and utilization at each center are

$$\lambda_F = 13.33 \text{ per hour} \quad \mu_F = 15 \text{ per hour} \quad \rho_F = 0.889$$
$$\lambda_P = 13.33 \text{ per hour} \quad \mu_P = 18 \text{ per hour} \quad \rho_P = 0.741$$
$$\lambda_I = 13.33 \text{ per hour} \quad \mu_I = 21 \text{ per hour} \quad \rho_I = 0.635$$

Utilization factors are now calculated using the effective arrival rates.
Step 2: Analyze each machine in the system independently.

$$W_F = 1/(\mu_F(1 - \rho_F)) = 1/(15(1 - 0.889)) = 0.6 \text{ hours} = 36 \text{ minutes}$$

$$L_F = \rho_F/(1 - \rho_F) = 0.889/(1 - 0.889) = 8 \text{ ingots}$$

$$W_P = 1/(\mu_P(1 - \rho_P)) = 1/(18(1 - 0.741)) = 0.214 \text{ hours} = 12.9 \text{ minutes}$$

$$L_P = \rho_P/(1 - \rho_P) = 0.741/(1 - 0.741) = 2.86 \text{ ingots}$$

$$W_I = 1/(\mu_I(1 - \rho_I)) = 1/(21(1 - 0.635)) = 0.124 \text{ hours} = 7.5 \text{ minutes}$$

$$L_I = \rho_I/(1 - \rho_I) = 0.635/(1 - 0.635) = 1.74 \text{ ingots}$$

Step 3: Combine the results from each center to analyze the performance of the entire system.
    The expected amount of time that a part spends in a queuing system waiting and being served is calculated on a per visit basis. In this system, a part may visit a machine more than once. Therefore, the expected number of visits (v_j) must also be calculated.

$$v_F = \lambda_F/\gamma_F = 13.33/12 = 1.111 \quad v_P = v_I = 1.111$$

$$L = L_F + L_P + L_I = 8 + 2.86 + 1.74 = 12.6 \text{ ingots}$$

$$W = v_F W_F + v_P W_P + v_I W_I = 1.111(36 + 12.9 + 7.5) = 62.7 \text{ minutes}$$

The rework loop increases the WIP and processing time by 71.9% and 70.9%, respectively.

The next example in our discussion of open network will evaluate a system with more elaborate routing in a four-machine network.

Example 13.6

Figure 13.12 shows a four-machine open network queuing system consisting of a furnace (*F*), a press (*P*), a rolling mill (*M*), and an inspection area (*I*). Each machine has a single server and the rejection rate is 5%. The part routing matrix is shown in Figure 13.13. Given that the external arrival rate is $\lambda = 12$, and $\mu_F = 15$, $\mu_P = 12$, $\mu_M = 18$, and $\mu_I = 21$, determine the WIP and processing time of the system.

Solution

Step 1: Calculate the effective arrival rate. From Figure 13.13, we develop the following equations:

$$\lambda_F = \lambda + 0.05\lambda_I$$

$$\lambda_P = 0.3\lambda_F + 0.4\lambda_M$$

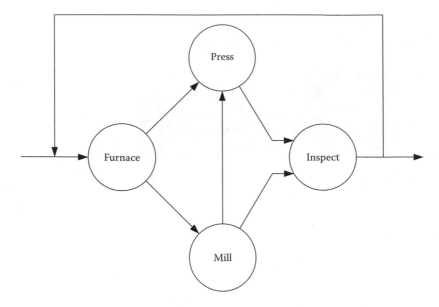

FIGURE 13.12   Four-machine system.

From/to	F	P	M	I	Exit
F	—	0.3	0.7	—	—
P	—	—	—	1.0	—
M	—	0.4	—	0.6	—
I	0.05	—	—	—	0.95

FIGURE 13.13   Part routing matrix.

$$\lambda_M = 0.7\lambda_F$$

$$\lambda_I = 1.0\lambda_P + 0.6\lambda_M = \lambda_F$$

From Example 13.5, we know that $\lambda_F = \lambda/P(A) = 12/0.95 = 12.63$. Solving these equations yields the effective arrival rates presented next. The service rates and utilizations at each machine center are also shown.

$\lambda_F = 12.63$	$\mu_F = 15$	$\rho_F = 0.842$
$\lambda_P = 7.33$	$\mu_P = 12$	$\rho_P = 0.611$
$\lambda_M = 8.84$	$\mu_M = 18$	$\rho_M = 0.491$
$\lambda_I = 12.63$	$\mu_I = 21$	$\rho_I = 0.602$

Step 2: Analyze each machine in the system independently. All the machines have only one server; therefore, we will use Table 13.1 for the operating characteristics of an M/M/1 queue.

$$W_F = 1/(\mu_F - \lambda_F) = 1/(15 - 12.63) = 0.422 \text{ h} = 25.33 \text{ min}$$

$$L_F = \lambda_F/(\mu_F - \lambda_F) = 12.63/(15 - 12.63) = 5.33 \text{ ingots}$$

$$W_P = 1/(\mu_P - \lambda_P) = 1/(12 - 7.33) = 0.214 \text{ h} = 12.84 \text{ min}$$

$$L_P = \lambda_P/(\mu_P - \lambda_P) = 7.33/(12 - 7.33) = 1.57 \text{ ingots}$$

$$W_M = 1/(\mu_M - \lambda_M) = 1/(18 - 8.84) = 0.109 \text{ h} = 6.6 \text{ min}$$

$$L_M = \lambda_M/(\mu_M - \lambda_M) = 8.84/(18 - 8.84) = 0.97 \text{ ingots}$$

$$W_I = 1/(\mu_I - \lambda_I) = 1/(21 - 12.63) = 0.119 \text{ h} = 7.2 \text{ min}$$

$$L_I = \lambda_I/(\mu_I - \lambda_I) = 12.63/(21 - 12.63) = 1.51 \text{ ingots.}$$

Step 3: Combine the results from each machine to analyze the performance of the entire system. Recall, that due to the *rework* loop, we must find the expected number of visits (v_j).

$$v_F = \lambda_F/\lambda = 12.63/12 = 1.05, \quad v_P = \lambda_P/\lambda = 7.33/12 = 0.61$$

$$v_M = \lambda_M/\lambda = 8.84/12 = 0.74, \quad v_I = \lambda_I/\lambda = 12.63/12 = 1.05$$

$$L = L_F + L_P + L_M + L_I = 5.33 + 1.57 + 0.97 + 1.51 = 9.38 \text{ ingots}$$

$$\begin{aligned}
W &= v_F W_F + v_P W_P + v_M W_M + v_I W_I \\
&= 1.05(0.422) + 0.61(0.214) + 0.74(0.109) + 1.05(0.119) \\
&= 46.8 \text{ minutes}
\end{aligned}$$

13.4.2 OPEN QUEUING NETWORK MODEL FOR GENERAL SYSTEMS

Models with generally distributed interarrival and service times are more realistic and we turn our attention to these models in this section. Consider a manufacturing system where multiple parts requiring multiple operations visit a required subset of machines to complete their processing. Sometimes, a part may visit the same machine for consecutive or nonconsecutive operations. Each operation requires a setup and parts are processed in batches called process batches. Material handling devices (MHDs) move batches of parts between the required pairs of machines. These are transfer batches. The transfer batch size need not be the same as the process batch size. Parts leave the system after the last operation.

A popular and effective approach to modeling such a complex system is called the parametric decomposition (PD) method. In contrast to the Jackson open queuing network modeling approach presented in the previous section, which is an exact analysis of a less realistic model, the PD method is an approximate analysis of a much more realistic model. Assuming that customers are served on an FCFS basis and there is infinite buffer space at each machine or workstation, it provides good estimates of key operational performance measures such as average cycle time, WIP inventory, and resource utilization for a specific machine, workstation, cell, or the entire system. The PD method is so named because it decomposes a large network consisting of multiple nodes into separate nodes and analyzes each separately, just as we did for the Jackson network. In the Jackson network, we assumed that the interarrival and service times can be described using the exponential distribution, which can be described using only one parameter. In a general network, we use the first two moments, mean and standard deviation (or to be more precise, the mean and the SCV), to partially characterize the two distributions. Like the Jackson network analysis in which the m-node network is treated as being made up of m independent queuing systems, the PD method also treats the

m nodes as being stochastically independent. However, in calculating the first two moments of the arrival and departure process at each node, it explicitly incorporates the effects of three network operations:

1. Departures
2. Merging of arrivals
3. Splitting of departures

13.4.2.1 Departures

As long as a server's utilization is less than 100%, the mean departure rate from the server is the same as the mean (effective) arrival rate. Calculation of the SCV is difficult however and can be approximated using the formula in the last column of the first row in Table 13.4. This formula assumes that the node consisting of S servers has utilization equal to ρ.

13.4.2.2 Splitting of Departures

Suppose that the combined customers leaving a station have a departure process that can be characterized by two moments λ and c_a^2. If these departures are split into k streams, the first two moments of the split streams can be calculated as shown in the second row of Table 13.4. In these formulae, p_i is the probability that a departure from the combined stream of customers will join stream i.

13.4.2.3 Merging of Arrivals

Suppose that multiple parts arrive into a node j with utilization ρ. Assume that the arrival processes can be approximated by independent general distributions with mean arrival rate and SCV of interarrival times given by λ_i and c_a^2, respectively, for $i = 1, 2, \ldots, k$. The two parameters—mean and SCV—of the distribution describing the merging or superpositioning of these arrivals can be approximately calculated as shown in the third row of Table 13.4. Note that the approximation is

TABLE 13.4

Effects of Three Basic Network Operations on the Arrival Rates

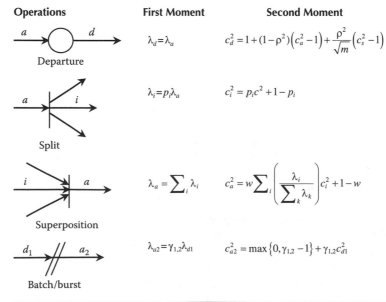

Operations	First Moment	Second Moment
Departure	$\lambda_d = \lambda_a$	$c_d^2 = 1 + (1-\rho^2)\left(c_a^2 - 1\right) + \dfrac{\rho^2}{\sqrt{m}}\left(c_s^2 - 1\right)$
Split	$\lambda_i = p_i \lambda_a$	$c_i^2 = p_i c^2 + 1 - p_i$
Superposition	$\lambda_a = \sum_i \lambda_i$	$c_a^2 = w \sum_i \left(\dfrac{\lambda_i}{\sum_k \lambda_k}\right) c_i^2 + 1 - w$
Batch/burst	$\lambda_{a2} = \gamma_{1,2} \lambda_{d1}$	$c_{a2}^2 = \max\{0, \gamma_{1,2} - 1\} + \gamma_{1,2} c_{d1}^2$

only for the second moment (SCV) calculations. Based on extensive empirical testing, Whitt (1983) has provided the SCV values. The parameter w is calculated using Equation 13.33.

$$w = \frac{1}{\left[1 + 4(1-\rho)^2(v-1)\right]} \tag{13.33}$$

where

$$v = \frac{1}{\sum_{i=1}^{k}\left(\lambda_i / \sum_{i=1}^{k}\lambda_i\right)^2}$$

ρ is the utilization of the server into which the arrivals merge

13.4.2.4 Combining the Effects of the Three Network Operations

Now that we have determined the effects of the three network operations, and because merging of arrivals, departures, or splitting of departures occur at every node, we can combine the results to determine the two parameters of the effective arrival into each node. This is done using the following equations,

$$\lambda_k = \lambda_{0k} + \sum_{i=1}^{m}\lambda_i q_{ik}, \quad k = 1, 2, ..., m \tag{13.34}$$

$$c_{ak}^2 = a_k + \sum_{i=1}^{m} c_{ai}^2 b_{ik}, \quad k = 1, 2, ..., m \tag{13.35}$$

where

$$a_k = 1 + w_j \left\{ \left(p_{0k}c_{0k}^2 - 1\right) + \sum_{i=1}^{m} p_{0k}\left[(1-q_{ik}) + q_{ij}\rho_i^2 x_i\right] \right\}$$

$$b_{ik} = w_j p_{ik} q_{ik}\left[v_{ik} + (1-v_{ik})(1-\rho_i^2)\right]$$

$$x_i = 1 + \frac{1}{S_i}\left[\left(\max\left\{c_{si}^2, 0.2\right\} - 1\right)\right]$$

$$w_k = \frac{1}{\left[1 + 4(1-\rho_k)^2(v_k-1)\right]}$$

$$v_k = \frac{1}{\sum_{i=0}^{m}\left(p_{ij}^2\right)}$$

$p_{ik} = \lambda_{ik}/\lambda_k$

$\lambda_{ik} = \lambda_k q_{ik}$

q_{ik} is the proportion of customers completing service at node i that proceed to node k

Thus, the heart of the PD method is to simultaneously solve Equations 13.34 and 13.35. Once the two moments of the effective arrivals into each node (now assumed to be stochastically independent due to the incorporation of the network operations into the second moment calculations in Equation 13.35) are known, the next step is to calculate the two moments of the effective service time at each node. This is easy to do and is shown in Equations 13.60 through 13.63, which incorporate setup and batch size information.

With the first two moments of the effective service and interarrival times, we can compute the average waiting time in queue of each product at each node, department, or the entire system.

Depending upon the number of (nodes) machines of type k, we analyze each node (machine type) as a GI/G/1 or GI/G/m queue and use Equations 13.24 and 13.25 to determine the expected time a part spends in queue at node k.

Because all products join the same queue to obtain service from node k, the average waiting time in queue for any product is given by Equations 13.24 or 13.25. However, the overall expected average sojourn time at node k—$E(W_k)$—depends upon the average waiting time and average processing time. The latter is known and thus the average sojourn time is obtained using the following formula:

$$E(W_k) = E(WQ_k) + (1/\mu_k) \tag{13.36}$$

Utilizing Little's law, the average number of products at node k as well as those waiting in queue can be readily determined:

$$L_k = \lambda_k E(W_k) \tag{13.37}$$

$$LQ_k = \lambda_k E(WQ_k) \tag{13.38}$$

To get the average number of parts of each type at each machine, the aforementioned results can be disaggregated as shown in Equations 13.67 and 13.68. If we need additional measures, for example, the variance of the flow times or number in the system, these can also be obtained rather easily as shown in Whitt (1983). We now have mean estimates of three important performance measures—machine utilization (given by $\lambda_k/S_k\mu_k$ for node k) as well as WIP inventory and product flow time (given by formulae (13.36) and (13.37), for each node).

Now that the building blocks for the PD method have been identified and its mechanism described, we can apply it to estimate key performance measures of complex manufacturing systems that have setup, process as well as transfer batches, and empty travel of the MHD. Such a model and software are discussed next. The only limiting assumptions are the FCFS service discipline and infinite buffer assumption. However, because we are using the open queuing network model to quickly evaluate and compare numerous system designs approximately and we will not design systems that have significant blocking, the two restrictive assumptions do not pose a serious problem for their intended purpose. Once we narrow down the range of designs being considered to a select few, we can always develop a detailed simulation model and analyze those in detail in the next (final) step of analysis. Simulation is an alternative tool for performance evaluation, but it is well known that it is expensive to build and run and thus must be used judiciously.

13.4.3 OPEN QUEUING NETWORK SOFTWARE FOR COMPLEX MANUFACTURING SYSTEMS

The Manufacturing system Performance Analyzer (MPA) is a useful software to evaluate the performance of a manufacturing system layout. It is described in detail in Meng (2002) and Meng and Heragu (2004) and is an extension of Whitt's (1983) Queuing Network Analyzer (QNA). MPA incorporates many realistic manufacturing considerations such as operational and transfer batch sizes, setup time, empty travel time of the MHD, and machine failure. MPA not only determines the commonly sought performance measures analytically (and therefore quickly), but also provides the user the option of automatically generating input data for the ProModel® simulation software.

MPA uses the PD method to analytically evaluate key performance measures of a queuing network. Given the first two moments—mean and SCV—of interarrival times of each customer type into the network and its routings as input, the PD method calculates the first two moments of interarrival time of an aggregate customer into each node. Each node is a queuing system and represents a server (machine, workstation, or MHD). Meng and Heragu (2004) identified an additional network operation—batch or burst—also listed in Table 13.4 and captured its effects on the aggregate arrival. When parts coming from a machine must wait at the next machine to be processed in larger batches, batching occurs. On the other hand, if products coming from one machine are processed in smaller batches at the next, "bursting" occurs. MPA assumes the following information is available:

- Number and types of machines
- Number and types of discrete MHDs in the material handling system (MHS)
- Number, types, volume, and routing of products to be manufactured in the specified production planning period
- First two moments of external arrival rate for each product
- First two moments of service time for each processing operation
- Setup time for each operation and transfer, for each product

Before we describe MPA, it is necessary to introduce the following notation:

Notation

i Product type index, $i = 1, 2, ..., p$

j, k Department index, $j, k = 1, 2, ..., n - 1$

n MHS department index

o^i Number of operations required by product i, $i = 1, 2, ..., p$

l Operation index, $l = 1, 2, ..., o^i$

λ_{0k}^i Arrival rate of product i from the external world into department k, $i = 1, 2, ..., p$; $k = 1, 2, ..., n - 1$

c_{0k}^{2i} SCV of interarrival time for product i from the external world into department k, $i = 1, 2, ..., p$; $k = 1, 2, ..., n - 1$

to_{kl}^i Natural service time for lth operation on product i at department k, $i = 1, 2, ..., p$; $k = 1, 2, ..., n - 1$; $l = 1, 2, ..., o^i$

c_{okl}^{2i} SCV of natural service time for lth operation on product i at department k, $i = 1, 2, ..., p$; $k = 1, 2, ..., n - 1$; $l = 1, 2, ..., o^i$

ts_{kl}^i Mean setup time for lth operation on product i at department k, $i = 1, 2, ..., p$; $k = 1, 2, ..., n - 1$; $l = 1, 2, ..., o^i$

c_{skl}^{2i} Setup time SCV for lth operation on product i at department k, $i = 1, 2, ..., p$; $k = 1, 2, ..., n - 1$; $l = 1, 2, ..., o^i$

tr_{jk} Travel time between departments j and k, $j, k = 1, 2, ..., n - 1$

c_{rjk}^{2i} SCV of travel time between departments j and k, $i = 1, 2, ..., p$; $k = 1, 2, ..., n - 1$

tp^i Mean setup time for loading product i on MHD in MHS department n, $i = 1, 2, ..., p$

c_t^{2i} SCV of setup time for loading product i on MHD in MHS department, $i = 1, 2, ..., p$

S_k Number of machines in department k, $k = 1, 2, ..., n$

Y_{kl}^i 1 if the lth operation of product i is done at department k, $i = 1, 2, ..., p$; $k = 1, 2, ..., n - 1$; $l = 1, 2, ..., o^i$; 0 otherwise

d_{jk} Distance between departments j and k

v Velocity of the MHD v

b_{kl}^i Batch size of product i at department k, for lth operation, $i = 1, 2, ..., p$; $k = 1, 2, ..., n - 1$; $l = 1, 2, ..., o^i$

bt_{jk}^i Transfer batch size of product i between departments j and k, $i = 1, 2, ..., p$; $k = 1, 2, ..., n - 1$; $l = 1, 2, ..., o^i$

Note that the setup time of an operation (ts_{kl}^i) is incurred only if the preceding operation on the machine is of a different type. We count the average number of the operations (N) needed to incur a setup (TS) and use the average setup time as the setup time for this operation ($ts_{kl}^i = TS/N$).

The MHS used to transfer parts between departments is modeled as another node of discrete MHDs. Notice that the nth department is assumed to be the MHS department with identical servers

(MHDs) in the aforementioned notation. It is convenient to have an indicator variable that tells us whether or not the two successive operations (say the $l-1$th and the lth) on a product i are done at a pair of departments j and k as shown in Equation 13.39 and to determine the total external arrival rate for product i as shown in Equation 13.40:

$$Y_{jkl}^i = Y_{j,l-1}^i Y_{kl}^i, \quad i = 1, 2, \ldots, p; \; j, k = 1, 2, \ldots, n-1; l = 2, 3, \ldots, o^i \tag{13.39}$$

$$\hat{\lambda}^i = \sum_{k=1}^{n-1} \lambda_{0k}^i, \quad i = 1, 2, \ldots, p \tag{13.40}$$

Because we are assuming deterministic routing for each product, λ_{0k}^i is greater than zero for only one k. Based on these, it is easy to get the total arrival rate of products into each department from the external world as well as out of each department to the external world. It is assumed that the intradepartment material transfer is accomplished by an MHD residing in that department and is treated like an "operation" for modeling purposes. However, the interdepartment material transfer is completed by the MHS department consisting of one or more identical MHDs and is treated as a transfer. Thus, because all the interdepartment transfers between departments must take place via the MHS department n, the arrival rate from a department j to department k for $j \neq k$ will be zero. The total arrival rate from a department to the MHS cell and vice versa is not zero however and can be easily determined:

$$\hat{\lambda}_{0k} = \sum_{i=1}^p \lambda_{0k}^i, \quad k = 1, 2, \ldots, n-1 \tag{13.41}$$

$$\hat{\lambda}_{kk} = \sum_{i=1}^p \sum_{l=2}^{o^i} \frac{\hat{\lambda}^i Y_{kkl}^i}{b_{k,l-1}^i}, \quad k = 1, 2, \ldots, n-1 \tag{13.42}$$

$$\hat{\lambda}_{jk} = 0, \quad j, k = 1, 2, \ldots, n-1, \, j \neq k \tag{13.43}$$

$$\hat{\lambda}_{kn} = \sum_{i=1}^p \sum_{\substack{j=1 \\ j \neq k}}^{n-1} \sum_{l=2}^{o^i} \frac{\hat{\lambda}^i Y_{kjl}^i}{b_{kl}^i}, \quad k = 1, 2, \ldots, n-1 \tag{13.44}$$

$$\hat{\lambda}_{nk} = \sum_{i=1}^p \sum_{\substack{j=1 \\ j \neq k}}^{n-1} \sum_{l=2}^{o^i} \frac{\hat{\lambda}^i Y_{jkl}^i}{bt_{jk}^i}, \quad k = 1, 2, \ldots, n-1 \tag{13.45}$$

$$\hat{\lambda}_{k0} = \sum_{i=1}^p \frac{\hat{\lambda}^i Y_{ko^i}^i}{b_{ko^i}^i}, \quad k = 1, 2, \ldots, n-1 \tag{13.46}$$

Observing that all interdepartment transfer takes place via the MHS department and that the MHS department never transfers parts to itself, the proportion of a batch of parts visiting department k after department j is easy to calculate:

$$p_{nn} = p_{jk} = 0, \quad j, k = 1, 2, \ldots, n-1, \, j \neq k \tag{13.47}$$

$$p_{jj} = \frac{\hat{\lambda}_{jj}}{\hat{\lambda}_{j0} + \hat{\lambda}_{jj} + \hat{\lambda}_{jn}}, \quad j = 1, 2, \ldots, n - 1 \tag{13.48}$$

$$p_{jn} = \frac{\hat{\lambda}_{jn}}{\hat{\lambda}_{j0} + \hat{\lambda}_{jj} + \hat{\lambda}_{jn}}, \quad j = 1, 2, \ldots, n - 1 \tag{13.49}$$

$$p_{nj} = \frac{\hat{\lambda}_{nj}}{\sum_{k=1}^{n-1} \hat{\lambda}_{nk}}, \quad j = 1, 2, \ldots, n - 1 \tag{13.50}$$

The various special features incorporated in MPA are discussed in the next several subsections.

13.4.3.1 Empty Travel Time of the MHD

The natural service time for a processing operation is the average time required for that operation and is usually readily available or can be obtained from the machine responsible for the processing operation. However, the natural service time for a transfer depends not only on the actual (loaded) travel from the originating station to the destination station, but also on the empty travel time from the station at which the material handling carrier is currently located to the flow originating station. While the empty travel time may be small and negligible for intracell transfers, it can have a significant impact on intercell transfers. We use expressions (13.51) through (13.53) to estimate empty travel time (Chow, 1987). This approach assumes a FCFS discipline. To make it more realistic, the loaded travel is time assumed to be stochastic and characterized by two moments—mean and SCV. Although it can be relaxed, we assume unloaded travel is negligible for intracell transfers. The probability that a material transfer takes place from cell j to cell k and the travel time per trip are given by

$$q_{jk} = \frac{\sum_{i=1}^{p} \sum_{l=2}^{o^i} \hat{\lambda}^i Y_{jkl}^i / b_{kl}^i}{\sum_{s=1}^{n-1} \sum_{t=1}^{n-1} \sum_{i=1}^{p} \sum_{l=2}^{o^i} \hat{\lambda}^i Y_{stl}^i / b_{tl}^i}, \quad j, k = 1, 2, \ldots, n - 1 \tag{13.51}$$

$$\hat{tr}_{jk} = \sum_{r=1}^{n-1} tr_{rj} \sum_{s=1}^{n-1} q_{sr} + tr_{jk}, \quad j, k = 1, 2, \ldots, n - 1 \tag{13.52}$$

$$tr_{jk} = \frac{d_{jk}}{v}, \quad j, k = 1, 2, \ldots, n - 1 \tag{13.53}$$

The first part of Equation 13.52 recognizes the fact that the material handling carrier dispatched to serve an intercell transfer request can be at any one of the $n - 1$ departments. It explicitly includes empty travel time from the department it is currently at to the cell where the transfer originates as well as the loaded travel time to the destination department. The SCV for the intracell and intercell transfer times is the corresponding variance divided by the square of the mean:

$$Var\left(\hat{tr}_{jk}\right) = E\left[\hat{tr}_{jk}^2\right] - E\left[\left(\hat{tr}_{jk}\right)\right]^2 \tag{13.54}$$

$$E\left[\left(\hat{tr}_{jk}\right)^2\right] = \left[\sum_{r=1}^{n}(tr_{rj})^2 \sum_{s=1}^{n} q_{sr} + (tr_{jk})^2 + 2tr_{jk} \sum_{r=1}^{n}(tr_{rj}) \sum_{s=1}^{n} q_{sr}\right] \tag{13.55}$$

13.4.3.2 Setup

We show how setup impacts the two moments of the effective service time of processing and transfer operations in this section. Assume that the natural process or transfer time is independent of the corresponding setup time for each operation or transfer. Obviously, the effective service time for an operation depends upon the batch size of the product. When x units of a product are processed as a batch, MPA assumes that the batch processing time is x times the processing time of each unit in the batch, but the SCV of the batch processing time is the same as that of the individual unit in the batch. The following expressions are derived on that assumption, but it is easy to modify the expressions to handle other batching assumptions.

$$\hat{t}e_{kl}^i = b_{kl}^i to_{kl}^i + ts_{kl}^i, \quad i = 1,2,\ldots,p; \; j,k = 1,2,\ldots,n-1; \; l = 1,2,\ldots,o^i \tag{13.56}$$

$$\hat{c}_{ekl}^2 = \frac{\left(b_{kl}^i\right)^2 \left(to_{kl}^i\right)^2 c_{okl}^{2i} + \left(ts_{kl}^i\right)^2 c_{skl}^{2i}}{\left(\hat{t}e_{kl}^i\right)^2}, \quad i = 1,2,\ldots,p; \; j,k = 1,2,\ldots,n-1; \; l = 1,2,\ldots,o^i \tag{13.57}$$

$$\hat{t}e_{rjk}^i = \left(\hat{t}r_{jk} + tp^i bt_{jk}^i\right), \quad i = 1,2,\ldots,p; \; j,k = 1,2,\ldots,n-1 \tag{13.58}$$

$$\hat{c}_{erjk}^2 = \frac{\left(\hat{t}r_{jk}\right)^2 c_{rjk}^{2i} + \left(bt_{jk}^i\right)^2 \left(tp^i\right)^2 c_t^{2i}}{\left(\hat{t}e_{rjk}^i\right)^2}, \quad i = 1,2,\ldots,p; \; j,k = 1,2,\ldots,n-1; \; l = 1,2,\ldots,o^i \tag{13.59}$$

The first and second moments for the service time distribution at each department and the MHS department can be obtained using the following expressions:

$$te_k = \frac{\sum_{i=1}^p \sum_{l=1}^{o^i} \hat{\lambda}^i \hat{t}e_{kl}^i Y_{kl}^i / b_{kl}^i}{\sum_{i=1}^p \sum_{l=1}^{o^i} \hat{\lambda}^i Y_{kl}^i / b_{kl}^i}, \quad k = 1,2,\ldots,n-1 \tag{13.60}$$

$$te_k^2 \left(\hat{c}_{erjk}^2 + 1\right) = \frac{\sum_{i=1}^p \sum_{l=1}^{o^i} \hat{\lambda}^i \, (\hat{t}e_{kl})^2 \left(c_{ekl}^{2i} + 1\right) Y_{kl}^i / b_{kl}^i}{\sum_{i=1}^p \sum_{l=1}^{o^i} \hat{\lambda}^i \hat{Y}_{kl}^i / b_{kl}^i}, \quad k = 1,2,\ldots,n-1 \tag{13.61}$$

$$te_n = \frac{\sum_{i=1}^p \sum_{\substack{j=1 \\ j \neq k}}^{n-1} \sum_{k=1}^{n-1} \sum_{l=2}^{o^i} \hat{\lambda}^i \hat{t}e_{rjk}^i Y_{jkl}^i / bt_{jk}^i}{\sum_{i=1}^p \sum_{\substack{j=1 \\ j \neq k}}^{n-1} \sum_{k=1}^{n-1} \sum_{l=2}^{o^i} \hat{\lambda}^i Y_{jkl}^i / bt_{jk}^i} \tag{13.62}$$

$$te_n^2 \left(\hat{c}_n^2 + 1\right) = \frac{\sum_{i=1}^p \sum_{\substack{j=1 \\ j \neq k}}^{n-1} \sum_{k=1}^{n-1} \sum_{l=2}^{o^i} \hat{\lambda}^i \, (\hat{t}e_{rjk})^2 \left(c_{erjk}^{2i} + 1\right) Y_{jkl}^i / bt_{jk}^i}{\sum_{i=1}^p \sum_{\substack{j=1 \\ j \neq k}}^{n-1} \sum_{k=1}^{n-1} \sum_{l=2}^{o^i} \hat{\lambda}^i \, \hat{Y}_{jkl}^i / bt_{jk}^i} \tag{13.63}$$

13.4.3.3 Batching

It is important to carefully consider batching because the same product can be batched differently at two successive stations, or different products visiting two machines for successive operations can be batched differently. The batch size calculation presented here uses the notion of relative batch size and calculates it for each pair of nodes. For a particular product i and operation l, let us define the relative batch size $\gamma_{jk}^{i,l} = b_{jl}^i / b_{k,l+1}^i$, as the ratio of the batch size of lth operation of product i done on machine j to the batch size of $(l+1)$th operation of that product on machine k. When multiple products flow from machine j to machine k, the relative batch size is as follows:

$$\gamma_{jk} = \frac{\sum_{i=1}^{n} \sum_{l=2}^{o^i} b_{jl-1}^i Y_{jkl}^i}{\sum_{i=1}^{n} \sum_{l=1}^{o^i} b_{kl}^i Y_{jkl}^i} \tag{13.64}$$

13.4.3.4 Combining the Effects of the Four Network Operations

The first two moments of the effective interarrival time into a node are obtained by solving the two systems of linear equations, which synthesize the effects of the four basic network operations (including superposition, departure, splitting, and batching) on the first and second moments of interarrival time (or arrival rates), respectively (see Table 13.4). To calculate the effective mean arrival rate into each node, we solve Equation 13.65 simultaneously to obtain a unique solution. Note that Equation 13.65 contains as many variables as equations and p_{jk} is the outgoing probability of node j, i.e., the proportion of customers leaving node j that go to k. The idea behind the linear equations is that arrival into a node is equal to arrivals from the outside world plus arrivals from other nodes. But because batch sizes at other nodes might be different from the batch size at the node under consideration, the batches departing from other nodes are transformed into batches at the node being considered by using the relative batch size formula shown in the following equation:

$$\hat{\lambda}_k = \sum_{i=1}^{p} \lambda_{0k}^i + \sum_{j=1}^{n} \hat{\lambda}_j p_{jk} \gamma_{jk}, \quad k = 1, 2, \ldots, n \tag{13.65}$$

The second set of linear equations to be solved simultaneously to yield the SCV of the interarrival time of the aggregate customer at a node synthesizes the effects of the four network operations, which characterize the propagation of variance of the flow through the network. Writing out the effects of this process on the SCV of machine k, we have the following equation:

$$c_{ak}^2 = a_k + \sum_{j=1}^{n} c_{aj}^2 b_{jk}, \quad k = 1, 2, \ldots, n \tag{13.66}$$

where

$$a_k = 1 + w_k \left\{ \begin{array}{l} \left(p_{0k}' \gamma_{0k} c_{0k}^2 - 1 \right) + \sum_{j=1}^{n} p_{jk}' \gamma_{jk} \left[\left(1 - p_{jk} \right) + p_{jk} \rho_j^2 x_j \right] \\ + \sum_{j=0}^{n} p_{jk}' \max \left\{ \gamma_{jk} - 1, 0 \right\} \end{array} \right\}$$

$$b_{jk} = w_k p_{jk}' p_{jk} \gamma_{jk} (1 - \rho_j^2)$$

$$w_k = \frac{1}{1 + 4(1 - \rho_k)^2 (u_k - 1)}$$

$$x_j = 1 + \frac{1}{\sqrt{m_j}} \left[\max\left\{ c_{sj}^2, 0.2 \right\} - 1 \right]$$

$$u_k = \frac{1}{\sum_{j=0}^{n} \left(p'_{jk} \right)^2}$$

$$p'_{jk} = \frac{\lambda_{ij}}{\lambda_j}$$

$$\gamma_{jk} = \frac{\sum_{i=1}^{p} \sum_{l=1}^{o^i} \hat{\lambda}_{jk}^i \gamma_{jk}^{i,l} Y_{jl}^i Y_{k,l+1}^i}{\sum_{i=1}^{p} \sum_{l=1}^{o^i} \hat{\lambda}_{jk}^i Y_{jl}^i Y_{k,l+1}^i}$$

With the first two moments of the effective service and interarrival times calculated, as before, we can analyze each node as a GI/G/1 or GI/G/m queue to determine the expected time a part spends in queue at node k and the average waiting time in queue of each product at each node using Equations 13.36 through 13.38. To get part-specific information at each node, the equations must be disaggregated as shown in Equations 13.39 and 13.40:

$$L_k^i = L_k \frac{\sum_{l=1}^{o^i} \lambda_{0k}^i}{\sum_{i=1}^{p} \sum_{l=1}^{o^i} \lambda_{0k}^i Y_{kl}^i}, \quad i = 1, 2, \dots, p; k = 1, 2, \dots, n \qquad (13.67)$$

$$LQ_k^i = LQ_k \frac{\sum_{l=1}^{o^i} \lambda_{0k}^i Y_{kl}^i}{\sum_{i=1}^{p} \sum_{l=1}^{o^i} \lambda_{0k}^i Y_{kl}^i}, \quad i = 1, 2, \dots, p; k = 1, 2, \dots, n \qquad (13.68)$$

The average time spent in the system and queue per visit for a batch of each product type are calculated as shown in the following equations:

$$E\left(WQ_k^i \right) = E(WQ_k) \qquad (13.69)$$

$$E\left(W_k^i \right) = E\left(WQ_k^i \right) + \frac{\sum_{l=1}^{o^i} te_{kl}^i Y_{kl}^i}{\sum_{l=1}^{o^i} Y_{kl}^i} \qquad (13.70)$$

Other performance measures may also be similarly obtained using the aforementioned information. We refer the reader to Meng and Heragu (2004), Meng et al. (2004, 2008).

13.4.4 CLOSED QUEUING NETWORK MODEL

So far in our discussion of the uses of queuing in the design of manufacturing models, we have used WIP and processing time as evaluation criteria. In all the examples, there has been no limit on the amount of WIP in the system. However, it is not typical for a manufacturing system to have an unlimited amount of WIP. (Recall that unlimited buffer at each node was one of the core assumptions in the PD method.) In most manufacturing systems, a low level of WIP is maintained. The desired level of WIP may be a function of storage space, cash investment, or various other factors.

In fact, in a just-in-time (JIT) lean manufacturing system, which focuses on the elimination or minimization of all forms of waste, WIP is seen as a major source of waste and managers strive hard to keep the WIP level to a minimum. Kanbans are used to control WIP. For modeling such problems, the closed queuing network is used. A queuing network in which the WIP or level of jobs is fixed at some level N is called a closed queuing network. A new part enters the network only when another leaves the network. Whereas WIP was an output statistic in an open network queuing model, it is a control parameter in the closed network. Another major difference is that each part i arrives at one or more machines from an external source with rate γ_i in an open queuing network. In fact, this is what makes an open network queuing system "open." In contrast, all the external arrival rates γ_i ($i = 1, 2, \ldots, m$) are equal to 0 in a closed queuing network.

To derive results for the closed network model, we proceed as we did for the open network model. We will first develop results for a simple two-node system. This helps us in understanding the computational difficulty in obtaining the usual performance measures. Then, we will show results for the multinode closed network. Because it is even more numerically difficult to compute these results on a computer, we will provide a method called the mean value analysis (MVA) method for arriving at performance measures. In fact, we will show how the MVA method can be used for closed networks with multiple nodes and multiple classes of jobs.

First let us consider a two-machine closed network queuing model in which we have two nodes (machines) connected in series and each job first goes through machine 1 and then machine 2 (Figure 13.14). If we assume that there are n pallets in the system each containing one part (job) and that a new job is introduced into the system (loaded onto a pallet) only as another leaves the system (unloaded from a pallet), then we have a closed two-node network that always has a fixed (n) number of jobs in it. As long as the queue in front of each machine can hold at least $n - 1$ jobs, there will be no blocking. Deriving results for the no-blocking situation is much easier than doing so for the blocking case. Hence, we will assume the queue capacity is at least $n - 1$ for both queues.

As before, let N_{t1}, N_{t2} be the number of jobs at nodes 1, 2 respectively, at time t. The stochastic process defining the number of parts at each of the nodes at time t, $t \geq 0$, is a Markov chain. Assuming we have reached steady state, let k and $n - k$ represent the average number of parts at nodes 1 and 2 at time t, respectively. Also, let $P(k, n - k)$ be the joint steady-state probability of having k and $n - k$ parts at nodes 1 and 2, respectively. The state changes when a service completion occurs at any of the two machines. It is easy to draw the partial steady-state diagram for the general state in which we have k and $n - k$ jobs at nodes 1 and 2, respectively (see Figure 13.15). When $k = 0$,

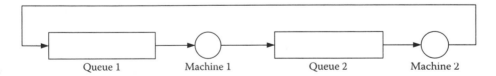

FIGURE 13.14 Closed network queuing system.

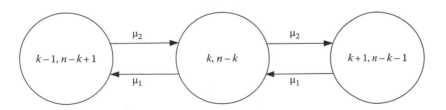

FIGURE 13.15 Partial rate diagram for a two-node closed network model.

it is obvious that the left state does not exist; similarly, when $k=n$, the right state does not exist. Therefore, the flow balance equations can be written as follows:

$$\mu_2 P(0, n) = \mu_1 P(1, n-1) \tag{13.71}$$

$$\mu_1 P(n, 0) = \mu_2 P(n-1, 1) \tag{13.72}$$

$$(\mu_1 + \mu_2) P(k, n-k) = \mu_1 P(k+1, n-k-1) + \mu_2 P(k-1, n-k+1) \tag{13.73}$$

Using Equation 13.73 and making appropriate substitutions, $P(1, n-1)$ can be written as

$$P(1, n-1) = \left(\frac{1}{\mu_1 + \mu_2}\right) \left[\mu_1 P(2, n-2) + \mu_2 P(0, n)\right]$$

$$= \left(\frac{1}{\mu_1 + \mu_2}\right) \left[\mu_1 P(2, n-2) + \mu_1 P(1, n-1)\right] \tag{13.74}$$

Hence

$$P(1, n-1) = \left(\frac{\mu_1}{\mu_1 + \mu_2}\right) \left[P(2, n-2) + P(1, n-1)\right] \tag{13.75}$$

Collecting the $P(1, n-1)$ terms on the left-hand side, we get

$$P(1, n-1) = \left(\frac{\mu_1}{\mu_2}\right) \left[P(2, n-2)\right] \tag{13.76}$$

Similarly, it can be shown that

$$P(2, n-2) = \left(\frac{\mu_1}{\mu_2}\right) \left[P(3, n-3)\right]$$

$$P(3, n-3) = \left(\frac{\mu_1}{\mu_2}\right) \left[P(4, n-4)\right]$$

$$\cdots$$

$$P(n-1, 1) = \left(\frac{\mu_1}{\mu_2}\right) \left[P(n, 0)\right]$$

Using the boundary condition that

$$P(0, n) + P(1, n-1) + P(2, n-2) + \cdots + P(n, 0) = 1 \tag{13.77}$$

we can then solve for each $P(k, n-k)$. To do so, note from Equation 13.77 that

$$1 - P(n, 0) = P(0, n) + P(1, n-1) + \cdots + P(n-1, 1)$$

$$= \left(\frac{\mu_1}{\mu_2}\right) \left[P(1, n-1) + P(2, n-2) + \cdots + P(n-1, 1)\right] \tag{13.78}$$

The second part of the right-hand side of Equation 13.78 is equal to $1 - P(0, n)$ due to Equation 13.77. Therefore, Equation 13.78 can be rewritten as

$$1 - P(n,0) = \left(\frac{\mu_1}{\mu_2}\right)\left[1 - P(0,n)\right] \tag{13.79}$$

Therefore,

$$P(n,0) = 1 - \left(\frac{\mu_1}{\mu_2}\right) + \left(\frac{\mu_1}{\mu_2}\right)\left[P(0,n)\right] \tag{13.80}$$

It is easy to see from the general expression for $P(k, n - k)$ that

$$P(0,n) = \left(\frac{\mu_1}{\mu_2}\right)^n \left[P(n,0)\right] \tag{13.81}$$

Making the substitution for $P(n, 0)$ from Equation 13.81 into Equation 13.80, we get the following equation:

$$P(n,0) = 1 - \left(\frac{\mu_1}{\mu_2}\right) + \left(\frac{\mu_1}{\mu_2}\right)^{n+1}\left[P(n,0)\right] \tag{13.82}$$

When we collect the $P(n, 0)$ terms on the left, it follows that

$$P(n,0) = \frac{1 - \left(\mu_1/\mu_2\right)}{1 - \left(\mu_1/\mu_2\right)^{n+1}} \tag{13.83}$$

From Equation 13.83, we can immediately obtain all other probabilities because

$$P(k,n-k) = \left(\frac{\mu_1}{\mu_2}\right)^{n-k}\left[P\left(n,0\right)\right] \tag{13.84}$$

Knowing the state probabilities $P(0, n)$ and $P(n, 0)$ is helpful in determining the utilization rates of the two nodes (machines) because utilization of the first machine is given by $1 - P(0, n)$ and utilization of the second machine is $1 - P(n, 0)$. From this, the throughput of the system, which is equal to utilization of the second server times its service rate μ_2, can be easily determined.

We now extend the preceding discussion to the multiple-node closed network. Assume that we have m nodes each having a single server. Let n_i represent the number of jobs at node i after steady state has been reached. Because the total number of pallets is n, the number of jobs in the system at anytime is also n. As in the two-node closed network, we let

$$\sum_{i=1}^{m} n_i = n$$

For the two-node case, the number of possible combinations of jobs at the two nodes or the number of states $(0, n)$, $(1, n - 1)$, ..., $(n, 0)$ was equal to $n + 1$. In the m-node case, the number of states is given by $(n + m - 1)!/(n!(m - 1)!)$. Clearly, the computation of performance measures is likely to be tedious, because of the explosion in the number of states. Before we present performance measures for this network, we provide two examples of how manufacturing systems may be modeled as closed m-node networks.

Consider a flexible manufacturing system in which a part is processed first on one of k machines and then on one of $m - k$ machines for the final processing step. After the first operation, the parts join a queue in front of a conveyor-robot MHS and are then transported to one of $m - k$ machines for the final processing step. After the last operation is completed, another conveyor-robot system is used to unload completed parts and load new ones onto the pallet. The probability of the parts visiting each of the m machines is known. This problem may be modeled as a $m+2$-node closed network as shown in Figure 13.16.

Now consider an extension of the preceding example. A flexible manufacturing cell processes r different types of jobs and each unit of each type undergoes N_r processing steps. There are max $\{N_i: i=1, 2, \ldots, r\}$ AGVs. After the first operation, all the part types are sent to the first AGV queue for transportation to the second operation (machine); similarly, after the second operation, all the part types are sent to the second AGV queue for transportation to the third operation (machine), and so on. After the last operation, all the completed part types are removed from the pallets and the empty pallets are sent to the last AGV queue for transportation to their respective first operation, where new parts are loaded. Thus, the first transportation for each part type is done by the first AGV, the second transportation by the second AGV, and so on. There are several machines capable of performing the intermediate operations for part type r and the probability that part type r will visit a specific machine for the N_rth operation is assumed to be known. Further, if the service time at each node and each server is exponential, this problem can again be appropriately modeled as a multiclass, multinode closed network.

As discussed earlier, deriving results for the general m-node network is quite involved. Hence, we only present some basic results for a general m-node network, and refer the reader to Viswanadham and Narahari (1992) for derivation of the results and further details. The system considered here assumes that there are m nodes in the network and r types of jobs are processed in it. The processing of job type j at node i (which has a single server) is exponential with parameter μ_i^j, $i=1, 2, \ldots, m$, $j=1, 2, \ldots, r$ and the routing probability p_{ij}^r that part type r will visit node j after operation at node i is known. The number of parts of type r at node i is n_i^r and N^r is the total number of parts of type r in the network. In other words, N^r is the number of pallets allotted to part type r. Obviously,

$$\sum_{i=1}^{m} n_i^r = N^r$$

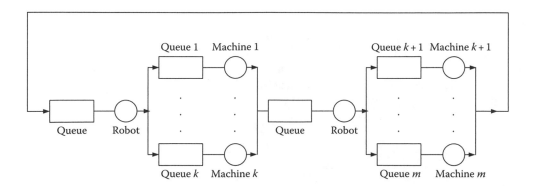

FIGURE 13.16 Closed network model of a simple automated manufacturing system.

Defining T_i as the vector $\left(n_i^1, n_i^2, \ldots, n_i^r\right)$ corresponding to node i and T as $(T_1, T_2, \ldots, T_m)$, we can get the steady-state joint probability of having all possible combinations of parts and part types at each node using the following product form solution:

$$P(T_1, T_2, \ldots, T_m) = \left(\frac{1}{NC}\right) \prod_{i=1}^{m} f(T_i)$$

where
  NC is a normalization constant
  $f(T_i)$ is a function of T_i

Calculating the normalization constant on a computer can pose serious numerical difficulties, especially as n and m become large. Hence, the preceding method of optimally determining the steady-state joint probability (from which other performance measures of the closed network can be obtained) is not desirable. Instead, we rely on the approximate MVA method. This is discussed in the next section.

13.4.4.1 Mean Value Analysis

MVA is computationally the least difficult among several techniques available to analyze closed network queuing systems. A discussion of alternative methods as well as MVA can be found in Viswanadham and Narahari (1992). The MVA procedure presented by Suri and Hildebrandt (1984) has three components: throughput times, production rates, and queue lengths. For our discussion, assume there are a total of n parts (pallets) in the system and

$$\sum_{j=1}^{r} N^j = n$$

As before, N^j is the number of pallets allotted to part type j and each node i has S_i servers with an exponential service rate for part type j with parameter μ_i^j, $i = 1, 2, \ldots, m$, $j = 1, 2, \ldots, r$.

The throughput time is the average time per visit of a part type r to machine i. The equation for the throughput time is as follows:

$$W_{ij} = \frac{1}{\mu_i^j} + \left(\frac{N^j - 1}{N^j}\right)\left(\frac{L_{ij}}{S_i \mu_i^j}\right) + \sum_{r \neq j}\left(\frac{L_{ir}}{S_i \mu_i^r}\right), \quad i = 1, 2, \ldots, m; \; j = 1, 2, \ldots, r \tag{13.85}$$

The first term in Equation 13.85 is the service time of part type j at machine i, the second term is the time spent in the queue waiting for other parts of type j, and the third term is the time spent in the queue waiting for parts of other types. L_{ij} is the average number of parts of type j at node i. Because the second term in Equation 13.85 is for calculating the time spent in node i's queue waiting for other parts of type j, we reduce L_{ij} by the quantity $(N^r - 1)/N^r$. When a part arrives at a machine, at most $N^r - 1$ other parts can be ahead of it in the queue. Hence, the term $[(N^r - 1)/N^r][L_{ir}]$ is the expected number of parts of type r in queue when a new one arrives. Dividing this by the service rate $S_i \mu_i^r$ then gives the time spent by the new part in the queue waiting for other parts of type r to be processed. Similarly, it follows that the last term in Equation 13.85 is the time spent by the new part in the queue waiting for parts of other types to be processed.

The second performance calculated by MVA is the production rate. In a closed network model, the production rate for part type r can be determined via Little's law ($L = \lambda W$) and is given by

$$X_j = \frac{N^j}{\left(\sum_{i=1}^{m} v_{ij} W_{ij}\right)} \qquad (13.86)$$

where v_{ij} is the expected number of visits of part type j to machine i.

The final component of MVA is the queue length. This calculation is again simply an application of Little's law. Because X_r is the equivalent of λ,

$$L_{ij} = X_j(v_{ij} W_{ij}) \quad \text{for all } i \text{ and } j$$

We now have all the components to utilize MVA to analyze closed network queuing systems. There are four iterative steps to the process.

MVA Algorithm

Step 1: The first step is to estimate an initial value of L_{ir}. The best method is to evenly distribute a part over all the stations it visits. For example, if a part type r visits the first three machines of a four machine loop, then $L_{1j} = L_{2j} = L_{3j} = N^j/3$, and $L_{4j} = 0$.

Step 2: Determine the throughput time W_{ir} using the following equation:

$$W_{ij} = \frac{1}{\mu_i^j} + \left(\frac{N^j - 1}{N^j}\right)\left(\frac{L_{ij}}{S_i \mu_i^j}\right) + \sum_{r \neq j}\left(\frac{L_{ir}}{S_i \mu_i^r}\right), \quad i = 1, 2, \ldots, m; \ j = 1, 2 \ldots, r$$

Step 3: Determine the production rate X_r for each machine r using the following equation:

$$X_r = \frac{N^j}{\sum_{i=1}^{m} v_{ij} W_{ij}}$$

Step 4: Determine the queue length L_{ij} using the equation $L_{ij} = X_j(v_{ij} W_{ij})$.

Step 5: Compare the L_{ij} value calculated in step 4 with the previous value. If the new value is within a desired range of the previous value, *stop*. If not, go to step 2.

The desired range can be arbitrarily chosen, but a 5% range is recommended. The procedure for MVA is utilized in the closed network queuing system described in Example 13.7. While the example is for a small system of three machines and two part types, it can be expanded to any number of machines and parts.

Example 13.7

Consider a machine shop that manufactures two parts. Part 1 visits machines 1 and 3, and part 2 visits machines 1 and 2. The service rates for machines 1, 2, and 3 are 10, 15, and 12 parts per hour, respectively. The service rates are not affected by the part being serviced. JIT/kanban considerations limit the WIP of parts 1 and 2 to 5 and 4 units, respectively.

1. Determine the amount of time each part spends waiting and being served at each machine.
2. Determine the number of each part waiting and being served at each machine.
3. Based on the solution to parts 1 and 2, what conclusions can be drawn about the system design?

Solution

Step 1: Determine initial values of L_{ij}. The number of each part type in the system is $N^1 = 5$, and $N^2 = 4$. Each part type visits only 2 of the 3 machines. Therefore, $L_{11} = L_{31} = N^1/2 = 2.5$ and $L_{12} = L_{22} = N^2/2 = 2$. Because part type 1 does not visit machine 2, and part type 2 does not visit machine 3, $L_{21} = L_{32} = 0$.

Step 2: Determine the throughput time W_{ij} using Equation 13.85.

$$W_{11} = 1/10 + [4/5][2.5/10] + 2/10 = 0.50 = 30 \text{ minutes}$$

$$W_{21} = 0$$

$$W_{31} = 1/12 + [4/5][2.5/12] = 0.25 = 15 \text{ minutes}$$

$$W_{12} = 1/10 + [3/4][2/10] + 2.5/10 = 0.50 - 30 \text{ minutes}$$

$$W_{22} = 1/12 + [3/4][2/15] = 0.25 = 0.167 = 10 \text{ minutes}$$

$$W_{32} = 0$$

Step 3: Determine the production rate X_r using the Equation 13.86. In this example, $v_{ij} = 1$ for all i and r. We get

$$X_1 = 5/[30 + 15] = 1/9 = 0.111 \text{ parts per minute}$$

$$X_2 = 4/[30 + 10] = 1/10 = 0.10 \text{ parts per minute}$$

Step 4: Determine the queue length L_{ij} using the equation $L_{ij} = X_r(v_{ij}W_{ij})$ for all i and j.

$$L_{11} = (0.111)(30) = 3.33 \text{ parts}$$

$$L_{21} = 0$$

$$L_{31} = (0.111)(15) = 1.67 \text{ parts}$$

$$L_{12} = (0.10)(30) = 3 \text{ parts}$$

$$L_{22} = (0.10)(10) = 1 \text{ part}$$

$$L_{32} = 0$$

Notice that $L_{11} + L_{31} = 5$, and $L_{12} + L_{22} = 4$.

Step 5: Compare the value of L_{ij} calculated in step 4 with the previous value. The values of L_{ij} warrant another iteration. Therefore, we return to step 2 and repeat until the new calculated values of L_{ij} are within the recommended range (5%) of the previous values. The following solution is found after four additional iterations.

$L_{11} = 4.28$ parts	$W_{11} = 47.8$ minutes
$L_{21} = 0$	$W_{21} = 0$
$L_{31} = 0.72$ parts	$W_{31} = 8.1$ minutes
$L_{12} = 3.60$ parts	$W_{12} = 47.5$ minutes
$L_{22} = 0.40$ parts	$W_{22} = 5.3$ minutes
$L_{32} = 0$	$W_{32} = 0$

The current design is creating a buildup in front of machine 1. Approximately 88% $[(4.28 + 3.60)/(5 + 4)]$ of the WIP is either waiting or being served at machine 1.

13.5 USE OF SIMULATION IN FACILITIES LAYOUT AND MATERIAL HANDLING

In various parts of this book, we have mentioned several times that simulation is a popular tool that can be used to evaluate system performance. We will now discuss how these tools could be used to analyze and design manufacturing and material handling systems. First, it should be emphasized that, like queuing network, simulation is not a prescriptive tool. Although it does not prescribe a course of action to follow to optimize a specified objective, it tells us how a given design—chosen perhaps using a prescriptive model, for example, the mathematical layout models and algorithms in Chapters 5, 10, and 11—will perform with respect to operational performance measures of interest to the user. Thus, it is a descriptive tool. That simulation is a descriptive and not a prescriptive model is not well understood in practice. For example, some practitioners incorrectly assume that simulation can develop an optimal or near-optimal facility layout or design! While simulation allows the user to input and test several designs or layouts, it does not generate a layout or facility design by itself. Thus, we have to resort to the models, algorithms, and approaches we discussed in Chapters 5, 10, and 11 to come up with a preliminary design. However, once we have a reasonably good and satisfactory preliminary design, we can then use simulation to see how some of the dynamic and probabilistic aspects of the problem (which were ignored at the preliminary design stage) impact the performance of the design with respect to chosen operational performance measures. For example, at the preliminary design stage, we were primarily concerned with minimizing the annual material handling costs and we did not consider the effects of MHD failure, vehicle blockages due to traffic congestion, variability in product demand, and other factors on system performance measures such as part waiting time, throughput rate, and MHD utilization. Typically, we use descriptive models—queuing network or simulation—to fine-tune the preliminary design so that we have a system design that is optimal or near-optimal not only with respect to design criteria but also operational performance measures.

Simulation is being increasingly used in place of other modeling tools for a variety of reasons.

- Due to the high level of automation we have in systems today, the problem being studied is dynamic, has a number of interacting probabilistic elements, is highly complex, and is therefore not amenable to analytical solution.
- Simulation is easy to understand, visualize, and communicate when compared to other modeling tools.
- With the increase in computing power and speed, it is relatively easy to develop and run simulation models.
- Whereas analytical queuing models only provide steady-state results, simulation can be used for transient as well as steady-state analysis.

Simulation may be defined as a computer technique that mimics the operation of a real manufacturing or service system over time. It involves generating an artificial history of the system being studied, observing the system history, and using these observations to estimate or predict the operating characteristics of the real system (Banks et al., 1995). Although it need not be computer based, we have included the term computer in the definition to emphasize that over the period of time for which simulation is being run, we need to keep track of a multitude of events and computations, and it is easier to do so using a computer-based tool.

The theory of simulation can be best understood via what is known as Monte Carlo simulation. The Monte Carlo method of simulation can be developed on a spreadsheet as illustrated in Example 13.9. Before we proceed with the example illustrating Monte Carlo simulation, it should be emphasized that this method of simulation estimates the system performance measure, e.g., average time in system, via sampling. Because interarrival and service times are typically not deterministic but are random variables that follow, in many cases, a known probability distribution, we must be able to sample data from the appropriate probability distribution. In order to do so, we must first be able to generate uniformly distributed random numbers in the interval [0, 1], i.e., all values between 0 and 1 must be equally likely to occur. Once we have done this, we can then use specialized methods to transform the random numbers into a random variate from the required distribution. Before further discussion on this, we explain how random numbers are generated on a computer. Although there are several methods available for generating random numbers, a popular method is called the linear congruential generator method. To produce a sequence of integers between 0 and $m - 1$, we use the following equation recursively:

$$x_{i+1} = (ax_i + c) \bmod m, \ i = 1, 2, \ldots$$

It is perhaps obvious to the reader that to generate the first integer in the sequence, we must have an initial "seed" value for x_0. In addition, we must specify values for a and c. Once we have done these, we can then generate the sequence of integers, x_1, x_2, ..., from which we can generate uniformly distributed random numbers in the interval [0, 1] using the following formula:

$$r_i = \frac{x_i}{m}, \quad i = 1, 2, \ldots$$

Random numbers generated for a starting seed of 30, $a = 19{,}866$, $c = 43$, and $m = 100$ are shown in Table 13.5. When the values for a, c, m, and x_0 are carefully selected and the preceding formula is used to generate random numbers, the numbers are truly random and meet all statistical properties. They are independent, uncorrelated, and uniformly distributed in the interval [0, 1]. However, because the random numbers can be produced on a computer again and again simply by specifying the values for initial seed and parameters a, c, and m and using a deterministic rule (formula), they are referred to as pseudorandom numbers. By properly selecting values for a, c, and m, not only can we generate truly random numbers but also generate enough of them so that we can carry out a full simulation run before the number sequence begins to repeat itself.

Just as there are several methods to generate random numbers, there are several techniques to transform random numbers into random variates for a specified distribution. We give a brief description of a popular method called the inverse transformation method. Suppose we want to generate random variates for an exponential distribution. We know that the pdf of an exponential distribution with parameter λ is given by

$$f(t) = \begin{cases} \lambda e^{-\lambda t} & \text{for } t > 0 \\ 0 & \text{for } t \le 0 \end{cases}$$

The cumulative probabilities for an exponential distribution are given by

$$F(x) = \int_0^x f(t)dt = \int_0^x \lambda e^{-\lambda t} \, dt = 1 - e^{-\lambda x}$$

Setting a uniformly distributed random number equal to the cumulative distribution function and solving for x yields a random variate for the exponential distribution as shown next:

$$1 - e^{-\lambda x} = r_i \Rightarrow e^{-\lambda x} = 1 - r_i$$

TABLE 13.5
Random Numbers

Iteration Number	Sequence of Integers	Random Number	Iteration Number	Sequence of Integers	Random Number
1	71	0.71	51	21	0.21
2	29	0.29	52	29	0.29
3	57	0.57	53	57	0.57
4	5	0.05	54	5	0.05
5	73	0.73	55	73	0.73
6	61	0.61	56	61	0.61
7	69	0.69	57	69	0.69
8	97	0.97	58	97	0.97
9	45	0.45	59	45	0.45
10	13	0.13	60	13	0.13
11	1	0.01	61	1	0.01
12	9	0.09	62	9	0.09
13	37	0.37	63	37	0.37
14	85	0.85	64	85	0.85
15	53	0.53	65	53	0.53
16	41	0.41	66	41	0.41
17	49	0.49	67	49	0.49
18	77	0.77	68	77	0.77
19	25	0.25	69	25	0.25
20	93	0.93	70	93	0.93
21	81	0.81	71	81	0.81
22	89	0.89	72	89	0.89
23	17	0.17	73	17	0.17
24	65	0.65	74	65	0.65
25	33	0.33	75	33	0.33
26	21	0.21	76	21	0.21
27	29	0.29	77	29	0.29
28	57	0.57	78	57	0.57
29	5	0.05	79	5	0.05
30	73	0.73	80	73	0.73
31	61	0.61	81	61	0.61
32	69	0.69	82	69	0.69
33	97	0.97	83	97	0.97
34	45	0.45	84	45	0.45
35	13	0.13	85	13	0.13
36	1	0.01	86	1	0.01
37	9	0.09	87	9	0.09
38	37	0.37	88	37	0.37
39	85	0.85	89	85	0.85
40	53	0.53	90	53	0.53
41	41	0.41	91	41	0.41
42	49	0.49	92	49	0.49
43	77	0.77	93	77	0.77
44	25	0.25	94	25	0.25
45	93	0.93	95	93	0.93
46	81	0.81	96	81	0.81
47	89	0.89	97	89	0.89
48	17	0.17	98	17	0.17
49	65	0.65	99	65	0.65
50	33	0.33	100	33	0.33

Taking natural logarithms on both sides and simplifying, we get

$$x = -\frac{1}{\lambda}\ln(1 - r_i) = \frac{1}{\lambda}\ln(r_i)$$

Notice that we have replaced $(1 - r_i)$ with r_i because $(1 - r_i)$ is also random. Because the parameter λ of the exponential distribution (i.e., its mean) is given and therefore known, repeating the procedure for other random numbers yields other random variates for this distribution. Using a similar approach, we can generate random variates for other distributions as well (see Banks et al., 1996; Law and Kelton, 1991). Now that we have a general idea of generating random numbers and random variates, we are ready to discuss Monte Carlo simulation. This is explained in Example 13.8.

Example 13.8

Consider a small machining center in which there is a drill, a cutter, and a finishing station. A part type arrives from an input conveyor in batches of size 1. The interarrival time follows an exponential distribution with a mean of 30 minutes. Each unit visits the drill first, cutter next, and finishing station before it is shipped to the warehouse via an output conveyor. The processing times follow known probability distributions and the relevant parameters are given in Table 13.6. (The drill and part type are referred to as Drill A and part type A for reasons that will become apparent later.) Develop a Monte Carlo simulation of the system and collect statistics on the idle time of Drill A and the average time in system for the first 20 parts.

Solution

Generate 20 sets of random variates on a spreadsheet using the linear congruential and inverse transformation methods for*

1. An exponential distribution with parameter (mean) of 30
2. An exponential distribution with mean of 60
3. An exponential distribution with mean of 30
4. A triangular distribution with parameters (minimum, mode, and maximum) of 20, 30, and 40

Record the random variables as shown in Table 13.7 (see the columns labeled IAT and ST). Set the simulation clock to zero. Use the first variate to correspond to the interarrival time of the first unit of part type A, i.e., the arrival of the first unit, the second to correspond to the service time

---

TABLE 13.6

Processing Time Parameters (in Minutes) for Part Type A

Machine	Part Type A
Drill A	Exponential (60)
Cutter	Exponential (30)
Finishing station	Triangular (20, 30, 40)

---

* Many simulation software packages have random number generators and methods to transform random numbers into random variates for several probability distributions.

TABLE 13.7
Monte Carlo Simulation

No.	Drill					Cutter					Finishing Station				
	IAT	TOA	ST	T SB	TSE	IT	ST	TSB	TSE	IT	ST	TSB	TSE	IT	TIS
1	35.02	35.02	69.04	35.02	104.06	35.02	3.67	104.06	107.74	104.06	27.09	107.74	134.82	107.74	99.80
2	28.15	63.17	55.30	104.06	159.36	0.00	15.62	159.36	174.98	65.13	34.19	174.98	209.16	61.04	145.99
3	40.87	104.04	80.75	159.36	240.11	0.00	30.10	240.11	270.20	0.00	34.09	270.20	304.30	12.26	200.25
4	2.57	106.61	4.13	240.11	244.24	0.00	46.36	270.20	316.56	94.31	26.86	316.56	343.42	84.46	236.81
5	83.82	190.42	166.63	244.24	410.87	0.00	17.01	410.87	427.88	74.35	28.22	427.88	456.10	88.46	265.67
6	46.18	236.60	91.36	410.87	502.23	0.00	42.32	502.23	544.55	12.98	32.70	544.55	577.25	0.00	340.64
7	28.15	264.76	55.31	502.23	557.54	0.00	3.55	557.54	561.09	0.00	29.55	577.25	606.80	0.00	342.04
8	2.08	266.84	3.17	557.54	560.70	0.00	21.12	561.09	582.21	121.82	32.04	606.80	638.84	73.90	372.00
9	72.16	339.00	143.32	560.70	704.03	0.00	8.72	704.03	712.74	66.59	24.50	712.74	737.25	47.02	398.24
10	38.15	377.16	75.31	704.03	779.34	0.00	4.93	779.34	784.27	135.76	31.86	784.27	816.13	135.73	438.97
11	70.85	448.01	140.70	779.34	920.03	0.00	31.83	920.03	951.86	116.52	30.74	951.86	982.60	111.50	534.59
12	74.67	522.68	148.34	920.03	1068.38	0.00	25.72	1068.38	1094.10	4.94	31.92	1094.10	1126.02	0.00	603.34
13	15.83	538.51	30.66	1068.38	1099.04	0.00	7.90	1099.04	1106.94	13.31	28.98	1126.02	1154.99	20.92	616.48
14	11.11	549.62	21.21	1099.04	1120.25	0.00	55.66	1120.25	1175.92	144.38	34.63	1175.92	1210.55	115.26	660.93
15	100.52	650.14	200.05	1120.25	1320.30	0.00	5.51	1320.30	1325.81	55.85	28.36	1325.81	1354.17	49.66	704.03
16	31.18	681.32	61.36	1320.30	1381.66	0.00	22.17	1381.66	1403.83	100.88	29.68	1403.83	1433.50	142.58	752.18
17	62.03	743.35	123.05	1381.66	1504.71	0.00	71.37	1504.71	1576.09	0.00	27.60	1576.09	1603.69	0.00	860.34
18	27.13	770.47	53.26	1504.71	1557.97	0.00	7.86	1576.09	1583.94	117.48	34.77	1603.69	1638.45	102.93	867.98
19	72.23	842.70	143.45	1557.97	1701.42	0.00	39.96	1701.42	1741.38	0.00	25.00	1741.38	1766.39	0.00	923.69
20	11.86	854.56	22.71	1701.42	1724.13	0.00	2.59	1741.38	1743.97	0.00	20.95	1766.39	1787.34	0.00	932.78
Total						35.02				1228.4				1153.5	10297

of the first unit of part type A on Drill A, the third to correspond to the service time of the first unit of part type A on the cutter, and the fourth to correspond to the service time of the first unit of part type A on the finishing station. Use the second, third, and other variates to correspond to the second, third, and other interarrival and service times, respectively. Record them as shown in Table 13.6. (The columns labeled IAT, TOA, ST, TSB, TSE, IT, TIS refer to the interarrival time, time of arrival, service time, time service begins, time service ends, idle time, and time in system, for the corresponding units of part type A and the three machines.)

From the data collected in Table 13.7, we see that the average time spent by a part in the system (based on a simulation run for 20 units) is 10,297/20 = 513.85 minutes. Similarly the percentage idle time for Drill A is 100 * 35.02/1724.13 = 2.03%. It is easy for the reader to see how other statistics, for example, average time spent in queue by the units, the percentage busy time for cutter, may be collected. Before we begin to draw inferences concerning the operating characteristics of the real system being studied (based on the output provided by the aforementioned simulation model), we must be sure that the simulation model is verified and validated. Verification is done to make sure that there are no unintended errors in the setting up of the model, e.g., wrong formulae used to determine the part waiting time. Validation is done to assure the users that the model behaves much like the real system.

Users would like to know that the output of the simulation model changes in a predictable direction when inputs are changed. For example, doubling the arrival rate should decrease the machine idle time, because the machine must now handle more parts. In addition to verification and validation, another difficult question that must be answered is how long should a simulation model be run before we can draw inferences concerning the operational characteristics of the real system being studied. Because simulation is a statistical experiment, we can have more confidence in our prediction of the system behavior if we have a large number of data points from which to draw an inference. This means running the simulation model longer. However, running the model longer costs precious computation time. Thus, we have to trade off between estimates with higher confidence and computation time. The 20 units for which we collected data is insufficient to make a reasonable estimate of the system performance. The simulation will have to be run longer. Because we wanted to briefly discuss the theory of simulation, we restricted our run to 20 units.

Assuming that the model has been verified and validated and that the appropriate run length has been determined, the model can then be used to examine design changes. For example, with the current service time of 60 minutes (exponentially distributed), we see that Drill A is utilized 98% of the time. Queuing theory tells us that higher levels of utilization (especially when the value approaches 1) cause the inventory buildup in the input queue of the drill to "explode". Even for the 20 unit simulation, we see that parts wait a significant amount of time in the input queue to Drill 13. Management may be interested in considering the impact of expending resources in order to minimize the process time variation. For example, by examining the process, making fixture changes, and performing periodic preventive maintenance, let us assume that management can reduce the service time variability to almost 0. The service time for Drill A can now be considered to be deterministic with a value of 60 minutes for each part. Using the simulation model, we can estimate the average time in system (see Exercise 33) for part type A to be 292 minutes—down from almost 515 minutes in the current setup! We can thus use simulation models to make estimates of system performance measures.

Even though the example considered is rather simple, the data collection efforts required in a simulation model are not trivial. When we expand the earlier problem to reflect additional considerations such as those found in real-world problems, the reader probably has an appreciation for the large amount of data tracking that must be done. Fortunately, we have several elegant software packages that do the data collection and tracking efficiently and allow the user to concentrate on modeling and output analysis. The newer packages are object-oriented even allow us to easily animate the simulation model.

Among the numerous simulation softwares available today, many have features that are particularly attractive for use in facilities design and material handling applications. Features of some 57 simulation softwares are compared in the December 2005 issue of *ORMS Today* magazine. Some of the popular simulation software packages include: *ARENA*, *AutoMod*, *PROMODEL*, SIMI0, SIMUL8, and WITNESS 2006. The reader interested in the aforementioned and other simulation software packages is strongly encouraged to refer to the simulation software buyer's guide and review in the December 2005 issue *ORMS Today* (pp. 44–55). In Example 13.9, we consider an expanded version of the problem described in Example 13.8 to give the reader an idea of some of the aspects that can be modeled using simulation.

Example 13.9

Consider Example 13.8. In addition to part type A, there is another one that is manufactured by the machining center. This part type B visits another drill (called B), and is then processed on the cutter and finishing station before it is shipped to the warehouse via the same output conveyor. The processing time for each unit of part type B on Drill B is normally distributed with a mean of 80 minutes and a standard deviation of 3 minutes. The processing time at the cutter and finishing station is the same as that of part type A (see Table 13.6). In addition, it is known that the interarrival time between any two part arrivals (whether they are type A or B) is normally distributed with a mean of 50 minutes, standard deviation of 5 minutes and that 50 percent of the arrivals are units of part type 13. The parts arrive in batches of size 1, 2, or 3 with known fixed probabilities. An automated guided vehicle (AGV) is used for material handling. It picks up a batch from the input conveyor and transports it to the input queue of the appropriate drill using the shortest path. Parts are taken from the output queue of the drills to the input queue of cutter, from the output queue of cutter to the input queue of the finishing station, and from the output queue of the finishing station to the output conveyor on an FCFS basis. The layout of the machines is shown in Figure 13.17, and the travel time, which is a function of the distance, is

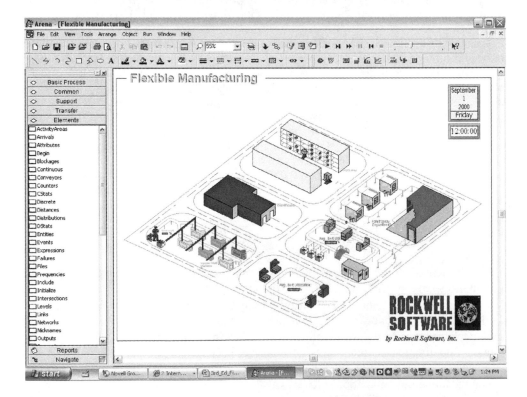

FIGURE 13.17   A sample simulation model. (Courtesy of Rockwell Software, Inc.)

assumed to be deterministic. Develop a simulation model for this example using any available software package. Modify the AGV dispatching rule so that, after dropping off a batch at an input queue, it travels to the nearest output queue that has parts waiting, transports them to the corresponding input queue, and repeats this procedure until all the parts (batches) in the input conveyor are transported. Run both models for 6 months and compare the AGV utilization as well as the part waiting times for the two systems.

In order to set up and solve the preceding problem using a simulation software package, we must explain modeling features of the simulation package. This is beyond the scope of this book, however, so we do not solve this numeric example. We do provide a graphical simulation model of the aforementioned problem (Figure 13.17) and a snapshot of an animation of the simulation (Figure 13.18), so the reader has a rough idea of what a simulation model developed using modern simulation software looks like and what the modeling capabilities of a simulation software are.

Simulation is an area that has exploded in terms of research, software development, and application in the last few decades. Complex models of real-world systems can now be easily captured in a PC-based software. These models are becoming easier and easier to develop. With the advances in PC hardware technology, the models and even their animations can be run in a reasonable time. For these and other reasons, there is abundant literature on this subject. There are entire books that are written on simulation modeling (e.g., Banks et al., 1996; Law and Kelton, 1991; Nelson, 1995) and simulation software (e.g., SLAM II—Pritsker (1986), SIMAN—Pegden et al. (1990), SIMAN V and ARENA–Banks et al. (1995)). Many industrial engineering programs offer a full semester course in simulation. It is virtually impossible to capture all the aspects of simulation in one section. What we have attempted to do in this section is therefore to explain to the reader that, in addition to the other models studied in this book, simulation is a very important tool that can be an extremely useful aid in facilities design and planning. As mentioned earlier, it is more useful in analyzing a system design than in designing the system itself. The reader interested in learning more about simulation is referred to Law and Kelton (1991) and Banks et al. (1996), among other sources.

13.6 SUMMARY

In this chapter, we introduced the reader to queuing, queuing network, and simulation modeling approaches. They have been applied to solve the manufacturing system design and planning problems rather successfully. In order to help the reader to understand how queuing models may be applied to system design, we presented detailed discussions on the basic types of queuing models. There are numerous queuing models and we presented or derived results only for the ones for which analytical solution is possible. In all the models studied, we were concerned primarily with the determination of average waiting time of a customer in the system and queue and also the average number of customers in both.

Discussion of the basic queuing models allowed us to understand development of results for queuing networks that find extensive use in manufacturing systems analysis and design. Because our objective was to introduce the reader to developments in queuing and queuing network theory, we covered only some basic open and closed networks. Only one solution technique was discussed for the open and closed networks. As mentioned previously, there are other techniques available and these are discussed in Viswanadham and Narahari (1992), Seidman et al. (1987), and Suri et al. (1993), amongst others. For the reader interested in exploring queuing network applications to manufacturing systems, we suggest extensive survey articles such as those by Buzacott and Yao (1986a,b) as well as a number of research articles including those by Solberg (1977), Shantikumar and Buzacott (1981), Stecke and Solberg (1981), to name a few. There is abundant literature on the queuing network subject. In this section, we have provided a partial list. For an extensive bibliography on this subject, we refer the reader to Viswanadham and Narahari (1992) and Suri et al. (1993).

ARENA Simulation Results

Department of Industrial Engineering

Summary for Replication 1 of 1

Project: Flexible Manufacturing   Run execution date :10/ 1/2007

Analyst: RSI   Model revision date:10 /1/2007

Replication ended at time   : 4800.0 Hours

Base Time Units: Hours

TALLY VARIABLES

Identifier	Average	Half Width	Minimum	Maximum	Observations
Time In System	2476.3	(Insuf)	262.50	4571.0	118
Yellow Flowtime	774.03	(Insuf)	261.00	1047.0	54
Green Flowtime	860.71	(Insuf)	516.00	1053.0	57
Blue Flowtime	903.37	(Insuf)	728.95	1059.0	41
Process Time	71.024	.45850	57.550	97.640	622
Overall Flowtime	841.42	(Insuf)	261.00	1059.0	152

DISCRETE-CHANGE VARIABLES

Identifier	Average	Half Width	Minimum	Maximum	Final Value
Yellow Paint Util	41.916	(Corr)	.00000	100.00	100.00
Green Paint Util	44.791	(Corr)	.00000	100.00	.00000
Blue Paint Util	48.958	11.927	.00000	100.00	.00000
GT Buffer 1 util	84.713	(Insuf)	.00000	100.00	100.00
GT Buffer 2 util	72.840	(Insuf)	.00000	100.00	.00000
GT Buffer 3 util	64.921	(Insuf)	.00000	100.00	100.00
GT Buffer 4 util	50.001	(Insuf)	.00000	100.00	.00000
GT Res 1 util	99.427	(Insuf)	.00000	100.00	100.00
GT Res 2 util	99.062	(Insuf)	.00000	100.00	100.00
GT Res 3 util	80.025	(Insuf)	.00000	100.00	100.00
GT Res 4 util	74.895	(Insuf)	.00000	100.00	100.00
Debur Util	46.354	(Corr)	.00000	100.00	100.00
Vac Clean Util	50.887	2.2020	.00000	100.00	100.00
Foam Gask Asy Util	54.964	(Corr)	.00000	100.00	100.00
Paint Conv 1 Util	11.970	(Insuf)	.00000	34.615	23.076
Paint Conv 2 Util	13.720	(Insuf)	.00000	17.741	15.591
GT Conv Util	26.353	(Corr)	.00000	52.000	28.000
InjMold Conv 2 Util	11.442	.24077	.00000	21.621	5.4054
InjMold Conv 1 Util	20.541	(Corr)	.00000	26.865	23.880
InjMold Worker Util	12.958	.43441	.00000	100.00	.00000
Foam Gask Conv Util	44.420	(Corr)	.00000	53.846	44.615

COUNTERS

Identifier	Count	Limit
Green	43	Infinite
Inj_Mold Parts	622	Infinite
Mach_Cell Parts	118	Infinite
Yellow	40	Infinite
Blue	47	Infinite

Simulation run time: 3.00 minutes.

Simulation run complete.

FIGURE 13.18   A sample simulation model output. (Courtesy of Rockwell Software, Inc.)

13.7 REVIEW QUESTIONS AND EXERCISES

1. List 10 examples of queuing systems you have seen. Explain the input and output process of each with the help of a diagram like Figure 13.1.
2. List five examples of queuing systems seen in automated manufacturing and material handling systems. Explain the input and output process of each with an illustration like Figure 13.1.
3. List the major performance measures that can be analyzed with a queuing system. Explain why these measures are important to a system designer and planner.
4. Suppose an independent, continuous random variable F follows an exponential distribution with parameter λ. Explain in your own words why the expected value of F is small. Use a graph showing the density function of an exponential distribution to explain your answer.
5. List three real-world examples of queuing systems where the exponential service time distribution is valid and not valid.
6. Show that the expected value of an independent, continuous random variable F that follows an exponential distribution with parameter λ is $1/\lambda$ and the variance is $1/\lambda^2$. [Hint: Solve $\int_0^\infty t\lambda e^{-\lambda t}\, dt$ to get $E(F)$ and the equation $E(F^2) - E(F)^2$ to get $V(F)$.]
7. (a) Explain the no-memory property in your own words.
   (b) If the service time follows an exponential distribution, what does that tell us about the number of service completions over a period of time. Explain.
8. Using the Kendall–Lee notation, list all the queuing models for which exact solutions can be found analytically.
9. Using the Kendall–Lee notation, classify the following queuing systems. Assume appropriate interarrival and service time distributions. State any other assumptions you make.
   (a) Bank with three tellers and a maximum capacity of 50 customers at any time
   (b) Airport runway
   (c) Restaurant with 25 tables
   (d) Mainframe computer with three processors, where jobs are processed according to priority codes assigned to each
   (e) Warehouse with two stocker cranes
   (f) A repair shop with three repair technicians
10. Explain the following:
    (a) Steady state
    (b) Infinite birth–death process
11. There are three states-0, 1, 2—in a birth–death process. The birth rates for states 0 and 1 are 2, state 2 is 0, and the death rate for states 1 and 2 is 3.
    (a) Construct a rate diagram similar to the one in Figure 13.2, write down the three flow conservation equations (for states 0, 1, and 2) and solve them to get steady-state probabilities.
    (b) Determine the values of L and W.
12. An automated teller machine (ATM) is located in a rest-stop along interstate highway I-87. Because the bank that services the ATM is 50 miles from the rest-stop, and the employee who is assigned to service the ATM has a flexible work schedule, the service times are not deterministic, rather exponentially distributed. When the ATM is empty, it takes the bank employee an average of 3 days before it is refilled. On average, an ATM has cash available in it for 20 days (exponentially distributed).*
    (a) Set up a two-state birth–death model for this problem.
    (b) Determine the percentage of time the ATM is not functional (i.e., has no cash in it).
13. Discuss the importance of Little's law.

---

* Based on a problem in Winston (1995).

FIGURE 13.19   Box scanning and tagging at a package shipping company.

14. Boxes arrive from an unloading dock at each receiving station of a package shipping company on a conveyor (see Figure 13.19) according to a Poisson distribution with a mean of 0.4 box per second. The boxes are scanned, tagged, and placed on another conveyor that feeds into a sortation system. Scanning and tagging together require 2 seconds per box (exponentially distributed). Determine the percentage of time the operator is busy, the average number of boxes in the first conveyor, and the throughput of the receiving station.

15. Suppose there are 10 receiving stations in the shipping company described in Exercise 14, and the conveyors from the 10 unloading docks merge into one conveyor. Boxes are picked up from the latter conveyor by operators at the 10 receiving stations and scanned and tagged as before. Assuming the arrival rate at the merge conveyor is 14,400 boxes per hour and the operator service times are 2 seconds per box (interarrival and service times exponentially distributed), determine the percentage of time an operator is busy, the average number of boxes in the merge conveyor, and the throughput of the system.

16. Suppose the 10 receiving stations in Exercise 14 are operating as 10 M/M/1 queues. Compare this system with the one described in Exercise 15. Which of the two do you prefer? Why?

17. Develop the results for the M/M/S/GD/∞/∞ system shown in Table 13.1.

18. A warehouse receives orders for storage according to a Poisson distribution with a mean of 64 orders per hour. It takes an employee an average of 3.75 minutes (exponentially distributed) to process the order and store it. Determine how many employees are required to minimize the labor cost and the waiting cost, assuming that each employee is paid $6 per hour and (system) waiting cost per hour is $3.75. Explain your answer.

19. A computer technical support center charges customers $10 per hour if they get through to the operator without waiting. A customer who has to wait is charged $7 per hour. The arrival rate of calls into the technical support center is 30 per hour (Poisson distribution) and the time to provide technical support is 5 minutes per customer (exponentially distribution).
    (a) Find the average price the technical support center charges its customers when there are
        (i) Three servers
        (ii) Four servers
        (iii) Five servers
    (b) How many operators must the center employ in order to maximize profits, assuming the operators' wages (including salary and benefits) are $20 per hour.

20. A freight truck has to be loaded with 70,000 cubic meters of a special material available at a certain location. Because the truck has no access to this location, a specially equipped loading device is used to load the material on self-powered platform trucks that then transport the material to

the freight truck parked at a nearby location that is accessible. Whereas the loading device costs $1000 each hour to rent, the platform trucks can be rented at only $100 per hour each. There is no limit on the number of platform trucks that can be rented. Each platform truck can hold 400 cubic meters of the material, and it takes an average of 5 minutes for it to haul the material, unload it on the freight truck, and return to the location for the next delivery. Due to space restrictions, loading equipment design, and other reasons, the loading itself requires an average of 12 minutes. Make appropriate assumptions about exponentiality and determine how many platform trucks should be rented in order to minimize the expected cost of hauling the material to the freight truck.*

21. Repeat Exercises 18 through 20 using the Sakasegawa (1977) W_q formula in Section 13.2.3. Are the answers you obtained different from those in Exercises 18 through 20? Explain.

22. Consider this simplified model of an emergency ambulance system. The only ambulance on an island is based at the only hospital on the island. The island is circular with radius R kilometers. The hospital is located at the center of the island. Emergency incidents arise at random uniformly over the island with density γ incidents/hour/square kilometer. The ambulance travels from the hospital to the incident, spends a random time S at the scene of the incident, and then returns to the hospital. The ambulance always travels at a fixed velocity V. Incidents that arise while the ambulance is out of the hospital are entered into an FCFS queue.

    (a) Assume the mean of the service time is μ and the variance of the service time is σ^2. Develop an expression for the mean delay before service. Define service as beginning when the ambulance leaves the hospital to respond to the incident and ending when it returns.

    (b) Develop an expression for μ in terms of the parameters of the problem.

    (c) Develop an expression for σ^2 in terms of the parameters of the problem.

    (d) Suppose the good people of the island want to see how a second ambulance would reduce the delay before service. Suggest two modeling strategies to estimate the delay when there are two ambulances. (Do not do more analysis; instead, describe in a few sentences how you would proceed to derive the average delay for these two approaches.)†

23. Consider an M/M/3 loss system (i.e., there are no queue positions). The arrival rate is λ. The servers are identical and have service rate μ.

    (a) Compute the probability that an arriving customer will not be served. Express your result as a function of $\rho = \lambda/\mu$.

    (b) Suppose that you observe these service times (in minutes) for the first five customers: 2.1, 10.8, 1.8, 4.7, 18.7. Discuss the validity of the assumption about the distribution of service times inherent in the M/M/3 model.

    (c) Suppose the arrival times of the first five customers studied in part (b) previously are 1.3, 1.5, 3.2, 9.0, and 10.6. What is the mean number of busy servers during the first 5 minutes of operation of the system?

    (d) When all servers are identical, we write the state transition diagram as a Markov chain. When we distinguish among servers, however, the state transition diagram becomes more complex, forming a multidimensional grid rather than a 1D chain. Consider the case of an M/M/2 loss system with nonidentical servers. Server 1 has service rate μ_1, and server 2 has μ_2. When both servers are idle, server 1 always serves the next arriving customer. The state-transition diagram in this case is shown in Figure 13.20. Using the principle of flow balance (i.e., flow into state = flow out of state), write the four simultaneous linear equations describing the steady-state distribution of state occupancies for this system.

    (e) Now extend these ideas to the M/M/3 loss system. Suppose that two of the servers are veterans with identical service rates μ, while the third server is a novice with service rate $\alpha < \mu$. We define the state of the system as a pair of numbers (number of veteran servers busy,

---

* Based on a problem in Winston (1995).
† The author gratefully acknowledges Professor T.R. Willemain, Rensselaer Polytechnic Institute, for providing this problem.

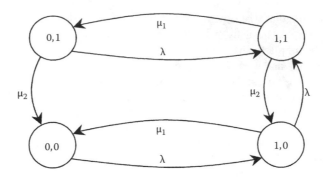

FIGURE 13.20   State diagram for an M/M/2 loss system.

number of novice servers busy). For instance, the empty-and-idle state is (0, 0). Suppose the system recognizes the desirability of having the veterans provide service when possible. In this case, the novice only serves a customer if the two veterans are busy. Draw the 2D state space describing this system, labeling the state transitions with the proper rates.

24. Explain the difference between a queuing model and a queuing network model. Which is more useful for modeling real word manufacturing design and planning problems? Why?

25. Derive the $P(n_1, n_2)$ and $P(0, 0)$ values for the two machine open network system depicted in Figure 13.9.

26. An AGV is used for transferring parts from one automated machine to another in a flexible manufacturing cell. Parts arrive from a conveyor to the first machine at the rate of nine pallets per hour where the processing takes an average of 5 minutes per pallet. The AGV loading, transfer, and unloading require 5 minutes per pallet, and processing on the second automated machine requires 6 minutes on average. Assuming all interarrival and service times to be exponential, and unlimited buffers, formulate the aforementioned problem as an open network model.
    (a) Determine the expected number of jobs waiting in front of the two machines and the AGV.
    (b) Under what conditions will results from the aforementioned model not be valid? Explain.

27. Suppose that in the situation described in Exercise 26, 5% of the pallets coming out of the second machine are sent to the first for rework. Answer the questions in Exercise 26 again.

28. A flexible manufacturing cell makes three part types on two machines. The first part type is processed on both, the second part on the first machine, and the third on the second. The service rates for the two machine types are 16 and 12 parts per hour, respectively. The number of pallets available for the three part types 1, 2, and 3 is 3, 6, and 4, respectively.
    (a) Determine the amount of time each part type spends waiting and being serviced at each machine.
    (b) Determine the number of each part type waiting and being served at each machine.

29. Obtain the answers to Exercises 14 through 16, 18 through 20, and 26 through 28 using any available software.

30. Consider a single-server, infinite capacity queuing system in which the arrival rate of customers is dependent upon the number of people in the system. Specifically, the arrival rate is $\lambda/(i+1)$, where i is the number of customers in the system. Answer the following three questions:[*]
    (a) Set up the steady-state equations and solve them $\left(\text{Hint} : \sum_{i=1}^{\infty} \frac{x^i}{i!} = e^x \right)$.
    (b) What does λ represent? You have been hired in an organization that has such an arrival rate and your first task is to measure λ. State one or two ways in which you can measure λ.
    (c) Set up an expression for the average time a customer is expected to spend in the line.

31. *Project*: Write a computer program for the PD method described in Section 13.4.2. Test its performance by generating a data set and comparing results of your program with those from simulation.
32. *Project*: Write a computer program for the PD method described in Section 13.4.3. Test its performance by generating a data set and comparing results of your program with those from simulation.
33. Develop a simulation model using any available simulation software for the problem in Example 13.8.
34. Develop a simulation model using any available simulation software for the problem in Example 13.9.
35. Consider the simulation model constructed in Example 13.8. Assuming the service time for drill A is deterministic with a value of 60 minutes per part, modify the simulation model appropriately, record the statistics, and
    (a) Show that the average time in system (expected) is 292 minutes per part type A
    (b) Calculate the average time in system for part type B
    (c) Compare the results obtained in Example 13.8 as well as (a) and (b) above and explain the importance of reducing process variability.
36. Discuss the advantages and disadvantages of analytical and simulation models. Which do you prefer? Why?

Bibliography

Adil, G.K., D. Rajamani, and D. Strong (1997), Assignment allocation and simulated annealing algorithms for cell formation, *IIE Transactions*, 29(1), 53–68.

Adrabinski, A. and M.M. Syslo (1983), Computational experiments with some heuristic algorithms for the traveling salesman problem, Technical Report No. 78, Wroclaw University, Wroclaw, Poland.

Akl, S.G. and G.T. Toussaint (1978), A fast convex hull algorithm, *Information Processing Letters*, 7, 219–222.

Alfa, A.S., M.Y. Chen, and S.S. Heragu (1991), A 3-opt based simulated annealing algorithm for vehicle routing problems, *Proceedings of the 13th Computers and Industrial Engineering Conference*, Orlando, FL, March 11–13, 21, pp. 635–639.

Allison, D.C.S. and M.T. Noga (1984), The rectilinear traveling salesman problem, *Information Processing Letters*, 18, 195–199.

Apple J.M. (1977), *Plant Layout and Material Handling*, 3rd ed., John Wiley & Sons, New York.

Arimond, J. and W.R. Ayles (1993), Phenolics creep up on engine applications, *Advanced Materials and Processes*, 143, 20.

Armour, G.C. and E.S. Buffa (1963), A heuristic algorithm and simulation approach to relative allocation of facilities, *Management Science*, 9(2), 294–300.

Asari, M. (1993), Electron beam hardening system, *Advanced Materials and Processes*, 143, 30–31.

Ashayeri, J. and L.F. Gelders (1985), Warehouse design optimization, *European Journal of Operational Research*, 21, 285–294.

Askin, R.G. and C.R. Standridge (1993), *Modeling and Analysis of Manufacturing Systems*, John Wiley & Sons, New York.

Avriel, M. (1976), *Nonlinear Programming: Analysis and Methods*, Prentice-Hall, Englewood Cliffs, NJ.

Ballou, R.H. (1988), *Business Logistics Management*, 3rd ed., Prentice-Hall, Englewood Cliffs, NJ.

Banks, J., B. Burnette, J.D. Rose, and H. Kozloski (1995), *Introduction to SIMAN V and CINEMA V*, John Wiley & Sons, New York.

Banks, J., J.S. Carlson, II, and B.L. Nelson (1996), *Discrete-Event Simulation*, Prentice-Hall, Englewood Cliffs, NJ.

Banks, J., J.S. Carlson, B.L. Nelson, and D.M. Nicol (2009), *Discrete-Event Simulation*, 5th Ed., Prentice-Hall, NJ.

Barnes, J.W. and M. Laguna (1990), Scheduling on parallel processors with linear delay penalties using tabu search, Technical Report, The University of Texas at Austin, Austin, TX.

Bazaraa, M.S. (1975), Computerized layout design: A branch and bound approach, *AIIE Transactions*, 7(4), 432–437.

Bazaraa, M.S. and O. Kirca (1983), A branch-and-bound-based heuristic for solving the QAP, *Naval Research Logistics Quarterly*, 30, 287–304.

Bazaraa, M.S. and H.D. Sherali (1980), Benders' partitioning scheme applied to a new formulation of the quadratic assignment problem, *Naval Research Logistics Quarterly*, 27(1), 29–41.

Bazaraa, M.S., H.D. Sherali, and C.M. Shetty (2006). *Nonlinear Programming, Theory and Algorithms*, 3rd ed., John Wiley & Sons, New York.

Bazaraa, M.S. and C.M. Shetty (1979), *Nonlinear Programming*. Wiley, New York, NY.

Bazaraa, M.S. and C. Shetty (1993), *Nonlinear Programming: Theory and Algorithms*, John Wiley & Sons, New York.

Beghin-Picavet, M. and P. Hansen (1982), Deux problems d'affectation non lineaires, *R.A.I.R.O Recherche Operationelle*, 16(3), 263–276.

Behnezhad, A.R. and B. Khoshnevis (1988), The effects of manufacturing progress function on machine requirements and aggregate planning, *International Journal of Production Research*, 26, 309–326.

Benders, J.F. (1962), Partitioning procedures for solving mixed-variables programming problems, *Numerische Mathematik*, 4, 238–252.

Benjafaar, S (2002), Modeling and analysis of congestion in the design of facility layouts, *Management Science*, 48(5), 679–704.

Benjafaar, S., S.S. Heragu, and S. Irani (2002), Next generation factory layouts: Research challenges and recent progress, *Interfaces*, 32(6), 58–76.

Black, J.T. (1983), Cellular manufacturing systems reduce setup time, make small lot production economical, *Industrial Engineering*, 36–48.

Bodin, L.D., B.L. Golden, A.A. Assad, and M.O. Ball (1983), *Routing and Scheduling of Vehicles and Crews: The State of the Art*, Computers and OR Special Issue, 10, 69–211.

Bozer, Y.A., R.D. Meller, and S.J. Erlebacher (1994), An improvement-type layout algorithm for single and multiple floor facilities, *Management Science*, 40(7), 918–932.

Bozer, Y.A., E.C. Schorn, and G.P. Sharp (1990), Geometric approaches to solve the Chebyshev traveling salesman problem, *IIE Transactions*, 22(3), 238–254.

Bozer, Y.A. and M.M. Srinivasan (1989), Tandem configurations for AGV systems offer simplicity and flexibility, *Industrial Engineering*, 21(2), 23–27.

Bozer, Y.A. and J.A. White (1990), Design and performance models for end-of-aisle order picking systems, *Management Science*, 26(7), 852–866.

Brooke, A., D. Kendrick, and A. Meeraus (1988), *GAMS: A User's Guide*, The Scientific Press, Redwood City, CA.

Brown, P.A. and D.F. Gibson (1972), A quantified model for facility site selection applied to a multipoint location problem, *AIIE Transactions*, 4, 1–10.

Buffa, E.S. (1955), Sequence analysis for functional layouts, *The Journal of Industrial Engineering*, 6, 12–13, 25.

Buffa, E.S., G.C. Armour, and T.E. Vollmann (1964), Allocating facilities with CRAFT, *Harvard Business Review*, 42, 136–158.

Buffa, E.S. and R.K. Sarin (1987), *Modern Production and Operations Management*, 8th ed., John Wiley & Sons, New York.

Bunday, B.D. and R.E. Scranton (1980), The G/M/r machine interference model, *European Journal of Operational Research*, 4, 399–402.

Burbidge, J.L. (1963), Production flow analysis, *The Production Engineer*, 42(12), 742.

Burbidge, J.L. (1992), Change to group technology: Process organization is obsolete, *International Journal of Production Research*, 30(5), 1209–1219.

Burkard, R.E. (1984), Locations with spatial interaction—Quadratic assignment problem, in R.L. Francis and P.B. Mirchandani (Eds.), *Discrete Location Theory*, Academic Press, New York.

Burkard, R.E. and K.H. Stratman (1978), Numerical investigations on quadratic assignment problems, *Naval Research Logistics Quarterly*, 25, 129–144.

Burke, P.J. (1956), The output of queuing systems, *Operations Research*, 4, 699–704.

Buzacott, J.A. and D.D. Yao (1986a), Flexible manufacturing systems: A review of analytical models, *Management Science*, 32(7), 890–905.

Buzacott, J.A. and D.D. Yao (1986b), On queuing network models of flexible manufacturing systems, *Queuing Systems*, 1, 5–27.

Carrie, A.S. (1973), Numerical taxonomy applied to group technology and plant layout, *International Journal of Production Research*, 11(4), 399–416.

Chandrasekharan, M.P. and R. Rajagopalan (1986a), MODROC: An extension of rank order clustering for group technology, *International Journal of Production Research*, 24(5), 1221–1233.

Chandrasekharan, M.P. and R. Rajagopalan (1986b), An ideal seed non-hierarchical clustering algorithm for cellular manufacturing, *International Journal of Production Research*, 24(2), 451–464.

Chandrasekharan, M.P. and R. Rajagopalan (1987), ZODIAC: An algorithm for concurrent formation of part-families and machine-cells, *International Journal of Production Research*, 25(6), 835–850.

Chang, T.C., R.A. Wysk, and H.P. Wang (1991), *Computer-Aided Manufacturing*, Prentice-Hall, Englewood Cliffs, NJ.

Chang, Y.L. and R.S. Sullivan (1991), *Quantitative Systems Version 2.0*, Prentice-Hall, Englewood Cliffs, NJ.

Chen, J. (1995), Cellular manufacturing systems design: Optimal solution of large scale real-world problems, Unpublished PhD dissertation, Rensselaer Polytechnic Institute, Troy, NY.

Chow, S.W.M. (1987), An analysis of automated storage and retrieval systems in manufacturing assembly lines, *IIE Transactions*, 18(2), 204–214.

Ciminowski, R. and E.L. Mooney (1995), Heuristics for a new model of facility layout, *Computers and Industrial Engineering*, 29(1–4), 273–277.

Ciminowski, R. and E.L. Mooney (1997), Heuristics for a proximity-based facility layout model, *Computers and Industrial Engineering*, 32(2), 341–349.

Cinar, U. (1975), Facilities planning: A systems analysis and space allocation approach, in C.M. Eastman (Ed.), *Spatial Synthesis in Computer-Aided Building Design*, John Wiley & Sons, New York.

Collins, N.E., R.W. Eglese, and B.L. Golden (1988), Simulated annealing: An annotated bibliography, *American Journal of Mathematical and Management Sciences*, 8, 209–307.

Conway, R.W. and W.L. Maxwell (1961a), A note on the assignment of facility location, *The Journal of Industrial Engineering*, 12, 34–36.

Conway, R.W. and W.L. Maxwell (1961b), A queuing model with state dependent service rate, *The Journal of Industrial Engineering*, 12, 132–136.

Currie, K.R. (1992), An intelligent grouping algorithm for cellular manufacturing, *Proceedings of the 14th Annual Computers and Industrial Engineering Conference*, Orlando, FL.

Das, C.S. and S.S. Heragu (1988), A transportation approach to locating plants in relation to potential markets and raw material sources, *Decision Sciences*, 19(4), 819–829.

Daskin, M. (2013), *Network and Discrete Location: Models, Algorithms, and Applications*, 2nd Ed., John Wiley & Sons, New York.

Davis, L. (1991), *Handbook of Genetic Algorithms*, Van Nostrand Reinhold, New York.

DePuy, G.W. (2007), *ExASRS v2.0: Excel Automated Storage/Retrieval Systems—A User's Manual for Configuring Preliminary Designs*, CICMHE/Material Handling Industry of America, Inc., Charlotte, NC.

Dilworth, J.B. (1989), *Production and Operations Management: Manufacturing and Non-Manufacturing*, Random House Publishing Company, New York.

Disney, R.L. (1981), Queuing networks, *Proceedings of the Symposia in Applied Mathematics*, 25, 53–83.

Donaghey, C.E. and V.F. Pire (1991), *BLOCPLAN-90, User's Manual*, Industrial Engineering Department, University of Houston, Houston, TX.

Dutta, K.N. and S. Sahu (1982), A multi-goal heuristic for facilities design problem: MUGHAL, *International Journal of Production Research*, 20(2), 147–154.

Editor (1990), AGVs in America: An inside look, *Modern Materials Handling*, 45(10), 56–60.

Editor (1993), Technology trends: Concepts for work and tool holding, *American Machinist*, 137, 10.

Editor (1996a), Buyer's guide: Lift trucks and AGVs, *IE Solutions*, 28(4), 44–49.

Editor (1996b), Buyer's guide: Simulation software, *IE Solutions*, 28(5), 54–63.

Editor (1996c), Northstar custom telephones rely on flexible conveyor systems, *Assembly Magazine Online*, May 1996.

Editor (2001), Is just-in-time returning to just-in-case? *Modern Materials Handling*, November 1.

Emmons, H., A.D. Flowers, C.M. Khot, and K. Mathur (1992), *Storm: Personal Version 3.0: Quantitative Modeling for Decision Support*, Prentice-Hall, Englewood Cliffs, NJ.

Faber, Z. and M.W. Carter (1986), A new graph theoretic approach for forming machine cells in cellular production systems, in A. Kusiak (Ed.), *Flexible Manufacturing Systems: Methods and Studies*, North Holland, New York, pp. 301–318.

Feller, W. (1957), *An Introduction to Probability Theory and Its Applications*, John Wiley & Sons, New York.

Fisher, E.L. and S.Y. Nof (1984), FADES: Knowledge-based facility design, *Proceedings of the 1984 Annual Industrial Engineering Conference*, Chicago, IL, May 1984, pp. 74–82.

Foulds, L.R. (1983), Techniques for facility layout: Deciding which pairs of activities should be adjacent, *Management Science*, 29, 1414–1426.

Foulds, L.R. and J.W. Giffin (1984), A graph theoretic heuristic for multi-floor building layout, *Proceedings of the IIE Conference*, Wayzata, Minnesota, April 13–15, pp. 202–205.

Francis, R.L. (1967), Sufficient conditions for some optimum-property facility designs, *Operations Research*, 15(3), 448–466.

Francis, R.L., L.F. McGinnis, and J.A. White (1992), *Facility Layout and Location: An Analytical Approach*, 2nd ed., Prentice-Hall, Englewood Cliffs, NJ.

Francis, R.L. and J.A. White (1974), *Facility Layout and Location: An Analytical Approach*, Prentice-Hall, Englewood Cliffs, NJ.

Fujine, M., T. Kaneko, and J. Okijima (1993), Aluminum composites replace cast iron, *Advanced Materials and Processes*, (6), 20.

Garey, M.R. and D.S. Johnson (1979), *Computers and Intractability: A Guide to the Theory of NP-Completeness*, W.H. Freeman and Company, New York.

Gavett, J.W. and N.V. Plyter (1966), The optimal assignment of facilities to locations by branch and bound, *Operations Research*, 14, 210–232.

Geoffrion, A.M. and G.W. Graves (1974), Multicommodity distribution system design by Benders decomposition, *Management Science*, 20(5), 822–844.

Ghiani, G., G. Laporte, and R. Musmanno (2004), *Introduction to Logistics Systems Planning and Control*, John Wiley & Sons, Chichester, U.K.

Gilmore, P.C. (1962), Optimal and suboptimal algorithms for the quadratic assignment problem, *Journal of the Society for Industrial and Applied Mathematics*, 10, 305–313.

Glover, F. (1989), Tabu search, part I, *ORSA Journal on Computing*, 1(3), 190–206.

Glover, F. (1990), Tabu search, part II, *ORSA Journal on Computing*, 2(1), 4–32.

Goetschalckx, M.P. (1983), Storage and retrieval policies for efficient order picking, Unpublished PhD dissertation, Georgia Institute of Technology, Atlanta, GA.

Goetschalckx, M.P. and H.D. Ratliff (1988), Sequencing picking operations in a man-aboard order picking system, *Material Flow*, 4(4), 255–263.

Goetschalckx, M.P. and H.D. Ratliff (1990), Shared storage policies based on the duration of stay, *Management Science*, 36(9), 1120–1132.

Goldberg, D. (1989), *Genetic Algorithms in Search, Optimization, and Machine Learning*, Addison-Wesley, Reading, MA.

Golden, B., L. Bodin, T. Doyle, and W. Stewart, Jr. (1980), Approximate traveling salesman algorithms, *Operations Research*, 28(3), 694–711.

Gongaware, T.A. and I. Ham (1981), Cluster analysis application for group technology manufacturing systems, *Proceedings of the Ninth American Metal Working Research Conference*, Dearborn, MI, pp. 503–508.

Gray, A.E., U.S. Karmarkar, and A. Seidmann (1992), Design and operation of an order-consolidation warehouse: Models and application, *European Journal of Operational Research*, 58, 14–36.

Groover, M.P. (1996) *Fundamentals of Modern Manufacturing: Materials, Processes and Systems*, Prentice-Hall, Englewood Cliffs, NJ.

Groover, M.P. and E.W. Zimmers, Jr. (1984), *CAD/CAM: Computer Aided Design and Manufacturing*, Prentice-Hall, Englewood Cliffs, NJ.

Gross, D. and C.M. Harris (1985), *Fundamentals of Queuing Theory*, John Wiley & Sons, New York.

Hales, S., S.S. Heragu, R.J. Graves, S. Jennings, and C.J. Malmborg (2007), A multimedia educational tool integrating materials handling technology, analysis and design using a virtual distribution center, *European Journal of Industrial Engineering*, 1(1), 93–110.

Ham, I. (1976), Introduction to group technology, SME Technical Paper MMR 76-093, Society of Manufacturing Engineers, Dearborn, MI.

Ham, I., K. Hitomi, and T. Yoshida (1985), *Group Technology*, Kluwer-Nijhoff Publishing Company, Boston, MA.

Hanna, S.R. and S. Konz (2004), *Facility Design and Engineering*, 3rd ed., Holcomb Hathaway Publishers, Scottsdale, AZ.

Hassan, M.M.D., G.L. Hogg, and D.R. Smith (1985), A construction algorithm for the selection and assignment of material handling equipment, *International Journal of Production Research*, 23(2), 381–392.

Hax, A.C. and D. Candea (1984), *Production and Inventory Management*, Prentice-Hall, Englewood Cliffs, NJ.

Heragu, S.S. (1988), Machine layout: An optimization and knowledge based approach, Unpublished PhD dissertation, Department of Mechanical and Industrial Engineering, University of Manitoba, Winnipeg, Canada.

Heragu, S.S. (1990), Modeling the machine layout problem, *Proceedings of the 12th Computers and Industrial Engineering Conference*, Orlando, FL, March 1990, Vol. 19(1–4), pp. 294–298.

Heragu, S.S. (1992a), A heuristic algorithm for identifying machine cells, *Information and Decision Technologies*, 18, 171–184.

Heragu, S.S. (1992b), Recent models and techniques for the layout problem, *European Journal of Operational Research*, 57(2), 136–144.

Heragu, S.S. (1994), Group technology and cellular manufacturing, *IEEE Transactions on Systems, Man and Cybernetics*, 24(2), 203–215.

Heragu, S.S. and A.S. Alfa (1992), Experimental analysis of simulated annealing based algorithms for the layout problem, *European Journal of Operational Research*, 57(2), 190–202.

Heragu, S.S. and C.S. Das (1994), Exact solution of the material handling equipment selection and assignment problems, DSES Technical Report #37-94-406, Rensselaer Polytechnic Institute, Troy, NY.

Heragu, S.S., L. Du, R.J. Mantel, and P.B. Schuur (2005), Mathematical model for warehouse design and product allocation, *International Journal of Production Research*, 43(5), 327–338.

Heragu, S.S. and Y.P. Gupta (1994), A heuristic for designing cellular manufacturing facilities, *International Journal of Production Research*, 32(1), 125–140.

Heragu, S.S. and S.R. Kakuturi (1997), Grouping and placement of machine cells, *IIE Transactions*, 29(7), 561–571.

Heragu, S.S. and J.S. Kochhar (1994), Material handling issues in adaptive manufacturing systems, in E.M. Malstrom and I.W. Pence, Jr. (Eds.), *The Materials Handling Engineering Division 75th Anniversary Commemorative Volume*, MH-2, Paper presented at *the 1994 International Mechanical Engineering Congress and Exposition*, American Society of Mechanical Engineers, pp. 9–13, Chicago, IL.

Heragu, S.S. and A. Kusiak (1987), Expert systems in manufacturing design, *IEEE Transactions on Systems, Man and Cybernetics*, SMC-17(6), 898–912.

Heragu, S.S. and A. Kusiak (1988), Machine layout problems in flexible manufacturing systems, *Operations Research*, 36(2), 258–268.

Heragu, S.S. and A. Kusiak (1991), Efficient models for the facility layout problem, *European Journal of Operational Research*, 53(1), 1–13.

Heragu, S.S. and C.M. Lucarelli (1996), Automated manufacturing system design using analytical techniques, in J.D. Irwin (Ed.), *The Industrial Electronics Handbook, A Volume of the Electrical Engineering Handbook Series*, CRC Press, Boca Raton, FL.

Heragu, S.S., B. Mazacioglu, and K.D. Fuerst (1994), Meta-heuristic algorithms for the order picking problem, *International Journal of Industrial Engineering*, 1(1), 67–76.

Heragu, S.S. et al. (2003), Multimedia tools for use in materials handling classes, *European Journal of Industrial Engineering*, 28(3), 375–393.

Heskett, J.L. (1963). Cube-per-order index—A key to warehouse stock location, *Transportation and Distribution Management*, 3(1), 27–31.

Heskitt, J.L. (1964), Putting the COI to work in warehouse layout, *Transportation and Distribution Management*, 4, 23–30.

Hicks, P.E. and T.E. Cowan (1976), CRAFT-M for layout rearrangement, *Industrial Engineering*, 8(5), 30–35.

Hillier, F.S., R.W. Conway, and W.L. Maxwell (1964), A multiple server queuing model with state dependent service rate, *The Journal of Industrial Engineering*, 15, 153–157.

Hillier, F.S. and G.J. Lieberman (1990), *Introduction to Stochastic Models in Operations Research*, McGraw-Hill, New York.

Hillier, F.S. and G.J. Lieberman (1995), *Introduction to Operations Research*, 6th ed., McGraw-Hill, New York.

Hillier, F.S., O.S. Yu, D. Avis, L. Fossett, F. Lo, and M. Reiman (1991), *Queuing Tables and Graphs*, Elsevier, New York.

Ho, Y.C. (1987), Performance evaluation and perturbation analysis of discrete event dynamic systems, *IEEE Transactions on Automatic Control*, AC-32(7), 563–572.

Hodgson, T.J. (1982), A combined approach to the pallet loading problem, *IIE Transactions*, 14(3), 175–182.

Hodgson, T.J. and T.J. Lowe (1982), Production lot sizing with material-handling cost considerations, *AIIE Transactions*, 14(1), 44–51.

Holland, J. (1975), *Adaptation in Natural and Artificial Systems*, The University of Michigan Press, Ann Arbor, MI.

Houshyar, A. and L.F. McGinnis (1990), A heuristic for assigning facilities to locations to minimize WIP travel distance in a linear facility, *International Journal of Production Research*, 28(8), 1485–1498.

Irani, S.A., H. Zhang, J. Zhou, H. Huang, T.K. Udai, and S. Subramanian (2000), Production flow analysis and simplification toolkit, *International Journal of Production Research*, 38(8), 1855–1874.

Iri, M. (1968), On the synthesis of loop and cutset matrices and related problems, in K. Kondo (Ed.), *RAAG Memoirs*, Vol. 4(A-XII), Research Association of Applied Geometry, Japan, pp. 376–410.

Jackson, J.R. (1957), Networks of waiting lines, *Operations Research*, 5, 518–527.

Jackson, J.R. (1963), Jobshop like queuing systems, *Management Science*, 10, 131–142.

Jacobs, R.F. (1984), A note on SPACECRAFT for multi-floor layout planning, *Management Science*, 30(5), 648–649.

Jaikumar, R. (1984), Flexible manufacturing systems: A managerial perspective, Working Paper, Harvard Business School, Boston, MA.

Jaiswal, N.K. (1968), *Priority Queues*, Academic Publishers, New York.

Johnson, D.S., C.R. Aragon, L.A. McGeoch, and C. Schevon (1989), Optimization by simulated annealing: An experimental evaluation. Part I, Graph Partitioning, *Operations Research*, 37(6), 865–892.

Johnson, M.E. and M.L. Brandeau (1992), Stochastic modeling for automated material handling system design and control, Working Paper, Owen Graduate School of Management, Vanderbilt University, Nashville, TN.

Johnson, R.V. (1982), SPACECRAFT for multi-floor layout planning, *Management Science*, 28(4), 407–417.

Kamath, M., S. Sivaramakrishnan, and G.M. Shirhatti (1995), RAQS: A software package to support instruction and research in queuing systems, *Proceedings of the Fourth Industrial Engineering Research Conference*, May 1995, Nashville, TN.

Kendall, D.G. (1953), Stochastic processes occurring in the theory of queues and their analysis by the method of Markov chains, *Annals of Mathematical Statistics*, 24, 338–354.

Khumawala, B.M. (1972), An efficient branch and bound algorithm for the warehouse location problem, *Management Science*, 18(12), 718–731.

Kind, D.A. (1975), Elements of space management, *Transportation and Distribution Management*, 15(2), 29–34.

King, J.R. (1980), Machine-component grouping in production flow analysis: An approach using a rank order clustering algorithm, *International Journal of Production Research*, 18, 213–219.

King, J.R. and V. Nakornchai (1982), Machine-component group formation in group technology: Review and extension, *International Journal of Production Research*, 20(2), 117–133.

Kleinrock, L. (1975), *Queuing Systems, Volume 1: Theory*, John Wiley & Sons, New York.

Kochhar, J.S., B.L. Foster, and S.S. Heragu (1998), HOPE: A genetic algorithm for the unequal area layout problem, *Computers and Operations Research*, 25(8), 583–594.

Kochhar, J.S. and S.S. Heragu (1998), MULTI-HOPE: A tool for multiple floor layout problems, *International Journal of Production Research*, 36(12), 3421–3435.

Koenig, J. (1980), Planning and justifying a material distribution center, *Proceedings of the Spring Annual IIE Conference*, Dallas, TX.

Koopmans, T.C. and M. Beckman (1957), Assignment problems and the location of economic activities, *Econometrica*, 25, 53–76.

Kouvelis, P. and M.W. Kim (1992), Unidirectional loop network layout problem in automated manufacturing systems, *Operations Research*, 40(3), 533–550.

Kulwiec, R.A. (1980), *Advanced Material Handling*, The Material Handling Institute, Charlotte, SC.

Kulwiec, R.A. (1982), *Material Handling Handbook*, John Wiley & Sons, New York.

Kumar, K.R., A. Kusiak, and A. Vannelli (1986), Grouping of parts and components in flexible manufacturing systems, *European Journal of Operational Research*, 24, 387–397.

Kusiak, A. and W.S. Chow (1987), An efficient cluster identification algorithm, *IEEE Transactions on Systems, Man and Cybernetics*, SMC-17(4), 696–699.

Kusiak, A. and W.S. Chow (1988), Decomposition of manufacturing systems, *IEEE Transactions on Robotics and Automation*, RA-4(5), 457–471.

Kusiak, A. and S.S. Heragu (1987a), Group technology, *Computers in Industry*, 9(2), 83–91.

Kusiak, A. and S.S. Heragu (1987b), The facility layout problem, *European Journal of Operational Research*, 29(3), 229–251.

Kusiak, A. and S.S. Heragu (1988), KBSES: A knowledge based system for equipment selection, *The International Journal of Advanced Manufacturing Technology*, 3(3), 97–109.

Kwo, T.T. (1958), A theory of conveyors, *Management Science*, 5(1), 51–71.

Lacksonen, T.A. (1994), Static and dynamic layout problems with varying areas, *Journal of Operational Research Society*, 45(1), 59–69.

Land, A.H. (1963), A problem of assignment with interrelated costs, *Operations Research Quarterly*, 14, 185–198.

Law, A.M. and W.D. Kelton (1991), *Simulation Modeling and Analysis*, 2nd ed., McGraw-Hill, New York.

Lawler, E.L. (1963), The quadratic assignment problem, *Management Science*, 9(4), 586–599.

Lawler, E.L., J.K. Lenstra, A.H.G. Rinnooy Kan, and D.B. Shmoys (1985), *The Traveling Salesman Problem*, John Wiley & Sons, New York.

Lee, A. (1966), *Applied Queuing Theory*, Macmillan, New York.

Lee, R. and J.M. Moore (1967), CORELAP—Computerized relationship layout planning, *The Journal of Industrial Engineering*, 18, 195–200.

Leung, J. (1992), A graph theoretic heuristic for designing loop-layout manufacturing systems, *European Journal of Operational Research*, 57(2), 243–252.

Lin, S. and B. Kernighan (1973), An effective heuristic algorithm for the traveling salesman problem, *Operations Research*, 21(2), 498–516.

Little, J.D.C. (1961), A proof for the queuing formula $L = \lambda W$, *Operations Research*, 9, 383–385.

Logendran, R. (1990), A workload based model for minimizing total intercell and intracell moves in cellular manufacturing, *International Journal of Production Research*, 28(5), 913–925.

Logendran, R. (1991), Impact of sequence of operations and layout of cells in cellular manufacturing, *International Journal of Production Research*, 29(2), 375–390.

Love, R.F., J.G. Morris, and G.O. Wesolowsky (1988), *Facilities Location: Models and Methods*, North-Holland, New York.

Love, R.F. and J.Y. Wong (1976), Solving quadratic assignment problems with rectilinear distances and integer programming, *Naval Research Logistics Quarterly*, 23, 623–627.

Malek, M., M. Guruswamy, M. Pandya, and H. Owens (1989), Serial and parallel simulated annealing and tabu search algorithms for the traveling salesman problem, *Annals of Operations Research*, 21, 59–84.

Malmborg, C.J. and K. Bhaskaran (1990), A revised proof of optimality for the cube-per-order index rule for stored item location, *Applied Mathematical Modelling*, 40, 87–95.

Malmborg, C.J., B. Krishnakumar, G.R. Simons, and M.H. Agee (1989), EXIT: A PC-based expert system for industrial truck selection, *International Journal of Production Research*, 27(6), 927–941.

Malmborg, C.J. and Y.C. Shen (1994), Heuristic dispatching models for multi-vehicle materials handling systems, *Applied Mathematical Modeling*, 18, 124–133.

Matson, J.O. and J.A. White (1982), Operational research and material handling, *European Journal of Operational Research*, 11, 309–319.

Maxwell, W.L. and J.A. Muckstadt (1982), Design of automated guided vehicle systems, *IIE Transactions*, 14(2), 114–124.

Mayer, H. (1960), Introduction to conveyor theory, *Western Electric Engineer*, 4(1), 43–47.

McAuley, J. (1972), Machine grouping for efficient production, *The Production Engineer*, 51(2), 53–57.

McCormick, W.T., P.J. Schweitzer, and T.W. White (1972), Problem decomposition and data reorganization by a clustering technique, *Operations Research*, 20, 992–1009.

Meller, R.D., A. Goel, and R.R. Kori (2005), Automated guided vehicles decision-support tool, Technical Report, Virginia Polytechnic Institute and State University, Blacksburg, VA.

Meng, G. (2002), Open queuing network performance analyzer for reconfigurable manufacturing systems, PhD thesis, Department of Decision Science and Engineering System, Rensselaer Polytechnic Institute, Troy, NY.

Meng, G. and S.S. Heragu (2004), Batch size modeling in a multi-item, discrete manufacturing system via an open queuing network, *IIE Transactions*, 36(8), 743–753.

Meng, G., S.S. Heragu, and H.W.M. Zijm (2009), Two-level manufacturing system performance analyser, *International Journal of Production Research*, 47(9), 2301–2326.

Miller, D.M. and R.P. Davis (1977), The machine requirements problem, *International Journal of Production Research*, 15(2), 219–231.

Miller, D.M. and R.P. Davis (1978), A dynamic resource allocation model for a machine requirements problem, *IIE Transactions*, 10(3), 237–243.

Miller, D.M. and J.W. Schmidt (1984), *Industrial Engineering and Operations Research*, John Wiley & Sons, New York.

Mitrofanov, S.P. (1983), *Group Technology in Industry*, Vols. 1 and 2, Mashinostroienie, Leningrad, USSR (in Russian).

Montreuil, B. (1988), From gross layout to net layout, Working Paper, School of Industrial Engineering, Purdue University, West Lafayette, IN.

Montreuil, B. and H.D. Ratliff (1989), Utilizing cut-trees as design skeletons for facilities layout, *IIE Transactions*, 21, 136–143.

Montreuil, B., H.D. Ratliff, and M. Goetschalckx (1987), Matching based interactive facility layout, *IIE Transactions*, 19(3), 271–279.

Montreuil, B. and U. Venkatadri (1991), Strategic and interpolative design of dynamic manufacturing systems layouts, *Management Science*, 37, 682–694.

Moore, J.M. (1971), Computer program evaluates plant layout alternatives, *Industrial Engineering*, 3(8), 19–25.

Murty, K.G. (1983), *Linear Programming*, John Wiley & Sons, New York.

Murty, K.G. (1995), *Operations Research: Deterministic Optimization Models*, Prentice-Hall, Englewood Cliffs, NJ.

Muth, E.J. (1975), Modelling and systems analysis of multistation closed-loop conveyors, *International Journal of Production Research*, 13(6), 559–566.

Muther, R. (1973), *Systematic Layout Planning*, Van Nostrand Reinhold Company, New York.

Nagi, R., G. Harhalakis, and J.M. Proth (1990), Multiple routings and capacity considerations in group technology applications, *International Journal of Production Research*, 28(12), 2243–2257.

National Research Council (1998), *Visionary Manufacturing Challenges for 2020*, National Academy Press, Washington, DC.

Nelson, B.L. (1995), *Stochastic Modeling: Analysis and Simulation*, McGraw-Hill, New York.

Nicol, L.M. and R.H. Hollier (1983), Plant layout in practice, *Material Flow*, 1, 177–188.

Noble, J.S. and J.M.A. Tanchoco (1993), A framework for material handling system design justification, *International Journal of Production Research*, 31(1), 81–106.

Nugent, C.E., T.E. Vollmann, and J. Ruml (1968), An experimental comparison of techniques for the assignment of facilities to locations, *Operations Research*, 16, 150–173.

Ogburn, J.E. (1984), Order picking economy: Batch vs. individual, *Proceedings of the International Material Handling Forum*, Houston, TX.

Or, I. (1976), Traveling salesman type combinatorial problems and their relation to the logistics of regional blood banking, Unpublished PhD dissertation, Department of Industrial Engineering and Management Sciences, Northwestern University, Evanston, IL.

Peck, L.G. and R.N. Hazelwood (1958), *Finite Queuing Tables*, John Wiley & Sons, New York.

Pegden, C.D., R.E. Shannon, and R.P. Sadowski (1990), *Introduction to Simulation Using SIMAN*, McGraw-Hill, New York.

Picard, J. and M. Queyranne (1981), On the one-dimensional space allocation problem, *Operations Research*, 29(2), 371–391.

Pierce, J.F. and W.B. Crowston (1971), Tree—Search algorithms for quadratic assignment problems, *Naval Research Logistics Quarterly*, 18, 1–36.

Powell, M.J.D. (1964), An efficient method for finding the minimum of a function of several variables without calculating derivatives, *Computer Journal*, 7, 155–162.

Press, W.H., B.P. Flannery, S.A. Teukolsky, and W.T. Vetterling (1986), *Numerical Recipes: The Art of Scientific Computing*, Cambridge University Press, New York.

Pritsker, A.A.B. (1986), *Introduction to Simulation and SLAM II*, John Wiley & Sons, New York.

Pritsker, A.A.B. and P.M. Ghare (1970), Locating new facilities with respect to existing facilities, *AIIE Transactions*, 2(4), 290–298.

Quinn, J.P. (2006), 3PLs hit their stride, *Logistics Management/Supply Chain Management Review*, July, 45(7), 3T–10T.

Rajagopalan, R. and J.L. Batra (1975), Design of cellular production systems—A graph theoretic approach, *International Journal of Production Research*, 13(6), 567–579.

Rajagopalan, S. and S.S. Heragu (1997), Advances in discrete material handling systems design, *Sadhana: Academy Proceedings in Engineering Sciences*, 22(Part 2), 281–292.

Ravindran, A., D.T. Phillips, and J.J. Solberg (1987), *Operations Research: Principles and Practice*, John Wiley & Sons, New York.

Reed, R. (1961), *Plant Layout: Factors, Principles, and Techniques*, Richard D. Irwin, Inc., Homewood, IL.

Reed, R. (1973), *Plant Location, Layout and Maintenance*, Irwin, Homewood, IL.

Ritter, H., T. Martinetz, and K. Schulten (1992), *Neural Computation and Self-Organizing Maps: An Introduction*, Addison Wesley Longman Publishing Co., Redwood City, CA.

Rosenblatt, M.J. (1979), The facilities layout problem: A multi-goal approach, *International Journal of Production Research*, 17, 323–332.

Rosenblatt, M.J. (1986), The dynamics of plant layout, *Management Science*, 32, 76–86.

Ross, S. (1970), *Applied Probability Models with Optimization Applications*, Holden-Day, San Francisco, CA.

Rouwenhorst, B., B. Reuter, V. Stockrahm, G.J. van Houtum, R.J. Mantel, and W.H.M. Zijm (2000), Warehouse design and control: Framework and literatures review, *European Journal of Operational Research*, 122, 515–533.

Russell, R.S. and B.W. Taylor, III (1995), *Production and Operations Management: Focussing on Quality and Competitiveness*, Prentice-Hall, Englewood Cliffs, NJ.

Sahni, S. and T. Gonzalez (1976), P-complete approximation problem, *Journal of Associated Computing Machinery*, 23(3), 555–565.

Sakasegawa, H. (1977), An approximate formula $L_q = \alpha \beta^\rho (1 - \rho)$, *Annals of the Institute for Statistical Mathematics*, 29, 67–75.

Schrage, L.E. (1986), *Linear, Integer and Quadratic Programming with Lindo*, Scientific Press, Palo Alto, CA.

Seehof, J.M. and W.O. Evans (1967), Automated layout design program, *The Journal of Industrial Engineering*, 18(12), 690–695.

Seidmann, A., P. Schweitzer, and S. Shalev-Oren (1987), Computerized closed queueing network models of flexible manufacturing systems: A comparative evaluation, *Large Scale System*, 12, 91–107.

Seifoddini, H. and P.M. Wolfe (1986), Application of the similarity coefficient method in group technology, *IIE Transactions*, 18, 271–277.

Sepponen, R. (1969), *CORELAP 7 User's Manual*, Unpublished working paper, Technical University, Helsinki, Finland.

Shantikumar, J.G. and J.A. Buzacott (1981), Open queuing network models of dynamic job shops, *International Journal of Production Research*, 19, 245–256.

Shubin, J.A. and H. Madeheim (1951), *Plant Layout*, Prentice-Hall, New York.

Simmons, D.M. (1969), One-dimensional space allocation: An ordering algorithm, *Operations Research*, 17, 812–826.

Simon, H.A. (1975), Style in design, in C.M. Eastman (Ed.), *Spatial Synthesis in Computer-Aided Building Design*, John Wiley & Sons, New York.

Singh, N. (1993), Design of cellular manufacturing systems, *European Journal of Operational Research*, 69(3), 284–291.

Sinriech, D. (1995), Network design models for discrete material flow systems: A literature review, *International Journal of Advanced Manufacturing Technology*, 10, 277–291.

Skorin-Kapov, J. (1990), Tabu search applied to the quadratic assignment problem, *ORSA Journal on Computing*, 2(1), 33–45.

Sneath, P.H.A. and R.R. Sokal (1973), *Numerical Taxonomy*, Freeman and Co., San Francisco, CA.

Solberg, J.J. (1977), A mathematical model of computerized manufacturing systems, *Proceedings of the Fourth International Research Conference on Production Research*, Tokyo, Japan, pp. 22–30.

Srinivasan, M.M., Y.A. Bozer, and M. Cho (1994), Trip-based material handling systems: Throughput capacity analysis, *IIE Transactions*, 26(1, January), 70–89.

Stecke, K.E. and J.J. Solberg (1981), Loading and control policies for a flexible manufacturing system, *International Journal of Production Research*, 19(5), 481–490.

Stewart, W., Jr. (1977), A computationally efficient heuristic for the traveling salesman problem, in *Proceedings of the 13th Annual Meeting of Southeastern TIMS*, Myrtle Beach, S.C., pp. 75–83.

Stidham, S. (1974), A last word on $L = \lambda W$. *Operations Research*, 22(2), 417–421.

Sule, D.R. (1991), *Manufacturing Facilities: Location, Planning and Design*, PWS Kent, Boston, MA.

Suri, R. and S. de Treville (1992), Rapid modeling: The use of queuing models to support time-based competitive manufacturing, in *Proceedings of a Joint German/U.S. Conference*, Hagen, Germany, June 25–26, pp. 21–30.

Suri, R. and R.R. Hildebrandt (1984), Modeling flexible manufacturing systems using mean-value analysis, *Journal of Manufacturing Systems*, 3, 27–38.

Suri, R., J.L. Sanders, and M.L. Kamath (1991), Performance evaluation of production networks, in S.C. Graves, A.H.G. Rinooy Kan, and P. Zipkin (Eds.), *Logistics of Production and Inventory: Handbook in Operations Research and Management Science*, North-Holland, New York.

Suri, R., J.L. Sanders, and M. Kamath (1993), Performance evaluation of production networks, in S.C. Graves et al. (Eds.), *Handbooks in Operations Research and Management Science*, Vol. 4: *Logistics of Production and Inventory*, Elsevier Science Publishers, B.V., Amsterdam, the Netherlands, pp. 194–286.

Suryanarayanan, J.K., B.L. Golden, and Q. Wang (1991), A new heuristic for the linear placement problem, *Computers and Operations Research*, 18(3), 255–262.

Suskind, P.B. (April 1989), IEs should play a critical role in office development, *Industrial Engineering*, 21(4), 53–60.

Syslo, M.M., N. Deo, and J.S. Kowalik (1983), *Discrete Optimization Algorithms with PASCAL Programs*, Prentice-Hall, Englewood Cliffs, NJ.

Taha, H.A. (1992), *Operations Research: An Introduction*, Macmillan, New York.

Takacs, L. (1962), *Introduction to the Theory of Queues*, Oxford University Press, Oxford, U.K.

Tam, K.Y. and S.G. Li (1991), A hierarchical approach to the facility layout problem, *International Journal of Production Research*, 29(1), 165–184.

Tamashunas, V.M., J. Labban, and D. Sly (1990), Interactive graphics offer an analysis of plant layout and material handling systems, *Industrial Engineering*, June 1990, 22(6), 38–43.

Tate, D.M. and A.E. Smith (1995), A genetic approach to the quadratic assignment problem, *Computers and Operations Research*, 22(1), 73–83.

Tompkins, J.A. and J.M. Moore (1984), *Computer Aided Layout: A User's Guide*, AIIE, Norcross, GA.

Tompkins, J.A. and R. Reed, Jr. (1976), An applied model for the facilities design problem, *International Journal of Production Research*, 14(5), 583–595.

Tompkins, J.A. and J.A. White (1984), *Facilities Planning*, 1st ed., John Wiley & Sons, New York.

Tompkins, J.A., J.A. White, Y.A. Bozer, and E.H. Frazelle (2003a), *Facilities Planning*, 3rd ed., John Wiley & Sons, New York.

Tompkins, J.A., J.A. White, Y.A. Bozer, and J.M.A. Tanchoco (2003b), *Facilities Planning*, 3rd ed., John Wiley & Sons, Hoboken, NJ.

Towill, D.R. (1984), A production engineering approach to robot selection, *OMEGA*, 12(3), 261–272.

Urban, T.L. (1993), A heuristic for the dynamic facility layout problem, *IIE Transactions*, 25, 57–63.

Vakharia, A.J. and U. Wemmerlov (1990), Designing a cellular manufacturing system: A materials flow based operations sequence, *IIE Transactions*, 22(1), 84–97.

van Camp, D.J., M.W. Carter, and A. Vannelli (1992), A nonlinear optimization approach for solving facility layout problems, to appear in special issue of *European Journal of Operational Research*, 57(2), 174–189.

Viswanadham, N. and Y. Narahari (1992), *Performance Modeling of Automated Manufacturing Systems*, Prentice-Hall, Englewood Cliffs, NJ.

Vollmann, T.E., C.E. Nugent, and Zartler (1968), A computerized model for office layout, *The Journal of Industrial Engineering*, 19, 321–327.

Vos, W.J. (1979), Group technology: What can it do for you? *Proceedings of the V International Conference on Production Research*, Amsterdam, the Netherlands, August 12–16, 1979, pp. 48–53.

Webster, D.B. and R. Reed (1971), A material handling system selection model, *AIIE Transactions*, 3(1), 13–21.

Weiszfeld, E. (1936), Sur le point pour lequel la somme des distances de n points donnés est minimum, *Tohoku Mathematics Journal*, 43, 355–386.

Wemmerlov, U. and N.L. Hyer (1986), Procedures for the part-family/machine group identification problem in cellular manufacturing, *Journal of Operations Management*, 6(2), 125–147.

Wemmerlov, U. and N.L. Hyer (1989), Cellular manufacturing in the U.S. industry: A survey of users, *International Journal of Production Research*, 27(9), 1511–1530.

Whitt, W. (1983), The queueing network analyzer, *The Bell System Technical Journal*, 62(9), 2779–2815.

Wimmert, R.J. (1958), A mathematical method for equipment location, *The Journal of Industrial Engineering*, 9, 498–505.

Winston, W.L. (1994), *Operations Research: Applications and Algorithms*, Duxbury Press, Belmont, CA.

Winston, W.L. (2004), *Operations Research: Applications and Algorithms*, 4th ed., Dubury Press, Belmont, CA.

Zhang, C. and H. Wang (1992), Concurrent formation of part families and machine cells based on the fuzzy set theory, *Journal of Manufacturing Systems*, 11(1), 61–67.

Index